SELOC®

Managing Partners Dean F. Morgantini
Barry L. Beck

Executive Editor Kevin M. G. Maher, A.S.E.

Production Specialists Melinda Possinger
Ronald Webb

Manufactured in USA

104 Willowbrook Lane
West Chester, PA 19382
ISBN-13: 978-089330-042-5
ISBN-10: 089330-042-X
3456789012 0987654321

www.selocmarine.com
1-866-SELOC55

SELOC'S

KAWASAKI PERSONAL WATERCRAFT

Volume IA

1992 - 1998

Tune-Up
and
Repair Manual

MARINE
MANUALS

Other books by Joan Lorton Coles and/or Clarence W. Coles

Glenn's Complete Bicycle Manual
*The Book of Harry (Limited Edition)
*Glenn's Flat Rate Manual -- American Cars
Seloc's OMC Stern Drive Tune-up and Repair Manual
Seloc's Marine Jet Drive Tune-up and Repair Manual
Seloc's Mercury Volume I Tune-up and Repair Manual
Seloc's Mercury Volume II Tune-up and Repair Manual
Seloc's Mercury Volume III Tune-up and Repair Manual
Seloc's Force Outboard Tune-up and Repair Manual
Seloc's Yamaha Volume I Tune-up and Repair Manual
Seloc's Yamaha Volume II Tune-up and Repair Manual
Seloc's Yamaha Volume III Tune-up and Repair Manual
Seloc's Mariner Volume I Tune-up and Repair Manual
Seloc's Mariner Volume II Tune-up and Repair Manual
Seloc's Chrysler Outboard Tune-up and Repair Manual
Seloc's OMC Cobra Stern Drive Tune-up and Repair Manual
Seloc's MerCruiser Stern Drive Volume I Tune-up and Repair Manual
Seloc's MerCruiser Stern Drive Volume II Tune-up and Repair Manual
Seloc's MerCruiser Stern Drive Volume III Tune-up and Repair Manual
Seloc's Mercury/Mariner Volume I Tune-up and Repair Manual
Seloc's Mercury/Mariner Volume II Tune-up and Repair Manual
Seloc's Mercury/Mariner Volume III Tune-up and Repair Manual
Seloc's Johnson/Evinrude Volume I Tune-up and Repair Manual
Seloc's Johnson/Evinrude Volume II Tune-up and Repair Manual
Seloc's Johnson/Evinrude Volume IIA Tune-up and Repair Manual
Seloc's Johnson/Evinrude Volume III Tune-up and Repair Manual
Seloc's Johnson/Evinrude Volume IV Tune-up and Repair Manual
Seloc's Johnson/Evinrude Volume V Tune-up and Repair Manual
Seloc's Personal Watercraft Volume I Tune-up and Repair Manual
Seloc's Personal Watercraft Volume IA Tune-up and Repair Manual
Seloc's Personal Watercraft Volume II Tune-up and Repair Manual
Seloc's Personal Watercraft Volume IIA Tune-up and Repair Manual
Seloc's Personal Watercraft Volume III Tune-up and Repair Manual
Seloc's Personal Watercraft Volume IIIA Tune-up and Repair Manual
Seloc's Volvo Penta Stern Drive Volume I Tune-up and Repair Manual
Seloc's Volvo Penta Stern Drive Volume II Tune-up and Repair Manual
Seloc's Volvo Penta Stern Drive Volume III Tune-up and Repair Manual
Briggs & Stratton Single Cylinder OHV Tune-up and Repair Manual
Briggs & Stratton Single Cylinder Horizontal Crankshaft Tune-up & Repair Manual
Briggs & Stratton Single Cylinder Vertical Crankshaft Tune-up and Repair Manual
*Out of print.

Cover
An original painting
by
Ray Goudey II
B.F.A. Art Center
Pasadena, California

Technical Director
James Tapp

Graphics
Barbara Flotow
Christopher Oliver

Photography
L.E. Coles

Inquiries should be addressed to:

Seloc Publications
10693 Civic Center Drive
Rancho Cucamonga, California 91730

ISBN 0-89330-042-X

Printed in the United States of America

ACKNOWLEDGMENTS

Seloc Publications wish to express their thanks and appreciation to Kawasaki Motors Corporation U.S.A. -- Irvine, California for the assistance during the preparation of this manual. Permission by Kawasaki to use certain exploded drawings and wiring diagrams is a positive contribution to this manual for which we are most grateful.

Seloc extends a "hearty" thanks to Pomona Valley Kawasaki, Ontario, California and the owner -- Ralph Whitlinger -- for providing us with certain units to complete our model coverage. Very special recognition is extended to Ken Stout for his time and expertise during the photographic sessions of the disassembling and assembling work on the Kawasaki units included in this manual. His expertise and skill certainly earn him the title of Kawasaki PWC mechanic -- EXTRAORDINAIRE.

Clarence W. Coles
Seloc Publications

FOREWORD

This is a comprehensive tune-up and repair manual for Kawasaki Personal Watercraft produced by Kawasaki Motors Corporation, U.S.A., Irvine, CA. from 1992 thru 1998.

The book has been designed and written to cover stock, factory "out the door" engines and jet pumps. Modifications for higher than manufacturer's rated performance are so extensive and varied, no attempt has been made to include them in this volume (Actually, for such coverage, a separate comprehensive book would be required. In such cases the publisher's recommendation is to follow the aftermarket instructions with the particular product or service.)

All Seloc manuals, including this title, are directed toward the professional mechanic, the do-it-yourselfer, and the student developing her/his mechanical skills.

TABLE OF CONTENTS

4- MAINTENANCE (CONTINUED)

5- TROUBLESHOOTING

6- FUEL AND OIL

7- IGNITION

8- ENGINE

9- ELECTRICAL

1
DESCRIPTION & OPERATION

1-1 BRIEF HISTORY

The jet drive system for propelling a craft through the water arrived on the scene in the mid 1960's with the jet drive boat. In those early days, the jet drive system was mated only with high performance powerplants -- engines in the 454 cu. in. class and larger. For this reason, during the "gas crunch" in the 1970's the jet drives were labeled as inefficient and as "gas hogs".

In addition to these two negative terms, they earned the reputation as "bad boy" boats due to their noisy "straight" exhaust, high rpm operation, and their almost unbelievable maneuverability. These combined factors did little to enhance their image and certainly restricted their popularity.

With new and improved technology, personal watercraft arrived on the scene about the mid 1970's. Personal watercraft, as we know them today, were developed using the same principles as the jet boats, and originally powered with a single cylinder two-stroke engine.

In order to meet the demand for more speed and the ability to carry more than just

A typical inner harbor summer weekend with scores of personal watercraft preparing to leave or just returning from a "fun day" on the water in the "outer harbor" or at sea close to shore. Just a reasonable amount of "TLC", will reward the owner and his/her friends with hours of trouble free enjoyment.

one person, watercraft manufacturers were quick to respond. Today, most modern craft are powered with a twin cylinder or three-cylinder two-stroke powerplant, coupled to a single stage pump.

Aftermarket shops and manufacturers have come into existence all across the United States and Canada. Their output of specialty products and services permit the owner to gain more speed and in over the competition at racing events wherever enough water is available.

As mentioned in the "Foreword", this book has been designed and written to cover stock, factory "out the door" engines and jet drives. Modifications for higher than manufacurer's rated performance are so extensive and varied, no attempt has been made to include them in this volume. (Actually, for such coverage, a separate comprehensive book would be required.) In such cases the publisher's recommendation is to follow the after market instruction with the particular product or service.

Series Covered

The following Kawasaki series produced from 1992 thru 1998 are covered in this manual. (Earlier models are covered in Seloc's Personal Watercraft -- Volume I -- Kawasaki -- Early days thru 1991.)

Model	Approx. Yr. of Production
JS550 Series	1992-1994
JF650 Series	1992-1996
JL650 Series	1992-1995
JS650 Series	1992-1993
JS750 Series	1992-1996
JH750 Series	1992 & On
JT750 Series	1994 & On
JH900 Series	1995 & On
JT900 Series	1997 & On
JH1100 Series	1996 & On
JT1100 Series	1997 & On

Other Seloc Personal Watercraft Manuals

Volume I - Kawasaki - early days thru 1991.

Volume II - Bombarier Sea Doo - early days thru 1991.

Volume IIA - Bombardier Sea Doo - 1992 thru 1997.

Volume III - Yamaha - early days thru 1991.

Volume IIIA -- Yamaha -- 1992 thru 1997.

Volume IV -- Polaris -- 1992 thru 1997.

Overall view of the popular Model 750 Series twin cylinder installation.

1-2 PRINCIPLES OF OPERATION

One of the first lessons to be learned in any elementary physics class is Newton's basic law: "For every force, there is an opposite and equal force". This statement is the basic principle of the jet pump. Water is "sucked" and "scooped" in from under the craft by a powerful pump rotating at incredible speed and then discharged, "blown" out, sternward in the opposite direction. In this manner the watercraft is propelled forward.

The personal water craft covered in this manual are all equipped with a twin or 3-cylinder water cooled two-stroke engine, matched with a single stage (one impeller) jet pump.

On a very few models, a reverse "gate" is swung down over the pump outlet nozzle forcing the exhausted water back in a forward direction thus moving the craft sternward.

Personal watercraft jet pumps may be classified as "axial flow" or "mixed flow".

Many times the Model 900ZXi Series engine is modified with aftermarket equipment and service for higher than factory "out-the-door" performance.

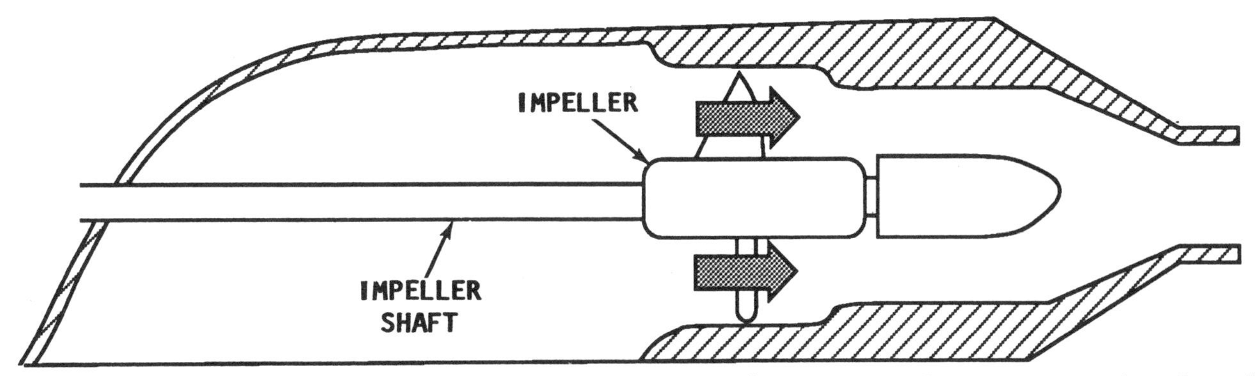

Cross-section line drawing to depict water flow through an axial flow jet pump. The water passes through parallel to the axis of impeller rotation.

Kawasaki watercraft are equipped with an axial flow jet pump, except the Model 550 which has a mixed flow pump -- see Chapter 10.

Axial Flow

Water in an axial pump moves on a single axis, as depicted in the adjacent illustration -- thus the word "axial" is used. In simple layman's terms -- water is ingested and discharged parallel to the axis of impeller rotation, as shown.

1-3 ENGINE AND JET DRIVE

The basic principle and arrangement of engine and pump are almost identical for all manufacturers.

The engine crankshaft is coupled either directly to the pump shaft through a coupler containing a rubber "shock absorber" or through a short driveshaft without any gear reduction. An impeller mounted on the pump shaft draws water into an opening in the hull through a suction and intake casting. Once the craft has attained forward motion the intake also serves as a "scoop" adding to the volume of water moved through the pump.

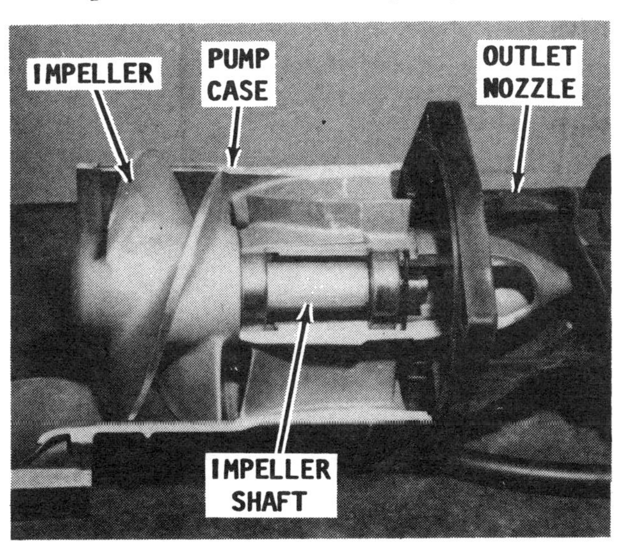

Photo of a cutaway "demonstration" type Kawasaki axial flow jet pump. A few major parts are identified.

Volume and Velocity

The outlet nozzle is slightly funnel shaped. This design coupled with the capacity of the pump (impeller) causes the volume of water entering the pump to exit the nozzle with increased velocity. This principle is similar to air passing through the venturi of a carburetor.

The amount of water ejected sternward from the nozzle at high velocity is the force propelling the craft through the water. This force can actually be measured and calculated in foot pounds or Newton meters. The

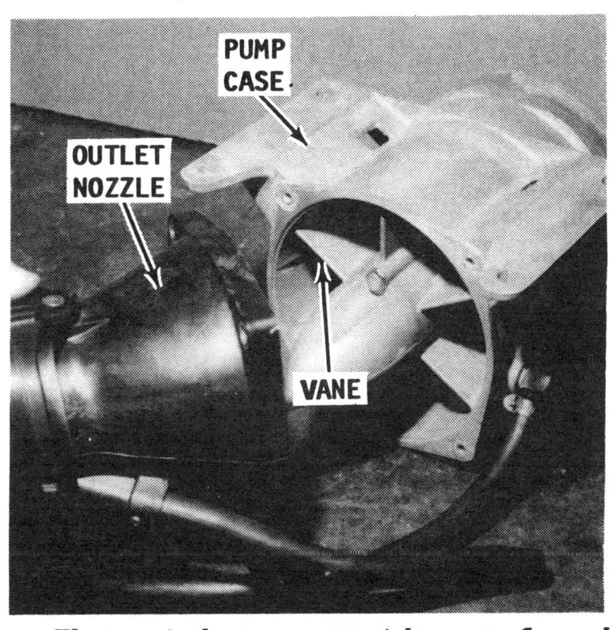

The vanes in the pump case straighten water flow and the conical shape of the outlet nozzle increases flow velocity in much the same way as air passing through a carburetor venturi.

greater the velocity of water mass moving through the nozzle, the greater the thrust to move the craft.

Cooling System

A fitting on the jet pump, aft of the impeller, syphens off cooling water. This water is delivered first to the exhaust manifold -- the hottest part of the engine. The water is then channeled around the cylinder walls, cylinder head and exhaust pipe.

After the exhaust pipe, some water is channeled overboard through a bypass hose.

The remaining water is mixed with exhaust gases in an expansion chamber and finally discharged from the exhaust outlet. This mixture of water and exhaust gases has the affect of cooling and somewhat quieting the exhaust emission.

On all models, a bypass hose connects the exhaust pipe water jacket to an outlet on the starboard side of the hull. When water is discharged as a "tattle-tale" stream, the operator is assured cooling water is circulating through the engine properly.

Cooling Water and Bilge Hoses

On most pumps two hoses are attached to the pump case and outlet nozzle. One hose channels some of the water flowing through the pump to the exhaust manifold -- the hottest part of the engine -- to cool the block during operation. The other hose siphons water out of the bottom of the hull,

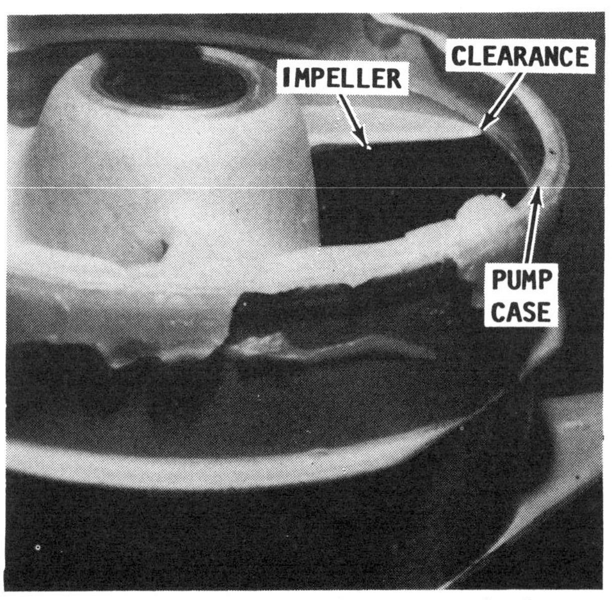

A feeler gauge may be used to measure the clearance between the impeller blades and the pump case.

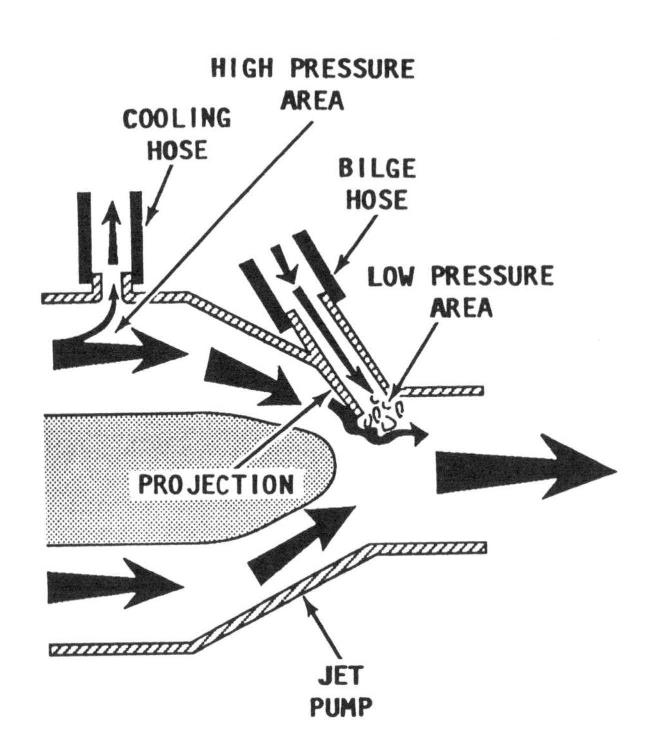

The direction of water flow through the cooling hose is opposite to the water flow moving through the bilge hose. The cooling hose routes water from the pump to the engine. Water is actually "vacuumed" from the bilge through the bilge hose to the pump where it is forced out through the nozzle.

and is referred to as the bilge system. The cooling hose has water flowing **FROM** the pump. The bilge hose has water flowing **TO** the pump. Both hoses attach to the pump in a similar manner.

What determines the direction of water flow?

The answer to this question is in a simple explanation of high and low pressure areas along the inside surface of the pump.

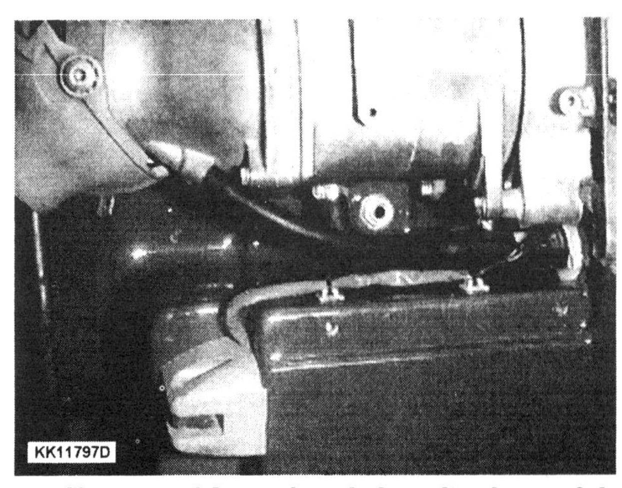

Close view of the nozzle with the cooling hose and the bilge hose clearly visible.

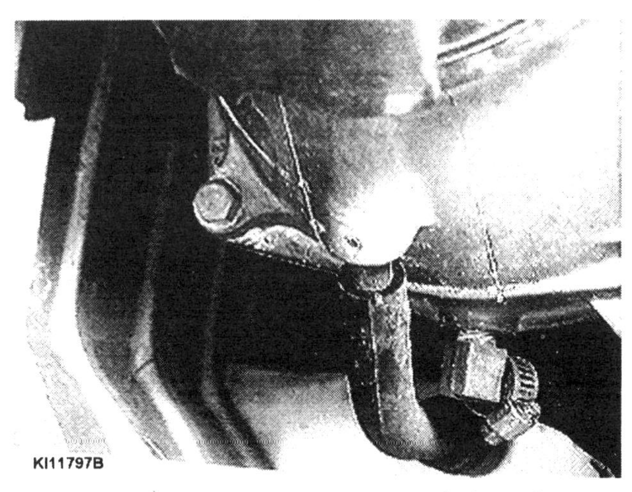

The bilge fitting protrusion extends into the water flow area. Water passing through causes a low pressure area a "vacuuming" -- siphoning -- effect -- sucking water out of the bilge.

Close inspection of the area inside the pump at the cooling water hose fitting reveals a smooth rounded shoulder with no obstruction or obstacle to impede the flow of water. A high pressure area develops here and draws the water down the hose attached to the fitting.

Further inspection of the area inside the hosing at the bilge hose fitting reveals a small protrusion around the opening. This protrusion causes disturbance to the water flow and creates an area of low pressure. This low pressure area will have the effect of emptying air and water from the hose aft with the impeller water flow -- similar to the action of a vacuum cleaner, as depicted in the accompanying illustration. If the other end of the hose is submerged in water inside the hull, the water will be "vacuumed" out. In this manner the bilge system drains the bilge.

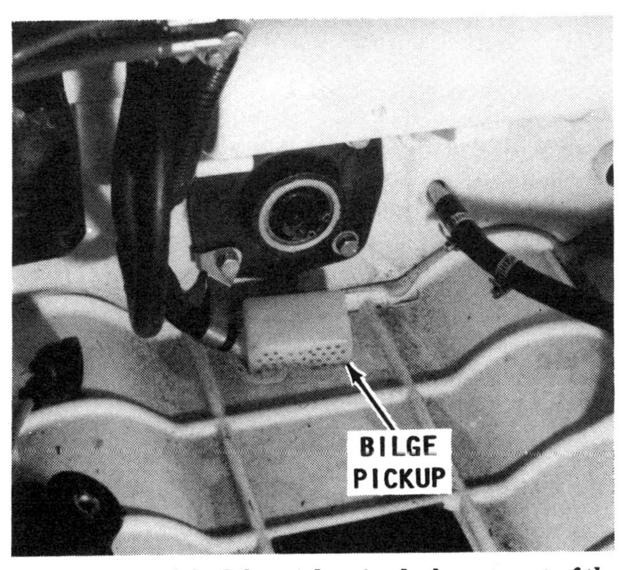

Location of the bilge pickup in the lowest part of the engine compartment. (Photograph taken with the engine removed for clarity.

Many an owner -- with good intentions -- thinking his action would smooth water flow and increase pump performance -- has filed the protrusion described in the previous paragraph smooth with the surrounding area of the pump. During operation of the water craft, the bilge system worked in **REVERSE.** Water was pumped from the impeller housing into the bilge area, quickly filling the bilge and possibly sinking the craft before the rider had time to realize what was happening.

Bilge Breather Fitting

A bilge breather fitting is installed at the highest point of the bilge line. This fitting has a small breather hole to prevent siphoning water from the pump into the bilge. If the engine was shut down with the bilge hose filled with water and the jet pump fitting was positioned higher then the end of the bilge line, water would flow back and fill the bilge area. The small breather hole in the fitting prevents this back-siphoning.

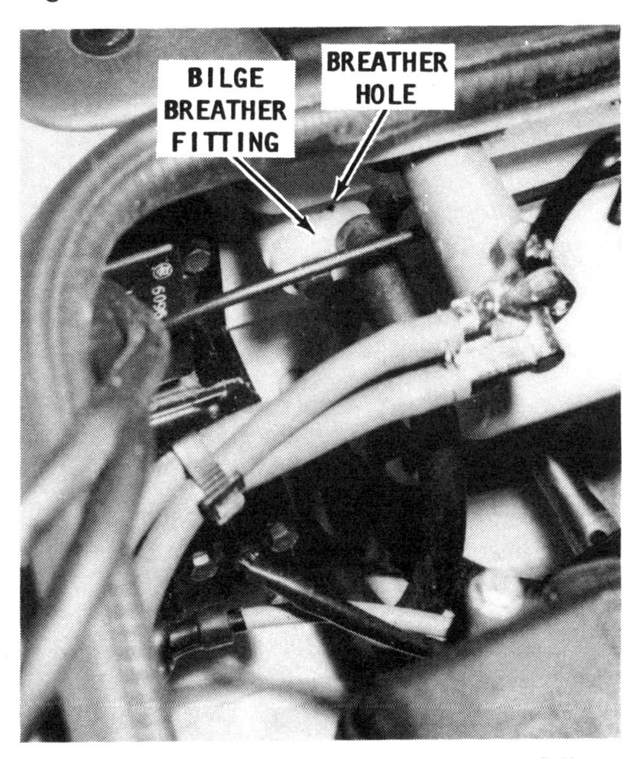

*A bilge breather fitting may be installed in different locations, depending on the engine model. However, it is **ALWAYS** installed at the highest point in the bilge hose. The fitting has a small "breather" hole which **MUST** be kept clean and unobstructed.*

1-4 IMPELLERS

Impellers provide thrust using a combination of water flow and water pressure in a closed environment.

The blades of an impeller overlap, thus water is trapped and forced through the impeller into the pump case and nozzle. The water moving past the impeller resists cavitation because it is under pressure. Blades in the pump case are angled exactly opposite to those of the impeller. These blades redirect the water to establish a concentrated flow through the nozzle.

The manufacturer recommends a radius of about 1/64" (0.3-0.5mm) for the leading edge of the impeller blade. A sharper radius will result in cavitation and a greater radius will reduce pump efficiency.

Cavitation Burns

Cavitation burns are the worst enemy of an impeller. These burns literally "eat away" material from the impeller, leaving holes and weakening the impeller structure. In extreme cases entire blades have been known to "depart" from the impeller hub because the cavitation burns at the base of the blade were so severe.

Cavitation burns are the result of imperfections (damage) on the impeller blades or air mixed with the water flow.

Cavitation burns may be caused by:

Wave jumping - air is sucked into the pump as the craft leaves the surface of the water.

Bad seal between the pump case and outlet nozzle -- allowing air to enter.

Leading edge of the impeller becoming damaged.

To explain exactly what causes cavitation burns, first, let us examine the last cause. If the impeller leading edge is damaged and "mushrooms" over the blade, a low pressure area will form under the mushroomed lip.

At atmospheric pressure, water boils at 212°F. Anytime there is an imperfection on the surface of an impeller blade and water passes over that imperfection, the water pressure is lowered. If the pressure is lowered, water will boil at a much lower temperature. Therefore, air bubbles will form in the water boiling under the mushroomed lip. The bubbles will creep down the surface of the blade and accumulate at the impeller hub. A high pressure area is formed at the base of the blade. Here, the air bubbles will collapse and reform back into water with a release of energy. This energy is absorbed by the impeller and results in material being eaten away.

The impeller should be dressed to a slightly rounded edge. It is impossible to achieve a perfectly straight sharp edge, as one side will always be convex and the other side concave. The concave side will form an area of low pressure and encourage cavitation as described above.

The other two causes of cavitation -- defective seals and wave jumping -- occurs when air is sucked in with the water flow. The same principles apply, as just described, in the previous paragraphs. Air bubbles will creep down the surface of the blade and accumulate at the impeller hub. A high pressure area is formed at the base of the blade. Here, the air bubbles will collapse and reform back into water with a release of energy. This energy is absorbed by the impeller and results in material being eaten away.

1-5 REPLACEABLE PARTS

Any time the engine or jet pump is disassembled, all manufacturers and the authors heartily recommend all parts included in an overhaul kit be installed. An engine kit usually includes all gaskets, seals, O-rings, and other parts required to rebuild the engine. Pistons, rings, bearings, crankshafts and even blocks should be inspected to determine if they are fit for further service. The pump kits usually include all gaskets, seals, O-rings, and other parts required to restore full efficiency and "like new" condition to the jet pump unit.

1-6 DEBRIS REMOVAL AND ENGINE OVERHEATING

As may be expected, a clogged impeller will slow the craft. The volume of water passing through the nozzle and water velocity will be reduced. A clogged impeller may cause restricted cooling water through the engine.

Many times the cooling water supply hose, if not properly secured with a hose clamp, may "blow off". The engine loses its supply of cooling water and internal damage quickly occurs.

Some units are equipped with an overheating warning horn. If the horn sounds, **SHUT DOWN** the engine immediately and remove the restriction or check the cooling hose connection.

Caution must be exercised when moving the craft off a beach to avoid sucking in mud, sand, pebbles, or rocks through the intake. Such ingestion could cause damage to the impeller, the vanes in the impeller duct, or clog engine cooling passages.

The engine cooling passages must be flushed with clear water on a regular basis. This is accomplished by simply connecting a garden hose to the cooling water hose or fitting on the manifold, if so equipped. Such an easy maintenance practice should become a habit and will prevent accumulation of debris which could restrict engine coolant water flow and cause overheating.

1-7 REVERSE CAPABILITY

At press time, only a few Kawasakl watercraft were equipped with a "reverse gate". This "gate" is installed in such a manner to swing down over the steering nozzle and forces the water from the jet pump nozzle to be directed forward -- thus moving the craft sternward. The gate is controlled by the operator through a cable.

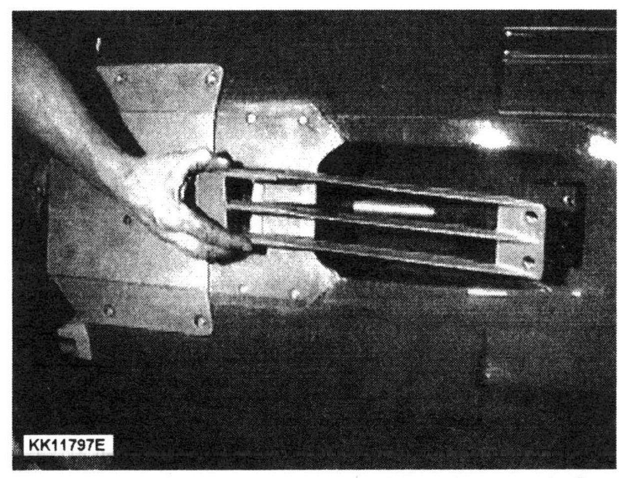

Any obstruction, such as a plastic bag, sucked up against the rock grate will drastically restrict water volume through the pump. A reduction in water flow affects craft performance and cooling water to the engine. Such a condition may quickly lead to a serious and expensive overheating problem.

With the "gate" is the fully raised position, all water is directed to the rear of the craft and the craft moves forward.

If the "gate" is lowered partially, some of the water is directed forward and some is allowed to move sternward -- the craft is "static" in the water and the "gate could be considered in a "neutral" position.

When the "gate" is allowed to move to the fully lowered position, almost all of the water is directed forward and the craft moves sternward -- the gate could be considered in a "reverse" position.

Adjustment of the reverse cable is covered in Chapter 11, Control Adjustments.

1-8 SPECIAL FEATURES

The Kawasaki Series covered in this manual have a couple of unique features worth mentioning at this time. One is an rpm limiter to prevent an almost "runaway" engine condition when the operator is wave jumping, using a ramp, or performing any number of other gyrations for fun or during competion.

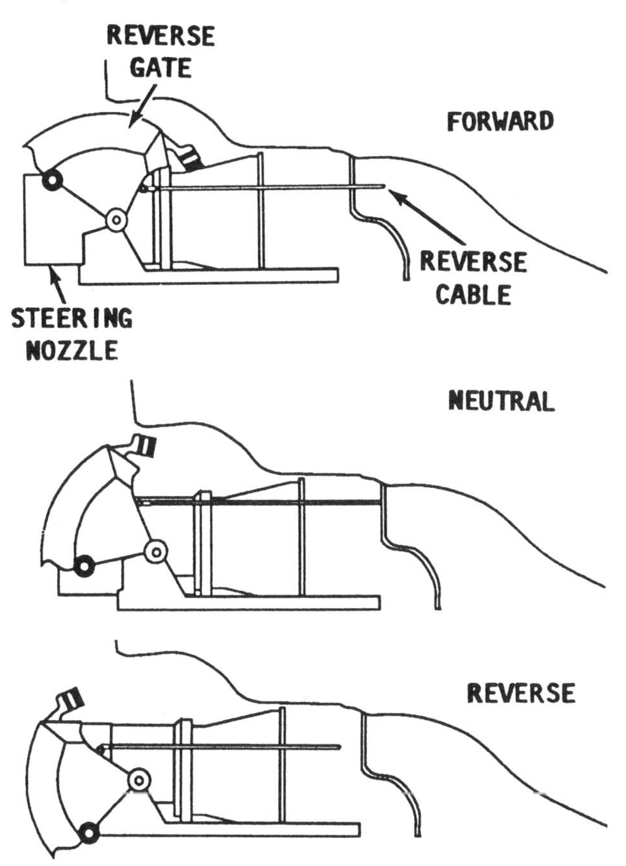

*Simple line drawings to illustrate water flow from the nozzle with the reverse "gate" in the full **FORWARD** position (top), in the **NEUTRAL** position (center), and in the full **REVERSE** position (bottom).*

A minor, but important feature --labeled "Throttle Opening Limiter" -- is a simple arrangement to limit engine rpm entirely. Such a device is important to outfits renting personal watercraft to novice users.

Another feature is a self-circling system enabling an operator who may be thrown from the craft to get back aboard and regain control of the craft.

RPM Limiter

Jumping waves can cause extensive internal engine stresses. As soon as the craft leaves the water, engine speed -- if left unchecked -- would rapidly increase to as much as 12,000 rpm. As soon as the craft re-enters the water, engine speed instantly decreases to about 6,000 rpm. This sudden increase and decrease in engine speed is dramatically hard on engine life. Repeated jumps have been known to cause bent crankshafts and/or broken connecting rods.

Therefore, the need certainly existed for an rpm limiter which would automatically reduce engine rpm under such conditions.

Since 1990, an electronic rev limiter has been installed on all models. This type limiter is, as the factory intended, very difficult to bypass. No! We are not going to reveal exactly how it functions or how it can be bypassed.

Throttle Opening Limiter

The engines covered in this manual are equipped with a device to limit throttle opening. If the owner wishes to limit craft speed while the unit is on loan -- or renting -- to a novice rider, a throttle limiter, next to the twist grip on the handle bar, as shown in the accompanying illustration, may be adjusted for this purpose.

The owner of the craft simply loosens the bolt -- slides the lever on the throttle limiter to the desired position -- tightens the bolt to hold the adjustment -- thus, engine rpm is limited and the craft's speed is restricted.

Self-Circling Mode

Some Kawasaki models are equipped with an emergency tether behind the engine "kill" switch. On all other craft, Kawasaki engineers incorporated a self-circling feature causing the craft to automatically follow a large circular course if the operator should fall or be thrown from the craft.

The theory behind this feature is similar to that of a free rolling automobile tire. When a single tire is set rolling freely, it will eventually slow down -- begin to wobble -- finally circle on its edge on one side -- eventually coming to rest on the ground.

With the personal watercraft -- once the operator leaves the craft, engine rpm is automatically reduced to an idle speed. Because the center of gravity for the craft shifts dramatically forward due to the loss of the rider, the bow drops and the craft begins to "plow" through the water instead of "planning" on an even keel.

Now, the action of waves against the hull will act to provide a continuous force preventing the craft from following a straight course. This wave action constantly pushing on the craft will cause it to follow a circular course. Without steering from an operator, the craft slowly makes a circular sweep to port -- or to starboard -- giving the operator the chance to catch the craft as it comes around -- climb aboard -- and continue his fun ride.

This self-circling feature will **ONLY** work if the engine idle speed is set correctly. If idle speed is set too high -- the craft will continue in a straight course.

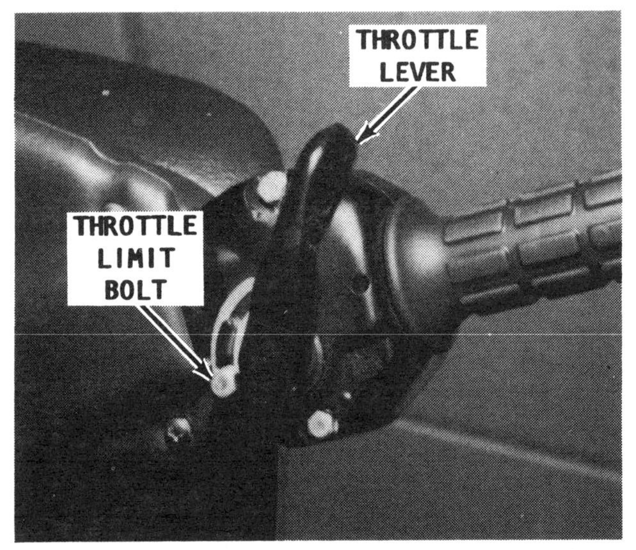

A throttle limiter device may be employed to limit engine rpm and thus craft speed. Rental outfits find this device most useful to restrict "customers" speed on the water.

2
SAFETY

2-1 INTRODUCTION

Personal watercrafting is one of the fastest growing leisure activites in the world. Riding a personal watercraft can be an exciting, exhilarating, and safe experience provided the operator exercises prudent behavior and remains in constant control of the vehicle.

Personal watercraft may be likened to a firearm. Weapons do not injure people, but the individual handling them is the true culprit. In a similar manner, if the personal watercraft is operated in an unresponsible manner the driver, passengers, and others using the same waterway may be in danger.

The greatest hazard in operating a personal watercraft stems not from the craft itself, but from the behavior of the operator. The Personal Watercraft Industry Association (PWIA) and the National Marine Manufacturers Association (NMMA) recognize there are a number of problems due to irresponsible craft operators contributing to tarnishment of the sport. Specifically, these problems range from alcohol and substance abuse to excessive noise, speed, and reckless maneuvering. Insensitive operators can disrupt an otherwise enjoyable day for other craft users through misconduct on the water. These individuals often endanger their own lives as well as the lives of others by losing full control of their craft.

Organized events, races, and exhibitions, properly planned and controlled have proven to be a spectactular event appealing to a wide range and number of participants and spectators.

Craft Classification

Any craft equipped with propulsion machinery is classified as a motorboat.

Personal watercraft are classified as "Class A inboard boats" by the Coast Guard and therefore are subject to most of the same laws and requirements as more conventional larger craft. A decal, usually affixed somewhere in the stern area, lists exceptions the Coast Guard has granted for this type of water vehicle.

Information

In 1989, several states recognized the need to impose specific regulations governing operation of personal watercraft. Rules and regulations differ from state to state and are constantly being revised. Therefore, it would be an impossible task to present current legislation on personal watercraft activites. Before the printing ink for this publication was dry, additions, deletions, and revisions would already be in effect.

For current information, the reader may call 1-800-336-BOAT for the name and ad-

*PWC owners and their friends prepare for a **SAFE** fun day on the water. By paying attention to U.S. Coast Guard safety requirements and using good judgement, their enjoyment will actually be increased many fold.*

dress of the local Boating Law Administrator. The reader may also contact the PWIA Government Relations Representative, Washington, D.C.

Regulation Enforcement

The U.S. Coast Guard and state enforcement officers have the authority to stop any craft to check for compliance with federal or state law. They can order a craft to return to the closest dock and remain out of the water until the hazardous condition of the craft is corrected.

Examples are:

Inadequate number of Personal Flotation Devices.

Missing/Inoperative fire extinguisher.

Overloading.

Operation at night without proper lights.

Fuel leakage.

Fuel transportation (other than approved fuel tank).

Failure to meet ventilation requirements.

Failure to meet carburetor backfire flame arrestor requirements.

Excessive leakage or accumulation of water in the bilge.

The U.S. Coast Guard and state enforcement officers have the authority to stop any craft if they determine the craft is being operated in a negligent manner. Negligent operation is defined as "the failure to excercise the degree of care necessary to prevent the endangering of life, limb, or property of any person".

Examples are:

Operating while under the influence of drugs or alcohol.

Excessive speed in a congested area.

Excessive speed in stormy or foggy conditions.

Coming too close to another vessel.

Operating in a swimming area where bathers are present.

Towing water skiers where obstructions exist or where a fall might cause skier to be struck by another vessel.

Operating in the vicinity of a dam.

Cutting through a marine parade or regatta.

2-2 MINIMUM LEGAL REQUIREMENTS FOR EQUIPMENT ONBOARD CRAFT

All personal watercraft owners **MUST** provide the following equipment:

One approved Type I, II, III or IV PFD (personal flotation device) for each person on board or being towed on water skis, etc.

An efficient sound producing device.

A B-I approved hand portable fire extinquisher.

Visual distress signals for nighttime use, if operated on coastal waters.

Personal Flotation Devices (PFDs)

The Coast Guard requires an approved life-saving device be worn by each person on board. Devices approved are identified by a tag indicating Coast Guard approval. Such devices may be life preservers, or buoyant vests. Ring buoys, or buoyant cushions are not acceptable.

Life preservers have been classified by the Coast Guard into five type categories. All PFDs presently acceptable on recreational craft fall into one of these five designations. Only four of the five catagories are stipulated by legal requirements. Type V is not acceptable for use on a personal watercraft.

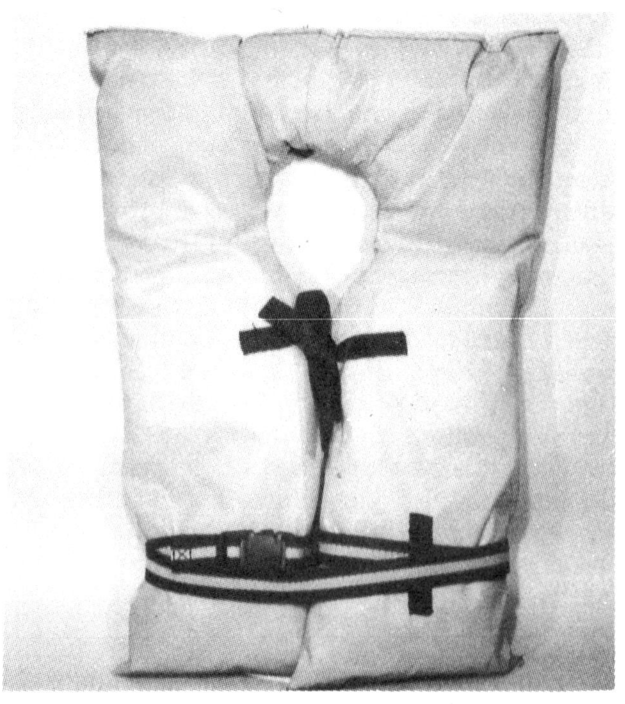

Type I PFD Coast Guard approved life jacket. This type flotation device provides the greatest amount of buoyancy.

A Type III PFD requires the wearer to be active to remain upright in the water. This type device is comfortable and allows the wearer to actually swim while in the water.

A PFD **MUST** be U.S. Coast Guard approved, in good and serviceable condition, and of an appropriate size for the person intending to wear it.

It is an accepted fact that most boating people own life preservers, but too few actually wear them. There is little or no excuse for not wearing one because the modern comfortable designs available today do not subtract from an individual's boating pleasure.

PFDs may have a long serviceable life, if proper care is exercised and the device is not abused. Wipe them clean and dry before storing in a well ventilated area during the off-season. **NEVER** use a PFD as a fender. A crushed PFD loses its buoyancy and becomes useless.

Type I PFD has the greatest required buoyancy and is designed to turn most **UNCONSCIOUS** persons in the water from a face down position to a vertical or slightly backward position. The adult size device provides a minimum buoyancy of 22 pounds and the child size provides a minimum buoyancy of 11 pounds. The Type I PFD provides the greatest protection to its wearer and is most effective for all waters and conditions.

Type II PFD is designed to turn its wearer in a vertical or slightly backward position in the water. The turning action is not as pronounced as with a Type I. The device will not turn as many different type persons under the same conditions as the Type I. An adult size device provides a minimum buoyancy of $15\frac{1}{2}$ pounds, the medium child size provides a minimum of 11 pounds, and the infant and small child sizes provide a minimum buoyancy of 7 pounds.

Type III PFD is designed to permit the wearer to place himself (herself) in a vertical or slightly backward position. The Type III device has the same buoyancy as the Type II PFD but it has little or no turning ability. Many of the Type III PFD are designed to be particularly useful when water skiing, sailing, hunting, fishing, or engaging in other water sports. Several of this type will also provide increased hypothermia protection.

Type IV PFD is designed to be thrown to a person in the water and grasped and held by the user until rescued. It is **NOT** designed to be worn. The most common Type IV PFD is a ring buoy or a buoyant cushion.

ONE LAST WORD

Common sense dictates the "cheap and cheerful" approach does not pay when applied to a life preserving device. Select and buy the very best. The life you save may be your own.

NOW, THESE WORDS

Due to the nature and maneuvering characteristics of personal watercraft, and from a "practical" standpoint, the following paragraph on bells, horns, and whistles hardly apply. The information is included to make the manual "complete" and because according to the U.S. Coast Guard they are required. Also, the authors feel the reader may "skipper" or "crew" on a larger craft and therefore, might be required to use the information presented here.

Bells, Horns and Whistles

All craft must carry some means of producing an efficient sound signal and the sound must be audible for a specific distance and duration, depending on the size of the craft. For personal watercraft, an audible sound must carry for one half mile and be of four to six second duration.

Law enforcement craft are the **ONLY** boats allowed to use sirens.

Personal watercraft operators are required to sound fog and maneuvering signals when underway in fog or rain as a means of communication between craft which are not visible. This system was devised to alert nearby craft of the presence of others in the water and as a means of communicating maneuvering intentions. Signals indicating maneuvering tactics are short one second blasts.

One blast indicates the craft is turning to starboard.
Two blasts indicate the craft is turning to port.

Three blasts indicate the craft is in reverse.
Five or more blasts indicate - stay away - danger.

A prolonged blast lasting 4 to 6 seconds duration every two minutes indicates a straight line course in poor visibility.

These signals should be memorised by every personal watercraft operator, not necessarily because the operator would give them, but in the event of a head on collision with say a large powerboat, the skipper may

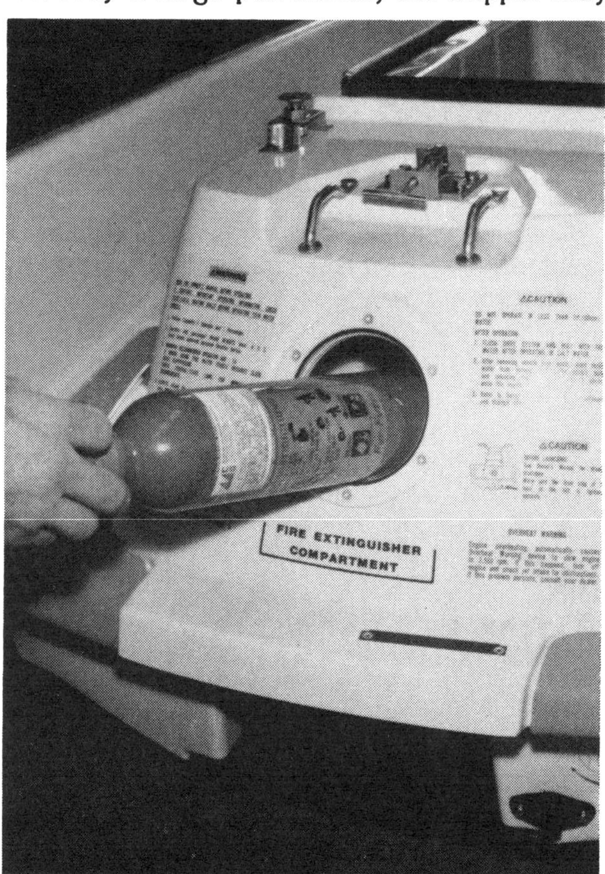

Personal Watercraft are required to carry a Type B-1 fire extinguisher aboard. On many craft, a compartment for the extinguisher is located in the stern where it is readily available.

signal his intention to turn either to port or starboard using this system of communication.

Fire Extinguishers

Personal watercraft are classified as "Class A inboard boats" by the Coast Guard and therefore are required to carry a B-I fire extinguisher.

All fire extinguishers must bear Underwriters Laboratory (UL) "Marine Type" approved labels. With the UL certification, the extinguisher does not have to have a Coast Guard approval number.

B-I contains 1-1/4 gallons foam, or 4 pounds carbon dioxide, or 2 pounds dry chemical agent, or 2-1/2 pounds Halon.

READ labels on fire extinguishers. If the extinguisher is U.L. listed, it is approved for marine use.

Any personal watercraft can be rolled over and held under water for a few seconds to extinguish the fire. Turning the craft over in the water will prevent oxygen from entering the engine compartment. Without oxygen, the fire will soon exhaust the remaining oxygen in the compartment and then go out. **DO NOT** open the engine compartment until the smoke has completely stopped, then open with great caution and have a fire extinguisher at hand.

Visual Distress Signals

Coast Guard Regulations require personal watercraft to carry night visual distress signals when used on coastal waters - even though it is illegal to operate a personal watercraft at night.

Coastal waters include: the Great Lakes, the territorial seas and those waters directly connected to the Great Lakes and the territorial seas, up to a point where the waters are less than two miles wide.

These signals assist rescuers in locating and aiding people in distress. This information is useful to every personal watercraft owner, to know how to respond to others in distress.

For night use: Pyrotechnic visual distress signaling devices (red flares, hand held or aerial) **MUST** be Coast Guard Approved, in serviceable condition and stowed to be readily accessible. If they are marked with a date showing the serviceable life, this date must not have passed.

Coast Guard Approved pyrotechnic devices carry an expiration date. This date can **NOT** exceed 42 months from the date of manufacture and at such time the device can no longer be counted toward the minimum requirements.

The only non-pyrotechnic visual distress signaling device approved by the coast guard for use at night is an electric distress light -- not a flashlight but an approved electric distress light which **MUST** automatically flash the international **SOS** distress signal (. . . - - - . . .)four to six times each minute.

2-3 MINIMUM LEGAL REQUIREMENTS FOR REGISTRATION OF CRAFT

Federal regulations require all personal watercraft to be issued a registration number and have the number displayed in a specific manner on the craft.

The "Certificate of Number" is issued by the state and must be carried on board at all times. The Hull Identification Number (HIN) is necessary to register the craft with the state.

The registration number must be displayed as follows:

The figures must be read from left to right.

The figures must be displayed on the forward half of each side of the bow of the craft.

The figures must be in bold, block letters of good proportion and must not be less than 3 inches high.

The figures must be of contrasting color to the craft hull, or background.

The figures must be as high above the waterline as practical.

No figures other than those assigned to the craft can be displayed on the forward half of the craft.

Letters must be separated from numbers by spaces or hyphens.

Validation decals (if required by state) must be displayed within 6 inches of the registration number.

Elaborate color schemes and custom detailing are very much a part of the fun of owning a personal watercraft. **HOWEVER,** the paint design must not obstruct the registration numbers or the validation decal. Like an automobile without a license plate, sooner or later a craft with no visible registration numbers will attract the attention of the law.

Hull Identification Number

Federal regulations require all personal watercraft to have a Hull Identification Number (HIN) permanently attached to the craft in two separate locations: on the starboard side of the transom, above the waterline. The HIN must also be displayed in an unexposed location, which is usually left to the discretion of the manufacturer.

2-4 SAFETY PRACTICES

MOST IMPORTANT WORDS IN THIS MANUAL

Under normal vehicle operation -- in an automobile, truck, motorcycle, tractor -- if an emergency situation arises in front of the vehicle, the immediate response is foot off the throttle **-- reducing SPEED** -- and hit the brake.

WRONG, WRONG, WRONG, in a personal watercraft. Water rushing through the jet pump and out the steering nozzle is the **ONLY** means of control. Once the throttle is fully backed off, direction of the craft can **NOT** be changed regardless of how drastic the operator may move the handle bars. The craft will continue on its course.

THEREFORE, if an emergency arises dead ahead, **DO NOT** back off the throttle completely. Reduce speed, but keep control.

Full Throttle Operation

Prolonged operation at full throttle will raise the cylinder head temperature by

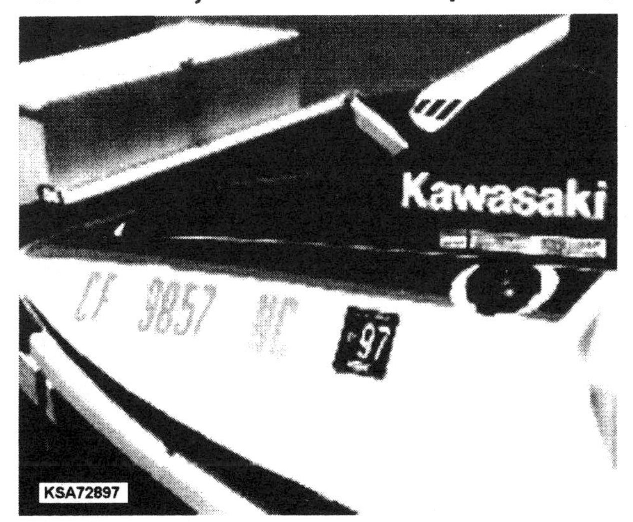

The validation decal and registration number are required by almost all states and Canadian provinces to be displayed port and starboard on the forward portion of the hull.

about 100° above normal. Some units are equipped with a thermo-switch engine temperature sensor. This sensor is connected directly to the CDI unit and grounded to the engine kill switch. If the engine temperature exceeds a maximum value specifed by the manufacturer, the ignition will be grounded and the engine shut down to prevent damage through overheating.

Jumping Waves

In the very early days of personal watercraft, jumping could cause extensive internal engine damage. When the craft left the water, unchecked engine rpm rapidly increased to as high as 12,000 rpm. As soon as the craft re-entered the water, engine rpm instantly decreased to about 6,000 rpm. This sudden increase and decrease in engine speed dramatically shortened engine life. In those days, repeated jumps were known to cause bent crankshafts, sheared flywheel keys, and broken connecting rods.

Yamaha engineers developed an rpm limiter system as an integral part of the CDI unit. This limiter prevents high engine rpm and an almost "runaway" condition, during wave or ramp jumping. On "stock", out the door factory units, this limiter restricts engine speed to about 7100 rpm. Thus, wave jumping, using a ramp, and a host of other almost unbelievable manuevers can be performed without undue stress on the engine. Today, with an rpm limiter system installed, a bent crankshaft or broken connecting rod is almost non-existent.

Kawasaki engineers developed the limiter system to make it very difficult to bypass, as was done many times in the past. No!, we shall not reveal operation or possibilities for bypass.

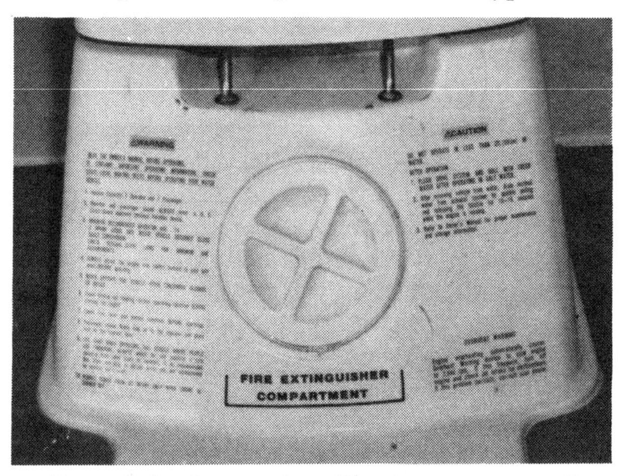

Several decals may be affixed to the stern of the hull to indicate compliance with or exceptions granted to normal Coast Guard regulations.

Alcohol and Substance Use

Operating a personal watercraft while under the influence of either alcohol or drugs is both illegal and life threatening. This fact cannot be stated too strongly. Such substances affect vision and slow reaction times. The danger is quadrupled in a craft on the water compared to on the shore.

Age Restrictions

Children under 14 years of age should not operate a watercraft unless accompanied on a two passenger craft by an adult experienced operator. Specific minimum age limits have been imposed by many states.

Speed Restrictions

Where no speed limits are posted, commonsense dictates the craft should be operated so it will pose no danger to other water users or the operator. Many states now have enforced speed limits for certain areas and conditions.

When entering "no wake" or 5 mph zones, such as fishing areas, swimming areas or marinas, the operator must slow to headway speed, the slowest speed at which appropriate steering capability can be maintained.

Individuals responsible for law enforcement on the water are keeping an ever vigilant eye on personal watercraft in many areas. Horseplay such as "wetting down" or "buzzing" other water users may result in a "ticket" for negligent operation.

Operation at Night

Personal watercraft are not equipped with lights and therefore, it is illegal to operate such a craft on the water after sunset.

Shallow Water Operation

Under normal conditions, personal watercraft may be operated in very shallow waters, and it is possible to maneuver the craft directly onto a sloping beach.

Caution must be exercised when moving the craft off a beach to avoid sucking in mud, sand, pebbles, or rocks through the intake. Such ingestion could cause damage to the impeller, the vanes in the pump bowl, or clog engine cooling passages.

WARNING
Expelled sand and small pebbles may cause INJURY to persons in the area.

A grated section set into the hull bottom helps prevent the ingestion of small rocks and pebbles.

Carry about a foot of stiff wire aboard for use in probing into the impeller area to dislodge debris where fingers cannot reach.

Engine Compartment Ventilation

All motorboats (personal watercraft included) built after April 25, 1940 and before August 1, 1980, powered by a gasoline engine or by fuels having a flashpoint of 110° F or less **MUST** have the following, which is quoted from a recent Coast Guard publication:

> At least two ventilation ducts fitted with cowls or their equivalent for the purpose of properly and efficiently ventilating the bilges of every engine and fuel tank compartment. Each duct must be at least two inches in internal diameter.
>
> There shall be at least one exhaust duct installed so as to extend from the lower portion of the bilge to cowls in the open air, and at least one intake duct installed so as to extend to a point at least midway to the bilge or at least below the level of the carburetor air intake.

Flame Arrestors

A flame arrestor, as the name suggests, safeguards against flame caused by engine backfire.

A gasoline engine installed in a motorboat or motor vessel after April 25, 1940, except outboard motors, must have a Coast Guard Approved flame arrestor fitted to the carburetor. This requirement applies to personal watercraft because the engine is enclosed.

Fuel System

All parts of the fuel system should be selected and installed to provide maximum service and protection against leakage.

The capacity of the fuel filter must be large enough to handle the demands of the engine as specified by the engine manufacturer.

All fittings and outlets must come out the top of the tank.

In order to obtain maximum circulation of air around fuel tanks, the tank should not come in contact with the hull except through the necessary supports. The supporting surfaces and hold-downs must fasten the tank firmly and they should be insulated from the tank surfaces. This insulation material should be non-abrasive and non-absorbent material.

Taking On Fuel

The fuel tank should be kept almost full to allow for expansion, but still prevent water from entering the system through condensation caused by temperature changes. Water droplets forming are one of the greatest enemies of the fuel system. By keeping the tank almost full, the air space in the tank is kept to an absolute minimum and there is less room for moisture to form. It is a good practice not to store fuel in the tank over an extended period, say for six months. Today, fuels contain ingredients that change into gums when stored for any length of time. These gums and varnish products will cause carburetor problems and poor spark plug performance. An additive (Sta-Bil) is available and can be used to prevent gums and varnish from forming.

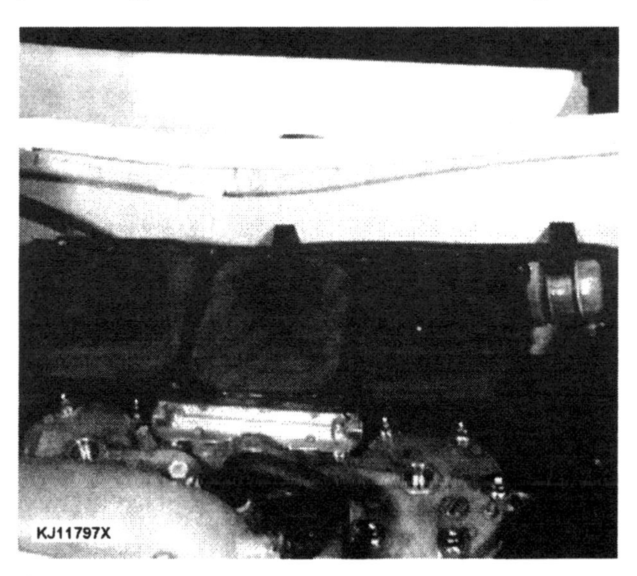

Carburetors on personal watercraft are required to be fitted with a flame arrestor because the engine compartment is an enclosed area.

Excessive Noise

Every manufacturer designs a stock exhaust system to keep noise to a minimum. A poorly maintained or modified exhaust system will produce excessive and disturbing noise. This is a form of air polution and is illegal in most states.

Automotive Replacement Parts

When replacing fuel, electrical, and other parts, check to be sure they are marine type and Coast Guard Approved. Automotive parts are not made to the high standards of marine parts. The carburetors must **NOT** leak fuel; electrical equipment **MUST** be able to operate in a gas fume enclosed area without exploding; etc. Automotive parts could cause a fire endangering the operator, passengers, and the craft. The part may look the same and even have a similar number, but if it is not **MARINE** it is not safe to use and will not pass Coast Guard inspection.

Overloading

Each personal watercraft is designed to carry a specific number of persons. Attempting to board more passangers than the craft is designed to carry, will upset weight distribution causing a marked decrease in performance and possible loss of control.

2-5 BOATING ACCIDENT REPORTS

New federal and state regulations require an accident report to be filed with the nearest state boating authority within 48 hours if a person is lost, disappears, or is injured to the degree of needing medical treatment beyond first aid. The time limit is reduced to 24 hours if the accident results in a loss of life.

Accidents involving only property or equipment damage **MUST** be reported within 10 days, if the damage is in excess of $500.00. Some states require reporting of accidents with propery damage less then $500.00 or a total boat loss. A $1,000.00 **PENALTY** may be assessed for failure to submit the report.

WORDS OF ADVICE

Take time to make a copy of the report to keep for your records or for the insurance company. Once the report is filed, the Coast Guard will not give out a copy, even to the person who filed the report.

The report must give details of the accident and include:

1- The date, time, and exact location of the occurrence.

2- The name of each person who died, was lost, or injured.

3- The number and name of the vessel.

4- The names and addresses of the owner and operator.

If the operator cannot file the report for any reason, each person on board **MUST** notify the authorities, or determine that the report has been filed.

In nautical terms, the front of the craft is the **bow;** the rear is the **stern;** the right side, when facing forward, is the **starboard** side; and the left side is the **port** side. One easy way to remember this basic fundamental is to consider the words "port" and "left" both have four letters and go together.

All directional references in this manual use this terminology. Therefore, the direction from which an item is viewed is of no consequence, because **starboard** and **port** **NEVER** change no matter where the individual is located, or standing, even on his or her head.

2-6 SECURITY

As mentioned in the opening paragraphs of this chapter, personal watercrafting is one of the fastest growing leisure activites in the world. Like most personal possessions, personal watercraft have value and valuables are subject to be lost to unscrupulous characters -- thieves. The best advice for a personal watercraft owner is "lock it or lose it".

Do not leave the craft unattended on the beach or dock. **ALWAYS** remove the kill switch tether from the switch and of course, the ignition key, if the unit is equipped with these features. Certain aftermarket security devices are available for installation to prevent theft of personal watercraft. These include: A "secret" ignition grounding switch, which grounds one side of the ignition coil, and a hidden fuel shut-off valve.

Most owners transport their craft on trailers or in the bed of a pickup truck. If loaded on a trailer, chain and lock the craft to the trailer. Invest in a hitch lock so the trailer cannot be detached from the towing vehicle. If the craft is loaded on a pickup truck, chain and lock it to the bed, or to a large and heavy object.

3
TUNING

3-1 INTRODUCTION

The efficiency, reliability, fuel economy and enjoyment available from engine performance are all directly dependent on having the engine tuned properly. The importance of performing service work in the sequence detailed in this chapter cannot be overemphasized. Before making any adjustments, check the specifications in the Appendix. **NEVER** rely on memory when making critical adjustments.

Before beginning to tune any engine, check to be sure the engine has satisfactory compression. An engine with worn or broken piston rings, burned pistons, or scored cylinder walls, cannot be made to perform properly no matter how much time and expense is spent on the tune-up. Poor compression must be corrected or the tune-up will not give the desired results.

A practical maintenance program that is followed throughout the year, is one of the best methods of ensuring the engine will give satisfactory performance at any time.

The extent of the engine tune-up is usually dependent on the time lapse since the last service. A complete tune-up of the entire engine would entail almost all of the work outlined in this manual. A logical sequence of steps will be presented in general terms. If additional information or detailed service work is required, the chapter containing the instructions will be referenced.

Each year higher compression ratios are built into modern marine engines and the electrical systems become more complex, especially with electronic (capacitor discharge) units. Therefore, the need for reliable, authoritative, and detailed instructions becomes more critical. The information in this chapter and the referenced chapters fulfill that requirement.

3-2 TUNE-UP SEQUENCE

During a major tune-up, a definite sequence of service work should be followed to return the engine to the maximum performance desired. This type of work should not be confused with attempting to locate problem areas of "why" the engine is not performing satisfactorily. This work is classified as "troubleshooting". In many cases, these two areas will overlap, because many times a minor or major tune-up will correct the malfunction and return the system to normal operation.

The following list is a suggested sequence of tasks to perform during the tune-up service work. The tasks are merely listed here. In most cases, procedures are

A clean, properly tuned, and adequately maintained power unit can multiply by many fold the enjoyment derived from owning and operating a personal watercraft.

given in subsequent sections of this chapter. For more detailed instructions, see the referenced chapter, for the unit being seviced.

1- Perform a compression check of each cylinder. See Ignition Chapter.
2- Inspect the spark plugs to determine their condition. Test for adequate spark at the plug. See Ignition Chapter.
3- Start the engine in a body of water or connect a "Flushing Kit" with a garden hose and check the water flow through the engine, See Engine Chapter.
4- Check the carburetor adjustments and the need for an overhaul. See Fuel Chapter.
5- Check the fuel pump for adequate performance and delivery. See Fuel Chapter.
6- Make a general inspection of the ignition system. See Ignition Chapter.
7- Test the cranking motor and the solenoid. See Electrical Chapter.
8- Check the internal wiring.

3-3 COMPRESSION CHECK

A compression check is extremely important, because an engine with low or uneven compression between cylinders **CANNOT** be tuned to operate satisfactorily. Therefore, it is essential that any compression problem be corrected before proceeding with the tune-up procedure.

If the engine shows any indication of overheating, such as discolored or scorched paint, inspect the piston skirts visually thru the transfer ports, using a mirror and flashlight, for possible scoring. It is possible for a cylinder with satisfactory compression to be scored slightly.

Checking Compression

Remove the spark plug wires. **ALWAYS** grasp the molded cap and pull it loose with a twisting motion to prevent damage to the connection. Remove both spark plugs and remember from which cylinder they came for evaluation later. Ground both spark plug leads to the engine to render the ignition system inoperative while performing the compression check.

Insert a compression gauge into the forward spark plug opening. Crank the engine with the cranking motor, through at least four complete revolutions of the crankshaft, with the throttle at the wide-open position, to obtain the highest possible reading. Repeat the test and record the compression for the other cylinders.

The manufacturer specifies the compression pressure for a new engine should be 142.2 psi (981 kpa) for two cylinder engines, 110 psi (759 kpa) for three cylinder engines. A variation between cylinders is far more important than the actual readings. A variation of more than 15 psi between the two cylinders indicates the lower compression cylinder is defective. The problem may be worn, broken, or sticking piston rings, scored pistons or worn cylinders.

Use of an engine cleaner will help to free stuck rings and to dissolve accumulated carbon. Several different brand names are available from the local marine or automotive parts house. Follow the directions on the can closely.

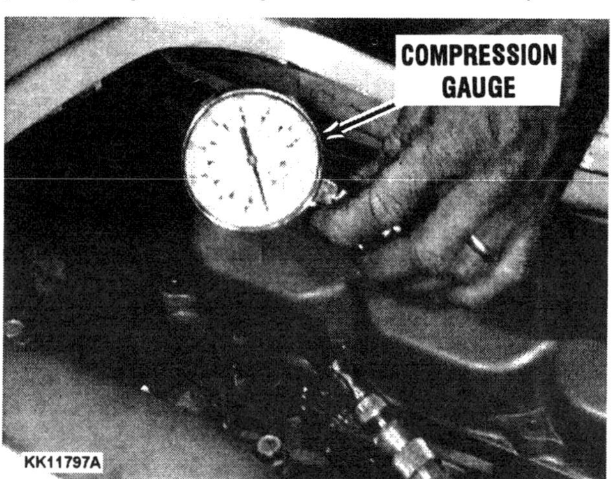

Performing a compression test on a three-cylinder engine. The spark plug high tension leads have been grounded in the rack on the electrical box -- shown in the next column -- to prevent placing an unnecessary extra load on the ignition coil.

Clear view of the rack for grounding the high-tension spark plug leads whenever special tests are conduncted on the ignition system.

3-4 SPARK PLUG INSPECTION

Inspect both spark plugs for badly worn electrodes, glazed, broken, blistered, or lead fouled insulators. Replace both the plugs, if either shows signs of excessive wear.

Make an evaluation of the cylinder performance by comparing the spark or "firing" condition with those shown in the Ignition Chapter. Because different engines may require different configurations, an increased number of spark plugs are manufactured to meet the specifications. Check each spark plug to be sure they are both of the same manufacturer and have the same heat range rating. If the new spark plugs are not of the same heat range, misfiring of the engine may occur.

When purchasing new spark plugs, **ALWAYS** ask the dealer if there has been a spark plug change for the engine being serviced. The spark plug gap standard is 0.030" for all series engines covered in this manual.

Crank the engine through several revolutions to blow out any material which might have become dislodged during cleaning.

Install the spark plugs and tighten them to a torque value of 14 ft lbs (20Nm). **ALWAYS** use a new gasket and wipe the seats in the head clean. The gasket must be fully compressed on clean seats to complete the heat transfer process and to provide a gas tight seal in the cylinder. If the torque value is too high, the heat will dissipate too rapidly. Conversely, if the torque value is too low, heat will not dissipate fast enough.

Broken Reed

A broken reed is usually caused by metal fatigue over a long period of time. The failure may also be due to the reed flexing too far because the reed stop has not been adjusted properly or the stop has become distorted.

If the reed is broken, the loose piece **MUST** be located and removed, before the engine is returned to service. The piece of reed may have found its way into the crankcase, behind the by-pass cover. If the broken piece cannot be located, the engine must be completely disassembled until it is located and removed.

An excellent check for a broken reed on an operating engine is to hold an ordinary business card in front of the carburetor. Under normal operating conditions, a very small amount of fine mist will be noticeable, but if fuel begins to appear rapidly on the card from the carburetor, one of the reeds is broken and causing the backflow through the carburetor onto the card.

A broken reed will cause the engine to operate roughly and with a "pop" back through the carburetor.

The reeds must **NEVER** be turned over in an attempt to correct a problem. Such action would cause the reed to flex in the opposite direction and the reed would break in a very short time.

3-5 ELECTRICAL POWER SUPPLY

Remove the battery from the engine compartment for service. Inspect and service the battery, cables and connections. Check for signs of corrosion. Inspect the battery case for cracks or bulges, dirt, acid, and electrolyte leakage.

Make sure the breather pipe is attached to the battery and not pinched by any part of the engine compartment.

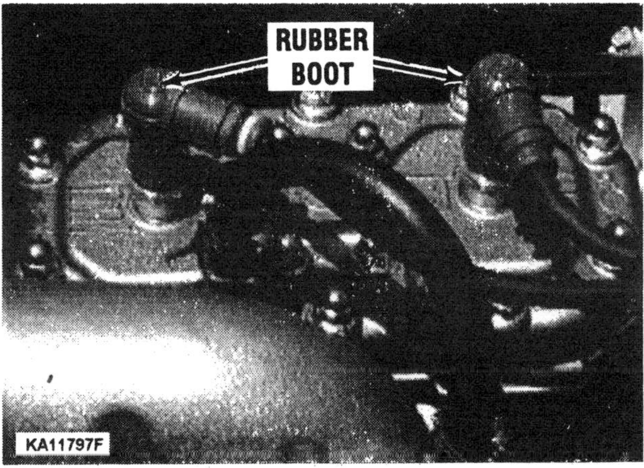

After a compression check or inspection of the spark plugs, the high-tension leads should be firmly connected to each plug, and then covered with the rubber boot.

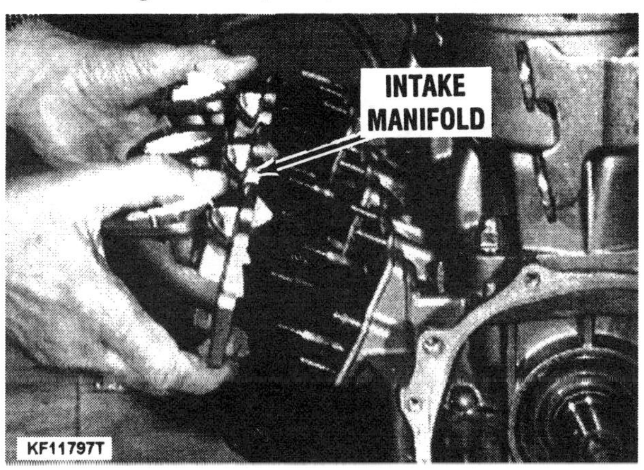

Removing/installing the intake manifold on a Model 900 Series engine prior to or following reed block service. The procedure is basically the same for other engines.

Clean the top of the battery. The top of a 12-volt battery should be kept especially clean of acid film and dirt, because of the high voltage between the battery terminals and prevent contact between the acid and the two rubber hold down bands. For best results, first wash the battery with diluted ammonia or baking soda solution to neutralize any acid present. Flush the solution off the battery with clean water. Keep the vent plugs tight to prevent the neutralizing solution or water from entering the cells.

"Maintenance Free Batteries"

Many batteries installed in personal watercraft are considered "Maintenance Free" batteries and do not require regular maintenance.

Standard Batteries

If the battery installed in the craft being serviced is not a "Maintenance Free" battery proceed with the following tasks.

Remove each filler cap using a pair of pliers. Fill each cell to the proper level with distilled water. The level of electrolyte should be between the upper and lower level marks during operation. When filling the battery, fill to the upper line and allow the battery to stand for twenty minutes. Check the level again and replenish as necessary.

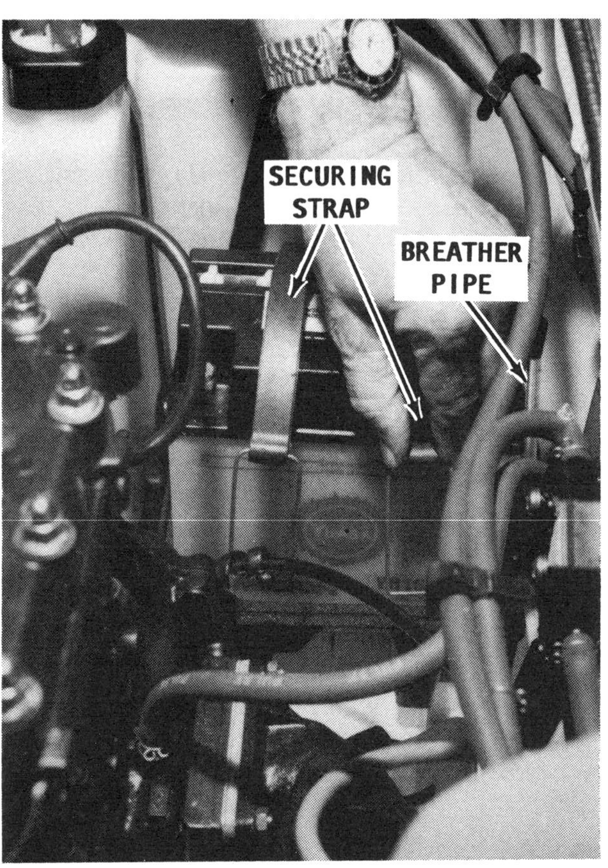

*Due to the violent maneuvers experienced by a personal watercraft -- more so than any other type craft -- the battery **MUST** be extremely well secured with **NO** movement possible.*

Use a temperature corrected hydrometer to test the specific gravity of the electrolyte. At 20°C (68°F) the hydrometer reading should be 1.26 on the scale. If it is necessary to charge the battery, leave the filler caps lightly resting on the cell openings to allow the gases to escape. The charging current should not exceed 1.9 amps for 10 hours.

After charging, push each filler cap into place and wipe the top of the battery clean before installation.

Clean the battery posts and battery cable ends with a wire brush to ensure good, clean connections. Clean the top surface of the battery and identify the "POS" or "+" and "NEG" or "-" embossed symbols. Correctly connect the battery cables to the battery observing polarity. If the cables are connected backwards, the ignition system **WILL** be destroyed the first time the engine is started.

Check to be sure the battery is fastened securely in position. The hold-down rubber straps should be tight enough to prevent any movement of the battery in the holder under the most violent maneuvers of the watercraft.

If the battery posts or cable terminals are corroded, the cables should be cleaned separately with a baking soda solution and a wire brush. Apply a thin coating of Multipurpose Lubricant to the posts and cable clamps before making the connections. The lubricant will help to prevent corrosion.

Jumper Cables

If booster batteries are used for starting an engine the jumper cables must be connected correctly and in the proper sequence to prevent damage to either battery, or the rectifier diodes.

ALWAYS connect a cable from the positive terminals of the dead battery to the positive terminal of the good battery **FIRST. NEXT,** connect one end of the other cable to the negative terminal of the good battery. Finally, connect the other end of the cable to the far side of the **ENGINE** for a good ground. By making the ground connection on the engine, any spark created will not be near the battery. An arc near the battery could cause an explosion, destroying the battery and causing serious personal injury.

If a trickle charger is used on a dead battery installed in the craft, one battery lead **MUST** be disconnected prior to connecting the charger, to prevent destroying the diodes in the rectifier.

NEVER use a trickle charger as a booster to start the engine because the diodes in the rectifier will be **DAMAGED.**

3-6 CARBURETOR ADJUSTMENT

Fuel and Fuel Tanks

Take time to check the fuel tank and all of the fuel lines, fittings, couplings, valves, flexible tank fill and vent. Turn on the fuel supply valve. If gas was not drained at the end of the previous season, make a careful inspection for gum formation. When gasoline is allowed to stand for long periods of time, particularly in the presence of copper, gummy deposits form. This gum can clog the filters, lines, and passageways in the carburetor.

Draining Fuel Tank

If the condition of the fuel is in doubt, drain, clean, and fill the tank with fresh fuel.

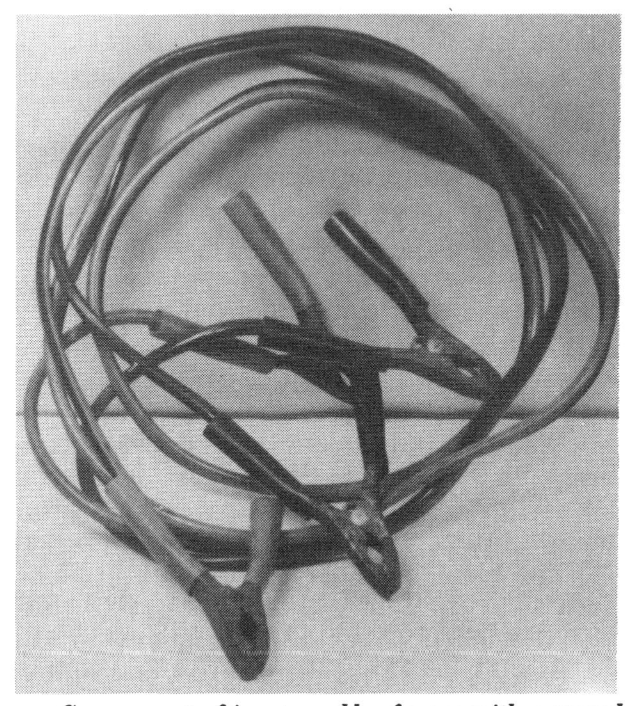

*Common set of jumper cables for use with a second battery to crank and start the engine. **EXTREME** care should be exercised when using a second battery, as explained in the text.*

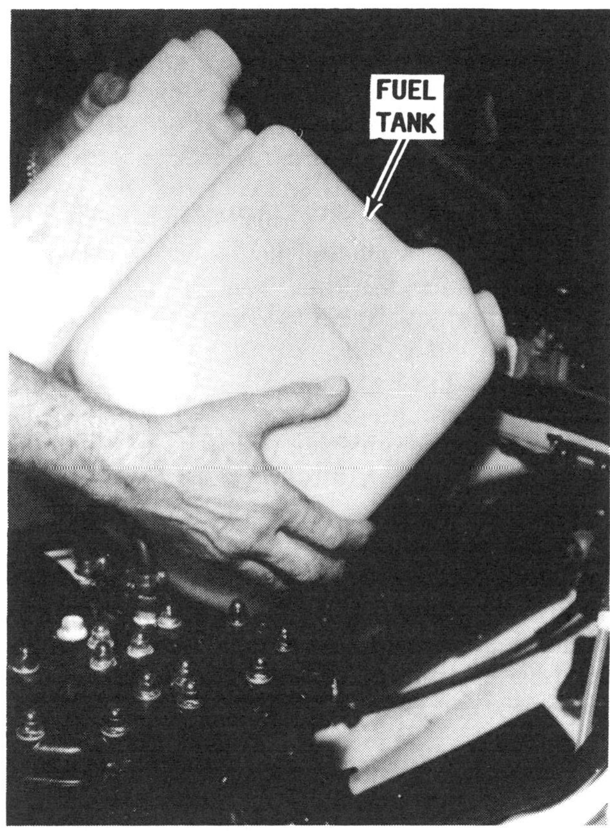

*Removing/installing the fuel tank in a Model 750SX stand-up craft in order to drain fuel which may have aged and "soured". **HOWEVER**, siphoning is more practical.*

The illustration at the top of this column shows a fuel tank being removed with the intention of draining the old fuel. Using a siphon method, as mentioned in the picture caption, is a much more practical method.

Add a 50:1 oil mixture to the first fuel at the beginning of a new season in addition to the oil supplied by the oil injection system.

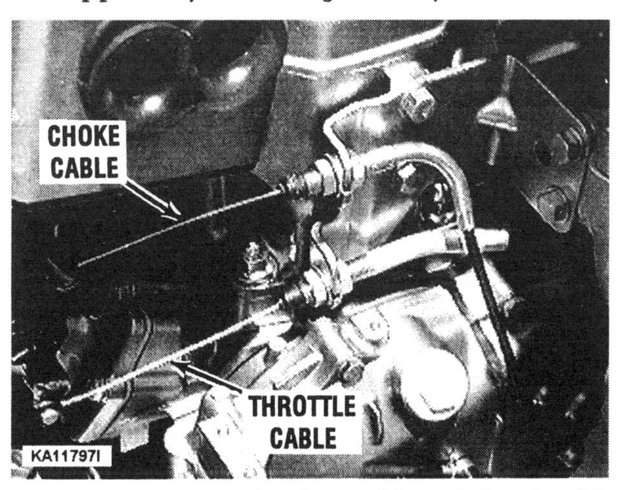

Typical setup for the choke cable and throttle cable adjustment through the locknut at each bracket. Chapter 6 -- Fuel and Oil covers adjustment of both cables. The bracket and routing of the cables will differ depending on the Model Series being serviced.

Low Speed and Idle Mixture Adjustment

The idle mixture and idle speed are set at the factory. Due to local conditions, it may be necessary to adjust the carburetor while the craft is operating in a test tank or secured in a body of water. For maximum performance, the idle mixture and the idle rpm should be adjusted under actual operating conditions.

Set the low speed adjustment screw at the specified number of turns open from a lightly seated position. Refer to the Appendix for the model being serviced.

Start the engine and allow it to warm to operating temperature.

CAUTION

Water must circulate from the jet pump to and from the engine anytime the engine is operating. Just a few minutes without cooling circulating water may cause extensive damage or seizure of the engine.

NEVER, AGAIN NEVER, operate the engine at high speed with a flush device attached. An engine operating at high speed with such a device attached, would **RUNAWAY** from lack of a load on the impeller shaft, causing extensive damage.

Connect a tachometer to the engine.

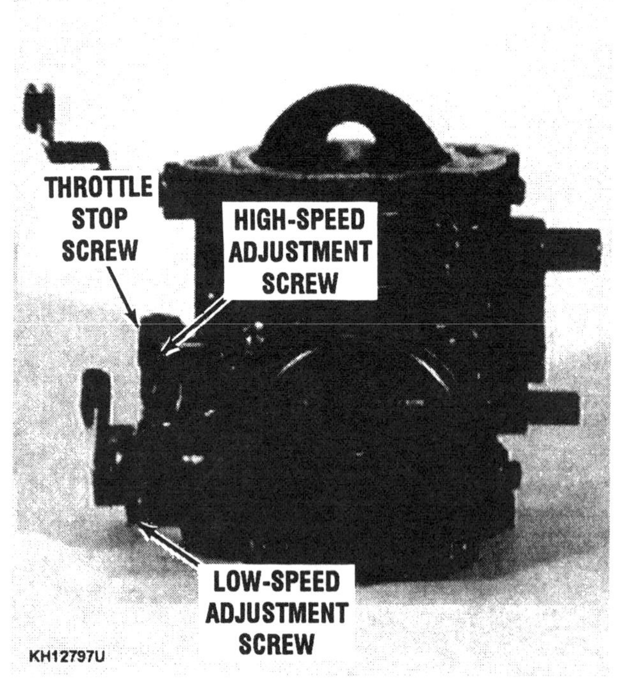

Open clear view of a Keihin CDK-40 carburetor with location of the low-speed and high-speed adjustment screws and the throttle stop screw clearly indicated.

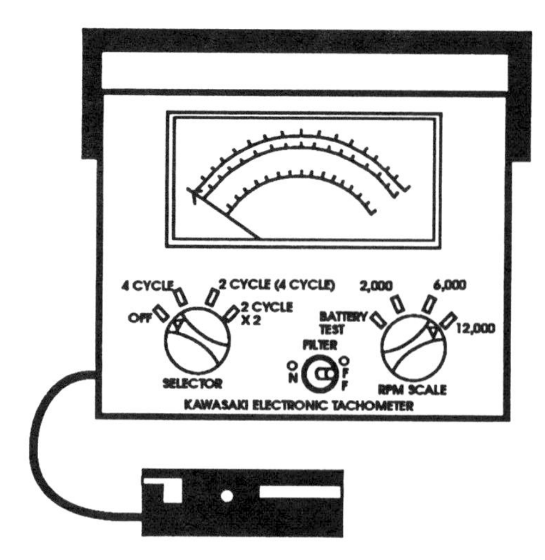

Maximum engine performance can only be obtained through proper tuning using a tachometer.

SPECIAL WORDS ON TACHOMETERS AND CONNECTIONS

A tachometer connected to the engine must be used to accurately determine engine speed during idle and high-speed adjustments.

Theoretically, it should be possible to connect a tachometer to any engine and receive an accurate count of crankshaft revolutions -- **RIGHT? WRONG!**

Unfortunately, this is not the case. Tachometer manufacturers admit, on certain two-stroke applications where points, and an ignition coil are **NOT** used, an attempt to connect a tachometer to the unit could "go up in smoke".

The general rules for connecting a tachometer to a marine two-stroke engine are:

1- The tachometer **MUST** be calibrated at the factory for two-stroke applications. Some tachometers have a dial indicator which may simply be switched from two-stroke to four-stroke and back again.

2- A tachometer may have as many as **FOUR** leads. The meter will always have a minimum of an input lead and a ground lead. The colors of these leads may vary between manufacturers, but the instructions provided with the meter will (or should), identify these leads. The input lead is connected to the primary **NEGATIVE** side of the ignition coil and the ground lead is connected to a suitable engine ground.

3- Some tachometers have a third wire which is considered the "hot" lead. The "hot" lead is attached to the **POSITIVE** battery terminal. On other tachometers, this "hot" lead is only used for an internal light bulb to illuminate the meter face. On still other tachometers this "hot" lead is necessary to the operation of the meter.

Once again, the instructions with the tachometer should identify the "hot" lead.

4- A fourth lead from the tachometer is used as the ground lead to the light bulb.

FIRST, last, and **ALWAYS**, check the tachometer manufacturer's instructions for use with a two-stroke unit.

Procedure

With the engine running, slowly turn the throttle stop screw until the engine idles at 1,500 rpm. Then rotate the low speed screw in or out 1/8th turn until the cylinders fire evenly and engine rpm increases. Readjust the throttle stop scew to reduce the rpm to idle speed.

High Speed Screw Adjustment

Set the high speed adjustment screw at the specified number of turns open from a lightly seated position. Refer to the Appendix for the model being serviced. This is the best setting for sea level. For operating at higher elevations, a leaner mixture is obtained by rotating the screw **CLOCKWISE** slightly, 1/8 turn at a time.

3-7 FUEL PUMPS

Many times, a defective fuel pump diaphragm is mistakenly diagnosed as a problem in the ignition system. The most common problem is a tiny pin-hole in the diaphragm. Such

Typical installation of the remote fuel pump installed on an outboard bulkhead in the engine compartment of a Model 650 Series watercraft with a Keihin CDK-34 carburetor.

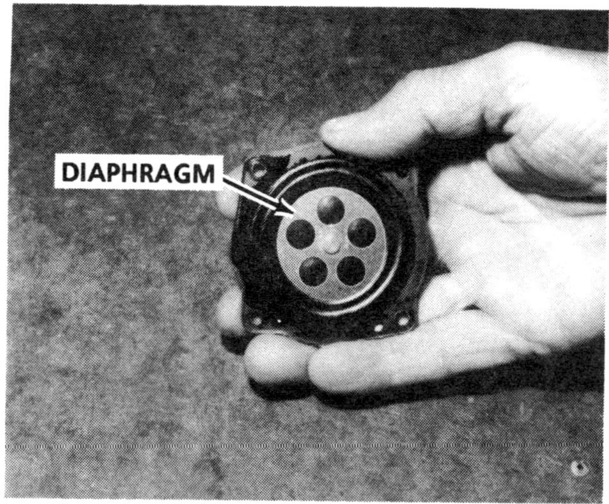

Close view of the fuel pump diaphragm from a Keihin carburetor with integral fuel pump. The diaphragm must be carefully inspected during a carburetor overhaul for the tiniest hole or tear. Such damage will "wet foul" the spark plug at idle speed.

a small hole will permit gas to enter the crankcase and set foul the spark plugs at idle-speed. During high-speed operation, gas quantity is limited, the plug is not fouled and will therefore fire in a satisfactory manner.

Remote Fuel Pump

At press time, the only engine covered in this manual with a remote fuel pump was the Model 650SX, with a Keihin CDK-34 carburetor. This remote fuel pump is shown in a typical installation on an outboard bulkhead in the engine compartment of a Model 650SX watercraft. To service a remote fuel pump, see Chapter 6 -- Fuel and Oil.

Integral Fuel Pump

Again, at press time, all other engines covered in this manual were equipped with Keihin

Back side -- fuel pump side -- of a rack of triple Keihin carburetors from a Model 900 Series watercraft.

carburetors -- all having an integral fuel pump with one exception -- on a triple carburetor installation, the center carburetor does not have the integral fuel pump. Fuel to this carburetor is supplied by the pump on the No. 1 and No. 3 carburetor.

Twin cylinder engines may have a single carburetor serving both cylinders -- Model 550 Series, Model 650 and some Model 750 Series engines or two carburetors -- each serving a single cylinder.

If the fuel pump fails to perform properly, an insufficient fuel supply will be delivered to the carburetor. This lack of fuel will cause the engine to run lean, lose rpm or cause piston scoring.

NEVER use liquid Neoprene on fuel line fittings. Always use Permatex when making fuel line connections. Permatex is available at almost all marine and hardware stores.

To service the fuel pump, see Chapter 6, Fuel and Oil.

3-8 CRANKING MOTOR AND SOLENOID

Cranking Motor Test

Check to be sure the battery is fully charged. Would you believe, many cranking motors are needlessly disassembled, when the battery is actually the culprit.

Lubricate the pinion gear and screw shaft with No. 10 oil.

Connect one lead of a voltmeter to the positive terminal of the cranking motor. Connect the other meter lead to a good ground on the engine. Check the battery voltage under load by depressing the ignition switch and observing the voltmeter reading.

If the reading is 9-1/2 volts or greater, and the cranking motor fails to operate, repair or replace the unit. See the Electrical Chapter.

Solenoid Test

An ohmmeter is required for this test. Select the Rx1000 ohm scale. Calibrate the meter by making contact with the meter leads to each other, and then rotating the calibration knob on the meter until the needle indicates zero ohms. Separate the meter leads.

Make contact with the Red meter lead to one of the large terminals on the solenoid. Make contact with the Black meter lead to the other large solenoid terminal. **NEVER** connect the battery leads to the large terminals of the solenoid, or the meter will be damaged.

Using battery jumper leads, connect the positive battery lead to the small terminal of the solenoid. Connect the negative battery lead to the small terminal of the solenoid. If the meter needle indicates continuity, the solenoid is serviceable. If the meter fails to indicate continuity, the solenoid is defective and **MUST** be replaced.

3-9 JET PUMP

Very few personal watercraft are equipped with a reverse gate. However, if the unit being serviced has a reverse gate, the gate **MUST** be properly adjusted to obtain maximum performance from the craft. When properly adjusted, the gate will permit the pump to deliver its full potential of thrust with no drag.

Gate Position

The shift rod lever adjustment should be checked from time to time to ensure the gate is firmly against the pump housing, when the unit is in the **FORWARD** position (wide open).

With the gate against the housing, any rattle noises will be avoided as the craft moves through the water. Proper positioning of the gate in forward gear will prevent wave action from accidently shifting the gate into reverse as the craft is operated through violent maneuvers.

A simple "tuning" task with the jet pump is to make a thorough check of the linkage and gate movement. Such a check will ensure maximum thrust and use of the horsepower developed by the engine.

Impeller

Excessively rounded jet impeller edges will reduce the efficiency of the jet drive.

The term cavitation "burn" is a common expression used throughout the world among people working with pumps, impeller blades, and forceful water movement.

"Burns" on the impeller blades are caused by cavitation air bubbles exploding with considerable force against the impeller blades. The edges of the blades may develop small "dime size" areas resembling a porus sponge, as the aluminum material is actually "eaten" by the condition just described.

4
MAINTENANCE

4-1 INTRODUCTION

The watercraft being prepared for service probably represents a sizeable investment In order to protect this investment and for the owner to receive the maximum amount of enjoyment from the craft it must be cared for properly while being used and when it is out of the water. Always store the watercraft with the bow higher than the stern and be sure to remove the inner hull drain plug/s. If you use any type of cover to protect craft, plastic, canvas, whatever, be sure to allow for some movement of air through the hull. Proper ventilation will assure evaporation of any condensation that may form due to changes in temperature and humidity.

GOOD WORDS

The authors estimate 75% of engine repair work can be directly or indirectly attributed to lack of proper care for the engine. This is especially true of care during the off-season period. There is no way on this green earth for a mechanical engine, particularly a marine engine, to be left sitting idle for an extended period of time, say for six months and then be ready for instant satisfactory service.

Imagine, if you will, leaving your automobile for six months, and then expecting to turn the key, have it roar to life, and be able to drive off in the same manner as a daily occurrence.

It is critical for a marine engine to be run at least once a month, preferably, in the water, but if this is not possible, then a flush attachment **MUST** be connected either directly to the engine block or in the hose line from the jet pump to the engine.

Move the craft to a body of water or connect a garden hose to the engine cooling water supply fitting or flush fitting on the cylinder head.

Start the engine and allow the rpm's to stabilize at idle speed **FOR JUST A FEW SECONDS**, and then turn the water on.

Adjust the water flow until a small trickle is discharged from the bypass outlet on the port side of the hull.

When the engine is to be shut down turn the water off **FIRST** -- raise the aft portion of the hull -- **WHILE THE ENGINE IS OPERATING AT IDLE** -- "rev" the engine just a **COUPLE** times to clear water from the exhaust system -- and then shut it down.

NEVER allow the engine to operate without cooling water for more than 15 seconds.

Operating the engine for more than 15 seconds without circulating water, will quickly lead to an overheat condition. If the engine is not shut down almost immediately, serious and expensive damage may be caused to internal engine components.

It is quite possible for the engine to "freeze" -- lock-up -- because of excessive heat. In such cases it becomes impossible to rotate the crankshaft and a complete removal, tear down, and overhaul is necessary.

A "squared away" 750 Series engine compartment. There is no substitute and few valid excuses for not developing good habits leading to regular maintenance practices. The results of a good program are always evident in top performance, maximum enjoyment, and "pride of ownership", all at minimum cost.

NEVER, AGAIN NEVER, operate the engine at high speed with a flush device attached. The engine, operating at high speed with such a device attached, would **RUNAWAY** from lack of load on the impeller, causing extensive damage.

At the same time, the shift mechanism, if the unit is so equipped, should be operated through the full range several times. The steering should also be operated from hard-over to hard-over.

Only through a regular maintenance program can the owner expect to receive long life and satisfactory performance at minimum cost.

The material presented in this chapter is divided into four general areas.

1- General information every owner should know.

2- Maintenance tasks that should be performed periodically to keep the craft operating at minimum cost.

3- Care necessary to maintain the appearance of the craft and to give the owner that "Pride of Ownership" look.

4- Winter storage practices to minimize deterioration during the off-season when the craft is not in use.

In nautical terms, the front of the craft is the **bow**; the rear is the **stern**; the right side, when facing forward, is the **starboard** side; and the left side is the **port** side. All directional references in this manual use this terminology. Therefore, the direction from which an item is viewed is of no consequence, because **starboard** and **port** **NEVER** change no matter where the individual is located or his position -- even standing on his/her head.

FORWARD
(BOW)

PORT
(LEFT SIDE)

STARBOARD
(RIGHT SIDE)

AFT
(STERN)

Common terminology used throughout the world, and of course extensively in this manual, for reference designation on watercraft of ALL sizes. ""Port", "Starboard", "Forward, and "Aft", never change even if an individual is standing on his/her head.

After a fun day on the water and before moving the craft out of the water, the owner of this TS650 should SLOWLY open the fuel tank cap to release built-up pressure in the tank.

After Use Tasks

At the end of a day's use, make a habit of opening the engine compartment to allow fumes to escape and sponging out all water in the bottom of the hull before trailering the craft. Leave the engine cover off when the craft is stored overnight to allow any moisture in the compartment to evaporate. Slowly open the fuel tank cap to release built up pressure. Pressure in the tank will increase considerably during hot weather. Releasing the pressure reduces the chance of an explosion and ensures smooth fuel flow when the engine is next started.

If the craft will remain unused for a week, spray storage oil inside the engine and a light film of lubricant, such as WD-40, Super Lube or Justice Bros. 80, on the engine exterior. Keeping the engine exterior lightly oiled at all times will result in water sliding off the engine and draining away rather than accumulating in casting webs -- resulting in corrosion.

Cooling

The engine cooling passages must be flushed with clear water on a regular basis. This is accomplished by simply connecting a "Flushing Kit" per the manufacturer's instructions **OR** by disconnecting the cooling hose from the pump to the engine and connecting a garden hose to the disconnected hose. Such an easy maintenance practice should become a habit and will prevent accumulation of debris which could restrict engine coolant water flow and cause overheating.

Connect a garden hose to the engine cooling water supply fitting or flush fitting on the cylinder head.

Start the engine and allow the rpm's to stabilize at idle speed **FOR JUST A FEW SECONDS,** and then turn the water on.

Adjust the water flow until a small trickle is discharged from the bypass outlet on the port side of the hull.

When the engine is to be shut down, turn the water off **FIRST** -- raise the aft portion of the hull -- **WHILE THE ENGINE IS OPERATING AT IDLE** -- "rev" the engine just a **COUPLE** times to clear water from the exhaust system -- and then shut it down.

NEVER allow the engine to operate without cooling water for more than 15 seconds.

Cooling System Flushing

Connect the garden hose to the engine cooling water supply fitting or flush fitting on the cylinder head.

The manufacturer recommends the engine be started and engine rpm allowed to stabilize at idle speed **FOR A FEW SECONDS,** before turning on the water. If the water is turned on before the engine is started, the water may flow back through the exhaust pipe into the engine. Should this occur the cylinder would fill with water and if the engine was cranking at the time -- trying to start -- the piston would be attempting to compress water and the connecting rod would buckle under the stress.

As soon as the engine stabilizes at idle rpm, turn on the cooling water and adjust the water flow until a small trickle is discharged from the bypass outlet on the portside of the hull.

Perform whatever adjustments or flushing is necessary and then turn off the water **WHILE THE ENGINE IS OPERATING AT IDLE.**

Work **QUICKLY** -- raise the aft section of the craft and "rev" the engine a few times to clear water from the exhaust system.

DO NOT allow the engine to operate without cooling water for more than 15 seconds.

Shut down the engine.

Controlling Corrosion

Since man first started out on the water, corrosion has been his enemy. The first form was merely rot in the wood and then it was rust, followed by other forms of destructive corrosion in the more modern materials.

The higher the concentration of salt in the water, the worse the corrosion which will take place.

One defense against corrosion is to use similar metals in the manufacture of the parts exposed to the water.

A second defense against corrosion is to insulate dissimilar metals. This can be done by using an exterior coating of Sea Skin or by insulating them with plastic or rubber gaskets.

Pump Impeller

The pump impeller is a precisely machined and dynamically balanced aluminum spiral. Observe the drilled recesses at exact locations to achieve the delicate balancing.

Excessive vibration of the jet pump may be attributed to an out-of-balance condition caused by the impeller being struck excessively by rocks, gravel or cavitation "burn".

The term cavitation "burn" is a common expression used throughout the world among people working with pumps, impeller blades, and forceful water movement.

"Burns" on the impeller blades are caused by cavitation air bubbles exploding with considerable force against the impeller

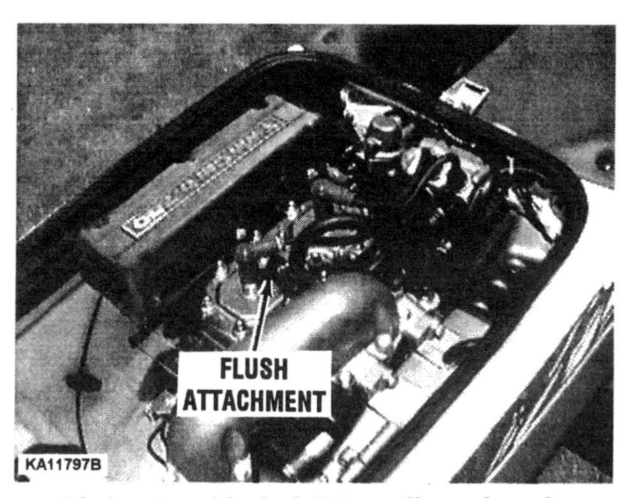

The location of the flush fitting will vary depending on the Series engine being serviced. First, the cap is removed, then a garden hose attached to the fitting, the engine started, and then the water turned on to flush the engine. This is a very important task, especially if the craft has been used in salt water. The water is to be turned off as soon as the engine is shut down.

blades. The edges of the blades may develop small "dime size" areas resembling a porous sponge, as the aluminum is actually "eaten" by the condition just described.

Excessive rounding of the impeller edges will reduce efficiency and performance. Therefore, the impeller should be inspected at regular intervals. If rounding is detected, the impeller should be removed from the housing, placed on a work bench and the edges restored to as sharp a condition as possible using a file. Refer to the Jet Pump chapter for detailed procedures on removal and installation of the impeller.

4-2 SERIAL NUMBERS

Most manufacturers use three different identification numbers stamped, embossed or decaled somewhere on the craft.

The engine serial number is the manufacturer's key to engine changes. This number identifies the year of manufacture, the qualified horsepower rating, and the parts book identification. If any correspondence or parts are required, the engine serial number **MUST** be used or proper identification is not possible.

Federal regulations require all personal watercraft to have a Hull Identification Number (HIN) permanently attached to the craft in two separate locations: one on the transom on the starboard side, above the waterline and the other in an unexposed location, which is usually left to the discretion of the manufacturer This "HIN" number can be most helpful when attempting to retrieve a stolen craft. The "HIN"

All states in the U.S. and provinces in Canada require a license identification number to be prominently displayed on the forward portion of the hull -- port and starboard.

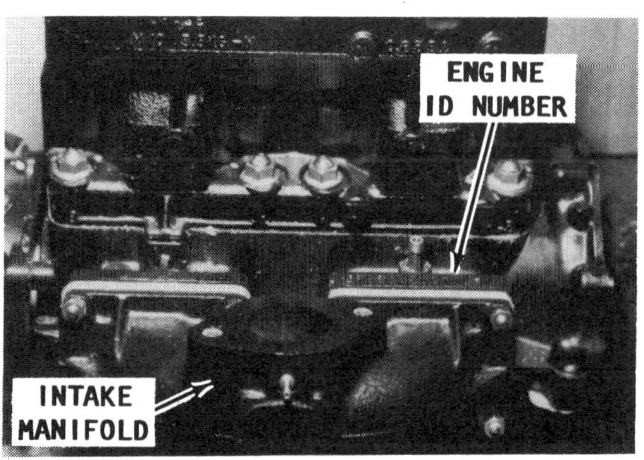

The engine ID number is embossed in different locations depending on the Model Series. Here the number is embossed on the block casting just above the intake manifold.

number is required in most states in order to properly register the craft.

Some manufacturers also use a Primary I.D. Number stamped on a plate attached to the inside of the engine compartment.

ONE MORE WORD

As a theft prevention measure, all metal plates with any type of identification number are especially made wafer thin. Any attempt to remove these plates will result in cracks across the serial number.

4-3 LUBRICATION - COMPLETE UNIT

As with every type mechanical invention with moving parts, lubrication plays a prominent role in operation, enjoyment, and longevity of the unit.

If a personal watercraft is operated in salt water the frequency of applying lubricant to fittings is usually cut in half as compared with the unit being used in fresh water. The few minutes involved in moving around the craft applying lubricant and at the same time making a visual inspection of its general condition will pay in rich rewards with years of continued service.

The first personal watercraft arrived on the scene in the United States in the early 1970s. It is not uncommon to see outboard units well over 20 years of age moving a boat through the water as if the unit had recently been purchased from the current line of models. An inquiry with the proud owner will undoubtedly reveal his main credit for its performance to be regular periodic maintenance. There is no reason this same longevity of performance cannot be true for personal watercraft.

The accompanying chart can be used as a guide to the frequency of maintenance while the craft is being used during the season. Unless otherwise stated use a good brand name multi-purpose water resistant lubricant. Detailed lubrication instructions are given below for locations requiring special attention.

In addition to the normal lubrication listed in the lubrication chart, the prudent owner will inspect and make checks on a regular basis as listed in the accompanying chart.

Throttle Cable Lubrication

Hold the throttle lever in the full open position against the grip. Use a small screwdriver and pry out the small seal from the groove of the throttle stop and a little way along the cable. Spray rust inhibitor along the exposed length of cable. Fill the groove with multi-purpose water resistant lubricant and then pull the seal down along the cable and back into place inside the groove. Wipe away any excess lubricant.

Steering Cable Lubrication

Pry the barrel of the steering cable free of the ball joint on the steering post arm. Coat the ball joint with lubricant and then snap the barrel back onto the ball joint.

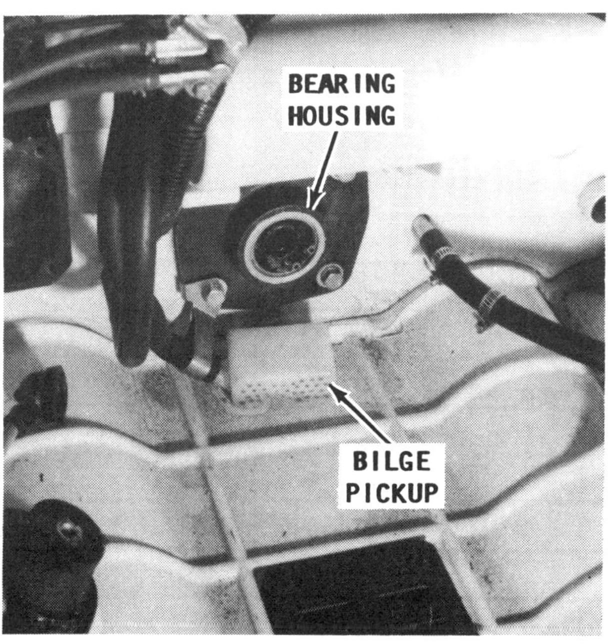

The bearing housing for the jet pump shaft is lubricated through the use of oil-impregnated bushings, which do not require maintenance. The bilge pickup holes should be checked regularly to ensure they are open, ready to pass bilge water when required.

The steering shaft ball joint requires periodic lubrication. Other model joints are different, but the principle and requirement for lubrication is the same.

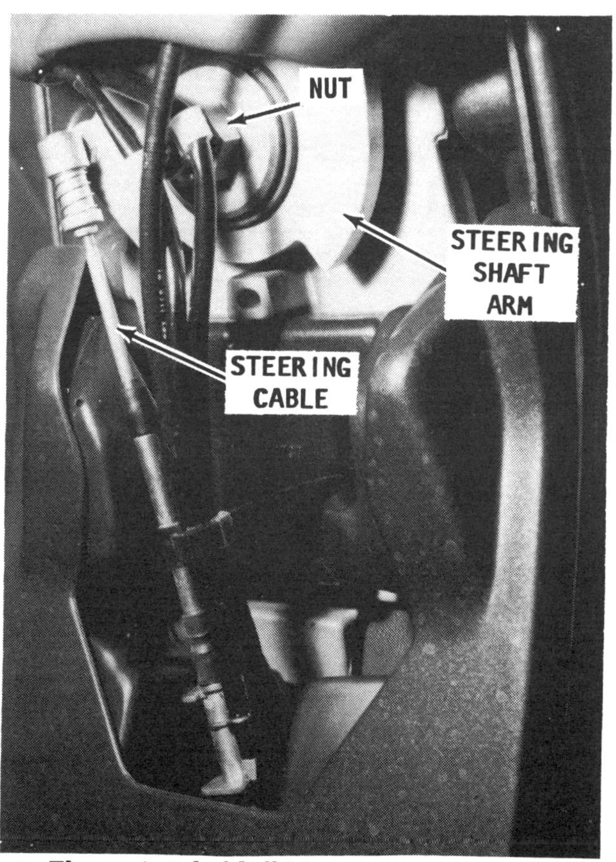

The steering shaft ball joint on a 650 series craft. The nut must be backed off; the electrical line disconnected; and then the joint lubricated. The exposed portion of the steering cable above the guide also requires lubrication. Other Model series have a similar arrangement.

Lubrication Points:

Lubrication Points	Initially every 10hrs	Initially every 50hrs	Initially every 100hrs	Then every 100hrs	Then every 200hrs
Carburetor throttle shaft			❖	❖	
Throttle cable - engine end			❖	❖	
Throttle cable - handlebar stop			❖	❖	
Choke cable - engine end			❖	❖	
Choke knob shaft			❖	❖	
Steering nozzle pivot shaft			❖	❖	
Steering post arm ball joint			❖	❖	
Reverse gate pivot pins			❖	❖	
Jet pump			❖	❖	
Seat latch			❖	❖	

Maintenance Tasks:

Maintenance Tasks	Initially every 10hrs	Initially every 50hrs	Initially every 100hrs	Then every 100hrs	Then every 200hrs
Check engine alignment					❖
Clean and gap spark plugs	❖	❖	❖	❖	
Adjust carburetor idle	❖		❖	❖	
Adjust carburetor mixture	❖		❖	❖	
Drain and inspect fuel tank					❖
Clean and inspect fuel filter	❖	❖	❖	❖	
Inspect fuel line			❖	❖	
Inspect check valve			❖	❖	
Adjust throttle control			❖	❖	
Adjust choke system			❖	❖	
Adjust steering system			❖	❖	
Inspect impeller		❖	❖	❖	
Check impeller to pump case clearance			❖	❖	
Inspect rubber coupling					❖
Inspect electrical connections	❖	❖	❖	❖	
Check battery electrolyte level	❖	❖	❖	❖	
Clean bilge strainer			❖	❖	
Inspect rubber water seal	❖		❖	❖	

Fuel/Oil Mixture

All engines covered in this manual are equipped with an oil injection system. Such a system replaces the age-old method of mixing oil with gasoline prior to filling the fuel tank. However, additional oil should be added to the fuel during a "break-in" period following an engine overhaul or following a long storage period.

"Break-in" Lubrication

In order to obtain extra lubrication during the first 10 hours of "break-in", the manufacturer recommends a premix of 25:1 mixture be used in the fuel tank. Therefore, any existing unused fuel in the tank should be removed before adding the premixed solution. This ratio will **ENSURE** adequate lubrication of moving parts which have been drained of oil during the storage period.

Professional mechanics who own oil injected units add an extra ounce of oil per each gallon of fuel. The addition of this extra oil may prevent engine seizure in the event of an oil pump failure or clog in an oil delivery line.

Use only outboard marine oil in the mixture, never automotive oils. Four-stroke automotive engine oil is not formulated to burn completely -- only to lubricate. Therefore, automotive oil, if used in a two-stroke engine will leave an undesirable residue.

Jet Pump Lubrication

The forward end of the pump shaft is supported by the bearing housing at the bulkhead. The aft end of the driveshaft is supported by the bearing and seal housing. Both of these bearing and supports are sealed units with "lifetime" lubrication ability. Therefore, no maintenance lubrication is necessary.

The jet pump of all Kawasaki watercraft covered in this manual have sealed forward and aft bearings. This means they do not require periodic maintenance.

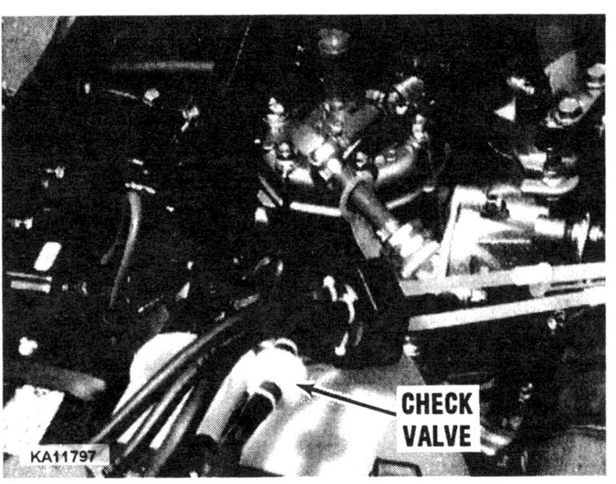

Typical location of the one-way check valve installed in the fuel tank vent line.

4-4 INSPECTION AND SERVICE

Fuel Tank Check Valve & Fuel Filter

A check valve is installed in the fuel tank vent line to prevent fuel from spilling during even the most violent maneuvers of the craft. This check valve should be checked about every 25 hours of engine operation. The valve will permit air to pass into the tank to replace the fuel drawn out, but will not allow fuel to flow in the opposite direction.

To check the valve, disconnect the hose from either side of the valve and lift the valve free. Now, attempt to blow air through the valve in the direction of the arrows. Air should pass through. Attempt to blow air through the valve in the opposite direction. The attempt should fail. If air will pass through the valve in either direction or if no air will pass through -- the valve is defective and must be replaced. When the valve is replaced, the arrows on the case **MUST** be directed **TOWARD** the fuel tank.

Typical installation of a water separation sediment bowl installed on some two and some three-cylinder units.

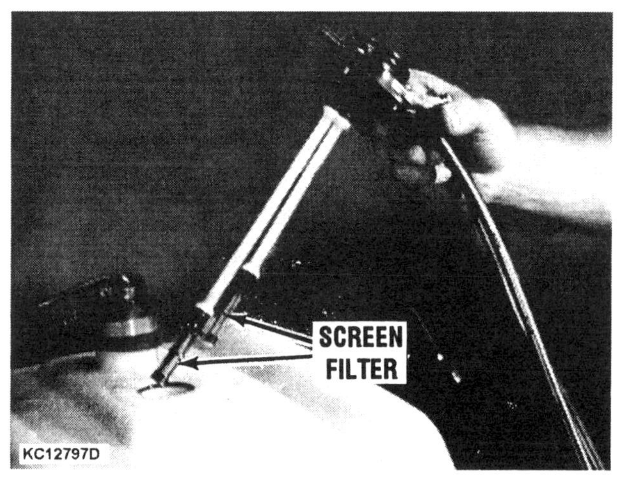

Fuel tank filters withdrawn from the fuel tank, after the tank has been removed from the craft. This is not considered a "routine" maintenance task.

Sediment Bowl

A water separation sediment bowl is installed on some two-cylinder and some three-cylinder model units. If water is observed in the sediment bowl, unscrew the ring nut, remove the bowl, discard the contents and replace the bowl -- tightening the ring nut just hand tight.

Fuel Tank Filters

A fuel filter assembly is installed on top of the fuel tank. An inlet line, an outlet line, and a vent line are connected to the assembly. The actual filters are screens attached to the lower end of two probes extending into the tank, as shown in the accompanying illustration on this page. In order to service -- clean -- the filters, the assembly must be removed from the tank. In order to accomplish this tank, in many cases the fuel tank must be removed from the craft. Such action requires the engine to be removed. Therefore, cleaning the filters is not considered a routine maintenance task.

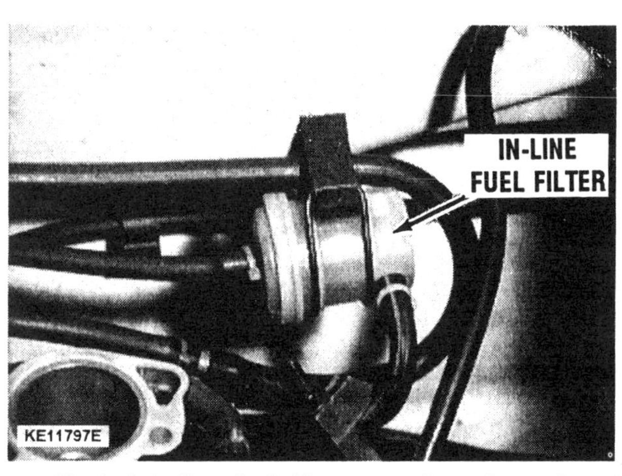

Typical in-line fuel filter secured to the outboard bulkhead in the engine compartment.

In-Line Filter

Many units covered in this manual have an in-line fuel filter installed on the outboard bulkhead in the engine compartment, as shown in the accompanying illustration.

To service the filter, first identify, and then pinch off the connected fuel lines with a some type of clamp. Disconnect the lines from the filter unit. Release the filter from the bulkhead through the attaching hardware.

Disassemble the filter, clean, and then blow the parts dry with compressed air. Secure the filter to the bulkhead with the attaching hardware in the same position from which it was removed. Connect the fuel lines to the same fitting from which there were removed. Remove the clamps.

Oil Filter

An oil filter is an integral part of the oil pump. This filter cannot be removed or serviced.

Drain Plug

A drain plug is installed in the transom to permit water in the bilges to be easily removed. Inspect the condition of the O-ring. Replace the plug as required. Clean any dirt or sand from the plug threads.

If the craft is to remain out of the water -- in storage but outside -- the plug may be left out to permit any rain water or melted snow which has found its way into the bilge to drain free of the craft. Remember to replace the plug before moving the craft into the water during the next outing or season. One method is to tie the plug to a handle bar or in a very visible location on the control panel, as a reminder.

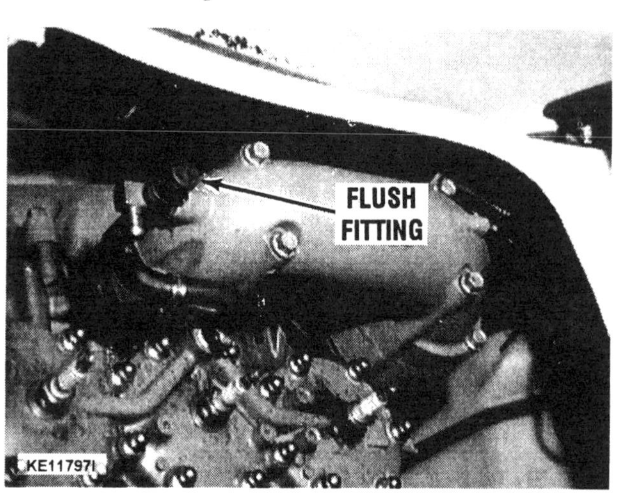

Location of the flush fitting at the aft end of the exhaust elbow on a Model 900 or Model 1100 engine.

FLUSHING -- COOLING AND BILGE SYSTEMS

Cooling System

A regular maintenance program should include flushing the cooling system to prevent sand or salt deposits from accumulating in the system. It is certainly worth the time and effort to flush after each outing and especially if the craft is used in salt water.

Remove the fitting cap, as shown in the accompanying illustration, and then connect a garden hose to the fitting.

Start the engine and allow it to idle for just a short time **BEFORE** turning the water on. The engine **MUST** be idling before the water is turned on to prevent water from back flowing through the exhaust pipe into the engine. Such action could cause severe and expensive damage to engine internal parts.

After the water is turned on, adjust the amount of flow until a "tattle-tale" stream of water is observed exiting from the bypass outlet on the starboard side of the craft.

Now, allow the engine to operate at idle speed for a few minutes with the water running, and then turn the water off, but allow the engine to continue idling. Increase engine rpm a couple of times for just a few seconds -- not more than 15 seconds -- to clear water from the exhaust system. Operation of the engine without water for a longer period could cause severe and expensive damage.

Shut down the engine, disconnect the garden hose and cover the fitting with the cap.

Bilge System

The bilge system must be kept in an operable condition at all times in order to perform properly in an emergency situation.

To **ENSURE** a "ready" condition of the bilge system and to prevent any clogging, the system should be flushed -- say every 25 hours of engine operation.

Disconnect both bilge hoses from the plastic breather fitting. Connect the bilge filter hose -- from the bottom of the hull -- to a garden hose. Turn on the water and flush the system for about a minute. During this time, water will flow into the engine compartment. Do not permit a large amount of water to accumulate in the compartment. Turn off the water.

Connect the other hose -- from the hull bulkhead -- to the garden hose. Turn the water

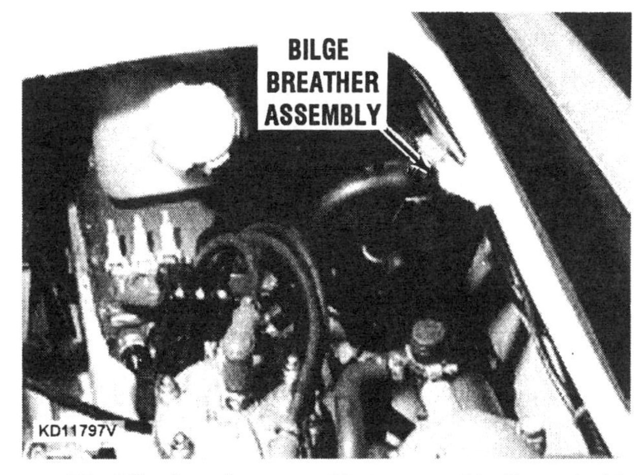

The bilge breather assembly is secured to the port side outboard bulkhead in the engine compartment. The text explains how the bilge system is to be flushed.

on and flush for a few minutes. Turn off the water and disconnect the garden hose.

Before connecting the hoses back to the fitting on the plastic breather, check to be sure the small hole on the top of the breather is clear. This small hole is essential to allow air to enter for proper operation of the bilge system.

4-5 IMPELLER-TO-PUMP CASE CLEARANCE

SPECIAL WORDS

The following procedure may be performed with the watercraft elevated on saw horses enabling a person to work underneath, or with the water craft on its side allowing access to the rock grate and impeller.

If the impeller-to-pump case clearance is to be determined with the water craft raised on sawhorses, the battery may be left secured in the craft. If the watercraft is to be positioned on its side, first disconnect the electrical leads at both

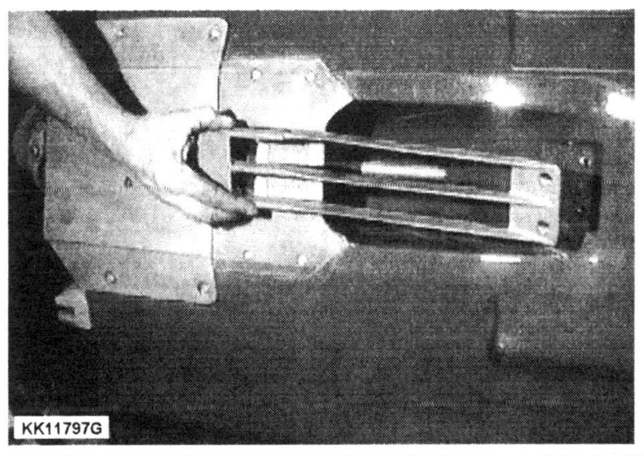

Removing the rock grate from the bottom of the hull in preparation to checking the impeller-to-pump case clearance. In this case the craft has been turned on its side, as described in the text.

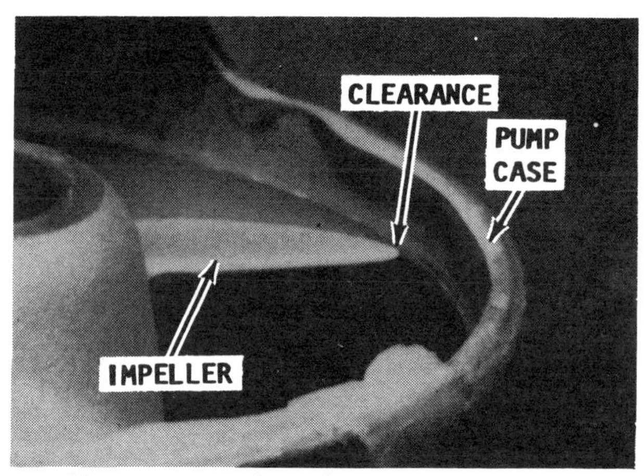

Close view to show clearance area between the impeller blades and the pump case, as explained in the text. Clearance table -- bottom of this column.

battery terminals, and then remove the battery from the craft.

Remove the bolts and washer securing the rock grate to the jet pump. If the seal between the hull and the pump housing is broken, or the foam packing is damaged, reseal the housing and intake area with silicone sealant to prevent impeller cavitation.

Insert a feeler gauge between each impeller blade and the pump case and determine the clearance. Make a note of the clearance, for each blade.

Determine the average of the clearances, by first adding them for a total, and then dividing by the number of impeller blades.

On a new craft the clearance in the following table will be what the manufacturer refers to as the "standard" clearance -- the first column. The second column is the "service limit". This value is the maximum allowable clearance.

IMPELLER-TO-PUMP CASE CLEARANCE

Series	Std. Clearance	Service Limit
550	0.008-0.012" (0.2-0.3mm)	0.024" (0.6mm)
650	0.008-0.012" (0.2-0.3mm)	0.024" (0.6mm)
750	0.008-0.012" (0.2-0.3mm)	0.024" (0.6mm)
900	0.006-0.012" (0.15-0.3mm)	0.024" (0.6mm)
1100	0.006-0.012" (0.15-0.3mm)	0.024" (0.6mm)

Clearance Results

If the clearance is less than the maximum value listed in the above table, no action is necessary.

If the clearance is more than the service limit, inspect the condition of the pump case. If the pump case has scratches deeper than 0.04" (1mm), replace the pump case. If the pump case is satisfactory, the problem must be with the impeller.

Visually inspect the impeller for nicks, scratches, pitting or a "mushroomed" edge. If the cause of the excessive clearance cannot be determined visually, the pump **MUST** be disassembled and the pump case measured and compared to specifications.

The jet pump must be removed from the craft to perform this work. Detailed illustrated procedures are presented in Chapter 10.

4-6 PRE-SEASON PREPARATION

Satisfactory performance and maximum enjoyment can be realized if a little time is spent in preparing the unit for service at the beginning of the season. Assuming the unit has been properly stored, as outlined in Section 4-10, a minimum amount of work is required to prepare the unit for use.

The following steps outline an adequate and logical sequence of tasks to be performed before using the craft the first time in a new season.

1- Lubricate the craft according to the manufacturer's recommendations. Refer to the lubrication table on Page 4-6. Remove, clean, inspect, adjust, and install the spark plugs with a new gasket (if they require a gasket). Make a thorough check of the ignition system. This

KL11797G

Just a small amount of time spent in pre-season preparation tasks will reward the owner with satisfactory performance and maximum enjoyment.

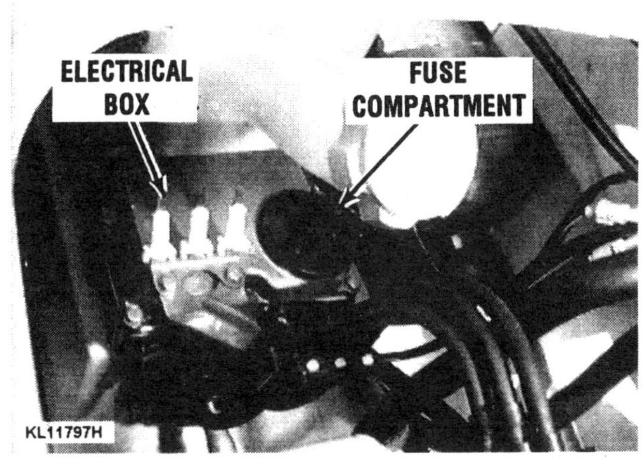

Location of the electrical box on a Model 900 Series craft. The box contains the rectifier/regulator, cranking motor relay, and the fuse container, which is easily accessible.

check should include: the wiring and the battery electrolyte level and charge. Many modern craft use a maintenance free battery, therefore this check is not necessary. If the exhaust outlet was plugged at the end of the last season, remove the plug.

2- Take time to check the gasoline tank and all fuel lines, fittings, couplings, valves, including the flexible tank fill and vent. If the fuel was not drained at the end of the previous season, it may have "soured" and give off an order of rotten eggs.

Check carefully for any gum deposits which may have formed. Such gum can clog the filters, lines, and passageways in the carburetor. To prevent gum from forming, a fuel additive such as "Sta-Bil" can be added to the fuel at the end of the season. Such an additive will prevent fuel from "souring" for up to twelve full months.

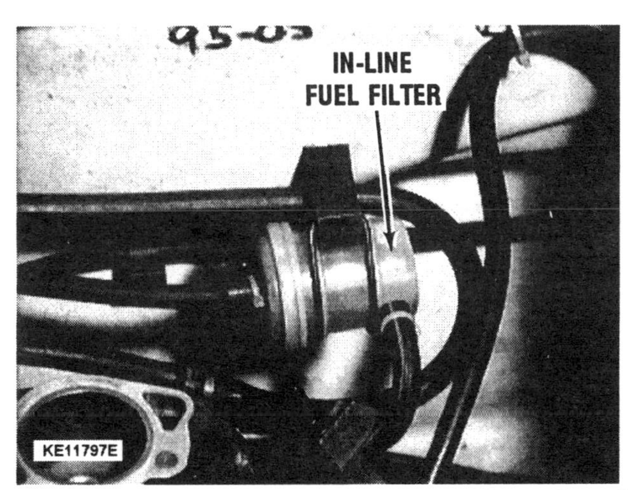

The in-line fuel filter secured to the outboard bulkhead in the engine compartment should be cleaned at the beginning of each new season.

The sediment bowl (water separation filter), if installed, should always be removed and drained as a pre-season preparation task.

GOOD WORDS

All manufacturers recommend the fuel filters be replaced or cleaned at the start of each season or at least once a year. An in-line filter attached to the outboard bulkhead is easily accessible. Many models have a water separation unit which is also easily accessible. The sediment bowl should be drained as a pre-season preparation task.

3- Close all water drains. Check and replace any defective water hoses. Check to be sure the connections do not leak. Replace any spring-type hose clamps, if they have lost their tension, or if they have distorted the water hose, with band-type clamps.

4- The engine can be run with the jet pump in water to flush it. If this is not

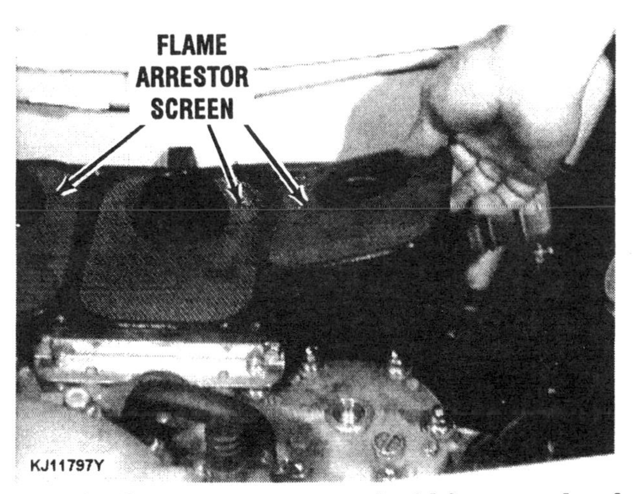

The flame arrestor screens should be removed and cleaned at the beginning of each season and on a regular basis as a maintenance task.

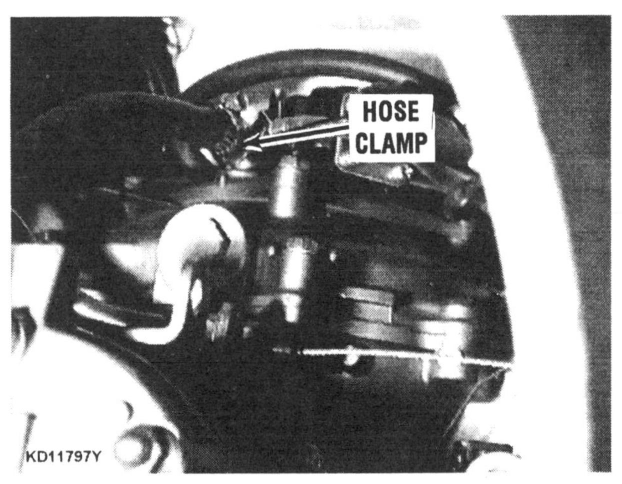

Check all hose clamps to be sure they are in good condition and tightened properly.

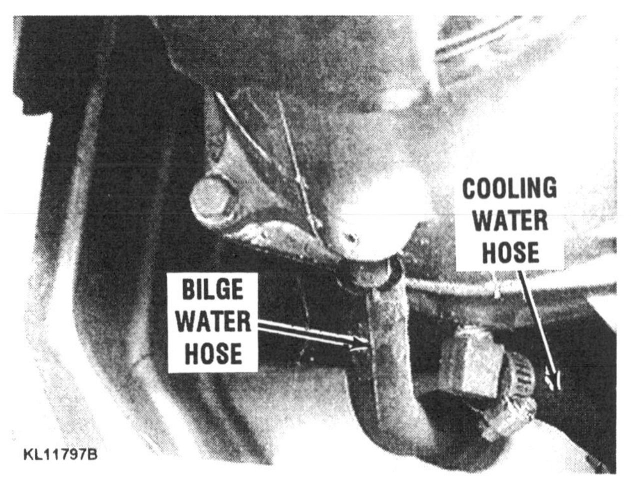

Close view of the cooling water connection and the bilge water system fitting at the aft end of the jet pump.

practical, a flush attachment may be used. This device is attached to the engine water supply hose, between the engine and the pump.

Connect a garden hose to the engine cooling water supply fitting or flush fitting on the cylinder head.

Start the engine and allow the rpm's to stabilize at idle speed **FOR JUST A FEW SECONDS,** and then turn the water on.

Adjust the water flow until a small trickle is discharged from the bypass outlet on the port side of the hull.

When the engine is to be shut down, turn the water off **FIRST** -- raise the aft portion of the hull -- **WHILE THE ENGINE IS OPERATING AT IDLE** -- "rev" the engine just a **COUPLE** times to clear water from the exhaust system -- and then shut it down.

NEVER allow the engine to operate without cooling water for more than 15 seconds.

Check the exhaust outlet for water discharge. Check for leaks.

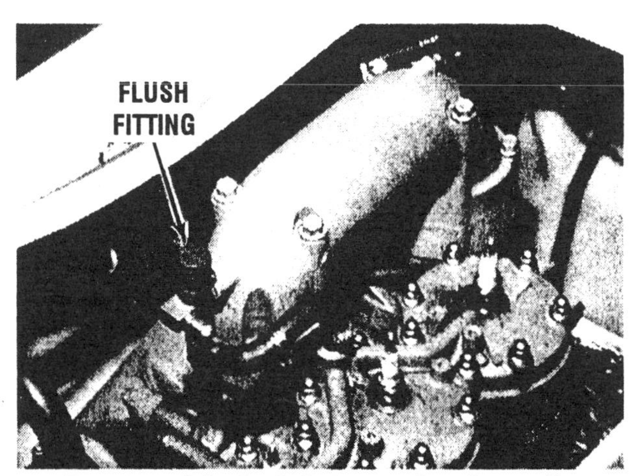

Location of the flush fitting on the exhaust elbow on a Model 900 or Model 1100 Series engine.

5- Check the electrolyte level in the battery and the voltage for a full charge. Clean and inspect the battery terminals and cable connections. **TAKE TIME** to check the polarity, if a new battery is being installed. Cover the cable connections with lubricant or special protective compound as a prevention to corrosion formation. Check all electrical wiring and grounding circuits.

6- Check all electrical parts on the engine and lower portions of the hull to be sure they are of a type to prevent ignition of an explosive atmosphere. Rubber caps help keep spark insulators clean and reduce the possibility of arcing. Electric cranking motors and high-tension wiring harnesses should be of a marine type to prevent an explosive mixture from igniting.

Many late model units are equipped with what is termed a "Maintenance Free" battery. As the name imples regular maintenance regarding electrolyte level and hydrometer reading is not necessary. The battery must be secured to prevent even the slightest movement during the most violent maneuvers of the the craft.

4-7 SEALANTS, ADHESIVES, LUBRICANTS, AND FUEL STABILIZERS

It is common practice for the larger manufacturers of personal watercraft to market their own line of products for use on their craft. Kawasaki chemical engineers have developed such a line available for use with Kawasaki outboard units, personal watercraft, snowmobiles, and motorcycles. Throughout this manual, the authors recommend the application of the manufacturer's products as a first choice. All products listed are alternatives of equal value, and may be used with confidence if the manufacturer's line is not available.

Sealants and Adhesives

Four sealants are recommended and they are **NOT** interchangeable. Each is designed to perform under a different set of conditions. Follow the directions on the package for cleaning and preparing surfaces. Sealants and adhesives **MUST** be applied **ONLY** to clean and dry parts. Apply sparingly -- excessive amounts may block oil passageways and cause serious damage.

*A gasoline stabilizer and conditioner may be used to prevent fuel from "souring" for up to twelve full months. However, draining fuel and adding new at the beginning of each season is still the best practice to **ENSURE** top performance.*

Loctite Lock N' Seal is a non-hardening, non-permanent locking compound. This material is recommended for application to the threads of load bearing fasteners. Loctite helps prevent loosening of the bolt due to vibration, thread wear, and corrosion.

Loctite Stud N' Bearing Mount compound is also a non-hardening, non-permanent locking compound. This material is recommended for application to the threads of load bearing fasteners submerged under water.

Loctite Superflex is a water resistant silicone sealer which provides a very effective flexible seal and is able to withstand high temperatures. This sealer is recommended for use where gaskets are required next to a metal surface at high temperature, for example, on the exhaust manifold cover gaskets. Loctite Superflex is blue in color.

Kawasaki Bond is a non-hardening material and is recommended for metal-to-metal joints such as the crankcase halves. This substance is highly resistant to oil and gasoline.

Lubricants

Different lubricants are recommended in the lubrication procedures presented in this manual. These lubricants are **NOT** interchangeable, each is designed to perform under varying conditions.

Shell Alvania EP1 is a general marine lubricant, chemically formulated to resist salt water. This lubricant is recommended for application to bearings, bushings, and oil seals.

Kawasaki Lubricant is a two-stroke engine oil. It has a petroleum base, and is considered a clean burning lubricant. Yamalube reduces carbon deposits and ensures maximum protection against engine wear. This lubricant also contains an ashless detergent to keep piston rings "free". Oil additives are usually not recommended by the manufacturer, and in some cases the use of such a substance may invalidate the warranty.

Fuel Stabilizer

"Sta-Bil" Fuel Conditioner and Stabilizer is recommended during engine operation and during the storage period. This fluid absorbs water in the fuel system and protects against corrosion.

If used during operation, this fuel additive will prevent the formation of gum and varnish deposits and greatly extend the period between required carburetor overhauls.

When added to the fuel during storage, the additive will prevent the fuel from "souring" for up to twelve full months.

4-8 FIBERGLASS HULLS

Fiberglass reinforced plastic hulls are tough, durable, and highly resistant to impact. However, like any other material they can be damaged. One of the advantages of this type of construction is the relative ease with which it may be repaired. Because of its break characteristics, and the simple techniques used in restoration, these hulls have gained popularity throughout the world.

A fiberglass hull has almost no internal stresses. Therefore, when the hull is broken or stove-in, it retains its true form. It will not dent to take an out-of-shape set. When the hull sustains a severe blow, the impact will be either absorbed by deflection of the laminated panel or the blow will result in a definite, localized break. Repairs are usually confined to the general area of the rupture.

Most modern marinas, especially those in or with salt water access, provide the boater with fresh water to rinse their craft before heading for home. Such a practice certainly helps to preserve the wax and paint job for that "Pride of Ownerhip" appearance.

Cleaning, Waxing, and Polishing

Any craft should be washed with clear water after each use to remove surface dirt and any salt deposits from use in salt water. Regular rinsing will extend the time between waxing and polishing. It will also give "pride of ownership", by having a sharp looking piece of equipment. Elbow grease, a mild detergent, and a brush will be required to remove stubborn dirt, oil, and other unsightly deposits.

Avoid harsh abrasives or strong chemical cleaners. A white buffing compound can be used to restore the original gloss to a scratched, dull, or faded area. The finish should be thoroughly cleaned, buffed, and polished at least once each season. Take care when buffing or polishing with a marine cleaner not to overheat the surface being worked, because the area will become "burned".

4-9 SUBMERGED ENGINE SERVICE

A submerged engine is always the result of an unforeseen accident. Once the craft is recovered, special care and service procedures **MUST** be closely followed in order to return the unit to satisfactory performance.

NEVER, again we say **NEVER** allow an engine that has been submerged to stand more than a couple hours before following the procedures outlined in this section and making every effort to get it running. Such delay will result in serious internal damage. If all efforts fail and the engine cannot be started after the following procedures have been performed, the engine should be disassembled, cleaned, assembled, using new gaskets, seals, and O-rings, and then started as soon as possible.

Submerged engine treatment is divided into three unique problem areas: submersion in salt water; submerged while powerhead was running; and a submerged unit in fresh water.

The most critical of these three circumstances is the engine submerged in salt water, with submersion while running, a close second.

Salt Water Submersion

NEVER attempt to start the engine after it has been recovered. This action will only result in additional parts being damaged and the cost of restoring the engine increased considerably. If the engine was submerged in salt water the complete unit **MUST** be disassembled, cleaned, and assembled with new gaskets, O-rings, and seals. The corrosive effect of salt water can only be eliminated by the complete job being properly performed.

Submerged While Running
Special Instructions

If the engine was running when it was submerged, the chances of internal engine damage is greatly increased. Remove the spark plugs to prevent compression in the cylinders. Use a socket wrench on the flywheel nut to rotate the crankshaft. If the attempt fails, the chances of serious internal damage, such as: bent connecting rod, bent crankshaft, or damaged cylinder, is greatly increased. If the crankshaft cannot be rotated, the engine must be completely disassembled.

CRITICAL WORDS

Never attempt to start an engine that has been submerged. If there is water in the cylinder, the piston will not be able to compress the liquid. The result will most likely be a bent connecting rod.

Submerged Engine — Fresh Water
SPECIAL WORDS

As an aid to performing the restoration work, the following steps are numbered and should be followed in sequence. However, illustrations are not included with the procedural steps because the work involved is general in nature.

In the United States, Coast Guard regulations require PWC's to contain sufficient built-in floatation -- usually styrofoam -- to prevent the craft from sinking when filled with water and the authorized number of persons aboard.

1- Recover the craft as quickly as possible.

2- Remove the engine cover and the spark plugs.

3- Remove the carburetor float bowl cover, or the bowl.

4- Flush the outside of the engine with fresh water to remove silt, mud, sand, weeds, and other debris. **DO NOT** attempt to start the engine if sand has entered any part of the engine. Such action will only result in serious damage to engine components. Sand in the engine means the unit must be disassembled.

CRITICAL WORDS

Never attempt to start an engine that has been submerged. If there is water in the cylinder, the piston will not be able to compress the liquid. The result will most likely be a bent connecting rod.

5- Remove as much water as possible from the engine. Most of the water can be eliminated by first holding the engine in a horizontal position with the spark plug holes **DOWN**, and then the flywheel with a socket wrench on the flywheel nut. Rotate the crankshaft through at least 10 complete revolutions. If you are satisfied there is no water in the cylinders, proceed with Step 6 to remove moisture.

6- Alcohol will absorb moisture. Therefore, pour alcohol into the carburetor throat and again rotate the crankshaft.

7- Pour alcohol into the spark plug openings and again rotate the crankshaft.

8- With assistance, rotate the craft into the horizontal position until the spark plug openings are facing **DOWN**. Pour engine oil into the carburetor throat and, at the same time, rotate the crankshaft to distribute oil throughout the crankcase.

9- With the craft upright, pour approximately one teaspoon of engine oil into each spark plug opening. Rotate the crankshaft to distribute the oil in the cylinders.

10- Install and connect the spark plugs.

11- Install the carburetor float bowl cover, or the bowl.

Move the craft to a body of water or connect a garden hose to the engine cooling water supply fitting or flush fitting on the cylinder head.

12- Obtain **FRESH** fuel and attempt to start the engine.

If using a garden hose, after the engine starts, allow the rpm's to stabilize at idle speed **FOR JUST A FEW SECONDS,** and then turn the water on.

Adjust the water flow until a small trickle is discharged from the bypass outlet on the port side of the hull.

Allow the engine to run for approximately an hour to eliminate any unwanted moisture remaining in the engine.

When the engine is to be shut down, turn the water off **FIRST** -- raise the aft portion of the hull -- **WHILE THE ENGINE IS OPERATING AT IDLE** -- "rev" the engine just a **COUPLE** times to clear water from the exhaust system -- and then shut it down.

NEVER allow the engine to operate without cooling water for more than 15 seconds.

CAUTION

Water must circulate through the jet pump -- to and from the engine, anytime the engine is operating. Circulating water will prevent overheating — which could cause damage to moving engine parts and possible engine seizure.

13- If the engine fails to start, determine the cause, electrical or fuel, correct the problem, and again attempt to get it running. **NEVER** allow an engine to remain unstarted for more than a couple hours without following the procedures in this section and attempting to start it. If attempts to start the engine fail, the unit should be disassembled, cleaned, assembled, using new gaskets, seals, and O-rings, as **SOON** as possible.

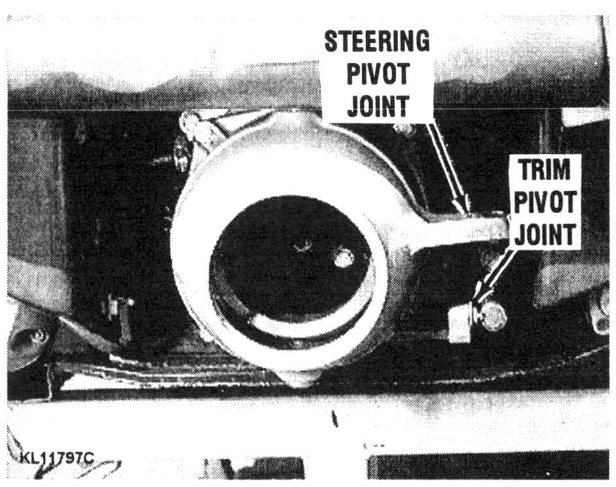

Close view of the steering pivot joint and the trim pivot joint -- if the trim system is installed.

4-10 WINTER STORAGE

Taking extra time to store the craft properly at the end of each season, will increase the chances of satisfactory service at the next season. **REMEMBER,** idleness is the greatest enemy of a marine engine. The unit should be run on a monthly basis. The steering and shifting mechanism should also be worked through complete cycles several times each month. The owner who spends a small amount of time involved in such maintenance will be rewarded by satisfactory performance, and greatly reduced maintenance expense for parts and labor.

Proper storage involves adequate protection of the unit from physical damage, rust, corrosion, and dirt.

The following steps provide an adequate maintenance program for storing the unit at the end of a season.

1- Empty all fuel from the carburetor.

For many years there has been the widespread belief simply shutting off the fuel at the tank and then running the engine until it stops is the proper procedure before storing the engine for any length of time. Right? **WRONG!**

First, it is **NOT** possible to remove all fuel in the carburetor by operating the engine until it stops. Considerable fuel is trapped in the float chamber and other passages and in the line leading to the carburetor. The **ONLY** guaranteed method of removing **ALL** fuel is to take the time to remove the carburetor, and drain the fuel.

Secondly, if the engine is operated with the fuel supply shut off until it stops, the fuel and oil mixture inside the block is removed, leaving bearings, pistons, rings, and other parts without any protective lubricant.

Remove the spark plugs and pour about one ounce of two-cycle engine oil into each spark plug hole. Wait a minute or two and then slowly rotate the flywheel/crankshaft through two complete revolutions to evenly distribute the oil. Install the spark plugs.

2- Drain the fuel tank and the fuel lines. Pour approximately one quart (0.96 liters) of benzol (benzine) into the fuel tank, and then rinse the tank and pickup filter with the benzol. Drain the tank. Store the craft with the engine compartment door open to allow air to circulate around the tank and with the fuel vent **OPEN** to allow air to circulate through the tank.

3- Clean the in-line fuel filter with benzol, see the Fuel chapter.

4- Lubricate the throttle and shift linkage. Lubricate the steering pivot shaft with water resistant multi-purpose lubricant.

5- Plug the exhaust outlet with a shop towel or similar item to prevent any type of contaminant from entering the exhaust system.

Battery Storage

Remove the battery from the craft and keep it charged during the storage period. Clean the battery thoroughly of any dirt or corrosion, and then charge the battery to the full specific gravity reading. After charging, store the battery in a clean cool dry place where it will not be damaged or knocked over.

NEVER store the battery with anything on top of it or cover the battery in such a manner as to prevent air from circulating around the filler caps. All batteries, both new and old, will discharge during periods of storage, more so if they are hot than if they remain cool. Therefore, the electrolyte level and the specific gravity should be checked at regular intervals. A drop in the specific gravity reading is cause to charge the battery back to a full reading.

One of the largest personal watercraft battery manufacturers recommends that a battery in storage should be checked every two weeks and charged if necessary. The electrolyte level can also be replenished at this time to prevent sulfation occurring.

In cold climates, **EXERCISE CARE** in selecting the battery storage area. A fully-charged battery will freeze at about 60° below zero. A discharged battery, almost dead, will have ice forming at about 19° above zero.

4-11 PRE-SEASON CHECK

Before attempting to start the craft after winter storage, check the steering and throttle action for smooth operation, without binding. If the battery was removed for storage, check to make sure it has a full charge.

Filling the Battery

If the battery is a "Maintenance Free" type battery no maintenance is necessary.

However, if the battery is not a "Maintenance Free" type, proceed as follows.

Remove each filler cap using a pair of pliers. Fill each cell to the proper level with distilled water. The level of electrolyte should be between the upper and lower level marks during operation. When filling the battery, fill to the upper line and allow the battery to stand for a while -- say twenty minutes. Check the level again and replenish, as necessary.

Use a temperature corrected hydrometer to test the specific gravity of the electrolyte. At 20°C (68°F) the hydrometer reading should be 1.26 on the scale. If it is necessary to charge the battery, leave the filler caps lightly resting on the cell openings to allow the gases to escape. The charging current should not exceed 1.9 amps for 10 hours.

After charging, push each filler cap into place and wipe the top of the battery clean before installation.

Clean the battery posts and battery cable ends with a wire brush to ensure good, clean connections. Clean the top surface of the battery and identify the "POS" or "+" and "NEG" or "-" embossed symbols. Correctly connect the battery cables to the battery observing polarity. If the cables are connected backwards, the ignition system **WILL** be destroyed the first time the engine is started.

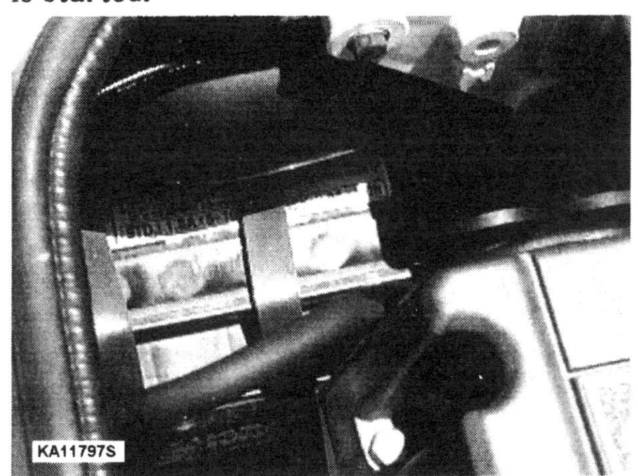

*A "Maintenance Free" battery well secured in a Model 650 Series watercraft. As mentioned several times in this manual, the battery **MUST** be so firmly in place, to prevent even the slightest movement during violent maneuvers of the craft. Battery life may be extended considerably if the battery is removed from the craft and stored in a warm dry area with proper air circulation during the off-season period.*

Make sure the breather pipe is in place on the battery and the pipe is not pinched when the battery is installed.

First Time Start

If the proper maintenance was observed at the end of the previous season, both the carburetor and fuel tank should be completely empty. Fill the gas tank with the proper mixture and open the fuel valve, if equipped. Open the choke. **DO NOT** attempt to crank the engine until it starts, **AT THIS TIME.** Considerable cranking time will be spent before the oil/fuel mixture reaches engine bearings and other moving parts. Expensive parts may be damaged due to lack of lubrication and the cranking motor will become excessively hot, which is also **BAD NEWS.**

If the craft cannot be moved to a body of water, connect a garden hose to the engine cooling water supply fitting or flush fitting on the cylinder head.

Start the engine and allow the rpm's to stabilize at idle speed **FOR JUST A FEW SECONDS,** and then turn the water on.

Adjust the water flow until a small trickle is discharged from the bypass outlet on the port side of the hull.

When the engine is to be shut down, turn the water off **FIRST** -- raise the aft portion of the hull -- **WHILE THE ENGINE IS OPERATING AT IDLE** -- "rev" the engine just a **COUPLE** times to clear water from the exhaust system -- and then shut it down.

NEVER allow the engine to operate without cooling water for more than 15 seconds.

CAUTION

Water must circulate through the jet pump -- to and from the engine, anytime the engine is operating. Circulating water will prevent overheating — which could cause damage to moving engine parts and possible engine seizure.

The correct procedure for starting is to obtain an **AEROSOL** can of WD-40. It must be an aerosol canister because the liquid must be atomized, a squirt can will not do the job. **DO NOT** use ether or a commercial starting fluid because they will evaporate almost instantly. These products work well with four-stroke engines equipped with an oil pump, but they do not contain a lubricant essential for a two-stroke engine. WD-40 is an extremely flammable oil.

Make sure both spark plugs are tight, so the mist from the canister will not ignite. Spray the WD-40 through the flame arrestor and continue spraying with **NO** throttle until the engine starts. As soon as the engine is operating, cease spraying the WD-40. The engine may emit some smoke for a few seconds after it starts. Watch the sediment bowl (if so equipped). As soon as it fills, the engine should operate on the regular oil/fuel mixture. Work the throttle to keep the engine at idle speed until it reaches operating temperature, then close the choke. Advance the throttle a few times and the engine should then idle smoothly.

5
TROUBLESHOOTING

5-1 INTRODUCTION

This chapter is divided into six main sections as follows:

5-2 Mechanical engine problems.
5-3 Fuel system problems.
5-4 Ignition system faults.
5-5 Cranking system faults.
5-6 Charging system malfunctions.
5-7 Troubleshooting Charts.

If a change in engine rpm is experienced at normal operating speed, such a change is usually a symptom of an engine problem or trouble in the jet pump. An increase or a decrease in rpm may be an indication of the area to be checked. The following generalizations may be helpful in isolating the problem.

A noticeable change in craft performance, as evidenced by one of the following symptoms, may be attributed to one or more of the areas listed:

Lower than Normal Engine Rpm

Mechanical problems in the engine:
- Plugged flame arrestor.
- Fouled spark plug/s, incorrect heat range or gap.
- Shorted or weak ignition coil.
- Contaminated fuel.
- Restricted exhaust system.
- Inadequate fuel pump pressure.
- Loss of compression due to blown head gasket.

Mechanical problems in the jet drive:
- Worn or damaged impeller.
- Main bearing failure.
- Rubber coupler failure.

Higher than Normal Engine Rpm

Mechanical problem in the engine:
- None.

Mechanical problems in the jet drive:
- Clogged intake grate -- possibly plastic bag or other debris.
- Line, sea weed, etc. entangled in impeller.
- Ingested sand.

If a reduction in speed is experienced, accompanied with no loss or gain of engine rpm, the loss of performance can fairly accurately be attributed to some type of drag on the craft or restriction of the water ejected from the jet pump.

Drag on the hull is usually caused by damage to the craft.

Many times, drag from the jet pump is caused by a partially lowered reverse gate, if so equipped, or from dragging debris caught in the intake grate.

A partially lowered reverse gate may be caused by incorrect cable adjustment or by binding around the gate pivot pins.

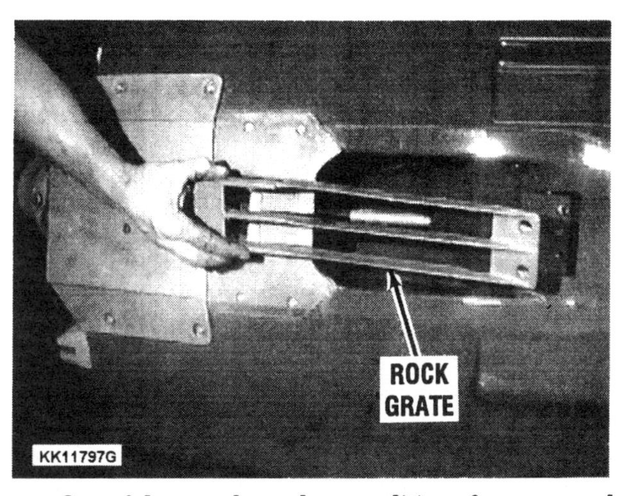

One of the most hazardous conditions for a personal watercraft is restriction of water through the rack grate caused by a plastic bag, plastic wrap, or by some type of sea weed. Even the smallest reduction in water volume will affect jet pump performance and could quickly lead to engine overheating.

Engine Troubleshooting

Troubleshooting must be a well thought-out procedure. Always attempt to proceed with the troubleshooting in an orderly manner. The "shot in the dark" approach will only result in wasted time, incorrect diagnosis, replacement of unnecessary parts, and frustration.

Obviously, if the instructions are to be of maximum benefit as a guide, they must be fully understood and followed in the proper sequence.

When an engine fails to start, the trouble must be localized to one of four general areas: cranking system; ignition system; fuel system; or compression. Each of these areas must be systematically inspected until the trouble is isolated. At that time, detailed tests of the system or area must be made to determine the part causing the problem. The last section of this chapter describes charging system troubleshooting.

Troubleshooting Check

When using this Troubleshooting Check, proceed sequentially through each test until the defect is uncovered. Then skip to the detailed testing procedure and check for that system. For example, if, when using the Troubleshooting Check procedure, the first two systems, cranking and ignition tests OK, but the third test shows there is trouble in the fuel system, then proceed to the detailed test under the Fuel System Troubleshooting Check in Section 5-3.

Cranking System Test

1- Turn the key switch to the **START** position or press the ignition button. The cranking motor should crank the engine at a normal rate of speed.

If the cranking motor cranks the engine slowly or fails to crank it at all, the trouble is either in the battery or in the cranking system. Test the battery first, then if the battery checks out, proceed directly to Troubleshooting the Cranking System in Section 5-7 for detailed testing procedure to isolate the problem.

Ignition System Test

2- Connect a standard spark tester (available at almost any automotive parts store at very nominal cost), between the spark plug and the spark plug lead. Hold the spark plug lead with an insulated holder (to

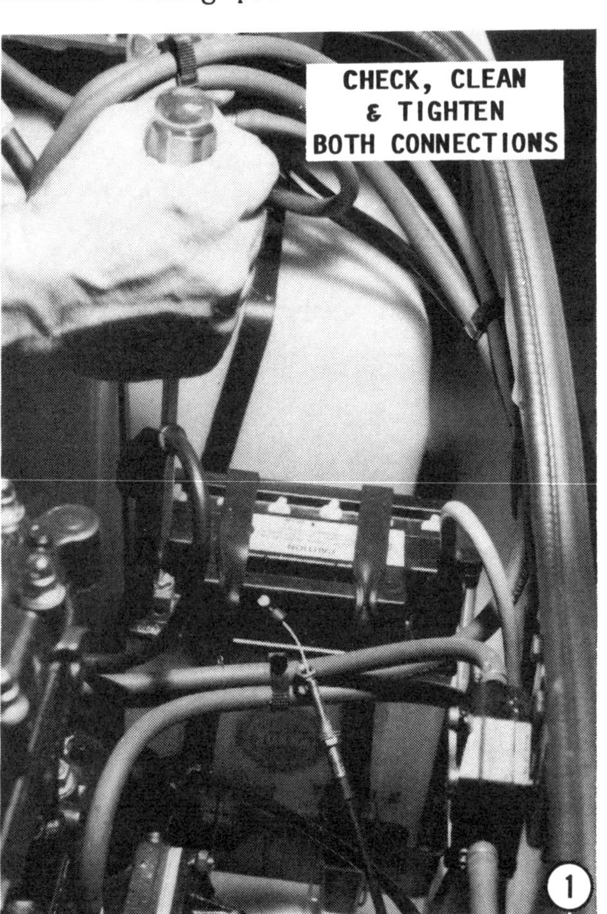

avoid receiving a "shock" when the current passes through). Crank the engine with the cranking motor. If there is **NO** spark, or if the spark is very weak, the trouble is in the ignition system. Proceed directly to the Ignition Troubleshooting in Section 5-4 for detailed testing procedures to uncover the problem.

3- Remove the flame arrestor cover and screens. Look down into the throat of each carburetor -- if so equipped -- and at the same time open and close the throttle valve assembly several times. Observe the throttle valves to be sure they move from the fully closed to the fully open positions. Perform the same check with the choke valve assembly.

Attempt to determine the age of the fuel in the fuel tank. Many fuels tend to "sour" (take on the odor of rotten eggs), over a three to four month duration, especially if an additive such as Sta-Bil has not been used. Other fuels take longer to "sour".

In no case, should an attempt be made to start the engine if the fuel in the tank is more than 12 months old. If the tank is drained (siphoned) of fuel, loosen the filler cap to prevent condensation.

Compression Test

4- Good compression is the key to proper engine performance. An engine with worn piston rings or a blown gasket cannot be made to perform satisfactorily until the mechanical defects contributing to low compression are corrected. Generally, a compression gauge is used to determine the cranking pressure within each cylinder.

To make a compression test, remove both spark plugs and remember or identify from which cylinder they were taken. Evaluation of the spark plug firing end can be most useful in determining how the cylinder is functioning. After the spark plugs have

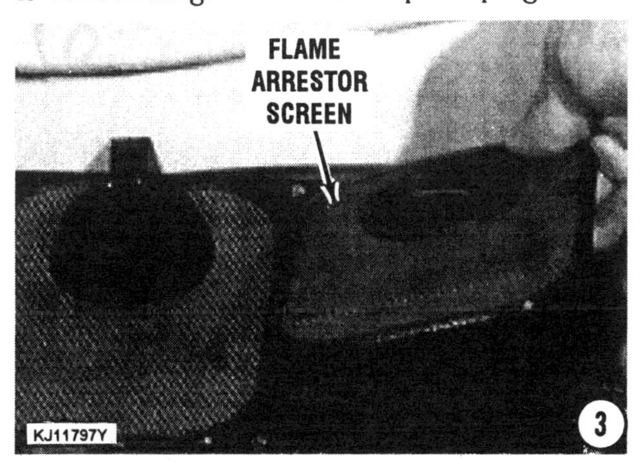

been removed, install a compression gauge into one cylinder and crank the engine.

The throttle valve and choke **MUST** be in the **WIDE OPEN** position in order to obtain maximum readings. Crank the engine through several revolutions and observe the highest reading on the compression gauge.

FACTORY ACCEPTABLE PRESSURES

MODEL/SERIES	PRESSURE RANGE
550SX	78 - 125 psi
650X-2	125 - 192 psi
750SX	121 - 187 psi
750ZXi	83 - 135 psi
750ST & STS	129 - 199 psi
750XiR	129 - 199 psi
900STX	83 - 135 psi
1100ZXi	95 - 151 psi

The differences of readings between cylinders is actually more important than the actual numerical number, but should not vary by more than 10%.

The manufacturer lists acceptable pressures for the various models, as given in the accompanying table. The significance in a compression

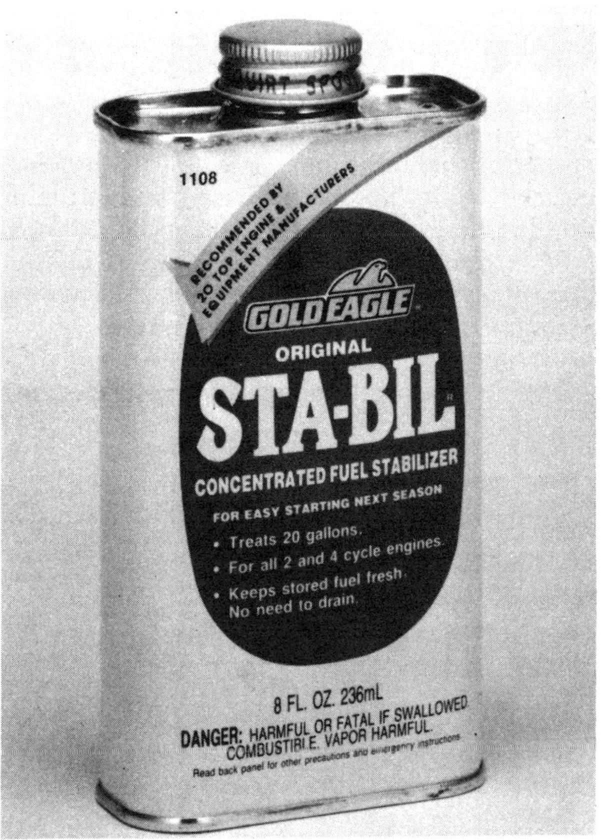

A gasoline stabilizer and conditioner may be used to prevent fuel from "souring" for up to twelve full months.

test is the variation in pressure reading between cylinders.

If a greater varation exists, then the lower reading cylinder should be checked by making a cylinder leak test. A simple and fairly effective leak test can be made by first inserting a teaspoonful of oil into the spark plug opening of the lower reading cylinder, and then cranking the engine a few times to distribute the oil.

Check the compression again to see if inserting the oil caused a change. The oil helps make a temporary seal around the rings and increases the compression. If the reading increased, the compression loss was probably due to worn rings. If the reading did not change appreciably, the loss may be due to a blown head gasket.

5-2 MECHANICAL ENGINE PROBLEMS

Jumping Waves

Even though "jumping waves" with a personal watercraft may be considered **GREAT** sport, consider the extensive internal engine damage which will surely occur if this type action is consistently practiced.

During wave jumping or using a wet ramp, when the craft clears the water, the engine rpm increases dramatically -- approaching the limit set by the rpm limiter in an unloaded condition.

As soon as the craft returns to the water surface, the sudden load on the impeller is transferred along the driveline. Speed of the rotating pump shaft and the crankshaft is suddenly reduced while the flywheel continues to spin rapidly due to its mass. This condition, especially when repeated over a period of time, often shears the Woodruff key and the flywheel becomes repositioned on the crankshaft.

A very small misalignment will result in a change in ignition timing. The craft operator will notice a drastic loss of power. A worse condition would be for the flywheel to shift enough engine start would not be possible.

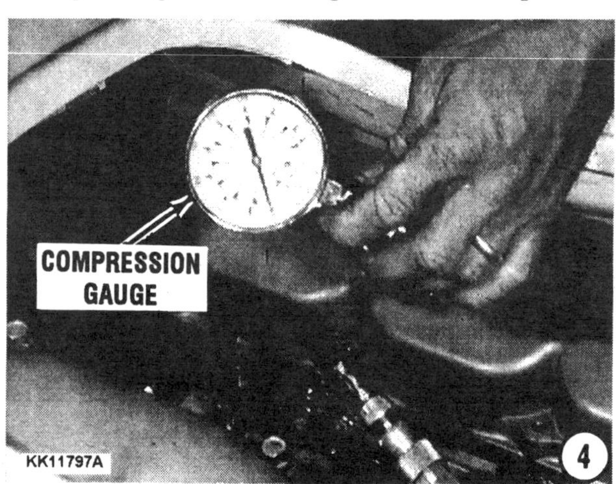

This sudden increase and decrease in engine speed dramatically shortens engine life. Bent crankshafts and broken connecting rods can usually be directly attributed to this type of "fun play".

Easy Check for Twisted Crankshaft

A twisted crankshaft may be detected by first removing the spark plugs and setting up a dial indicator in one of the spark plug openings. If a dial indicator is not available, a pencil, straw, or other small diameter straight object may be used.

With the indicator, or other device, in place, determine TDC (top dead center) of the No. 1 piston. Now, make two matching marks -- one on the flywheel and the other adjacent on the stator. Next, rotate the flywheel **EXACTLY** 180° -- preferably using a degree wheel on the flywheel. Make a second mark on the stator adjacent to the mark on the flywheel.

NOW, check for TDC of the other piston/s. The dial indicator should indicate the same reading as for the first piston **AND** the two marks on the stator should be **EXACTLY** opposite each other through the center of the crankshaft. If the indicator is off or if the marks on the stator are not aligned, as described -- the crankshaft is twisted and **MUST** be replaced.

Engine Noises

Engine noises can be generally classified as knocks, slaps, clicks, or squeaks. These noises are usually caused by loose bearings, sloppy pistons, or other moving parts of the engine.

A main bearing knock is usually identified by a dull thud which is noticeable under engine load. Operating the engine while the craft, secured to a dock, will usually reveal a main bear-

Removal of the crankshaft is not necessary to check for a "twisted" condition, as explained in the text.

ing knock. If the noise disappears when one of the spark plug wires is removed, then the noise is probably coming from that cylinder. The cause may be either the rod bearing, piston pin, or the piston, which has quieted down because the load has been removed from the cylinder. If a rod bearing has failed, the noise will be loudest when the engine is decelerating. Piston pin noise and piston slap are generally louder when a cold engine is first started.

Carbon build up in the combustion chamber can cause interference with a piston.

Many times the source of an unusual sound may be isolated by using a stethoscope or other listening device. One such device is a long shank screwdriver. Allow one hand to extend over the handle of the screwdriver. Then, place the clenched fist against an ear, and make firm contact with the other end of the screwdriver on the engine block, at each cylinder or noise area. Take care and use good judgement when using this method of attempting to detect the cause of a problem because the noise will travel through other metallic parts of the engine and could lead to a false interpretation of what is being hear hearing.

LEAK DOWN PROCEDURE

On a two-stroke engine, the crankcase has to be completely sealed to prevent pressure from escaping during primary compression. This same pressure sealing is also necessary to prevent a change in the air/fuel mixture -- resulting in a lean mixture -- being delivered to the cylinder.

First take measures to seal off the intake and exhaust manifolds Next, obtain a pressure/vacuum gauge and a "Mini-Vac" device.

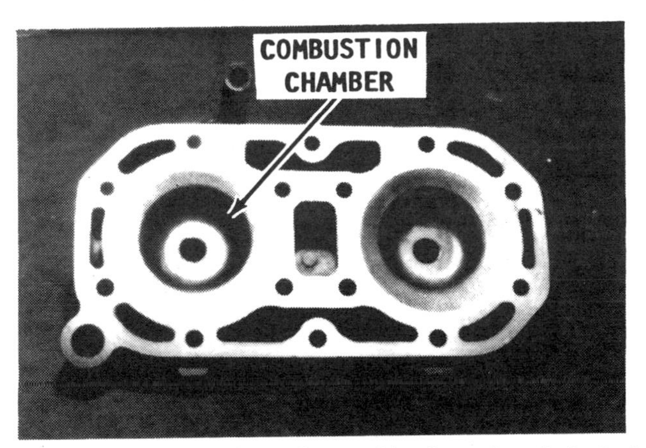

This head has been thoroughly cleaned and checked and is ready for installation. Carbon buildup in the combustion chamber can cause interference with a piston.

Now, plug all vacuum lines at the crankcase. Remove a spark plug. Using a spark plug adapter threaded into the spark plug opening, pressurize the crankcase through the spark plug opening with the "Mini-Vac" device to 6 psi. After six minutes, the pressure loss should not exceed one inch per minute for six minutes.

After the pressure test has been completed, perform a vacuum test. Using the "Mini-Vac" device, draw vacuum to 6-inches. Again, the loss of vacuum should not exceed one inch per minute for six minutes.

If the crankcase passes both tests, the unit is in satisfactory condition. If the crankcase failed either test, prepare a solution of soapy water in a "squirt" bottle. Pressurize the crankcase as described above, and then spray the soapy water solution on suspected areas of pressurization loss, such as the head gasket, cylinder base gasket, intake manifold, crankcase mating surface, rear crankshaft seal and the forward crankcase seal. If air bubbles form in the soapy water, even the smallest amount of pressure is being lost at the location.

The forward seal would require the removal of some items in order to use the soapy water, but it is an important spot for possible pressure leak.

5-3 FUEL SYSTEM PROBLEMS

The following paragraphs provide an orderly sequence of tests to pinpoint problems in the system. It is very rare for the carburetor by itself to cause failure of the engine to start.

Many times fuel system troubles are caused by a plugged fuel filter, a defective fuel pump, or by a leak in the fuel line from the fuel tank to the fuel pump or by a vacuum leak in the line between the pump and the crankcase.

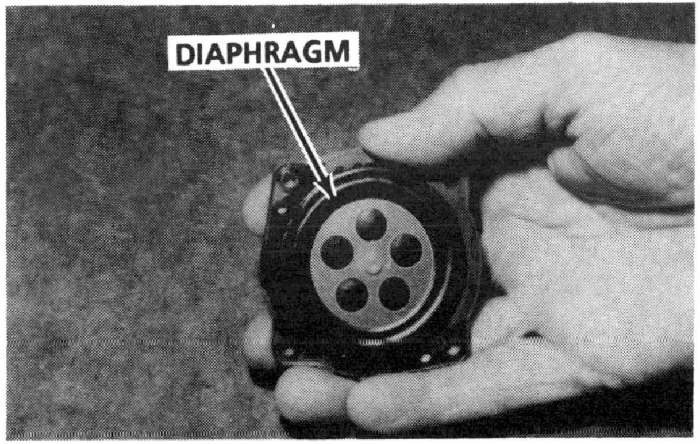

The tiniest "pin" hole in the carburetor diaphragm will seriously affect carburetor performance.

Fuel will begin to sour in three to four months and will cause engine starting problems. A fuel additive such as Sta-Bil may be used to prevent gum from forming during storage or prolonged idle periods.

When the engine is hot, the fuel system can cause starting problems. After a hot engine is shut down, the temperature inside the fuel bowl may rise to 200°F and cause the fuel to actually boil. All carburetors are vented to allow this pressure to escape to the atmosphere. However, some of the fuel may percolate over the high-speed nozzle and overflow into the intake manifold.

In order for this raw fuel to vaporize enough to burn, considerable air must be added to lean out the mixture. Therefore, the only remedy is to open the throttle as wide as possible and to crank the engine until enough air is drawn in to provide the proper mixture for the engine to start. **NEVER** move the throttle lever back-and-forth in an attempt to start a hot engine. This action will only compound the problem by adding more fuel to an already too-rich mixture.

A leak between the fuel tank and the pump many times will not appear when the engine is operating because the suction created by the pump sucking fuel will not allow the fuel to leak. Once the engine is turned off and the suction no longer exists, fuel may begin to leak.

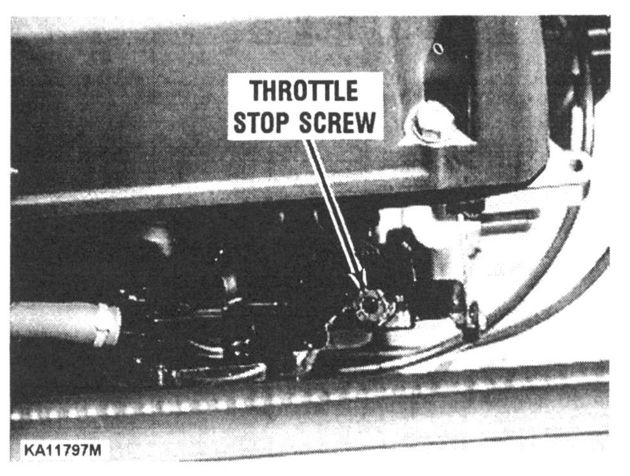

Close view of the throttle stop screw which must be set properly or the self-circling feature of the craft will be lost, if the operator is thrown into the water.

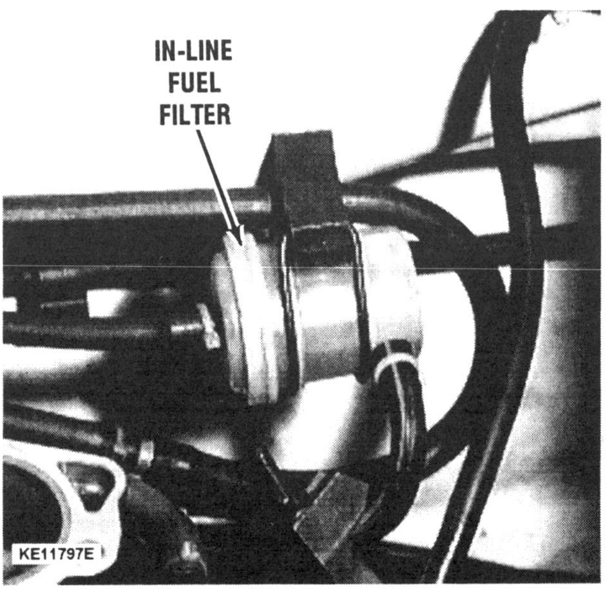

A leaking sealing ring at the in-line fuel filter will draw air into the system. The leak may not be evident until the engine is shut down -- fuel will then leak out past the damaged sealing ring.

Engine Surge

If the engine operates as if the load on the craft was being constantly increased and decreased, even though the operator attempts to maintain a constant engine speed, the problem can most likely be attributed to the fuel pump.

A small leak between the carburetor and the intake manifold, caused by a defective carburetor mounting gasket, will result in rough engine performance at idle speed.

Rough Engine Idle

Problems preventing an engine from running smoothly include: a fouled spark plug; an air leak in the intake manifold; and uneven compression between the cylinders.

Of course any problem in the carburetor affecting the air/fuel mixture will also prevent the engine from operating smoothly at idle speed. These problems usually include: leaking needle valve and seat; a dirty flame arrestor; defective choke; and improper adjustments for idle mixture or idle speed.

5-4 IGNITION SYSTEM FAULTS

WARNING

Check to be sure the engine compartment is well ventilated and free of any gasoline vapors before starting any of the following tests. A spark will be generated creating a potential fire hazard if fuel vapors are present.

1- Check to be sure the battery has a full charge. If not, correct the condition by charging the battery or making a substitution. Many times the battery is found to be the culprit.

2- Check all the terminal connections inside the electrical box. Check all primary and sec-

1

2

ondary ignition wiring, the engine stop or "kill" switch and the ignition switch.

3- Use a spark tester and check for spark. If a spark tester is not available, hold the plug wire about 1/4" (6.4mm) from the engine. Crank the engine through a few revolutions using the cranking motor and check for spark. A strong spark over a wide gap must be observed when testing in this manner, because under compression a strong spark is necessary in order to ignite the air/fuel mixture in the cylinder. This means it is possible to think a strong spark is present, when in reality the spark will be too weak when the plug is installed. If

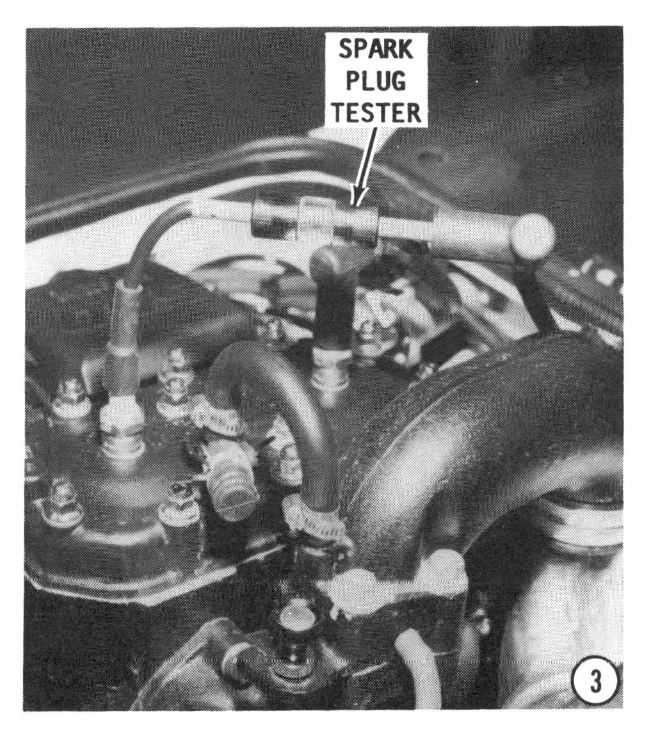

3

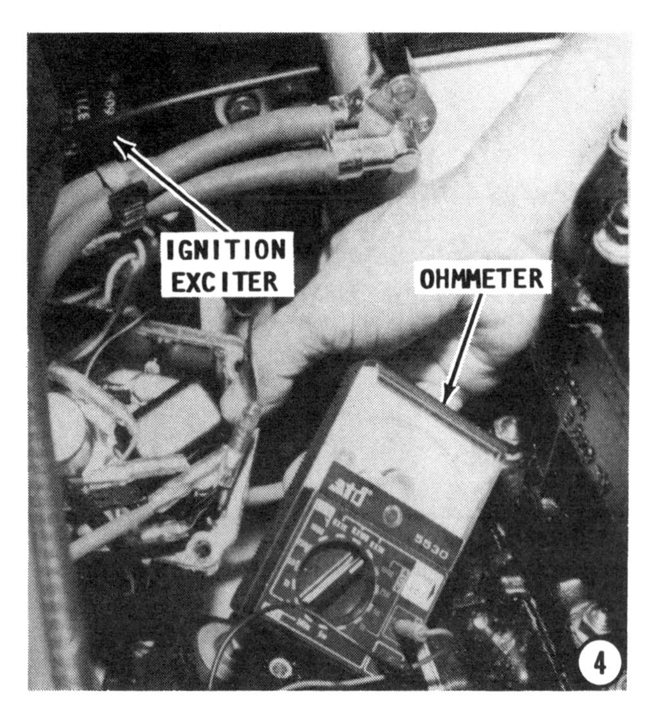

there is no spark, or if the spark is weak, the trouble is most likely under the flywheel in the stator assembly.

4- Perform the resistance test procedures given in the Ignition chapter to identify the faulty component.

Intermittent or Multiple Problems

Many ignition problems occur only during engine operation, when the component is subject to vibration and/or an increase in temperature. Many of these defective components return to normal once the engine is shut down and allowed to cool.

There is also the possibility of more than one defective component in the ignition system. One component failure may have caused a surge of power and "fried" another component.

Make every attempt to avoid the **BUY** and **TRY** method of troubleshooting. In **MOST** cases electrical components are not returnable, once the item leaves the store.

SPARK PLUG EVALUATION

Removal: Remove the spark plug wires by pulling and twisting on only the molded cap. **NEVER** pull on the wire or the connection inside the cap may become separated or the boot damaged. Remove the spark plugs and identify from which cylinder they were taken. **TAKE CARE** not to tilt the socket as you remove the plug or the insulator may be cracked.

Examine: Carefully examine the firing end to determine the firing condition in each cylinder.

Correct Color: A proper firing plug should be dry and powdery. Hard deposits inside the shell indicate the engine is starting to use some oil, but not enough to cause concern. The most important evidence is the light gray color of the porcelain, which is an indication this plug has been running at the correct temperature. This means the plug is one with the correct heat range and also that the air/fuel mixture is correct.

Overheating: A dead white or gray insulator, which is generally blistered, is an indication of overheating and pre-ignition.

Example of overheating, heavy load. Use a spark plug with a lower heat rating.

Spark plug fouled by oil. The engine may be in need of an overhaul.

The electrode gap wear rate will be more than normal and in the case of pre-ignition, will actually cause the electrodes to melt. Overheating and pre-ignition are usually caused by overadvanced timing, detonation from using too-low an octane rating fuel, an excessively lean air/fuel mixture, or problems in the cooling system.

Rich Mixture: A black, sooty condition on both the spark plug shell and the porcelain is caused by an excessively rich air/fuel mixture, both at low and high speeds. The rich mixture lowers the combustion temperature so the spark plug does not run hot enough to burn off the deposits.

Deposits formed only on the shell is an indication the low-speed air/fuel mixture is too rich. At high speeds with the correct mixture, the temperature in the combustion chamber is high enough to burn off the deposits on the insulator.

Too Cool: A dark insulator, with very few deposits, indicates the plug is running too cool. This condition can be caused by low compression or by using a spark plug of an incorrect heat range. If this condition shows on only one plug it is most usually caused by low compression in that cylinder. If both plugs have this appearance, then it is probably due to the plugs having a too-low heat range.

Fouled: A fouled spark plug may be caused by the wet oily deposits on the insulator shorting the high-tension current to ground inside the shell. The condition may also be caused by ignition problems which prevent a high-tension pulse to be delivered to the spark plug.

Carbon Deposits: Heavy carbon-like deposits are an indication of excessive oil consumption. This condition may be the result of worn piston rings.

Electrode Wear: Electrode wear results in a wide gap and if the electrode becomes carbonized it will form a high-resistance path for the spark to jump across. Such a condition will cause the engine to misfire during acceleration. If both plugs are in this condition, it can cause an increase in fuel consumption and very poor performance at high-speed operation. The solution is to replace the spark plugs with a rating in the proper heat range and gapped to specification.

Red rust-colored deposits on the entire firing end of a spark plug can be caused by water in the cylinder combustion chamber. This can be the first evidence of water entering the cylinders through the exhaust manifold. This condition **MUST** be corrected at the first opportunity.

5-5 CRANKING SYSTEM FAILURES

Regardless of how or where the solenoid is mounted, the basic circuits of the starting system on all makes of cranking motors are the same and similar tests apply. In the

Powdery deposits have melted and shorted-out this spark plug.

Red, Brown, or Yellow deposits, are by-products of combustion from fuel and lubricating oil.

following testing and troubleshooting procedures, the differences are noted.

WARNING
ALWAYS TAKE TIME TO VENT THE BILGE WHEN CONDUCTING ANY OF THE TESTS AS A PREVENTION AGAINST IGNITING ANY FUMES ACCUMULATED IN THAT AREA.

Faulty Symptoms

If the cranking motor spins, but fails to crank the engine, the cause is usually a corroded or gummy Bendix drive . The drive should be removed, cleaned, and inspected for signs of corrosion or wear.

If the cranking motor cranks the engine too slowly, the following conditions are possible causes, with corrective actions that may be taken:

- a- Battery charge is low. Charge the battery to full capacity.
- b- High resistance connections at the battery, solenoid, or motor. Clean and tighten all connections.
- c- Undersize battery cables. Replace cables with sufficient size.

If the cranking motor must be removed for inspection and/or servicing, most engines covered in this manual require the exhaust elbow, manifold and chamber to be removed first.

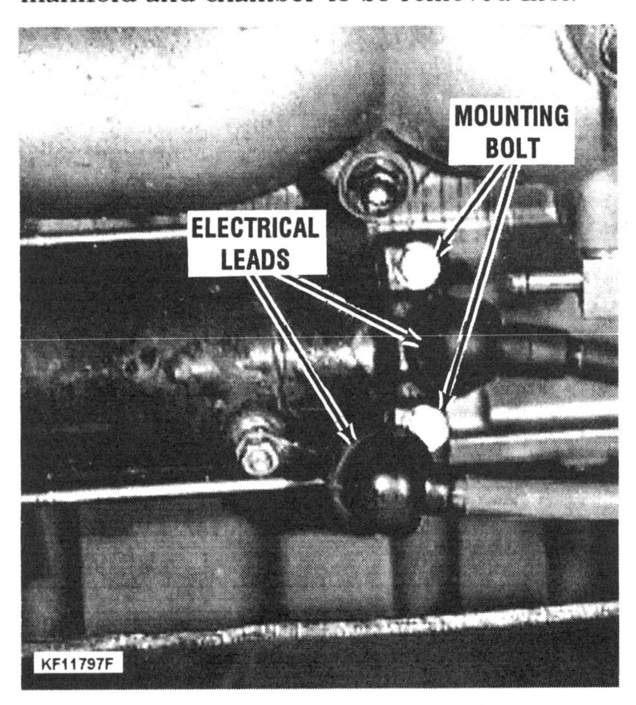

Close look at the cranking motor with the two mounting bolts and the electrical line clearly visible.

All engine cranking problems fall into one of three problem areas:

1- The cranking motor fails to turn.
2- The cranking motor spins rapidly, but does not crank the engine.
3- The cranking motor cranks the engine, but too slowly.

The following paragraphs provide a logical sequence of tests designed to isolate a problem in the cranking system.

Before wasting too much time troubleshooting the cranking motor circuit, the following checks should be made. Many times, the problem will be corrected.

- a- Battery fully charged.
- b- Main fuse is "good" (not blown).
- c- All electrical connections clean and tight.
- d- Wiring in good condition -- insulation not worn or frayed.

The following troubleshooting procedures are presented in a logical sequence.

Do not operate the cranking motor for more than 15 seconds. Prolonged cranking motor operation will cause overheating and damage the motor.

After each test, allow the cranking motor to cool for a minute or so.

Never depress the **START** button to activate the cranking motor while the engine is operating. Such action will damage the pinion and/or flywheel gears.

Cranking Circuit Tests

1- Remove both spark plug leads from the plugs and ground the ends to the engine

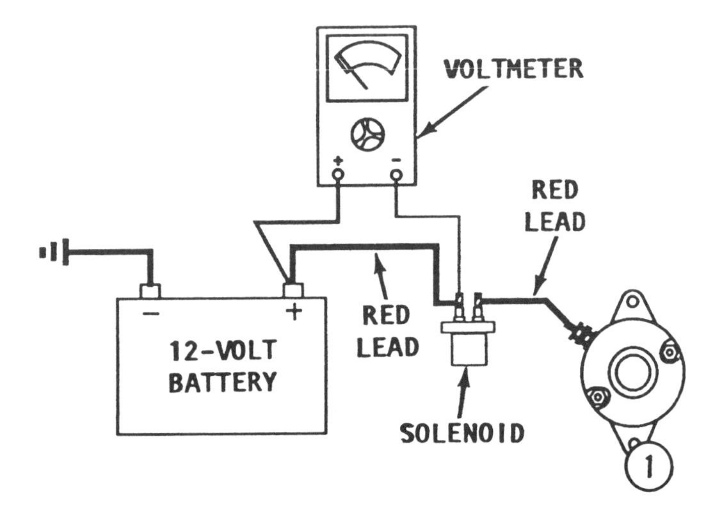

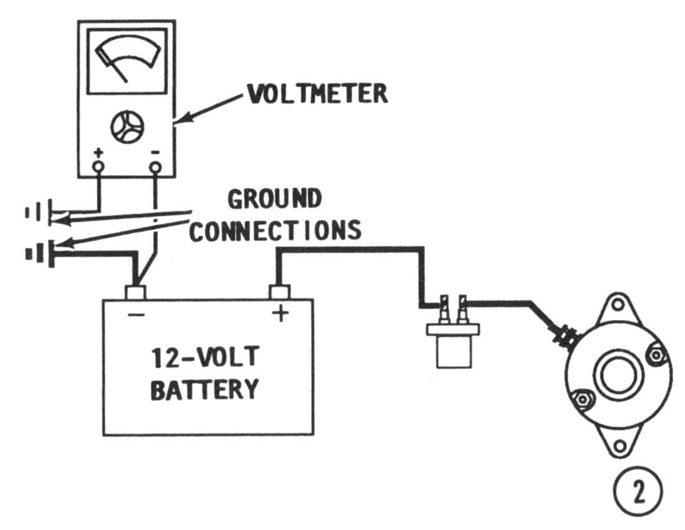

to prevent accidental engine start. Keep the leads grounded for the following tests.

Obtain a voltmeter and select the Vx1 DC scale. Observe the starter solenoid. Both large terminals have Red leads connected. One terminal has the battery cable connected and the other terminal has the lead from the cranking motor connected. Make contact with the negative Black meter lead to the terminal on the starter solenoid with the battery cable connected.

Make contact with the positive Red meter lead to the positive battery terminal. The voltmeter should register less then 0.25V DC. If the reading exceeds 0.25V DC, clean the connections at both ends of the positive battery cable and repeat the test. If the reading does not change, replace the cable.

2- Keep the meter on the same scale. Make contact with the negative Black voltmeter lead to the negative terminal on the battery. Make contact with the positive Red voltmeter lead to a good engine ground.

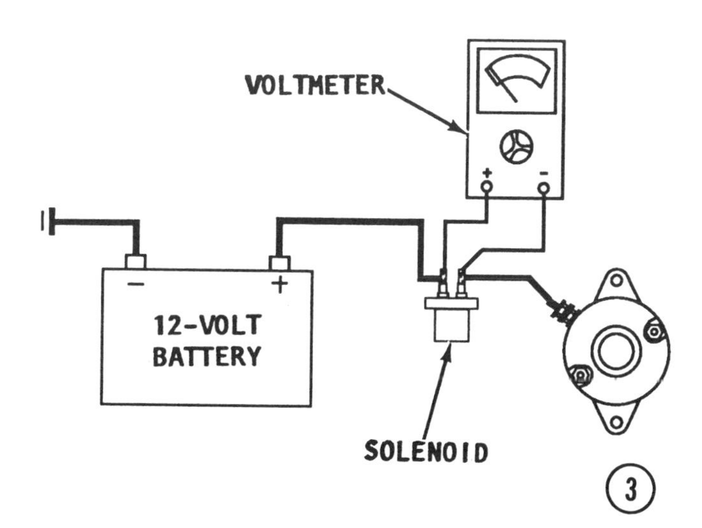

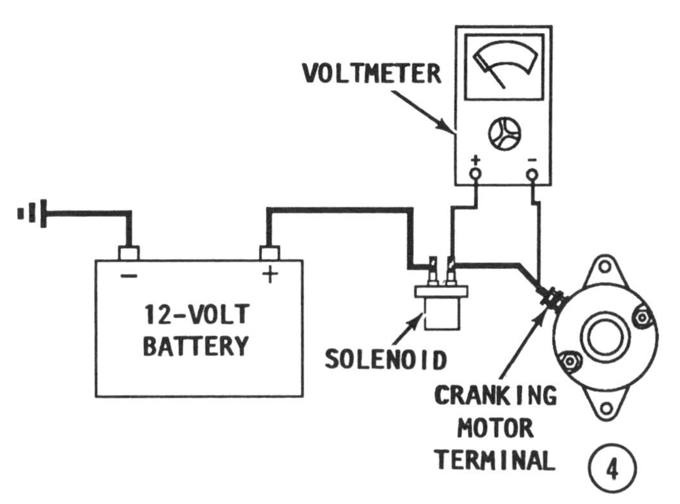

The voltmeter should register less then 0.25V DC. If the reading exceeds 0.25V DC, clean the connections at both ends of the negative battery cable and repeat the test. If the reading does not change, replace the cable.

CRITICAL WORDS

During the following tests -- Step 3 and Step 4, both meter leads must be placed on the solenoid terminals **WHILE** the engine is being cranked. **ALSO,** both meter leads must be removed from the solenoid terminals **WHILE** the engine is being cranked and **BEFORE** the engine has stopped cranking. If these precautions are not followed, the voltmeter may be damaged.

3- Keep the meter on the same scale. Obtain the services of an assistant.

Signal the assistant to start cranking the engine. While the engine is being cranked, make contact with the negative Black voltmeter lead to the solenoid terminal which has the lead from the cranking motor connected, and then quickly make contact with the positive Red voltmeter lead to the solenoid terminal which has the battery cable connected.

If the meter reading exceeds 0.25V DC, the starter solenoid must be replaced. Remove the meter leads and signal the assistant to cease cranking.

4- Keep the meter on the same scale and the assistant at the **START** button. Observe the same precautions as before in Step 3. While the engine is being cranked, make contact with the negative Black voltmeter lead to terminal on the cranking motor, and then quickly make contact with the positive Red voltmeter lead to the solenoid terminal which has the Red cranking motor connected.

If the meter reading exceeds 0.25V DC, the connections at both ends of the solenoid-to-cranking motor lead should be cleaned and the test repeated. If the reading does not change, the lead must be replaced.

CRANKING MOTOR RELAY REMOVAL FOR TESTING

Description

The cranking motor relay is actually a switch between the battery and the engine. The switch cannot be serviced. Therefore, if testing indicates the switch is faulty, it **MUST** be replaced. The relay is housed inside the electrical box. The box must first be removed from the engine, and then the relay removed from the box for testing purposes.

Electrical Box Removal Model 550 Series Only

1- Disconnect the battery ground cable from the battery. Remove the spark plug lead/s from the spark plug/s. Remove the two bolts securing the connector cover to

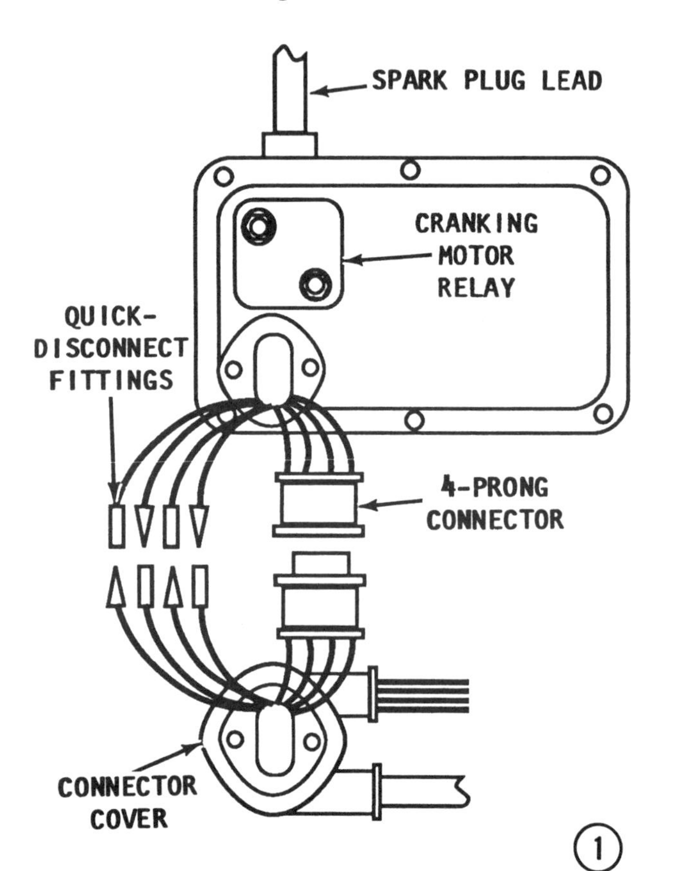

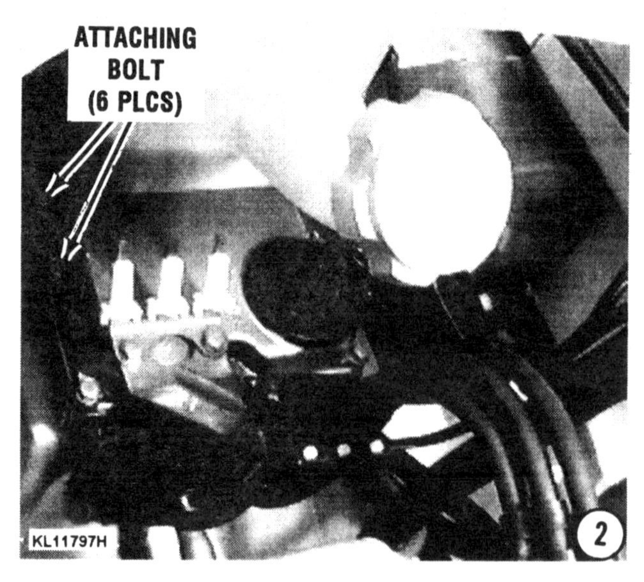

the electrical box. Pull off the cover. Disconnect the 4-pin connector and the four free leads at their quick disconnect fittings.

Pull back the two boots over the large cables at the cranking motor relay. Disconnect both large cables from the relay terminals. Remove the six bolts securing the electrical box to the engine compartment bulkhead. Remove the two bolts securing the two halves of the box together, and then open the box.

Electrical Box Removal All Models Except 550

2- Disconnect the battery ground cable from the battery. Remove the spark plug leads from the spark plugs. Pull back the large boot over the large cables at the cranking motor relay. Disconnect the cables from the relay terminals. Remove the bolts securing the cover to the

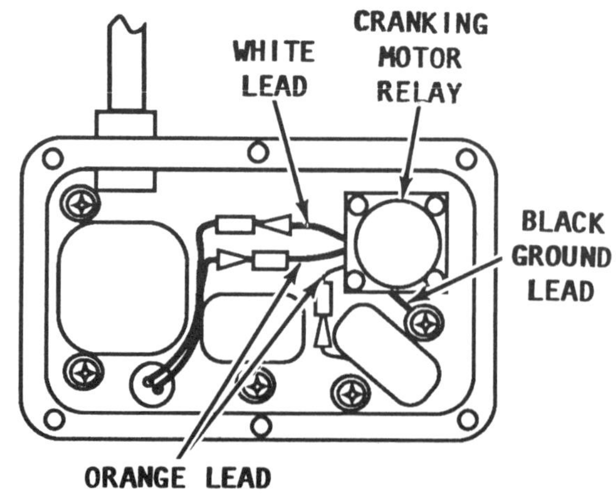

Line drawing to depict the cranking motor relay on the Model 550 Series.

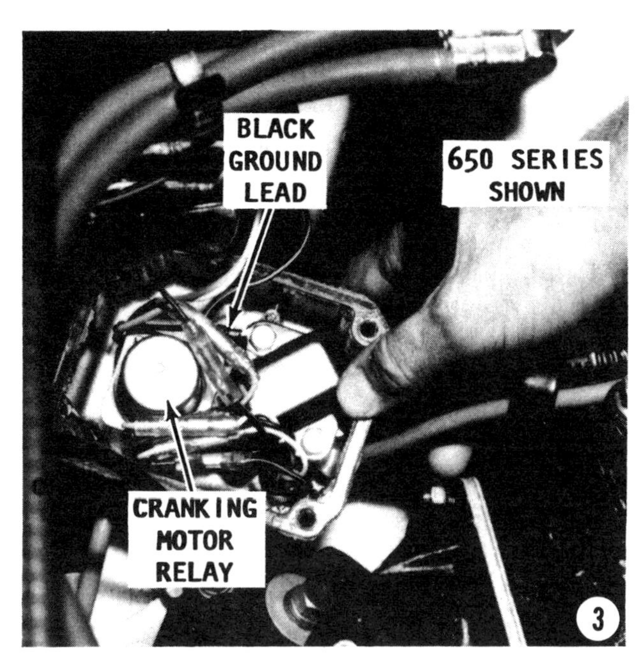

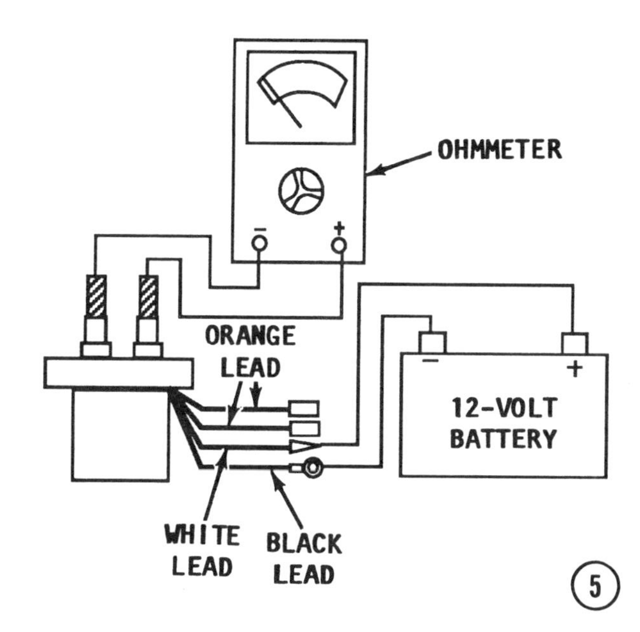

bulkhead or side of the hull, depending on the model. Note there are two different bolt sizes.

Cranking Motor Relay Removal
All Models

3- Inside the box -- the CDI unit must be removed from high performance 750 Models or all Model 900 and 1100 Series. Disconnect the ground switch from the relay at the rectifier/ regulator. Disconnect the two Orange leads and the White lead at their quick-disconnect fittings. Outside the box -- remove the nuts from the two large relay terminals. Pull the relay out of the box, taking care to observe the arrangement of washers and insulating grommets.

Cranking Relay Testing

4- Select the Rx1 ohm scale on the ohmmeter. Make contact with the two meter leads across the two large relay terminals. The meter should read an infinite resistance. If the meter shows a resistance of less than infinity, the relay is defective and must be replaced.

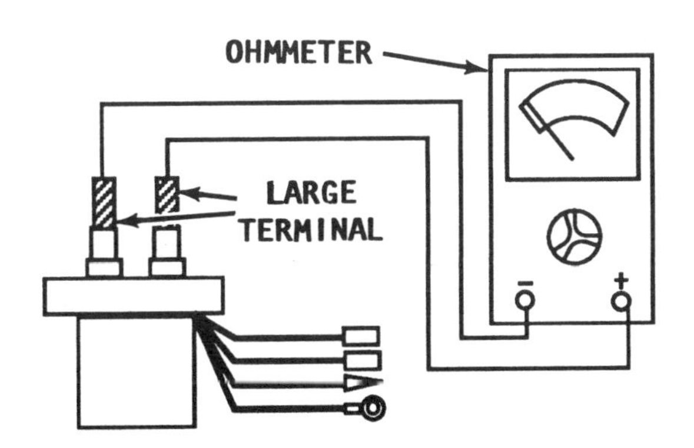

5- Keep the ohmmeter on the same scale. Obtain a 12V battery. Connect the negative battery terminal to the small Black ground lead from the relay. Connect the positive battery terminal to the small White lead from the relay.

Make contact with the two meter leads across the two large relay terminals. If the relay emits a "click" and the ohmmeter registers zero ohms, the relay is good. If no "click" is heard, or if the meter registers a high or infinite resistance, the relay is defective and must be replaced.

The cranking motor relay is housed inside the electrical box -- under the CDI unit on high performance Model 750 Series and all Model 900 and 1100 Series.

Cranking Relay Installation

6- Apply a coat of multi-purpose water restistant lubricant to all the insulating washers and grommets. Slide a large flat metal washer over each of the two large terminals, followed by a flat white plastic insulator. Then, insert the relay into the electrical box.

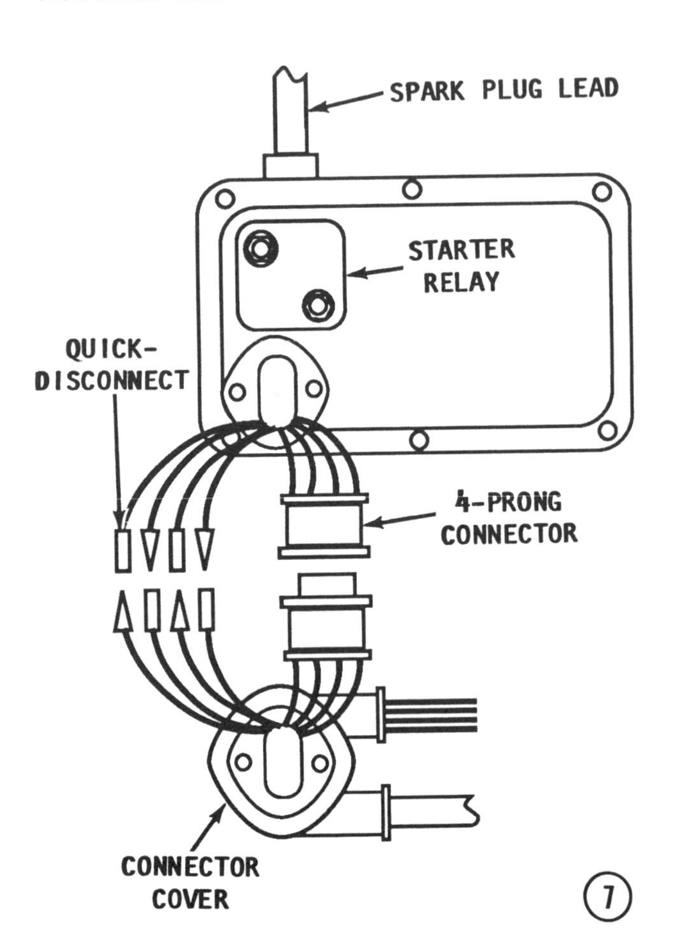

Install the following items over each of the relay terminals protruding from the box: the large Black plastic insulator, large White plastic insulator, small metal washer, and finally the locknut.

Tighten both locknuts securely. Connect the Orange and White leads, matching color-to-color and secure the relay Black ground lead under the rectifier/regulator mounting screw.

Electrical Box
Installation
Model 550 Series Only

7- Secure the halves of the box with the two bolts and washers. Position the box against the engine compartment wall and secure it with the six bolts. Connect the two large cables to the relay terminals, one from the battery and one to the cranking motor. Cover the terminals with the boots, one on each cable.

Connect the two halves of the 4-prong connector. Connect the four free leads, matching color-to-color, at their quick-disconnect fittings. Install the connector cover over the electrical box and secure it in place with the two bolts.

Install the spark plug leads and connect the battery ground cable.

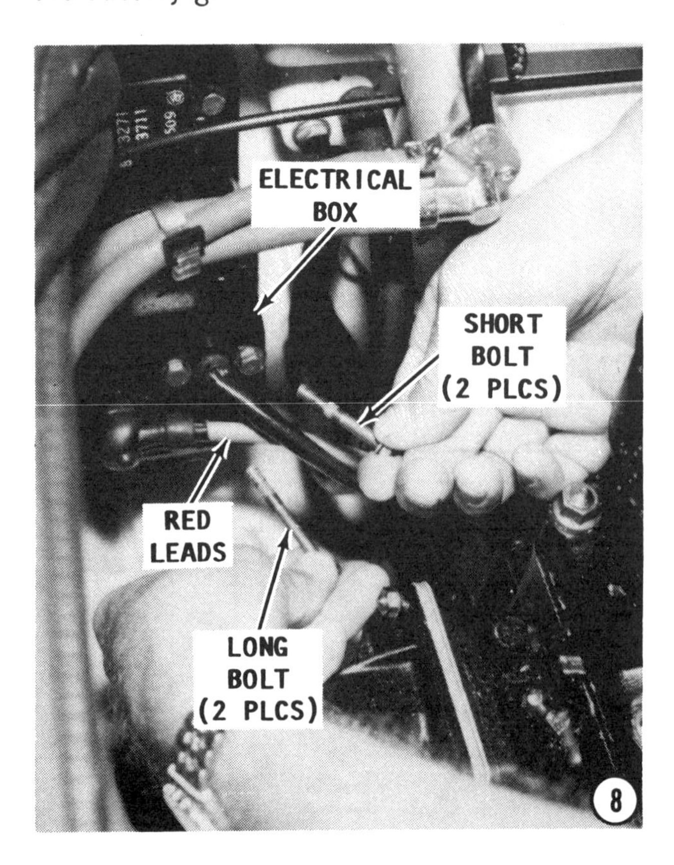

Electrical Box Installation
All Models Except 550

8- Install the cover over the electrical box mounted in the engine compartment. Secure the cover to the box with the four bolts and washers. Ensure the attaching bolts are positioned in the proper location and tightened securely. Connect the two large cables to the relay terminals, one from the battery and one to the cranking motor. Cover the terminals with the single large "boot".

Install the spark plug leads and connect the battery ground cable.

5-6 CHARGING SYSTEM MALFUNCTIONS

The lighting coil, rectifier/regulator, battery, and the necessary wiring to connect it all together comprise the charging system.

Before the charging system is blamed for battery problems, consider other areas which may be the cause:

1- Operating at low speed for short periods.

2- Voltage losses due to high resistance.

3- Corroded battery cables, connectors, and terminals.

4- Low electrolyte level in the battery cells.

5- Prolonged disuse of the battery causing a self-discharged condition.

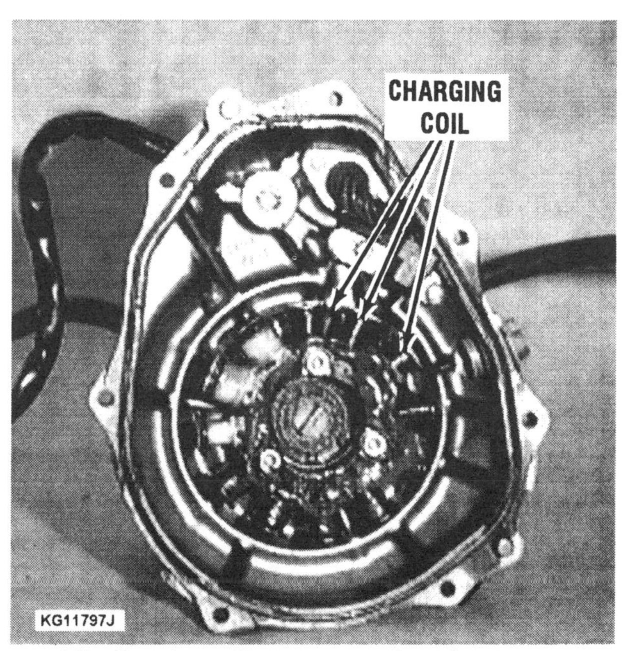

The charging coils are mounted on the stator assembly, attached to the inside of the flywheel cover on some Model 750 Series, and all Model 900 and 1100 Series.

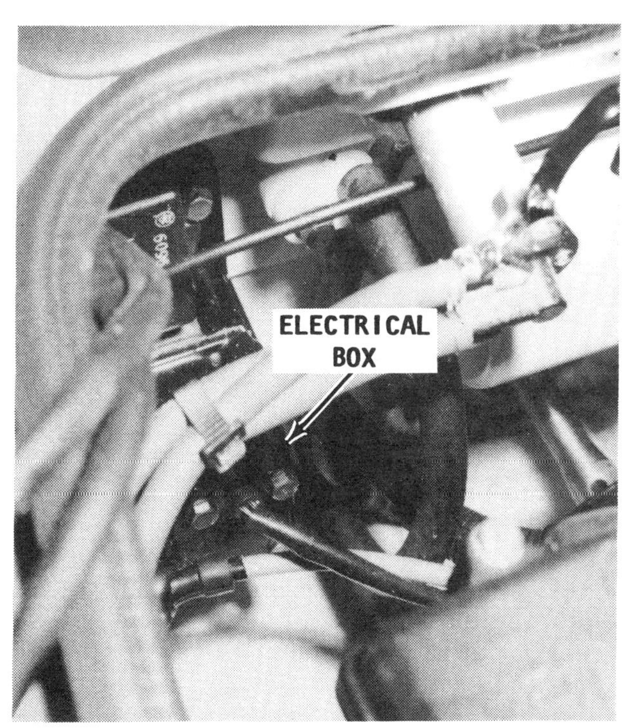

Location of the electrical box -- on a "stand-up" version watercraft -- which houses and protects the rectifier/regulator and the cranking motor relay. The box is watertight with grommets surrounding the harness leads where they enter the box.

Detailed procedures to service the battery is presented in Section 5-3. If, after charging, the battery fails to give the required specific gravity reading, then the battery will never hold a charge and must be replaced. This problem may have originated with the battery, a shorted cell perhaps, or may be attributed to the failure of the rectifier/regulator or charging coil. Refer to Chapter 7 -- Ignition -- for resistance tests on these two components.

Location for the electrical box on high performance Model 750 Series, and all Model 900 and 1100 Series watercraft. The rectifier/regulator and cranking motor relay are located under the CDI unit, within the electrical box.

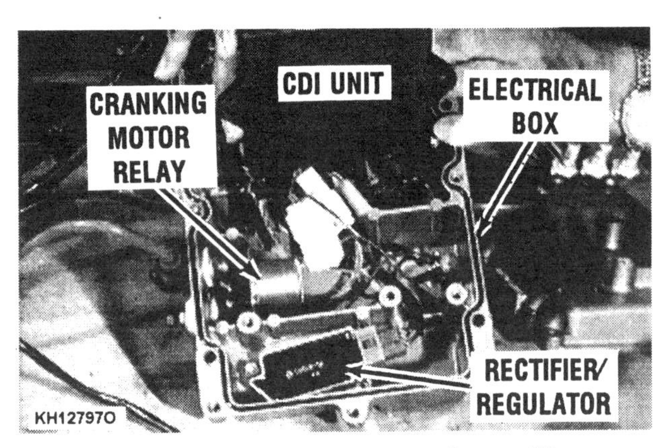

The CDI unit raised -- exposing the cranking motor relay and the rectifier/regulator underneath.

5-7 TROUBLESHOOTING CHARTS

The following six charts are presented as an aid to troubleshooting the engine and jet pump. The uppermost box identifies a specific problem. Start with the problem, then work downward. The probable causes are listed in columns under a specific system, for example: Fuel system, or Electrical system. Once a possible cause is identified, refer to the specific chapter for procedures on testing and service.

When using the charts -- proceed step-by-step through each of the tests or checks until the defect or cause of the problem is uncovered.

FIRST CHART -- Page 5-17 -- Engine fails to start.

SECOND CHART -- Page 5-18 -- Engine starts, then shuts down.

THIRD CHART -- Page 5-19 --Engine misfires.

FOURTH CHART -- Page 5-20 -- Engine lacks proper power.

FIFTH CHART -- Page 5-21 -- Abnormal engine noises.

SIXTH CHART -- Page 5-22 -- Jet pump problems.

ONE FINAL WORD

Before going out on the water, **TAKE TIME** to verify the drain plug is installed. Countless number of excursions have had a very sad beginning because the craft was eased into the water only to have the hull begin filling with water.

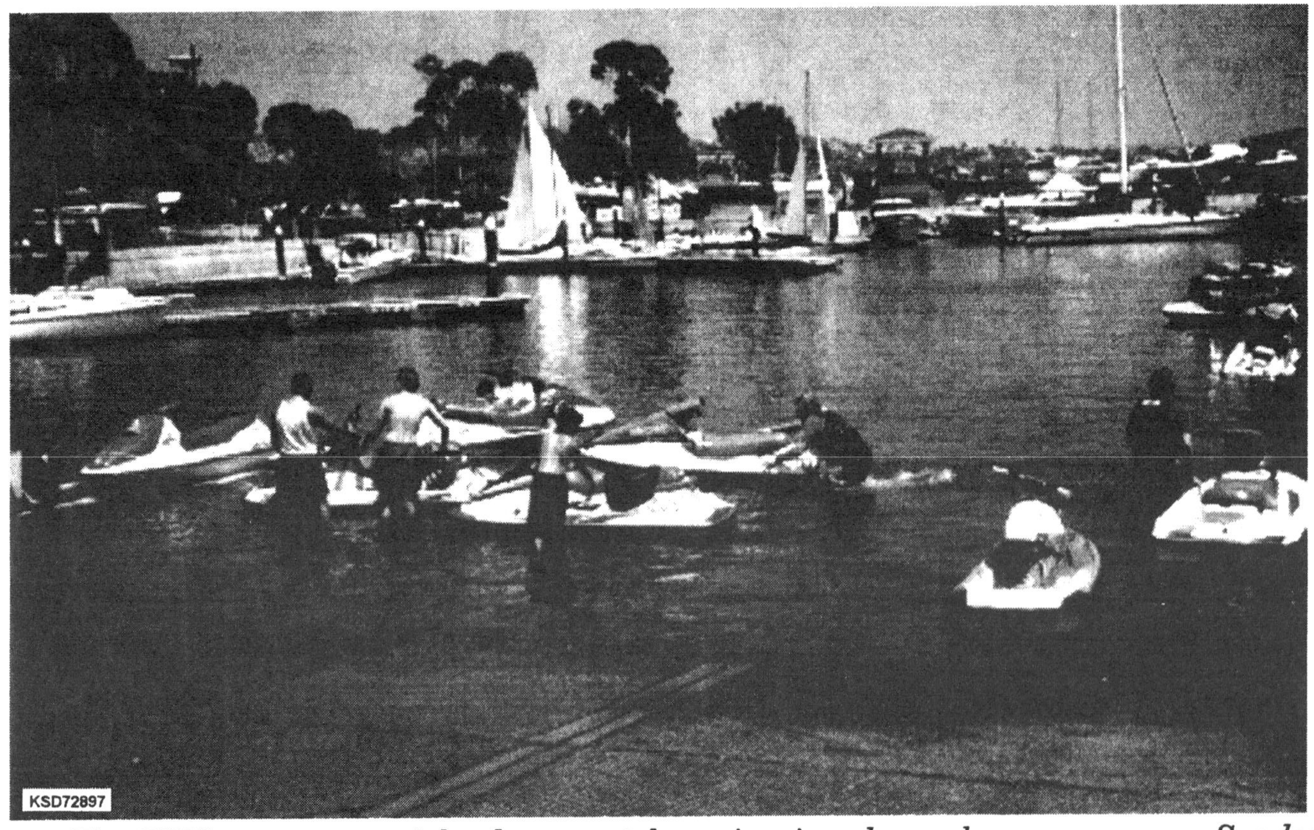

The PWC ramp at an inland or coastal marina is a busy place on a sunny Sunday afternoon.

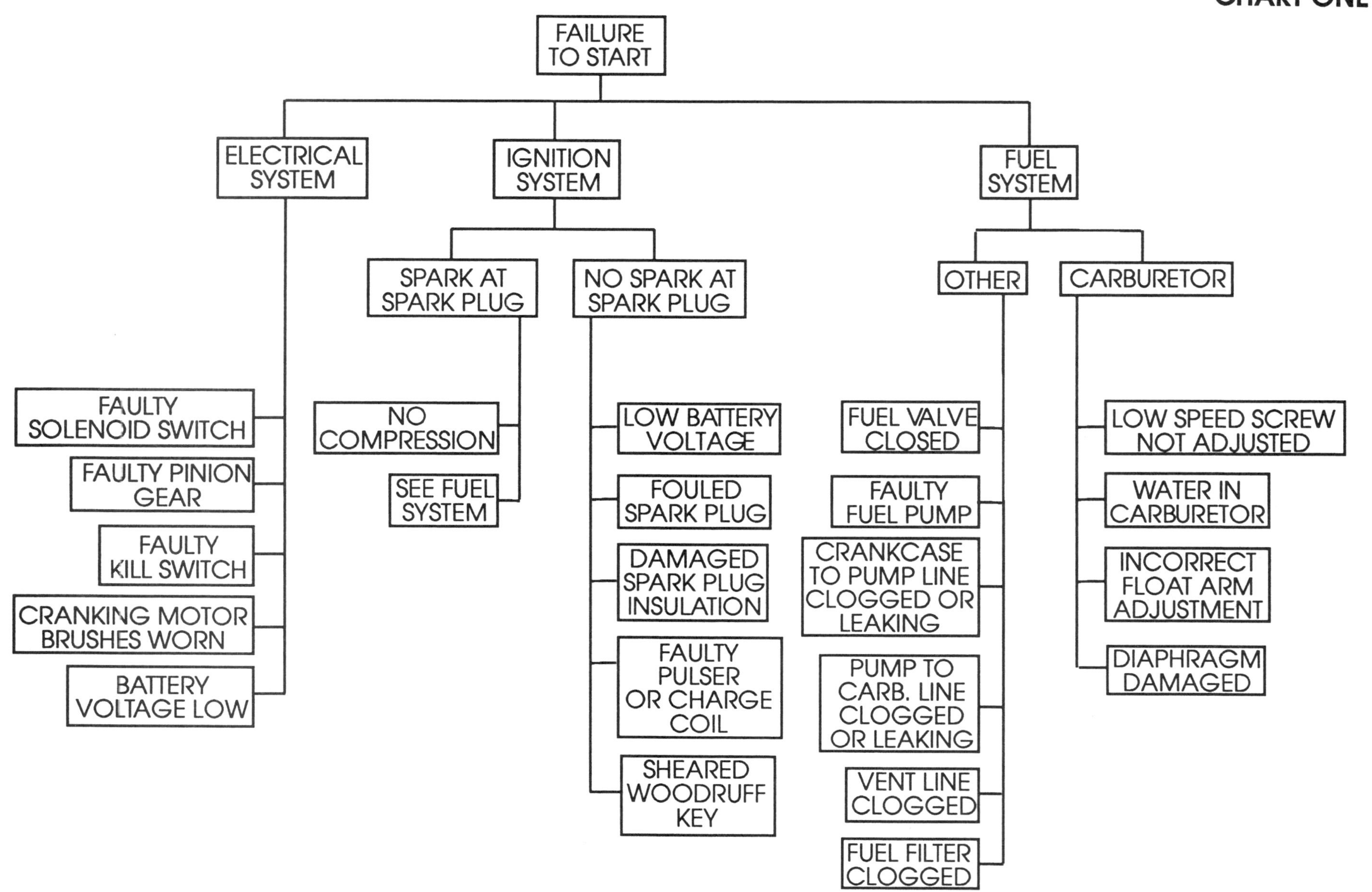
CHART ONE
FAILURE TO START
ELECTRICAL SYSTEM
FAULTY SOLENOID SWITCH
FAULTY PINION GEAR
FAULTY KILL SWITCH
CRANKING MOTOR BRUSHES WORN
BATTERY VOLTAGE LOW
IGNITION SYSTEM
SPARK AT SPARK PLUG
NO COMPRESSION
SEE FUEL SYSTEM
NO SPARK AT SPARK PLUG
LOW BATTERY VOLTAGE
FOULED SPARK PLUG
DAMAGED SPARK PLUG INSULATION
FAULTY PULSER OR CHARGE COIL
SHEARED WOODRUFF KEY
FUEL SYSTEM
OTHER
FUEL VALVE CLOSED
FAULTY FUEL PUMP
CRANKCASE TO PUMP LINE CLOGGED OR LEAKING
PUMP TO CARB. LINE CLOGGED OR LEAKING
VENT LINE CLOGGED
FUEL FILTER CLOGGED
CARBURETOR
LOW SPEED SCREW NOT ADJUSTED
WATER IN CARBURETOR
INCORRECT FLOAT ARM ADJUSTMENT
DIAPHRAGM DAMAGED

CHART TWO

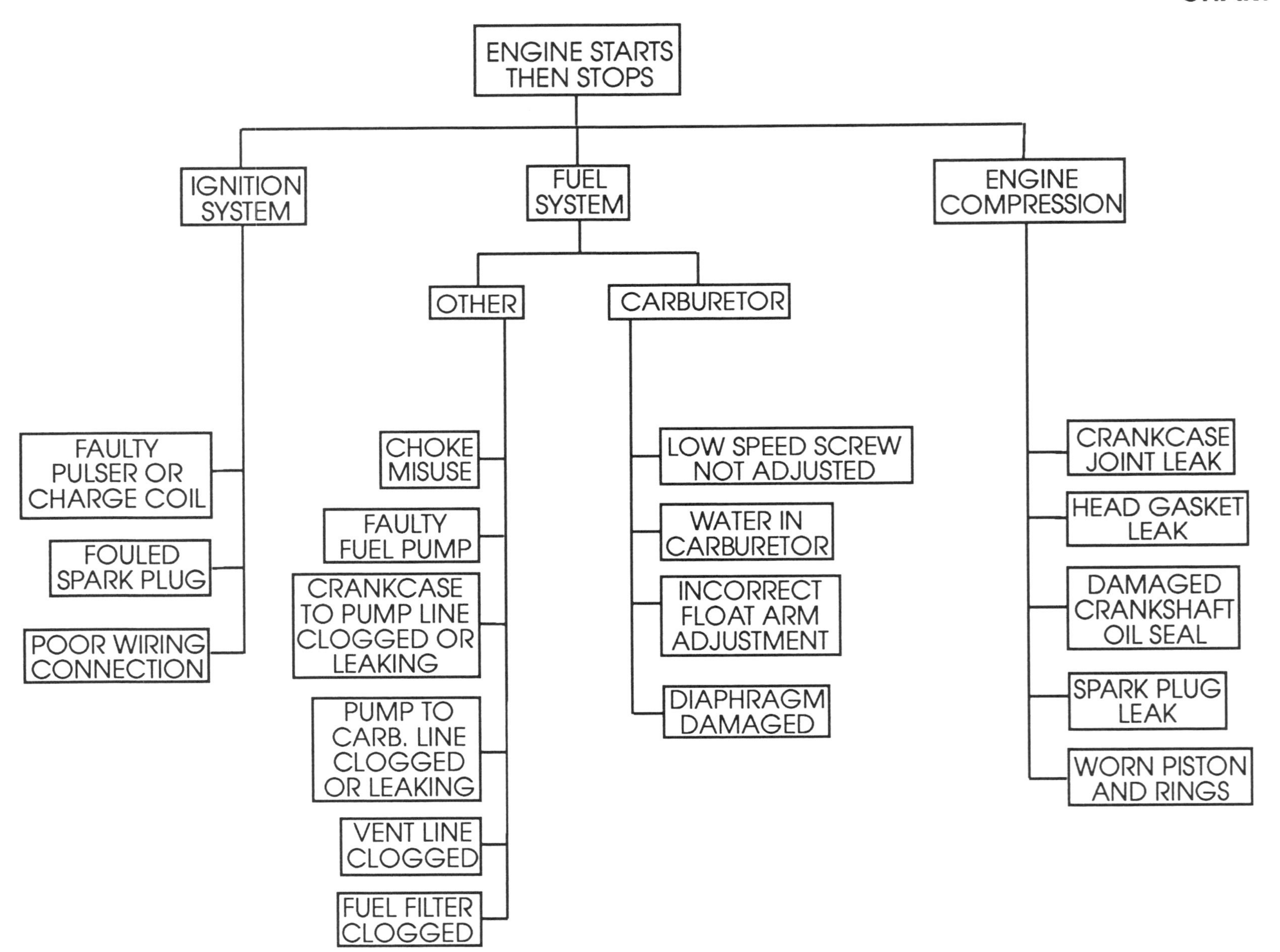
ENGINE STARTS THEN STOPS
IGNITION SYSTEM
FAULTY PULSER OR CHARGE COIL
FOULED SPARK PLUG
POOR WIRING CONNECTION
FUEL SYSTEM
OTHER
CHOKE MISUSE
FAULTY FUEL PUMP
CRANKCASE TO PUMP LINE CLOGGED OR LEAKING
PUMP TO CARB. LINE CLOGGED OR LEAKING
VENT LINE CLOGGED
FUEL FILTER CLOGGED
CARBURETOR
LOW SPEED SCREW NOT ADJUSTED
WATER IN CARBURETOR
INCORRECT FLOAT ARM ADJUSTMENT
DIAPHRAGM DAMAGED
ENGINE COMPRESSION
CRANKCASE JOINT LEAK
HEAD GASKET LEAK
DAMAGED CRANKSHAFT OIL SEAL
SPARK PLUG LEAK
WORN PISTON AND RINGS

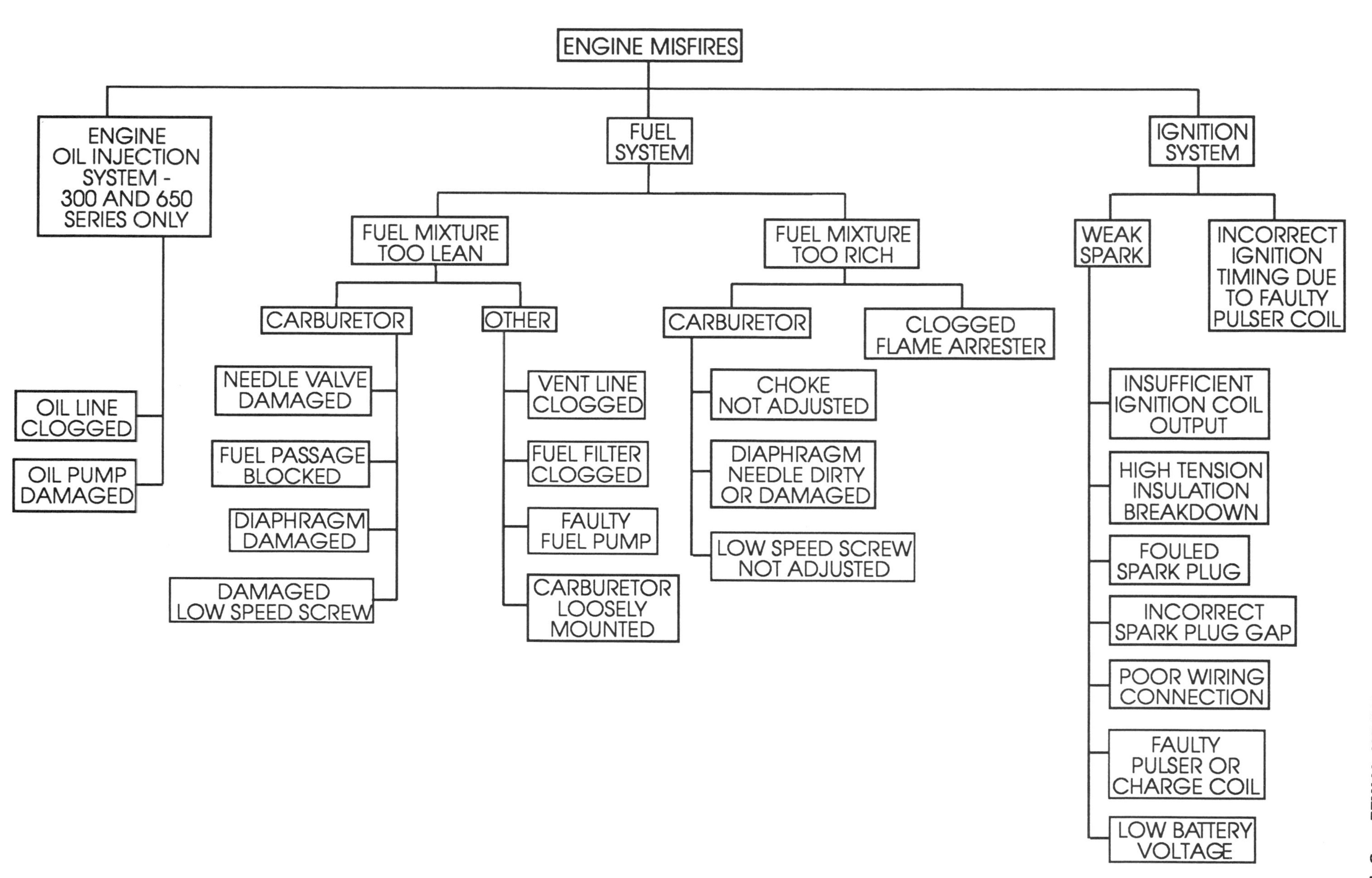
CHART THREE
ENGINE MISFIRES
ENGINE OIL INJECTION SYSTEM - 300 AND 650 SERIES ONLY
OIL LINE CLOGGED
OIL PUMP DAMAGED
FUEL SYSTEM
FUEL MIXTURE TOO LEAN
CARBURETOR
NEEDLE VALVE DAMAGED
FUEL PASSAGE BLOCKED
DIAPHRAGM DAMAGED
DAMAGED LOW SPEED SCREW
OTHER
VENT LINE CLOGGED
FUEL FILTER CLOGGED
FAULTY FUEL PUMP
CARBURETOR LOOSELY MOUNTED
FUEL MIXTURE TOO RICH
CARBURETOR
CHOKE NOT ADJUSTED
DIAPHRAGM NEEDLE DIRTY OR DAMAGED
LOW SPEED SCREW NOT ADJUSTED
CLOGGED FLAME ARRESTER
IGNITION SYSTEM
WEAK SPARK
INSUFFICIENT IGNITION COIL OUTPUT
HIGH TENSION INSULATION BREAKDOWN
FOULED SPARK PLUG
INCORRECT SPARK PLUG GAP
POOR WIRING CONNECTION
FAULTY PULSER OR CHARGE COIL
LOW BATTERY VOLTAGE
INCORRECT IGNITION TIMING DUE TO FAULTY PULSER COIL

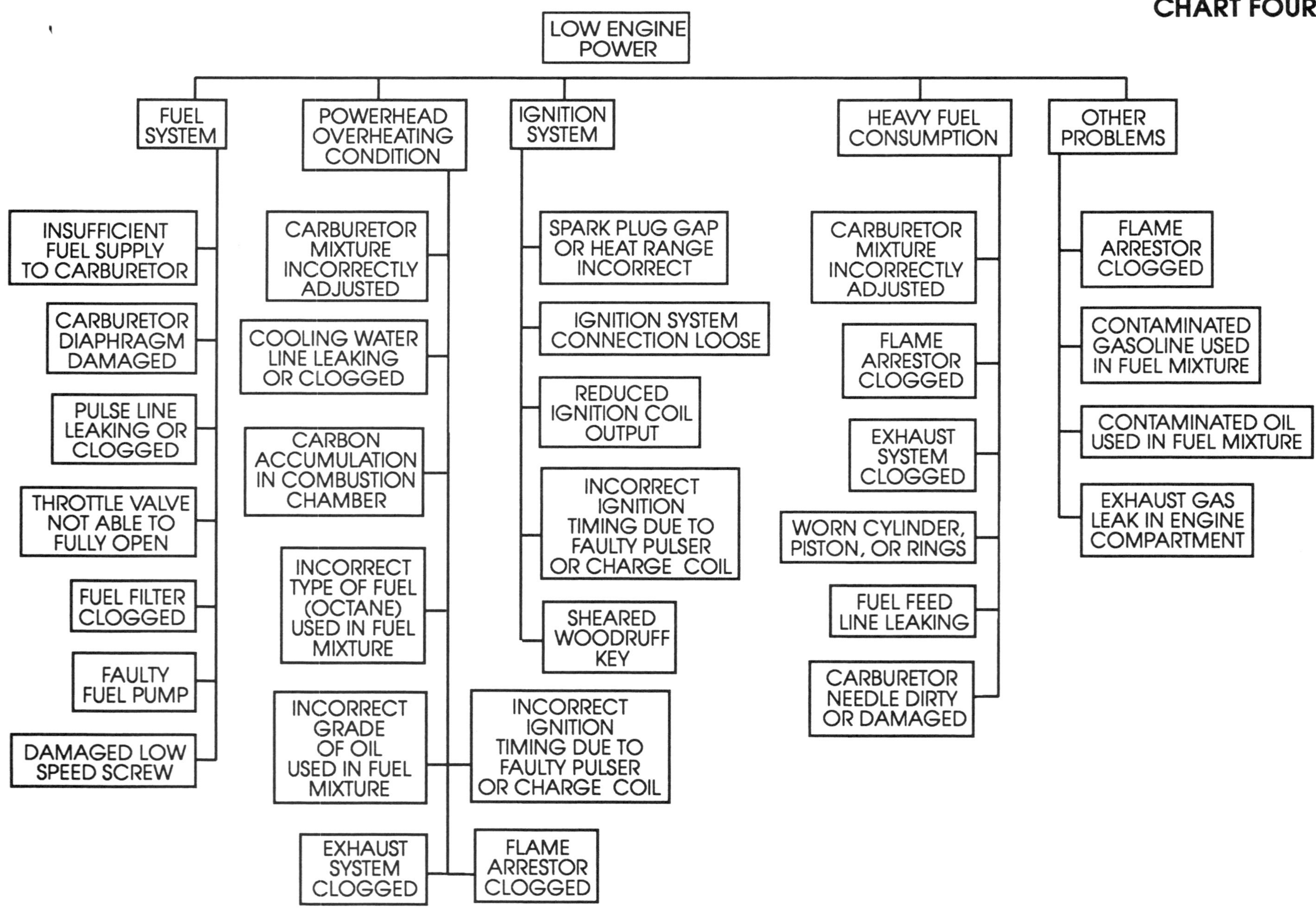
CHART FOUR
LOW ENGINE POWER
FUEL SYSTEM
INSUFFICIENT FUEL SUPPLY TO CARBURETOR
CARBURETOR DIAPHRAGM DAMAGED
PULSE LINE LEAKING OR CLOGGED
THROTTLE VALVE NOT ABLE TO FULLY OPEN
FUEL FILTER CLOGGED
FAULTY FUEL PUMP
DAMAGED LOW SPEED SCREW
POWERHEAD OVERHEATING CONDITION
CARBURETOR MIXTURE INCORRECTLY ADJUSTED
COOLING WATER LINE LEAKING OR CLOGGED
CARBON ACCUMULATION IN COMBUSTION CHAMBER
INCORRECT TYPE OF FUEL (OCTANE) USED IN FUEL MIXTURE
INCORRECT GRADE OF OIL USED IN FUEL MIXTURE
INCORRECT IGNITION TIMING DUE TO FAULTY PULSER OR CHARGE COIL
EXHAUST SYSTEM CLOGGED
FLAME ARRESTOR CLOGGED
IGNITION SYSTEM
SPARK PLUG GAP OR HEAT RANGE INCORRECT
IGNITION SYSTEM CONNECTION LOOSE
REDUCED IGNITION COIL OUTPUT
INCORRECT IGNITION TIMING DUE TO FAULTY PULSER OR CHARGE COIL
SHEARED WOODRUFF KEY
HEAVY FUEL CONSUMPTION
CARBURETOR MIXTURE INCORRECTLY ADJUSTED
FLAME ARRESTOR CLOGGED
EXHAUST SYSTEM CLOGGED
WORN CYLINDER, PISTON, OR RINGS
FUEL FEED LINE LEAKING
CARBURETOR NEEDLE DIRTY OR DAMAGED
OTHER PROBLEMS
FLAME ARRESTOR CLOGGED
CONTAMINATED GASOLINE USED IN FUEL MIXTURE
CONTAMINATED OIL USED IN FUEL MIXTURE
EXHAUST GAS LEAK IN ENGINE COMPARTMENT

CHART FIVE

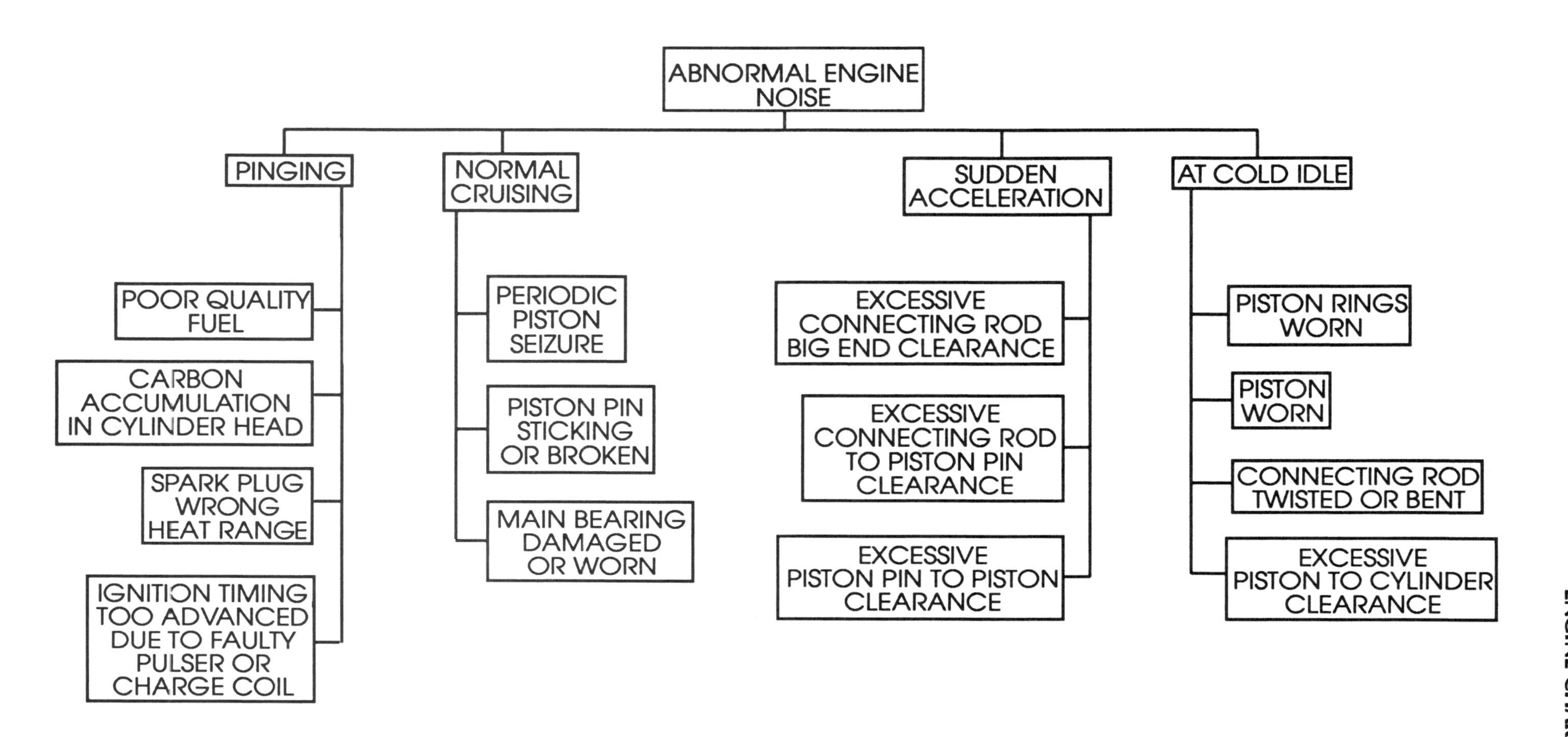
ABNORMAL ENGINE NOISE
PINGING
POOR QUALITY FUEL
CARBON ACCUMULATION IN CYLINDER HEAD
SPARK PLUG WRONG HEAT RANGE
IGNITION TIMING TOO ADVANCED DUE TO FAULTY PULSER OR CHARGE COIL
NORMAL CRUISING
PERIODIC PISTON SEIZURE
PISTON PIN STICKING OR BROKEN
MAIN BEARING DAMAGED OR WORN
SUDDEN ACCELERATION
EXCESSIVE CONNECTING ROD BIG END CLEARANCE
EXCESSIVE CONNECTING ROD TO PISTON PIN CLEARANCE
EXCESSIVE PISTON PIN TO PISTON CLEARANCE
AT COLD IDLE
PISTON RINGS WORN
PISTON WORN
CONNECTING ROD TWISTED OR BENT
EXCESSIVE PISTON TO CYLINDER CLEARANCE

CHART SIX

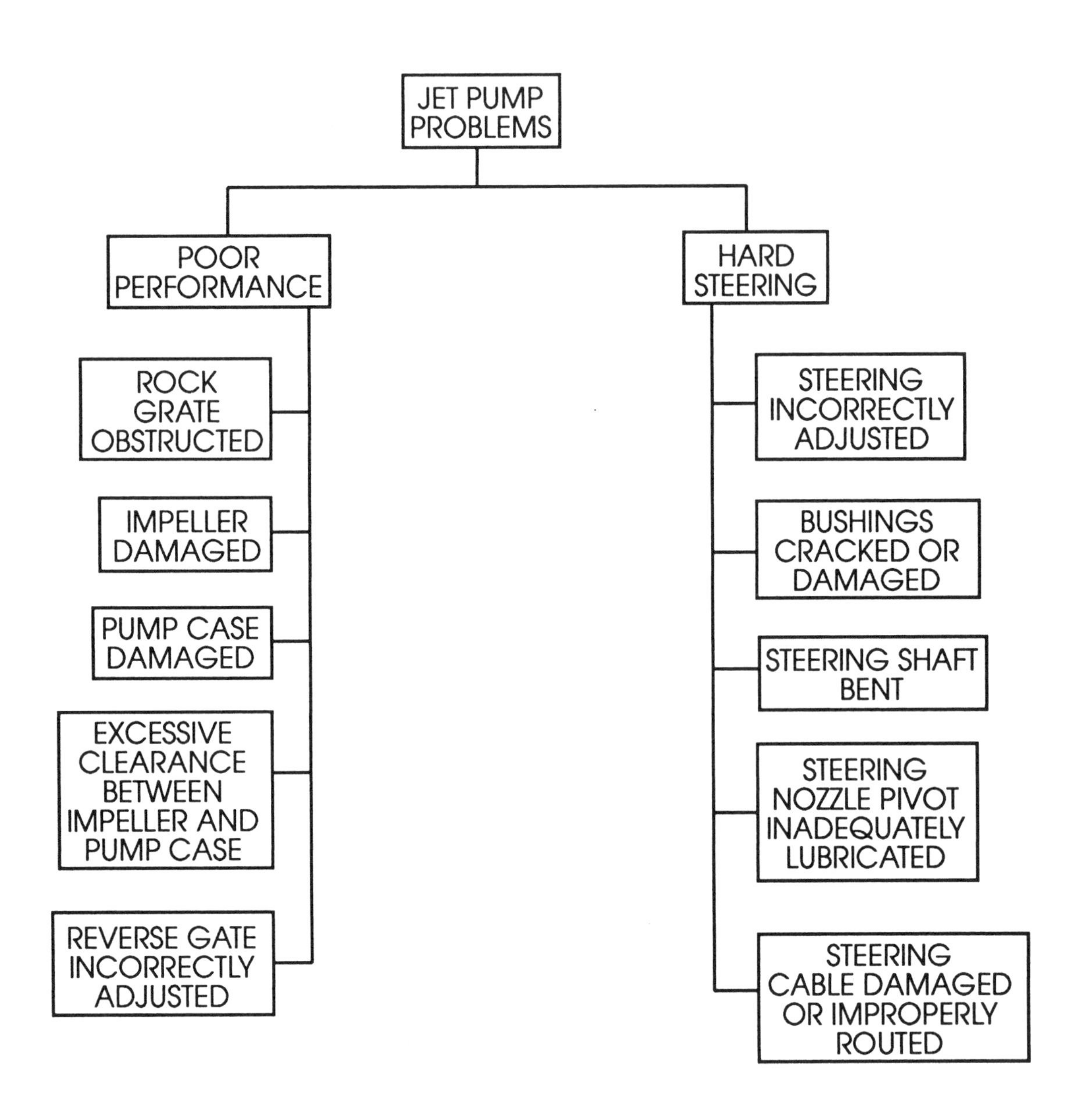
JET PUMP PROBLEMS
POOR PERFORMANCE
ROCK GRATE OBSTRUCTED
IMPELLER DAMAGED
PUMP CASE DAMAGED
EXCESSIVE CLEARANCE BETWEEN IMPELLER AND PUMP CASE
REVERSE GATE INCORRECTLY ADJUSTED
HARD STEERING
STEERING INCORRECTLY ADJUSTED
BUSHINGS CRACKED OR DAMAGED
STEERING SHAFT BENT
STEERING NOZZLE PIVOT INADEQUATELY LUBRICATED
STEERING CABLE DAMAGED OR IMPROPERLY ROUTED

6
FUEL AND OIL

6-1 INTRODUCTION

The carburetion and ignition principles of two-cycle engine operation **MUST** be understood in order to perform a proper tune-up on a marine engine.

If you have any doubts concerning your understanding of two-cycle engine operation, it would be best to study the Introduction section in the first portion of Chapter 8, before tackling any work on the fuel system.

The fuel system includes the fuel tank, fuel pump, fuel filters, carburetor, and the associated parts to connect it all together. Regular maintenance of the fuel system to obtain maximum performance, is limited to changing the fuel filters at regular intervals and using fresh fuel.

If a sudden increase in gas consumption is noticed, or if the engine does not perform properly, a carburetor overhaul, including boil-out, or replacement of the fuel pump may be required.

6-2 GENERAL CARBURETION INFORMATION

The carburetor is merely a metering device for mixing fuel and air in the proper proportions for efficient engine operation. At idle speed, a marine engine requires a mixture of about 8 parts air to 1 part fuel. At high speed or under heavy duty service, the mixture may change to as much as 12 parts air to 1 part fuel.

DIAPHRAGM CARBURETORS

Description

The two carburetors used on the Kawasaki PWC's covered in this manual are Keihin diaphragm type carburetors. One model Keihin has an integral fuel pump. When three carburetors are installed on an engine -- the center carburetor does not have the fuel pump. No. 1 and No. 3 carburetors supply the fuel.

If the design of the carburetor does not include the integral fuel pump a separate remote fuel pump is mounted in the engine compartment.

A diaphragm carburetor has no float bowl or float. Instead, a diaphragm takes the place of the float.

A hinged arm, still called a float arm, rests against the diaphragm and opens and closes the needle and seat assembly.

Operation

Following the fuel through its course, from the fuel tank to the combustion chamber of the cylinder, will provide an appreciation of exactly what is taking place.

After the engine starts, the fuel passes through the pump to the carburetor. All systems have some type of filter installed

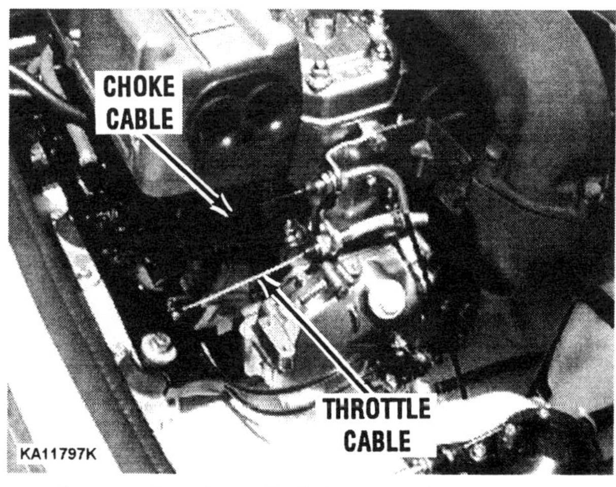

A properly adjusted choke cable and throttle cable along with clean flame arrestor screens and the use of fresh fuel will do much toward maximum performance.

The Keihin CDK-34 carburetor installed on the Model 650 Series engine is supplied with fuel from a remote fuel pump, shown below.

somewhere in the line between the tank and the carburetor.

Fuel is delivered to an inlet valve under pressure from the integral or remote fuel pump. When the choke valve is closed and the engine is cranked, or when the piston is on its upward stroke, the pressure in the carburetor throat orifice and the crankcase is less than atmospheric. The fuel chamber above the diaphragm is connected to the crankcase through a vacuum tube. A nozzle channel connecting the carburetor throat with the fuel chamber also allows the fuel chamber to be at less than atmospheric pressure.

Typical installation of a remote fuel pump mounted to the outboard bulkhead to service a CDK-34 carburetor, shown above.

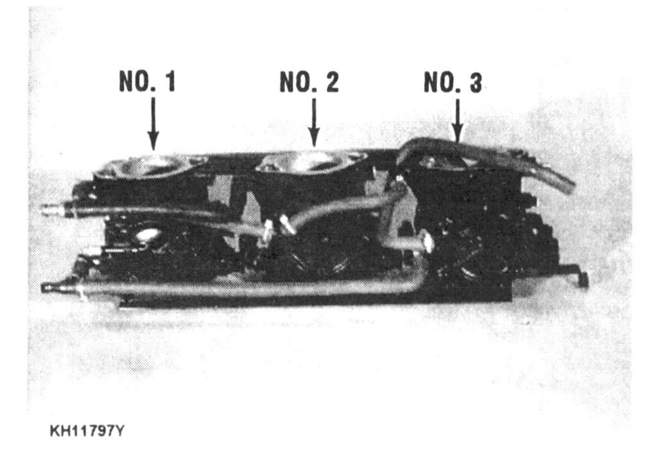

Rack of triple Keihin CDK-38 carburetors from a Model 900 ZXI watercraft. The No. 1 and No. 3 carburetors have an integral fuel pump. The No. 2 does not have the pump.

The chamber on the other side of the diaphragm is vented to atmosphere, and therefore, is always at atmospheric pressure.

With the two pressure differentials, the diaphragm will flex toward the lowest pressure -- towards the fuel chamber. As the diaphragm flexes, it raises the float arm which, because of a return spring, is in constant contact with the diaphragm. The float arm is hinged. As one side goes up, the other side comes down and allows the inlet valve to open. At the same time, the needle valve falls away from the valve seat. As soon as the inlet valve opens, fuel flows under pressure from the fuel pump into the fuel chamber.

When the piston is on its downward stroke, the crankcase, the carburetor throat, and the fuel chamber are all at a higher pressure than atmosphere. The fuel chamber filled with fuel at a higher pressure now flexes toward the side with the lower pressure -- the side vented to atmosphere.

The return spring pushes the float arm down against the diaphragm. The other side of the float arm raises up and pushes the needle valve against its seat closing the inlet valve.

On the next piston downward stroke, the diaphragm flexes back again forcing the fuel in the fuel chamber to be ejected through a jet into the carburetor throat. On the same stroke the fuel chamber is replenished with fresh fuel through the open needle valve. This cycle repeats itself, sending a squirt of fuel into the carburetor throat each downward stroke of the piston.

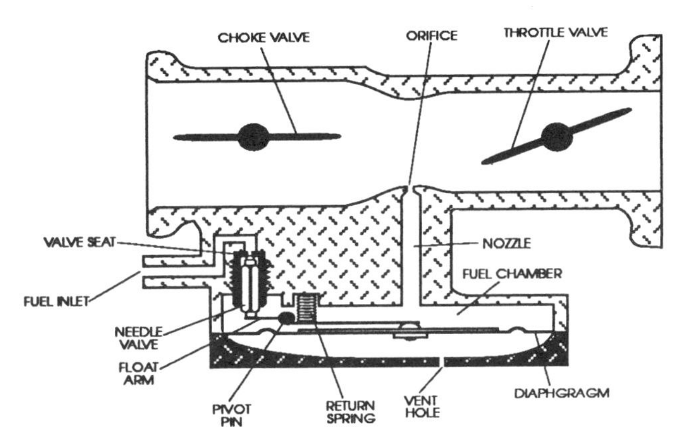

Simple cross-section drawing of a diaphragm type carburetor used on all series and model years covered in this manual.

In order to obtain the proper air/fuel mixture for all engine speeds, some models have high and low speed jets. These jets have adjustable needle valves which are used to compensate for changing atmospheric conditions. In almost all cases, the high-speed circuit has fixed high-speed jets and are not adjustable.

A throttle valve controls the flow of air/fuel mixture drawn into the combustion chambers. A cold engine requires a richer fuel mixture to start and during the brief period it is warming to normal operating temperature. A choke valve is placed ahead of the metering jets and venturi. As this valve begins to close, the volume of air intake is reduced, thus enriching the mixture entering the cylinders.

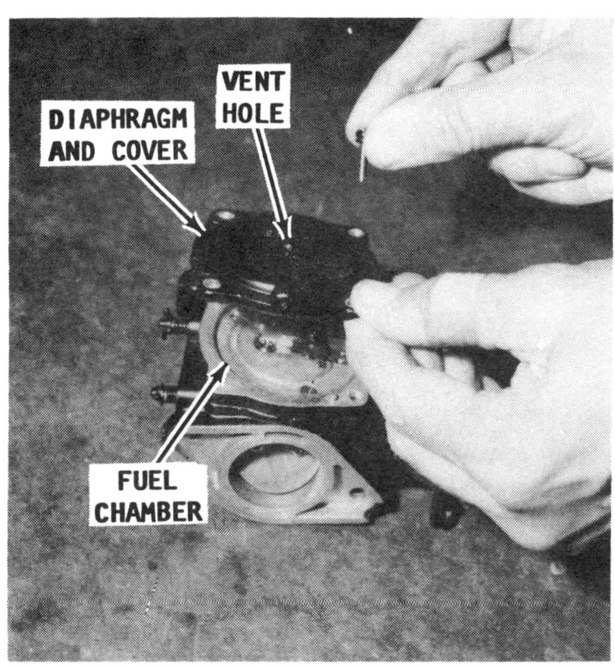

The vent hole for the fuel chamber above the diaphragm must be kept clear to the atmosphere at all times for the diaphragm to function efficiently.

When this choke valve is fully closed, a very rich fuel mixture is drawn into the cylinders.

6-3 FUEL COMPONENTS AND AVAILABLE GAS

The fuel system includes the fuel tank, fuel pump, fuel filters, carburetor, connecting fuel lines, and the associated parts to connect it all together. Regular maintenance of the fuel system to obtain maximum performance, is limited to changing the fuel filters, cleaning the flame arrestor screens and using fresh fuel.

Even with the high price of fuel, removing gasoline that has been standing unused over a long period of time, is still the easiest and least expensive preventive maintenance possible. In most cases this old gas, even with some oil mixed with it, can be used without harmful affects in an automobile using regular gasoline.

If a sudden increase in gas consumption is noticed, or if the engine does not perform properly, a carburetor overhaul, including boil-out, or replacement of the fuel pump may be required.

Leaded Gasoline

The manufacturer of the units covered in this manual recommend the engines be operated using regular unleaded gasoline with a minimum octane rating of 84 or higher.

In the United States, The Environmental Protection Agency (EPA) phased-out leaded fuel, "Regular" gasoline in 1993. Lead in gasoline boosts the octane rating (energy). Therefore, if the lead is removed, it must be replaced with another agent. Unknown to the general public, many refineries are adding alcohol in an effort to hold the octane rating.

Alcohol in gasoline can have a deteriorating affect on certain fuel system parts. Seals can swell, pump check valves can swell, diaphragms distort, and other rubber or neoprene composition parts in the fuel system can be affected.

Since about 1981, every effort has been made to use materials that will resist the alcohol being added to fuels.

Fuel containing alcohol will slowly absorb moisture from the air. Once the moisture content in the fuel exceeds about 1%, it will separate from the fuel taking the alcohol with it.

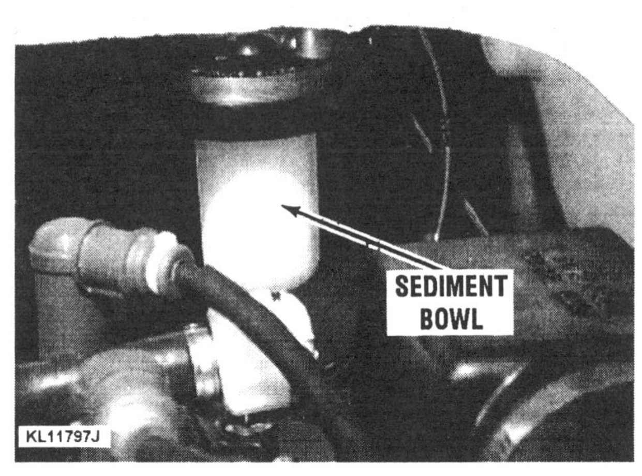

Usual installation of a water separation filter with sediment bowl. Easy access and removal through a ring nut is a big help to regular maintenance.

This water/alcohol mixture will settle to the bottom of the fuel tank. The engine will fail to operate. Therefore, storage of this type of gasoline is not recommended for more than just a few days.

A lead substitute additive -- octane booster -- can help prevent detonation when unleaded gasoline is used in a personal watercraft.

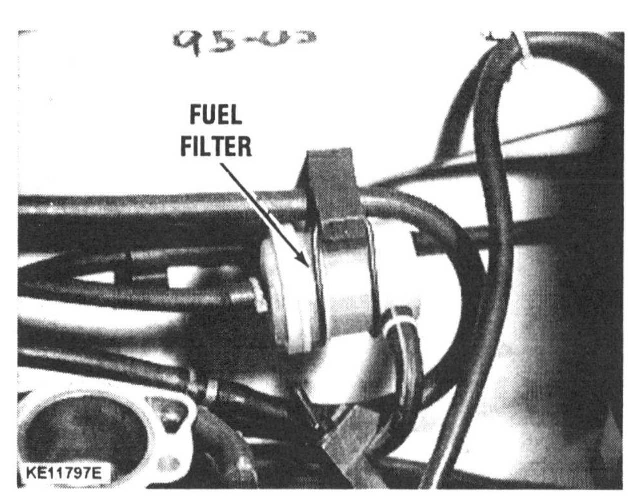

In-line fuel filter mounted to the outboard bulkhead in the engine compartment. This filter is used on many, but not all Kawasaki models covered in this manual.

Fuel Filter and Sediment Bowl

Some models are equipped with a fuel filter and sediment bowl mounted to the outside bulkhead in the engine compartment. The bowl is quickly removed through a ring nut at the top.

In-Line Fuel Filter

Most models have an in-line fuel filter mounted on the outside bulkhead in the engine compartment. The location affords easy access for maintenance.

Fuel Tank Screen Filters

Some models have a set of screen filters installed in the fuel tank, as shown in the accom-

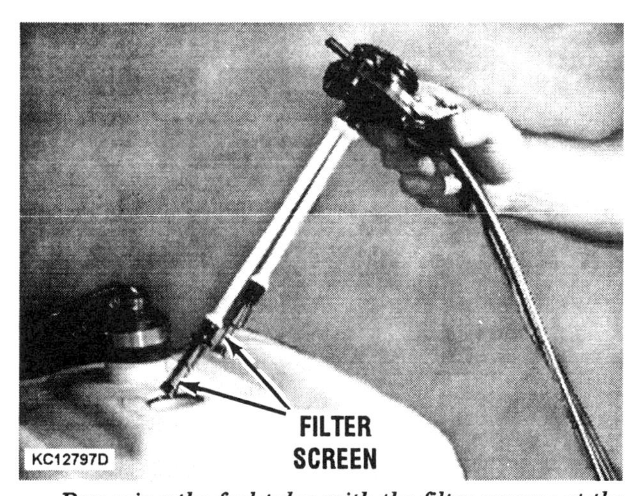

Removing the fuel tubes with the filter screens at the lower end from the fuel tank. As mentioned in the text, the tank must be removed to withdraw the tubes, and in most cases the engine must be removed in order for the tank to clear. Therefore, this is not considered a routine maintenance task.

panying illustration. Service -- cleaning -- of these filters is not considered a routine maintenance task because the fuel tank must be removed in order to withdraw the filters from the tank **AND** the engine must be "pulled" from the craft for the fuel tank to clear. Therefore cleaning these filter screens is not considered a routine maintenance task.

Air/Fuel Mixture

A suction effect is created in the intake manifold each time the piston moves upward in the cylinder. This suction draws air through the throat of the carburetor. A restriction in the throat, called a venturi, controls air velocity and has the effect reducing air pressure at this point.

The difference in air pressures at the throat and in the fuel chamber, causes the fuel to be pushed out of the metering jet extending down into the fuel chamber. When the fuel leaves the jet, it mixes with the air passing through the venturi. This air/fuel mixture should then be in the proper proportion for burning in the cylinders for maximum engine performance.

Throttle and Choke Valves

A throttle valve controls the flow of air/fuel mixture drawn into the combustion chambers. A cold engine requires a richer fuel mixture to start and during the brief period it is warming to normal operating temperature. A choke valve is placed ahead of the metering jets and venturi. As this valve begins to close, the volume of air drawn in is reduced, thus enriching the mixture entering the cylinders.

When this choke valve is fully closed, a very rich fuel mixture is drawn into the cylinders aiding engine start.

FUEL PUMP

Many times, a defective fuel pump diaphragm is mistakenly diagnosed as a problem in the ignition system. The most common problem is a tiny pin-hole in the diaphragm. Such a small hole will permit gas to enter the crankcase and set foul the spark plugs at idle-speed. During high-speed operation, gas quantity is limited, the plug is not fouled and will therefore fire in a satisfactory manner.

Remote Fuel Pump

At press time, the only engine covered in this manual with a remote fuel pump was the Model 650SX, with a Keihin CDK-34 carburetor. This remote fuel pump is shown in a typical installation on an outboard bulkhead in the engine compartment of a Model 650SX watercraft at the bottom of Page 6-2. Troubleshooting, testing and service procedures for the remote fuel pump are presented in Section 6-8, beginning on Page 6-33.

Integral Fuel Pump

Again, at press time, all other engines covered in this manual were equipped with Keihin carburetors -- all having an integral fuel pump with one exception -- on a triple carburetor installation, the center carburetor does not have the integral fuel pump, as indicated in the illustration at the top of Page 6-2. Fuel to this carburetor is supplied by the pump on the No. 1 and No. 3 carburetor.

Twin cylinder engines may have a single carburetor serving both cylinders -- Model 550 Series. Model 650 and some Model 750 Series engines have a single carburetor or two carburetors -- each serving a single cylinder.

If the fuel pump fails to perform properly an insufficient fuel supply will be delivered to the carburetor. This lack of fuel will cause the engine to run lean, lose rpm or cause piston scoring.

If the fuel pump diaphragm tears or is punctured, fuel will enter the crankcase through the pulse line causing an over-rich mixture. This mixture will foul the plugs and in extreme cases flood the crankcase.

NEVER use liquid Neoprene on fuel line fittings. Always use Permatex when making fuel line connections. Permatex is available at almost all marine and hardware stores.

Detailed illustrated procedures to service the integral fuel pump on a Keihin carburetor may be found in Section 6-7, beginning on Page 6-25.

OIL INJECTION

Briefly, oil from a remote oil tank feeds fuel to an oil pump mounted to the outside of the flywheel cover. This pump is driven by a short shaft directly off the forward end of the crankshaft. Therefore, when the demand for more oil is required due to increased rpm, the pump is able to deliver added oil to meet the demand, because the crankshaft is rotating faster.

Oil from the pump is routed through individual hoses directly to each carburetor where it is mixed with the fuel.

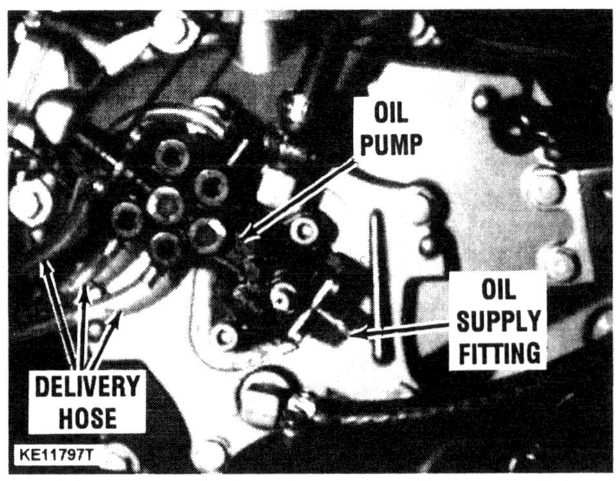

Close view of the oil pump and associated hoses installed on a Model 900 engine. A separate hose serves each carburetor.

Service of the oil pump and system is covered in detail in Section 6-9, beginning on Page 6-38.

Fuel/Oil Mixture

All engines covered in this manual are equipped with an oil injection system, replacing the age-old task of mixing oil with the gas before adding it to the tank.

However, additional oil should be added to the fuel during a "break-in" period following an engine overhaul or following a long storage period.

"Break-in" Lubrication

In order to obtain extra lubrication during the first 10 hours of "break-in", the manufacturer recommends a premix of 25:1 mixture be used in the fuel tank. Therefore, any existing unused fuel in the tank should be removed before adding the premixed solution. This ratio will **ENSURE** adequate lubrication of moving parts which have been drained of oil during the storage period.

Professional mechanics who own oil injected units add an extra ounce of oil per each gallon of fuel. The addition of this extra oil may prevent engine seizure in the event of an oil pump failure or clog in an oil delivery line.

Use only outboard marine oil in the mixture, never automotive oils. Four-stroke automotive engine oil is not formulated to burn

Close view of the oil injection system on a Model 650 Series engine. The system on a 550 Series engine is almost identical.

On a two-cylinder engine, the oil is injected through a nozzle into the intake manifold under the carburetor. Here, the carburetor has been removed to expose the oil injection nozzle.

completely -- only to lubricate. Therefore, automotive oil, if used in a two-stroke engine will leave an undesirable residue.

Removing Fuel From the System

For many years there has been the widespread belief that simply shutting off the fuel at the tank and then operating the engine until it stops is the proper procedure before storing the engine for any length of time. Right? **WRONG!**

It is **NOT** possible to remove all fuel in the carburetor by operating the engine until it stops. Some fuel is trapped in passages. The **ONLY** guaranteed method of removing **ALL** fuel is to take the time to remove the carburetor, and drain the fuel.

If the engine is operated with the fuel supply shut off until it stops the fuel and oil mixture inside the engine is removed, leaving bearings, pistons, rings, and other parts with little protective lubricant, during long periods of storage.

Proper Procedure

a- Shut off the fuel supply at the tank.

b- Disconnect the fuel line at the tank.

c- Start and operate the engine until it begins to run **ROUGH**.

d- Shut down the engine. Some fuel/oil mixture will remain inside the engine.

e- Remove and drain the carburetor.

By having disconnected the fuel supply, all **SMALL** passages are cleared of fuel even though some fuel is left in the carburetor.

f- Apply a light oil into the combustion chamber through the spark plug opening, as instructed in the Owner's manual.

NOTE

For short periods of storage, simply running the carburetor dry may help prevent severe gum and varnish from forming in the carburetor. This is especially true during hot weather.

ENGINE REVOLUTION LIMITER

In 1990, Kawasaki engineers developed an electronic revolution limiter This type limiter is, as the factory intended, very difficult to bypass. No!, we are not going to tell you exactly how it functions or how to bypass the limiter.

However, we can say -- the revolution limiter is an integral part of the CDI igniter and is now used on all Kawasaki PWC's. **ANY** attempt to bypass the limiter could actually result in serious and expensive consequences.

6-4 TROUBLESHOOTING

The following paragraphs provide an orderly sequence of tests to pinpoint problems in the fuel systems. It is very rare for the carburetor by itself to cause failure of the engine to start.

Fuel Problems

Many times fuel system troubles are caused by a plugged fuel filter, a defective fuel pump, or by a leak in the line from the fuel tank to the fuel pump. A defective choke may also cause problems. **WOULD YOU BELIEVE**, a majority of starting troubles which are traced to the fuel system are the result of an empty fuel tank, clogged fuel filter, or aged "sour" fuel.

WORDS OF CAUTION

Be sure to loosen the fuel filler cap on the fuel tank **BEFORE** opening any portion of the fuel system, to relieve any pressure built up inside the fuel system. On a hot summer's day, pressure in the fuel system can build up in a short time.

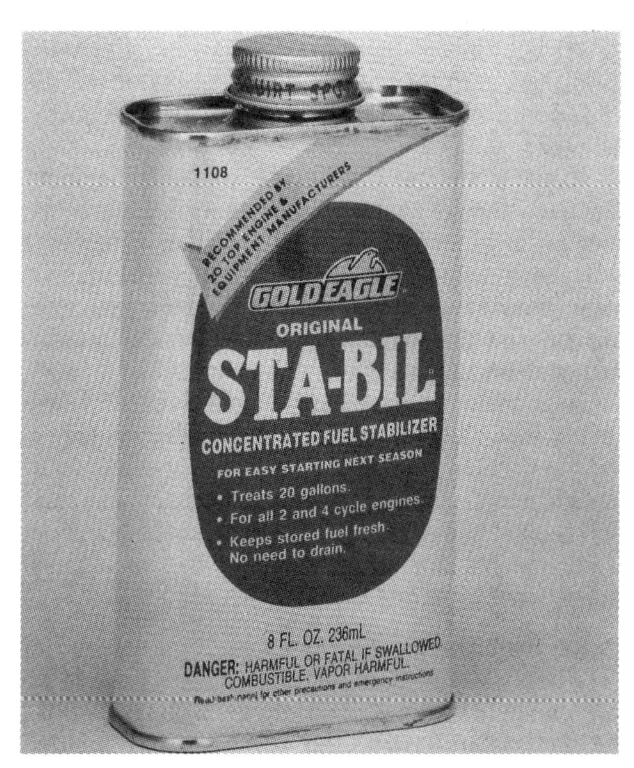

STA-BIL gasoline Stabilizer and Conditioner -- or equal substitute — may be used to prevent fuel from "souring" for up to twelve full months.

Fuel Filter and Sediment Bowl Service

Have a shop cloth handy to catch any spilled fuel. Un-thread the ring nut and pull down the sediment bowl with the fuel filter inside. A large O-ring seals the bowl to the filter cap and ring nut. The function of this O-ring is to prevent fuel and air leaks

If the O-ring becomes defective, the fuel pump will draw air into the sysetm and the engine will either not start, faulter, or die. Sometimes, this problem is not apparent with a full fuel tank, because the fuel is gravity flow to the carburetor. When the fuel level drops in the tank, the pump takes over, but its effectiveness is impaired by the air leak at the sediment bowl O-ring.

Clean out the sediment bowl and rinse the filter screen in solvent every 25 hours of engine operation.

"Sour" Fuel

Under average conditions (temperate climates), fuel will begin to breakdown in about four months. A gummy substance forms in the bottom of the fuel tank and in other areas. The filter screen between the tank and the carburetor and small passages in the carburetor will become clogged. The gasoline will begin to give off an odor similar to rotten eggs. Such a condition can cause the owner much frustration, time in cleaning components, and the expense of replacement or overhaul parts for the carburetor.

Even with the high price of fuel, removing gasoline that has been standing unused over a long period of time is still the easiest and least expensive preventative maintenance possible. In most cases, this old gas can be used without harmful effects in an automobile using regular gasoline.

A gasoline preservative additive, such as Stabil Fuel Conditioner and Stabilizer for 2 and 4-Cycle Engines, will keep the fuel "fresh" for up to twelve months. If this particular product is not available in your area, other similar additives are produced under various trade names.

Choke Problems

When the engine is hot, the fuel system can cause starting problems. After a hot engine is shut down, the temperature inside the fuel bowl may rise to 200°F and cause the fuel to actually boil. All carburetors are vented to allow this pressure to escape to the atmosphere. However, some of the fuel may percolate over the high-speed nozzle.

If the choke should stick in the open position, the engine will be hard to start. If the choke should stick in the closed position, the engine will flood making it very difficult to start.

In order for this raw fuel to vaporize enough to burn, considerable air must be added to lean out the mixture. Therefore, the only remedy is to remove the spark plugs; ground the leads; crank the engine through about ten revolutions; clean the plugs; install the plugs again; and start the engine.

Rough Engine Idle

If an engine does not idle smoothly, the most reasonable approach to the problem is to perform a tune-up to eliminate such areas as: fouled or faulty spark plugs.

Other problems that can prevent an engine from running smoothly include: An air

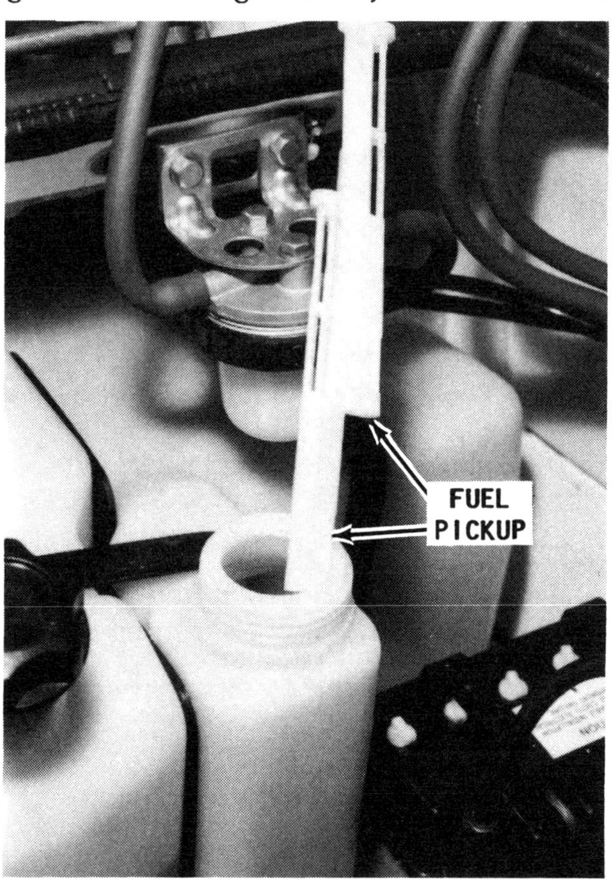

Withdrawing the fuel pickup tubes from the fuel tank on a Model 650 Series. The manufacturer suggests the screens at the bottom of the tubes be cleaned every 25 hours. This task is possible on a Model 650, but on other models, the tank must be "pulled" and the engine removed to provide clearance for the tank.

leak in the intake manifold; uneven compression between the cylinders; and sticky or broken reeds.

Of course any problem in the carburetor affecting the air/fuel mixture will also prevent the engine from operating smoothly at idle speed. These problems usually include: a weak float arm return spring; leaking needle valve and seat; defective choke; and improper adjustments for idle mixture or idle speed.

Excessive Fuel Consumption

Excessive fuel consumption can be the result of any one of two conditions, or both may apply.

1- Inefficient engine operation.

2- Poor boating habits of the operator.

If the fuel consumption suddenly increases over what could be considered normal, then the cause can probably be attributed to the engine or watercraft and not the operator.

Check the fuel system for possible leaks. A leak between the tank and the pump many times will not appear when the engine is operating, because the suction created by the pump drawing fuel will not allow the fuel to leak. Once the engine is turned off and the suction no longer exists, fuel may begin to leak.

If a minor tune-up has been performed and the condition still exists, then the problem most likely is in the carburetor and an overhaul is in order.

Engine Surge

If the engine operates as if the load on the craft is being constantly increased and decreased, even though an attempt is being made to hold a constant engine speed, the problem can most likely be attributed to the fuel pump, or a restriction in the fuel line between the tank and the carburetor.

6-5 CARBURETOR MODELS

Two different designs of Keihin carburetors are used on the engines covered in this manual. One model has an integral fuel pump and the other does not have the pump. The following short table lists the engine model, the number of carburetors installed, and the Keihin identification number.

MODEL	NUMBER	KEIHIN
550	Single	CDK-38
650	Single	CDK-34
Std 750	Single	CDK-40
H.P. 750	Double	CDK-40
900	Triple	CDK-38
1100	Triple	CDK-38

H.P. = High Performance

Remember, a remote fuel pump is used with the CDK-34 carburetor and when a triple carburetor installation is used, the center carburetor does not have the integral fuel pump.

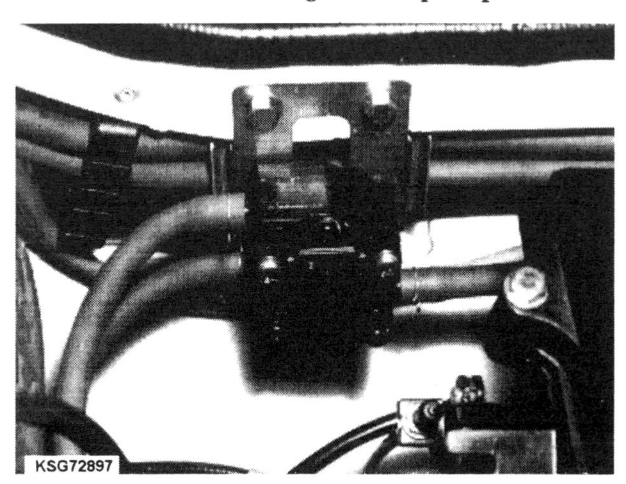

Remote fuel pump mounted to the outside bulkhead in the engine compartment to supply fuel to a Keihin CDK-34 carburetor -- in this case on a Model 650 watercraft. Location of the pump provides easy access to inspect the hose line connections or to remove and install the pump.

6-6 SERVICE KEIHIN CDK-34 CARBURETOR

This section provides complete detailed illustrated procedures to remove, disassemble, clean, inspect, assemble -- including bench adjustments -- install and operating adjustments for the Keihin CDK-34 carburetor.

SAFETY WORDS

Any time the fuel system is "opened" to disconnect a fuel line or to remove a component in the system, such as a carburetor, the battery terminals should be disconnected as a precaution against a possible spark. Be prepared to clamp off the disconnected fuel line to prevent fuel from spilling into the bilge or siphoning from a small reservoir in the system.

The Keihin carburetor covered in this section uses a remote fuel pump and is installed on the Model 650 Series engines.

The Keihin carburetor covered in this section is identified by the manufacturer as the CDK-34 with a 28mm venturi. The unit is installed on the twin cylinder Model 650 Series engine. This carburetor does not have an integral fuel pump.

Complete detailed illustrated procedures to service the remote fuel pump installed to serve the CDK-34 carburetor are presented in Section 6-8 of this chapter, beginning on Page 6-33.

REMOVAL AND DISASSEMBLING

Good shop practice dictates a carburetor repair kit be purchased and new parts be installed any time the carburetor is disassembled.

The remote fuel pump used with all Keihin carburetors on the 650 Series prior to 1991 and on all JetMates.

Make an attempt to keep the work area organized and to cover parts after they have been cleaned. This practice will prevent foreign matter from entering passageways or adhering to critical parts.

Begin by removing the flame arrestor cover, and then the flame arrestor "box".

1- Loosen the gas tank filler cap slightly to relieve any pressure built up in the fuel system. Disconnect the fuel supply and return hoses from the carburetor. Take time to plug the disconnected ends of both hoses with a golf tee, pencil end, or similar item to prevent contaminants from entering the fuel system.

2- Loosen the two locknuts on the choke cable -- one on each side of the cable support bracket.

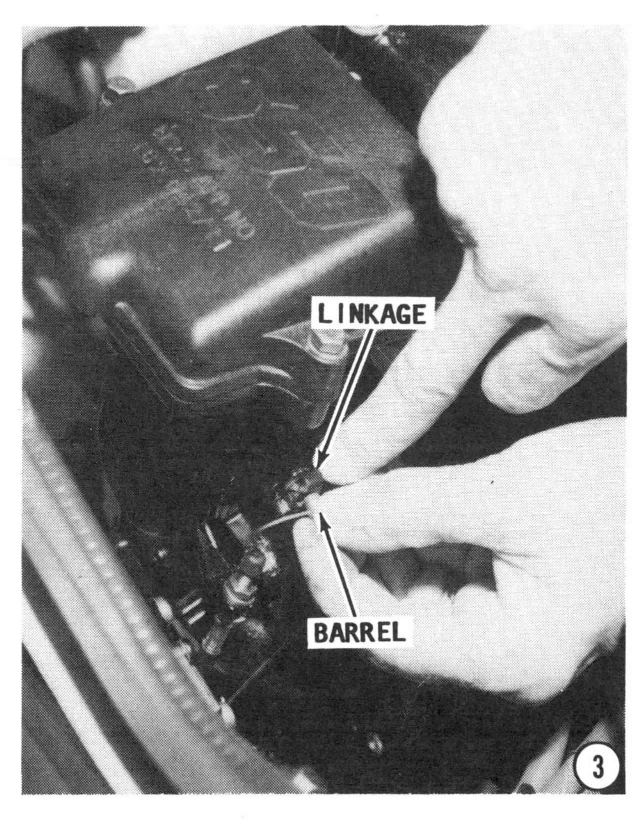

3- Depress the choke linkage and unhook the barrel from the hole in the linkage.

Repeat these last two steps to disconnect the throttle cable.

4- Remove the bolts securing the flame arrestor cover. Lift off the cover and the arrestor screen.

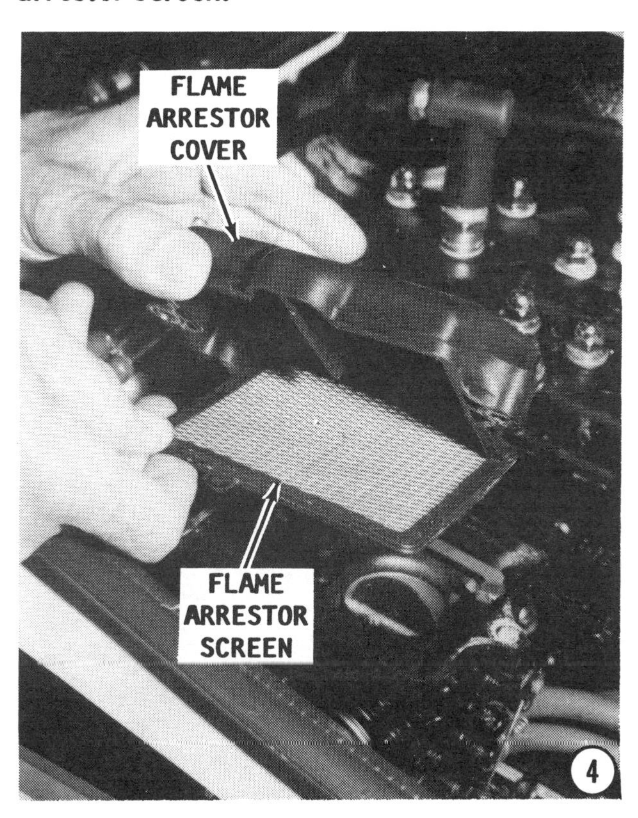

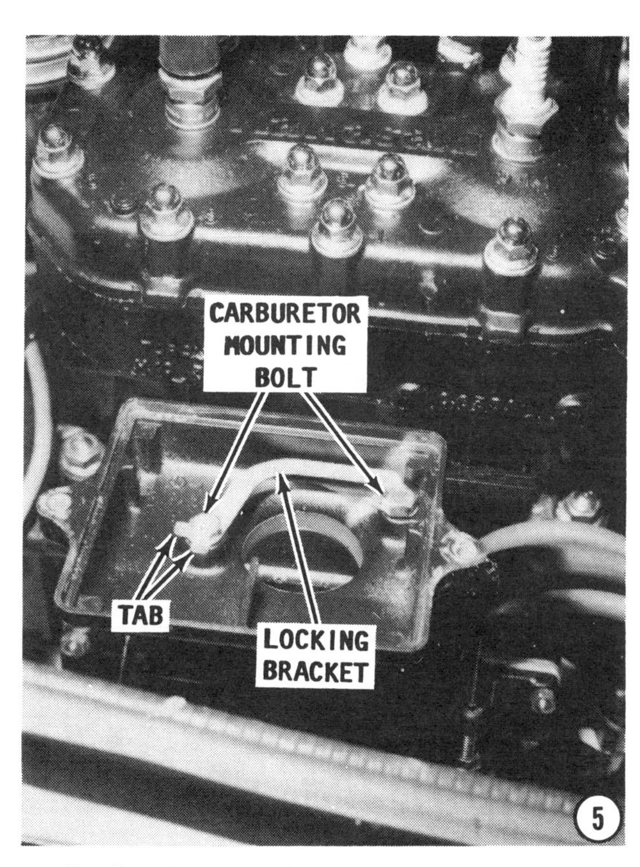

5- Bend down the two tabs against the sides of the two carburetor mounting bolts. Loosen the two bolts.

6- Lift off the lower air intake cover with the locking bracket and bolts still in place, and then remove the carburetor.

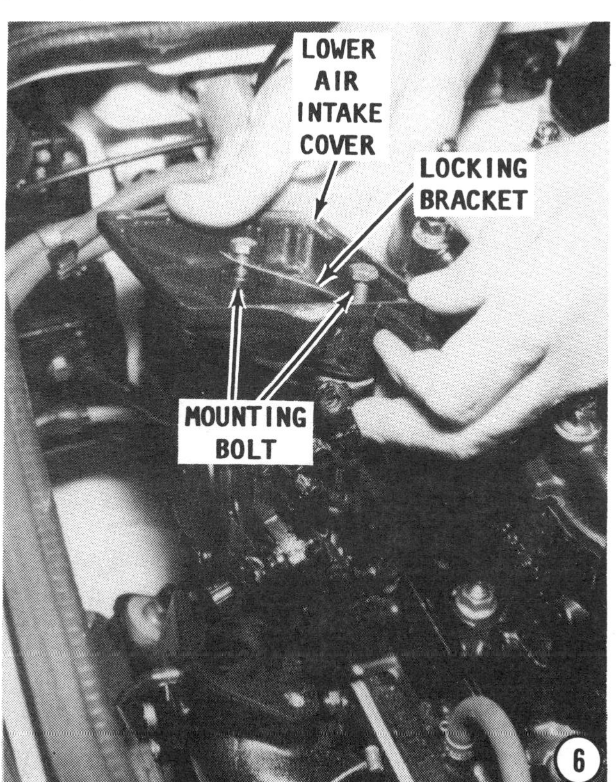

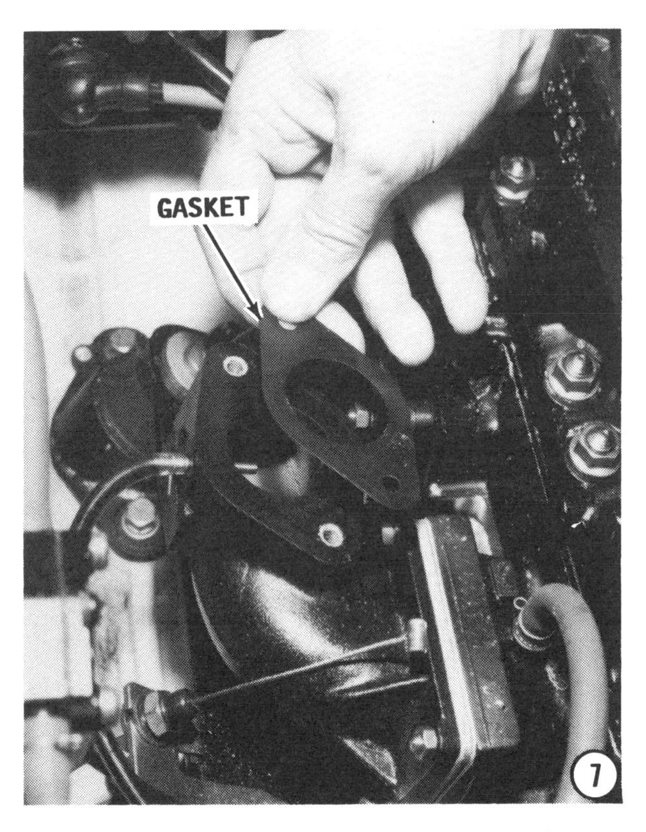

7- Remove the carburetor base gasket.

Carburetor Disassembling

8- Remove the four Phillips head screws securing the diaphragm cover to the carburetor and lift off the diaphragm cover.

Remove the diaphragm from the fuel chamber.

9- Remove the screw securing the diaphragm housing to the carburetor.

10- Lift off the diaphragm housing.

11- Remove and discard the sealing ring.

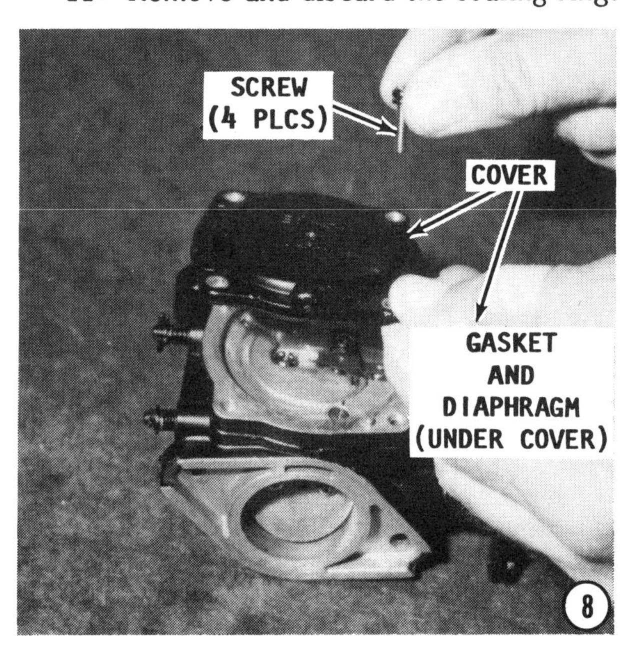

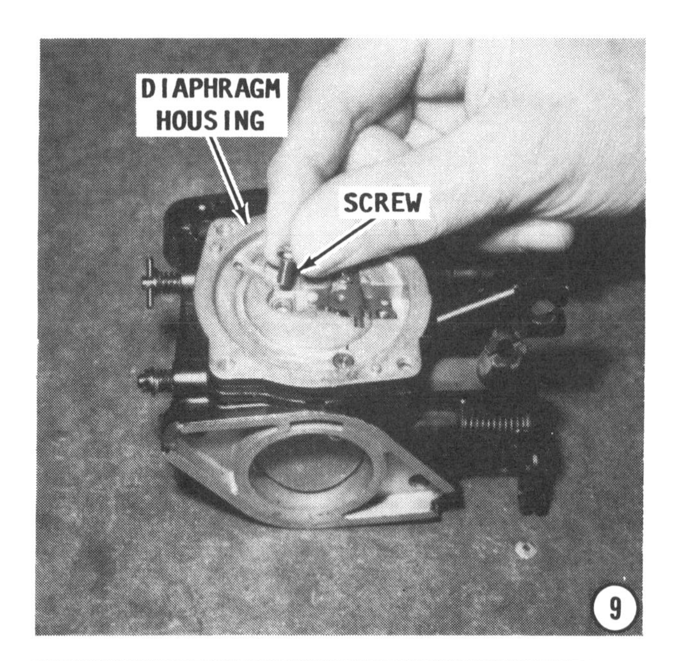

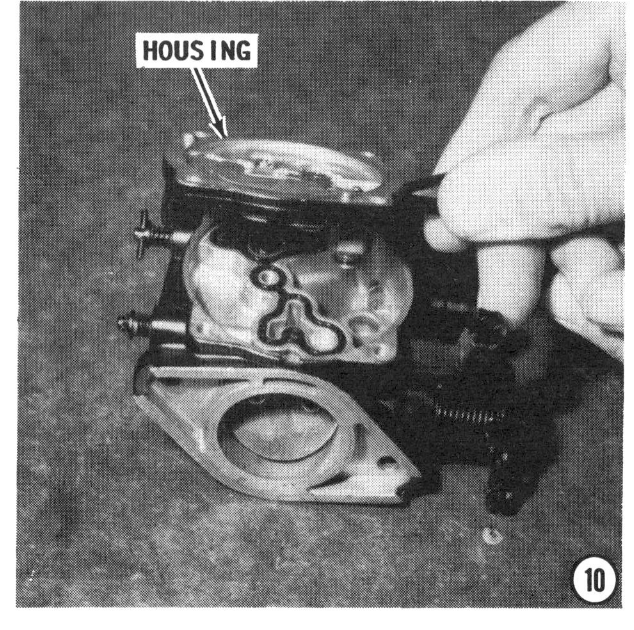

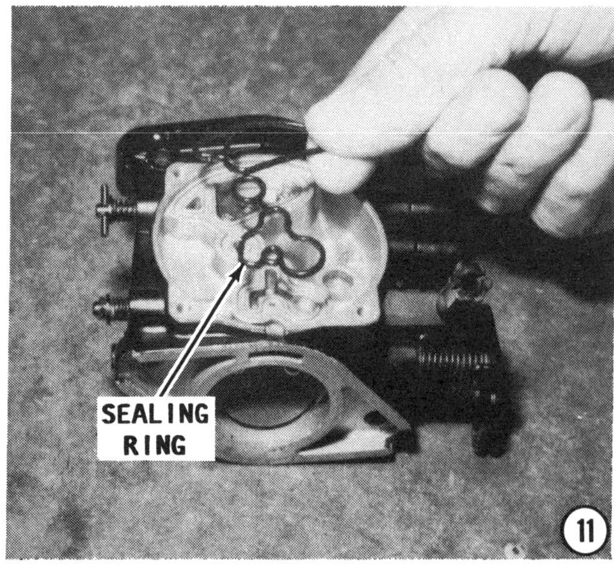

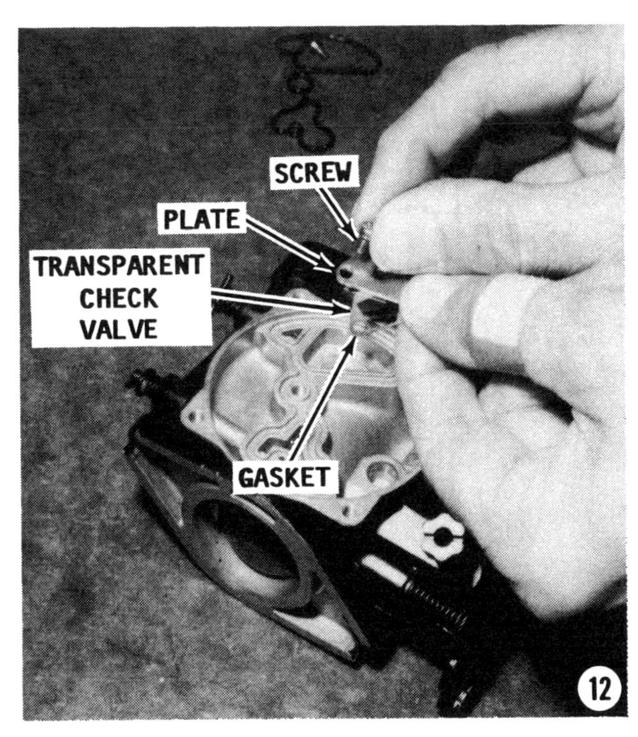

12- Obtain a stubby Phillips head screwdriver and a hammer. Now, tap lightly on the end of the screwdriver while attempting to remove the two screws securing the high speed check valve plate. Do not attempt to remove these screws "cold" -- without tapping the screwdriver -- because the heads will almost certainly be stripped! Remove the plate.

CAREFULLY lift off the transparent check valve. **TAKE CARE** not to lose this item. This valve controls the flow of fuel in one direction and prevents the engine from "faultering" when the throttle is suddenly

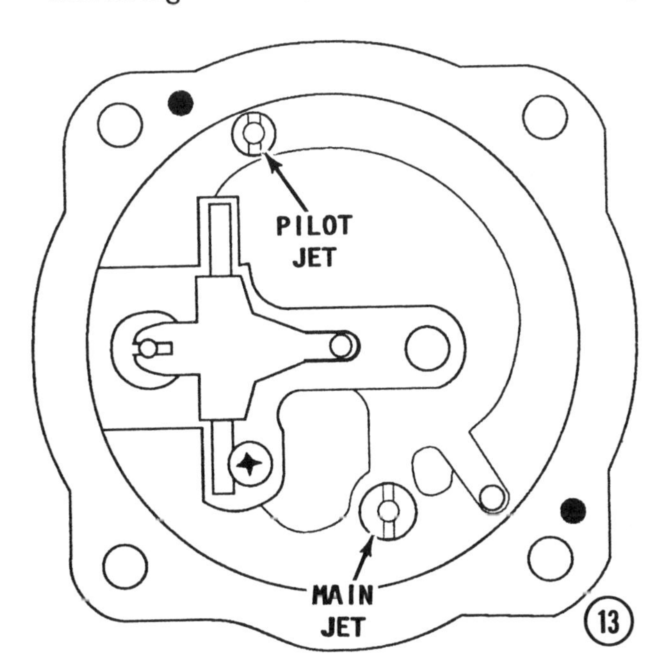

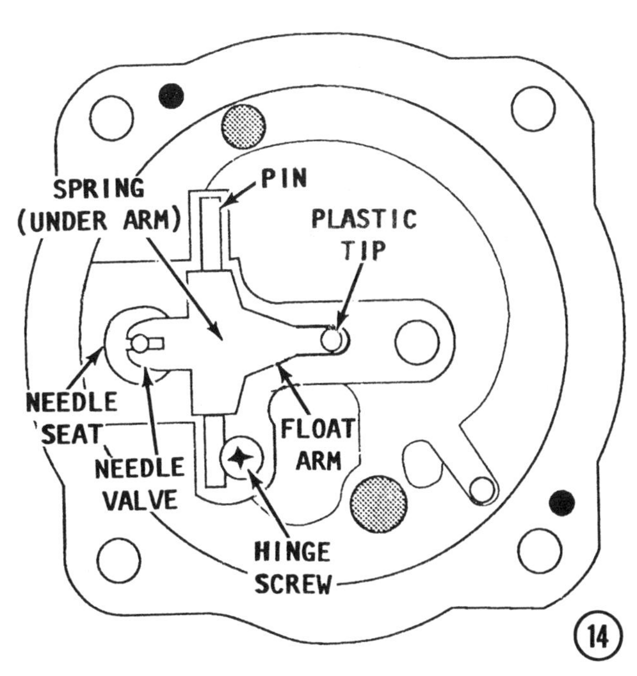

changed from wide open, to idle, and then back to wide open again. Remove the gasket under the check valve.

13- Use the correct size screwdriver to remove the main jet -- the larger of the two jets -- from the fuel chamber. Again, use the correct size screwdriver and remove the pilot jet from the chamber.

14- Use the stubby Phillips head screwdriver and a hammer from Step 12 and lightly tap the end of the screwdriver while attempting to remove the hinge screw. Do not attempt to remove this screw "cold" -- without tapping the screwdriver -- because the head will almost certainly be stripped!

Lift out the float arm and float pin. The needle valve will remain hanging on the pin by the clip. Lift the valve from the seat.

Remove the spring from its recess in the fuel chamber. Lift out the needle valve from the needle seat. The needle seat is pressed into the carburetor and **CANNOT** be easily removed.

15- Turn the housing over and remove and discard the O-ring around the seat.

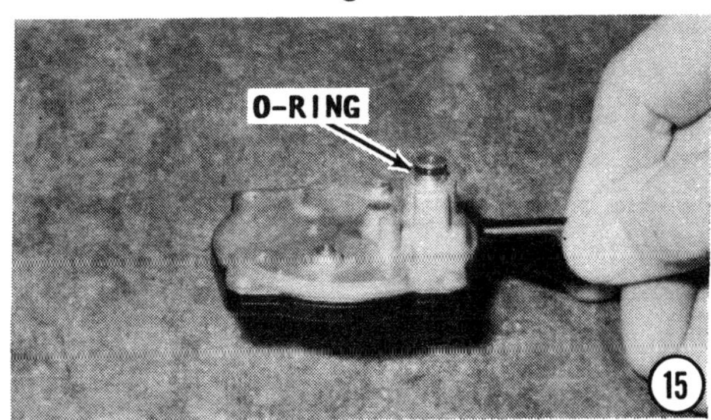

SPECIAL WORDS
ADJUSTING SCREWS

The low and high speed mixture screws must **NOT** be removed unless damage to the tapered end is suspected and the screws are to be replaced. Under normal service these two screws should remain undisturbed.

Each Keihin carburetor is individually adjusted at the factory using a flow meter. After the adjustment is completed, a limiter cap is installed over the head of the screw to prevent accidental tampering.

Under normal service these two screws should remain undisturbed.

CRITICAL WORDS

Perform Steps 16 and 17, **ONLY** if the high speed or low speed mixture cannot be adjusted properly. **ALSO**, perform the steps only if replacement screws have been obtained and are on hand. Failure to adjust the low speed or high speed mixture would suggest the tapered end of one -- but not necessarily both -- of the screws has become damaged. In almost all cases, "damage" is caused by the screw being threaded in too tightly and the tapered end of the needle becoming "grooved", because it passed through the orifice. Once the needle is "grooved", proper mixture cannot be obtained.

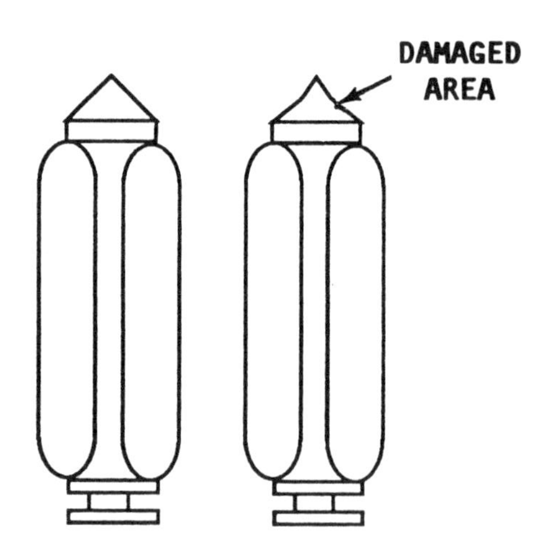

Line drawing of a new valve needle (left), with a smooth tapered surface compared with a damaged needle (right) -- unfit for further service.

16- Pull off the limiter cap on the low speed mixture adjusting screw -- the screw next to the embossed **"L"**. Back out the low speed mixture screw.

17- Pull off the limiter cap on the high speed mixture adjusting screw -- the screw next to the embossed **"H"**. Back out the high speed mixture adjusting screw.

CLEANING AND INSPECTING

NEVER dip rubber parts, plastic parts, or diaphragms in carburetor cleaner. These parts should be cleaned **ONLY** in solvent, and then blown dry with compressed air.

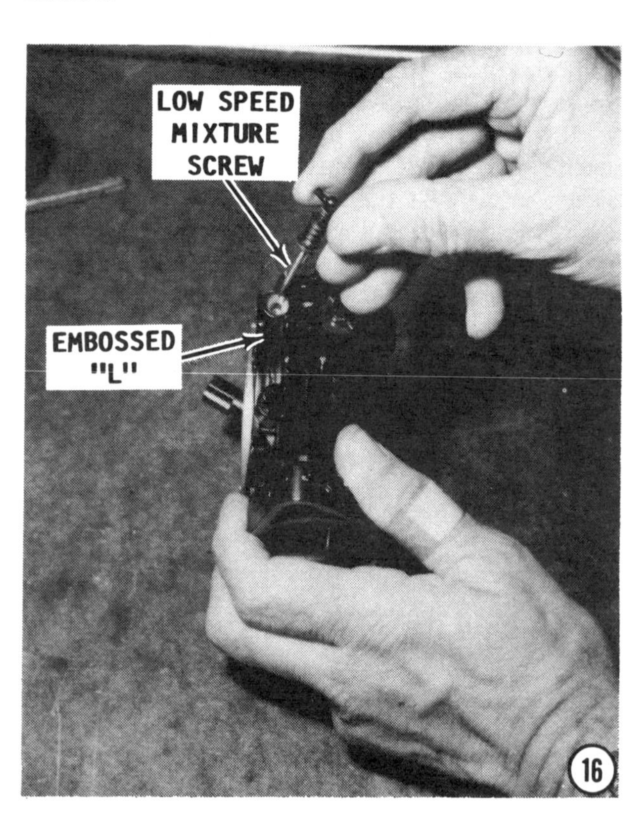

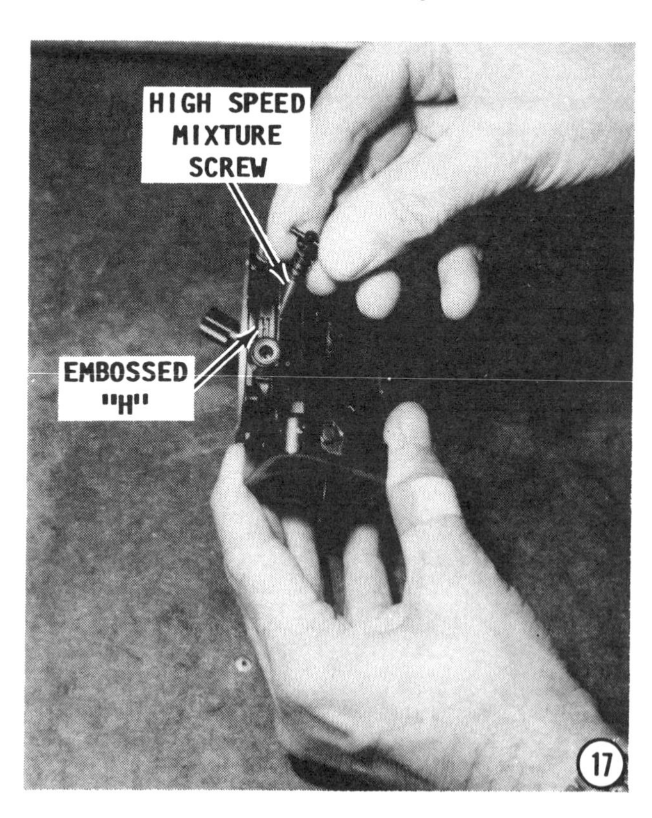

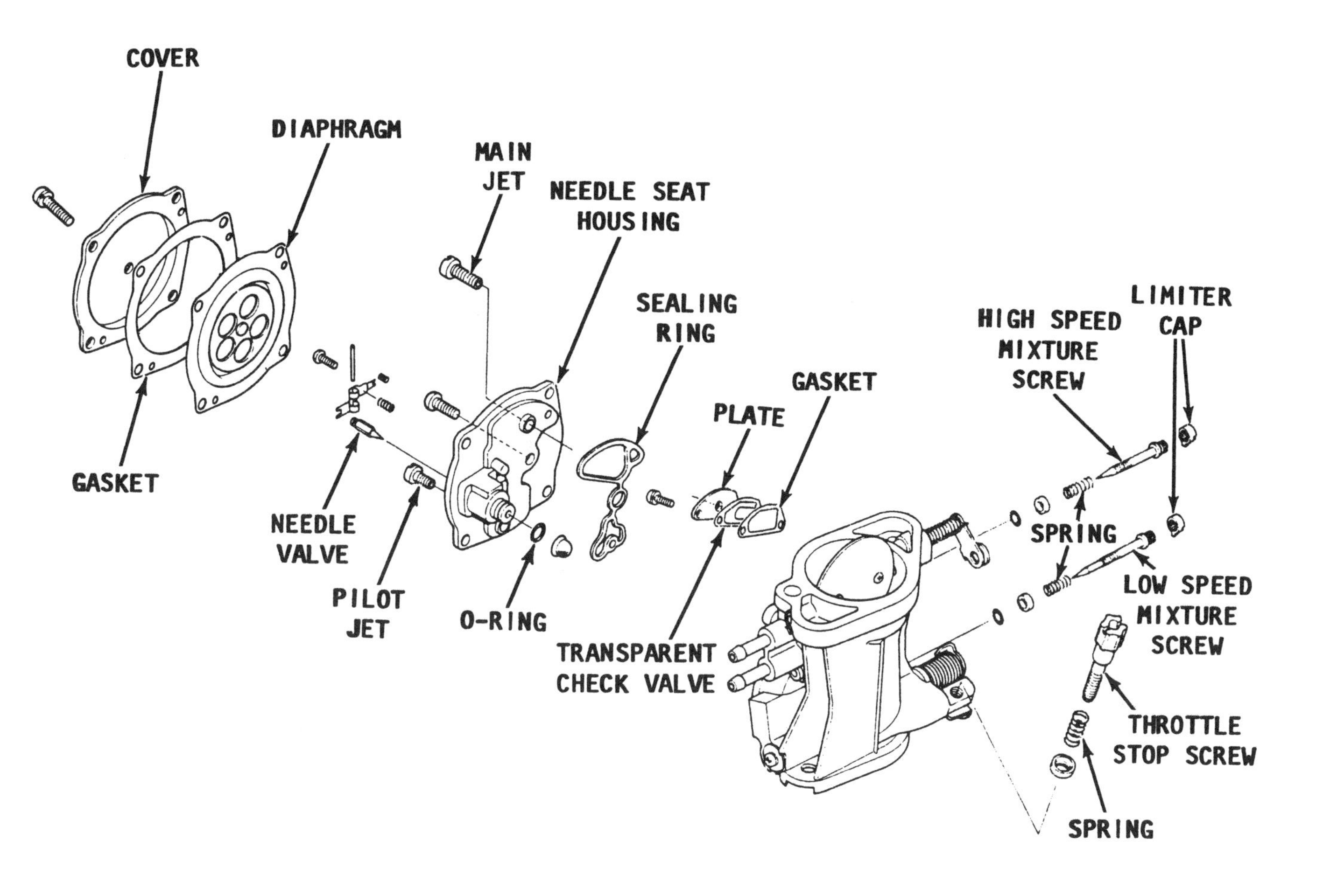

Exploded drawing of the Keihin CDK-34 diaphragm carburetor used on the Model 650 Series engine covered in this manual. Major parts have been identified.

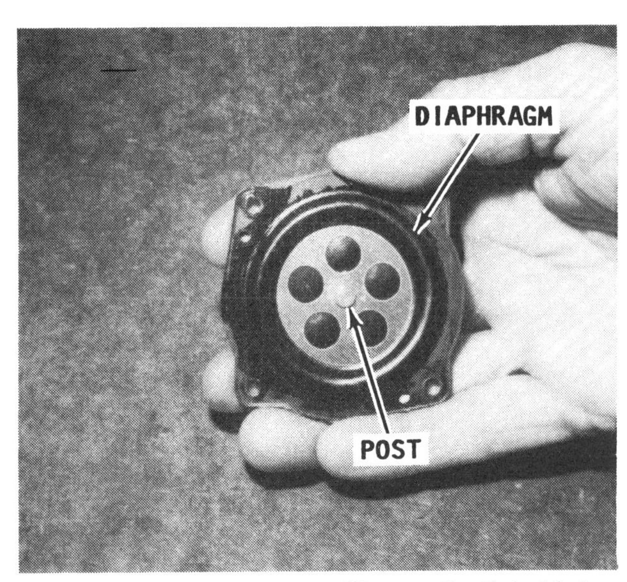

*A damaged diaphragm -- with even the tiniest hole --**MUST** be replaced to obtain full engine performance.*

Place all metal parts in a screen-type tray and dip them in carburetor cleaner until they appear completely clean, then blow them dry with compressed air.

Blow out all passages in the castings with compressed air. Check all parts and passages to be sure they are not clogged or contain any deposits. **NEVER** use a piece of wire or any type of pointed instrument to clean drilled passages or calibrated holes in a carburetor.

Move the throttle shaft back and forth to check for wear. If the shaft appears to be too loose, replace the complete throttle body because individual replacement parts are **NOT** available.

CRITICAL AREA

Inspect the condition of both the choke and throttle shafts. Most shaft designs incorporate a cutout portion to enable the butterfly valve to lay flat against the shaft. Two small holes are drilled in this cutout portion to accommodate the small screws securing the valve to the shaft.

The weakest point of the shaft is at the screw holes. Small cracks form at the holes due to metal fatigue. These cracks elongate the threaded holes and the small screws shake loose. Once loose, they are sucked into the carburetor, then the intake manifold, through the reed valves, and into the crankcase -- eventually finding their way into the combustion chamber. One of these small screws can then cause very extensive and expensive damage to reed valves, pistons, rings, and cylinder walls. Therefore,

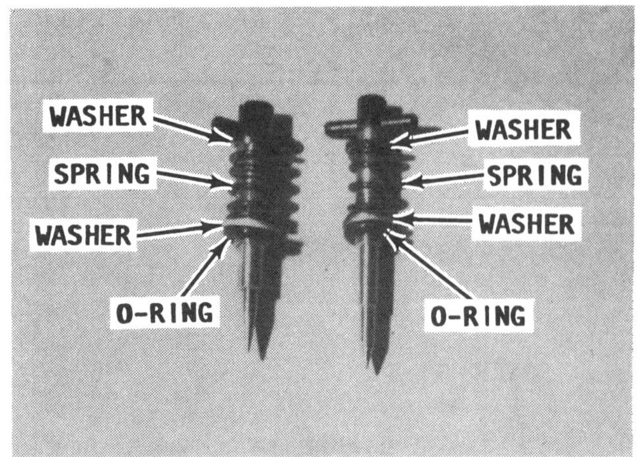

Check to be sure the low speed adjustment screw and the high speed adjustment screw are properly assembled, prior to installation.

check the cutout in the shaft **CLOSELY** for any evidence of a small crack.

Inspect the main body, airhorn, and venturi cluster gasket surfaces for cracks and burrs which might cause a leak. Check to be sure the float arm return spring has not been stretched. Check the float arm needle contacting surface and replace the diaphragm if this surface has a groove worn in it.

As previously mentioned, most of the parts which should be replaced during a carburetor overhaul are included in overhaul kits available from your local marine dealer. One of these kits will contain a fuel inlet needle and seat. The needle should be replaced each time the carburetor is disassembled as a precaution against leakage.

Make a thorough inspection of the fuel chamber and fuel diaphragms for the tiniest pin hole. If one is discovered, the hole will only get bigger. Therefore, the diaphragm must be replaced in order to obtain full performance from the engine.

ASSEMBLING

Special Words on Mixture Screws

Observe the low and high speed mixture screws. They appear to be very similar, but they are **NOT** interchangeable. To identify the correct screw, place them side by side as shown in the accompanying illustration. The low speed screw is shorter than the high speed screw. Both screws are threaded into the carburetor at the base of the fuel pump. The low speed screw is installed next to the embossed letter **"L"**, and the high speed screw is installed next to the embossed letter **"H"**.

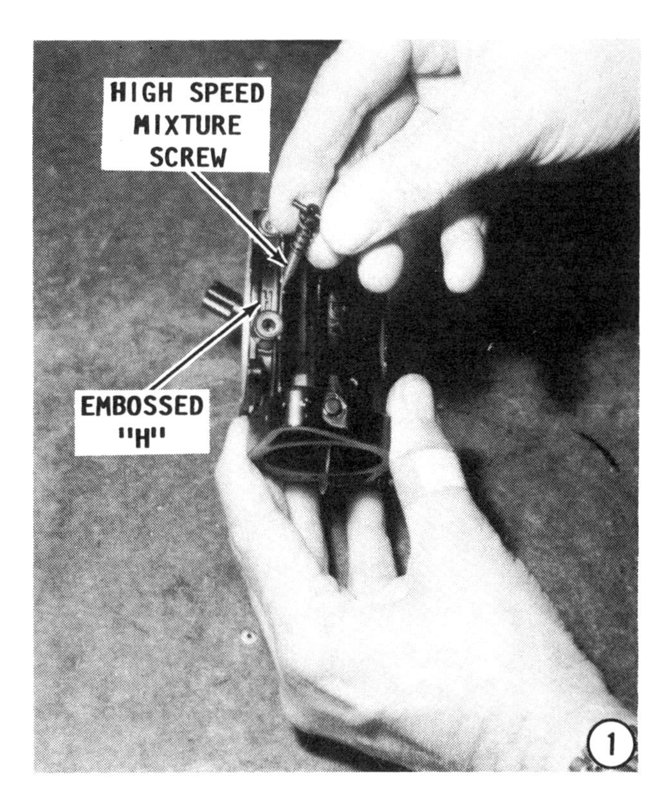

Both screws are adjusted later at the completion of the assembly procedures. Therefore, at this time, thread each screw in just a few turns. Check to make sure both screws are equipped with a washer, a spring, another washer, and an O-ring.

1- Install the high speed screw into the opening embossed with the letter "H".

2- Install the low speed screw into the opening embossed with the letter "L".

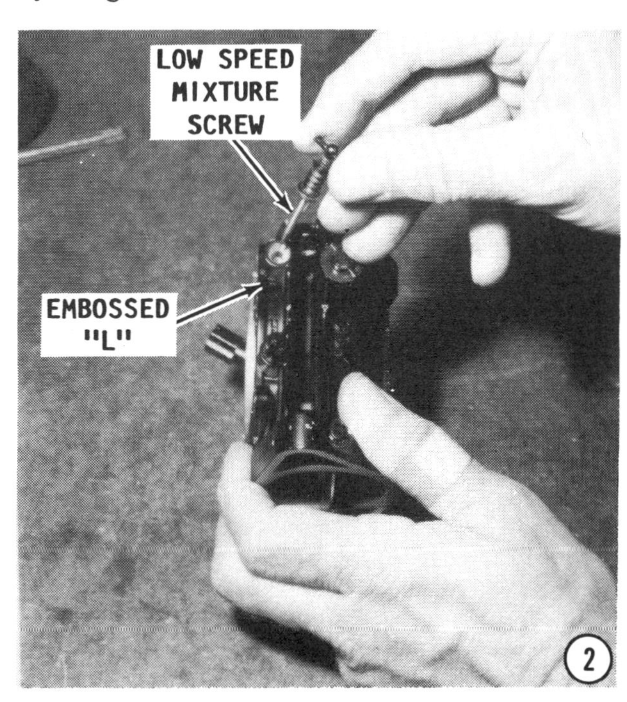

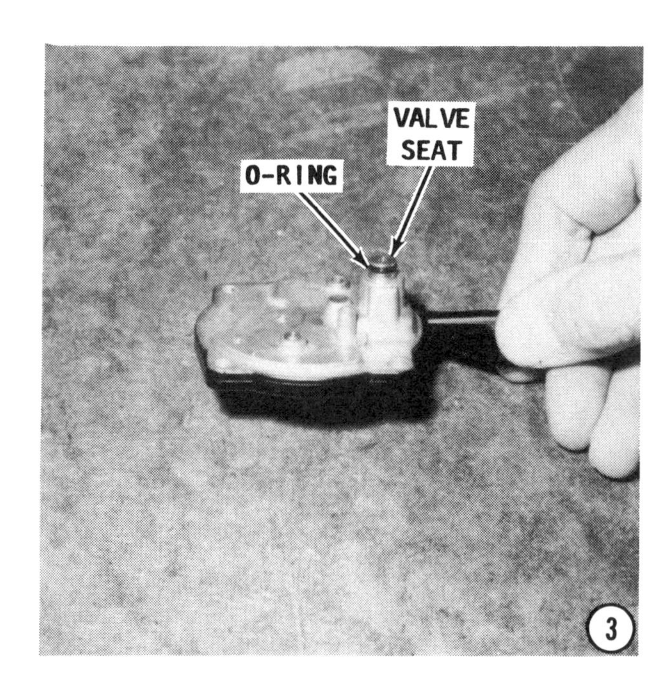

3- Install a new O-ring around the valve seat on the underside of the float chamber housing.

4- Install the small spring into the hole in the center of the float chamber. Slide the hinge pin into the float arm and hook the needle valve into the slot in the arm.

Lower the needle valve into the needle seat and index the dimple in the float arm over the spring. Position the hinge pin into place in the elongated slot in the float chamber and secure it with the small Phillips head screw. Tighten this screw very securely.

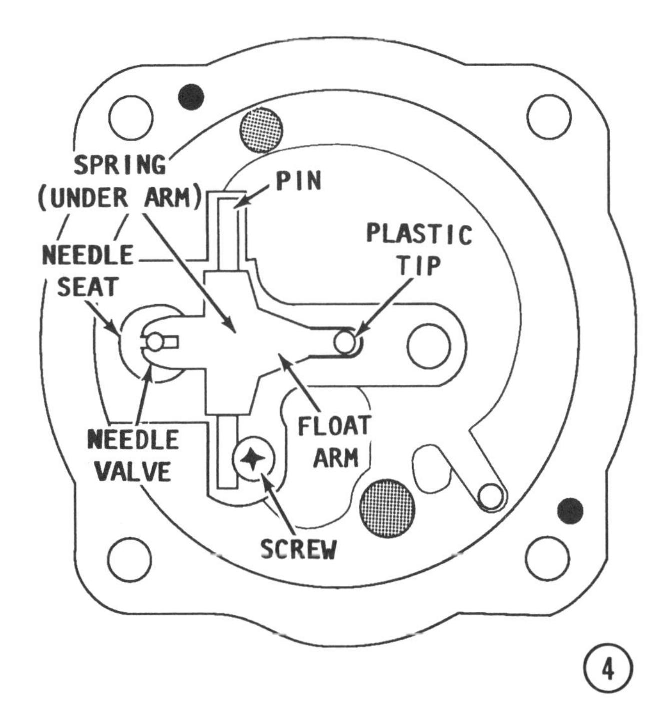

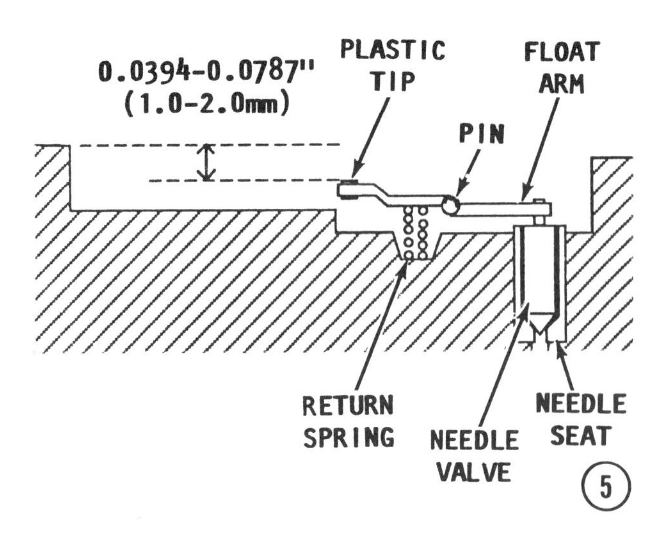

Float Arm Adjustment

5- Three conditions must be satisfied in order to achieve correct float arm adjustment.

a) The top of the float arm must be parallel to the fuel chamber surface.

b) The distance between the top surface of the float arm and the diaphragm contact surface of the carburetor must be between 0.0394-0.0787" (1.0-2.0mm).

c) The float arm must make contact with the needle valve, without applying any pressure to the valve.

If any of these conditions are not satisfied, then replace the float arm and/or return spring. **DO NOT** attempt to bend the float arm or to change the characteristics of the old return spring.

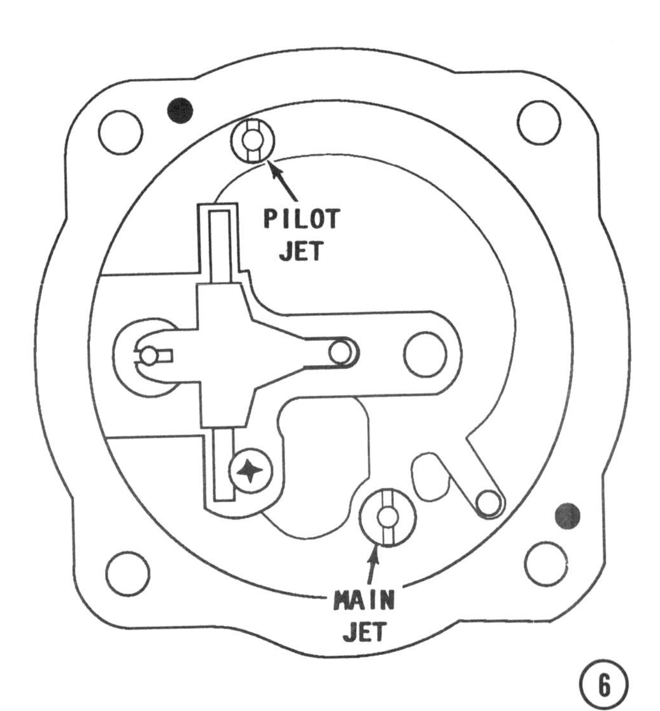

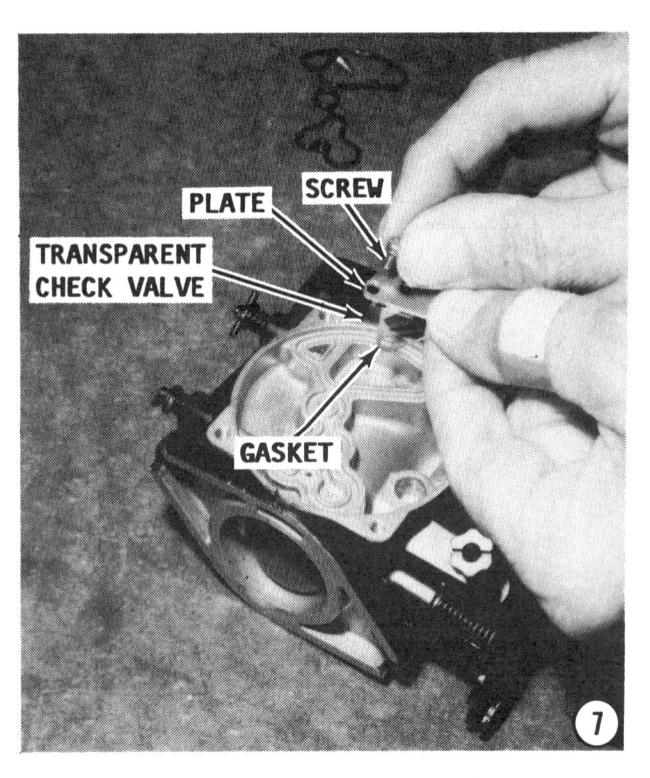

Depress the rounded end of the float arm. The needle valve should lift smoothly and return inside the seat when the arm is released.

6- Install the pilot jet -- the smaller of the two -- into the jet opening in the fuel chamber. Tighten the jet securely using the proper size screwdriver.

Install the main jet into the jet opening in the fuel chamber. Again, using the proper size screwdriver, tighten the jet securely.

7- Position the high speed check valve gasket into the recess in the fuel chamber, followed by the transparent check valve. Install the plate over these two items and secure it in place with two Phillips head

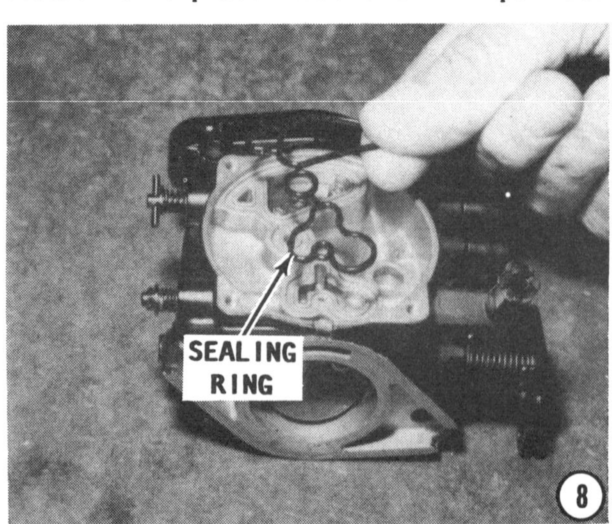

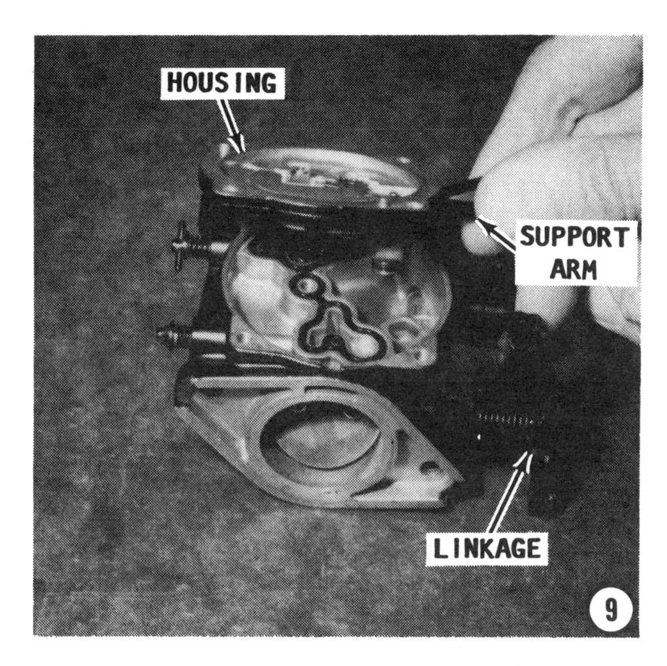

screws. Tighten the screws to a torque value of 1.5 ft lb (2Nm).

8- Position a **NEW** sealing ring over the fuel chamber.

9- Lower the housing over the fuel chamber, with the choke cable support arm on the same side as the linkage.

10- Install the Phillips head screw securing the housing to the chamber. Tighten the screw securely.

11- Place the diaphragm over the fuel chamber with the float post on the underside facing **DOWN** to rest against the float arm. Position the diaphragm cover over the diaphragm. Secure the cover to the carburetor with the four Phillips head screws.

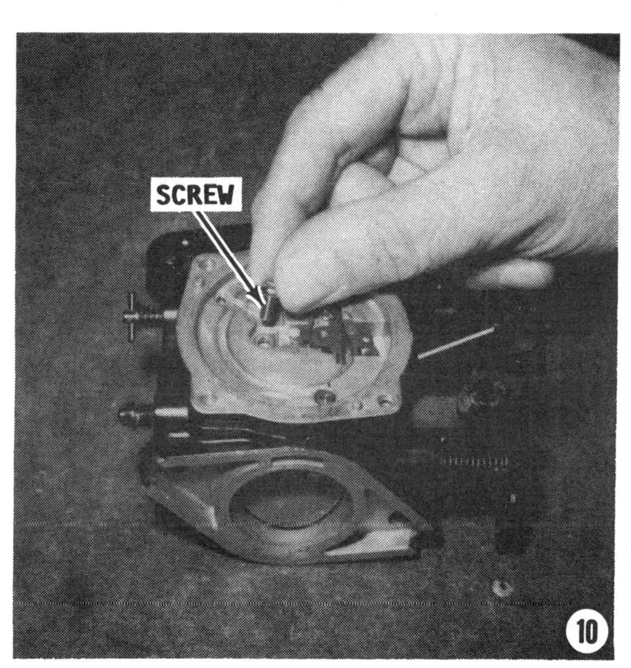

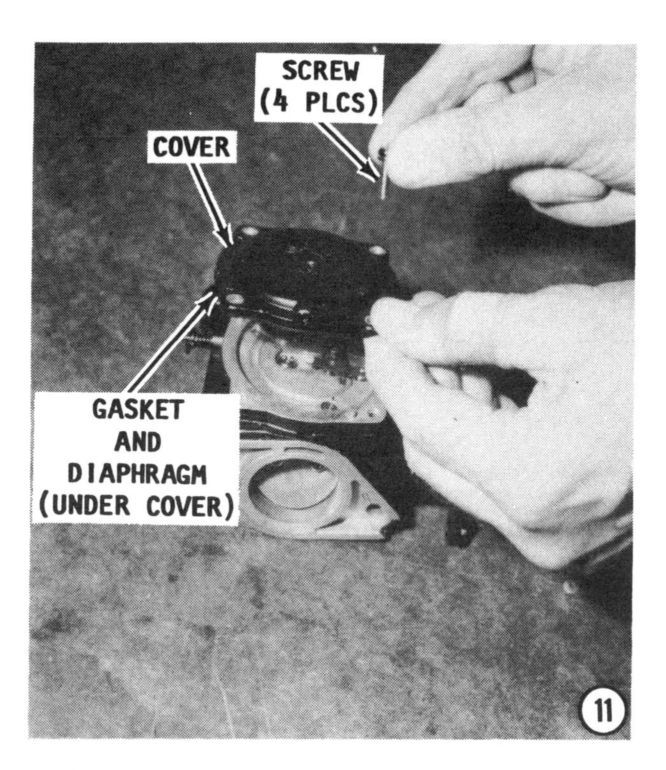

Tighten the screws alternately and evenly to a torque value of 3 ft lb (4Nm).

Carburetor Installation

12- Place a new carburetor mounting gasket over the intake manifold.

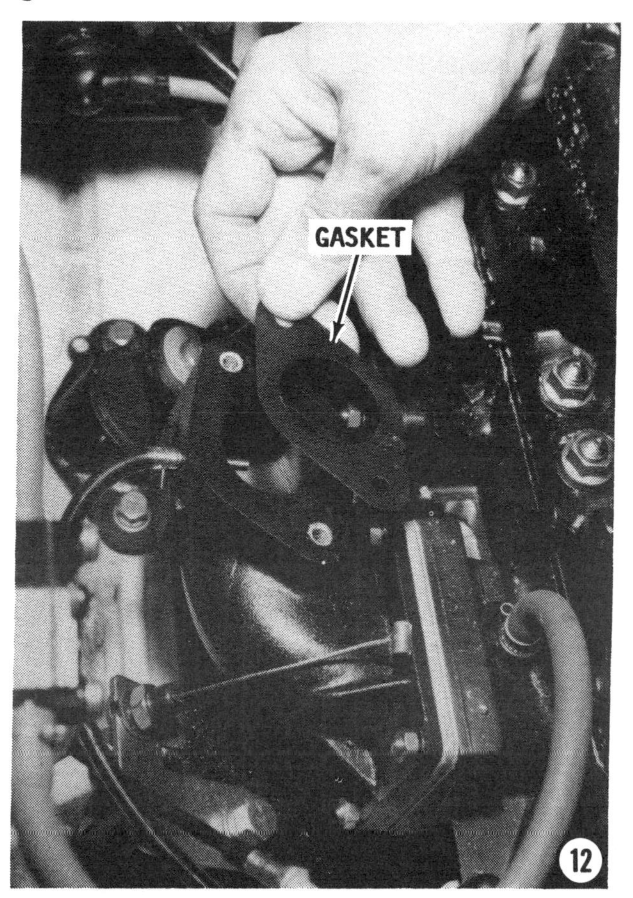

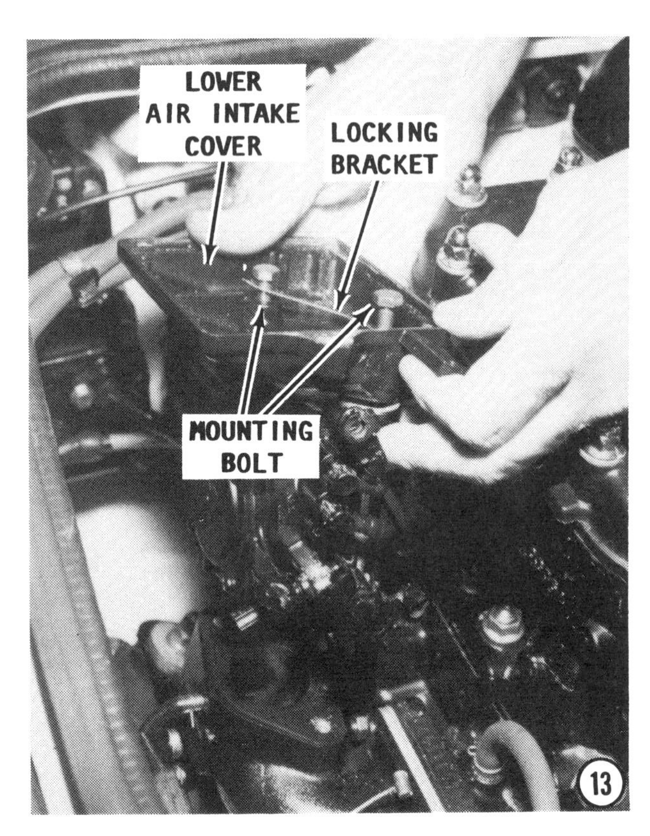

13- Place the carburetor over the intake manifold and gasket, aligning all bolt holes. Position the assembled lower air intake cover over the carburetor and guide the two bolts through the carburetor and gasket as the cover is lowered into place.

14- Tighten the two carburetor mounting bolts. After the bolts have been installed and tightened securely, bend the tab on the locking bracket up against the side of the bolts to prevent the bolts from backing out. If the tab does not align with the side of a bolt, loosen the bolt very slightly until a tab does align.

15- Position the flame arrestor screen over the lower cover with the cutout/radius edge facing **UPWARD.** Install the top cover and secure the two halves together with the two bolts.

Choke and Throttle Cable Installation

16- Slide the choke cable into the cable support arm on the carburetor, with a lock-

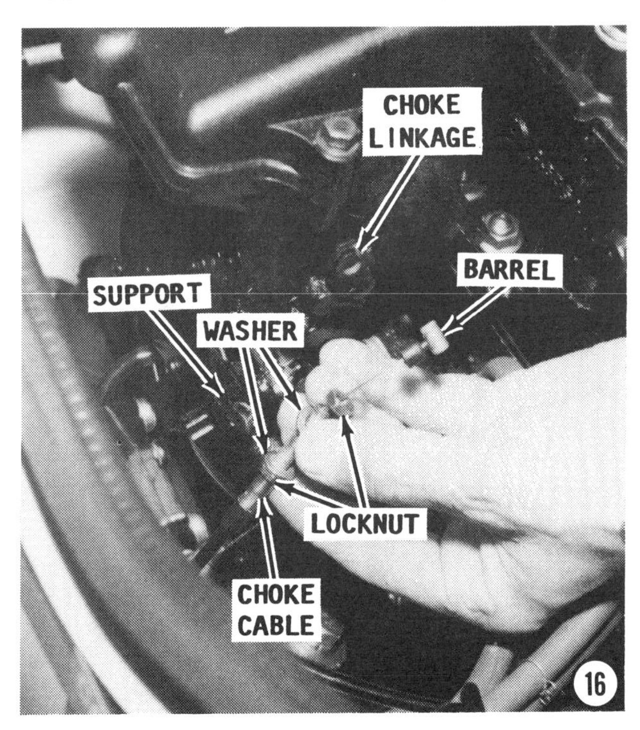

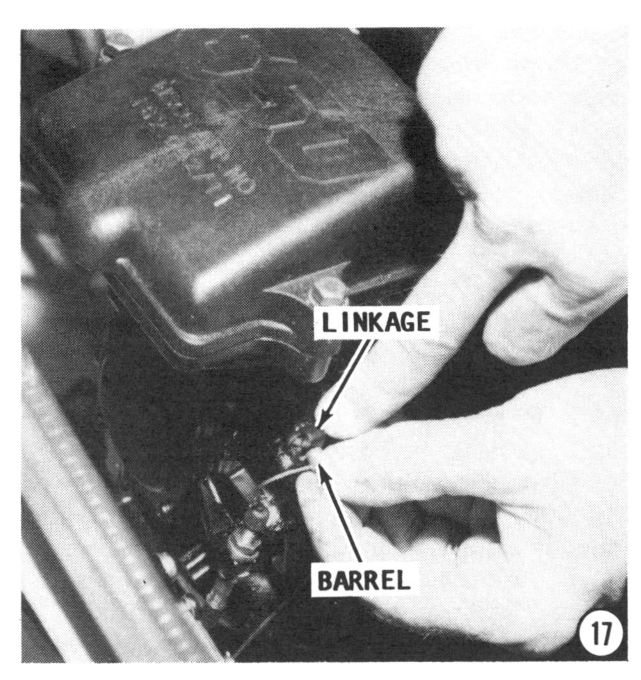

nut and washer on each side of the support. Temporarily, hand tighten the two locknuts against each other.

17- Depress the choke linkage on the carburetor. Push the barrel at the end of the choke cable into the slotted hole in the choke linkage.

18- Adjust the position of the two locknuts to give an acceptable amount of slack in the cable, and then tighten the locknuts against each other to hold the cable in place on the support bracket.

Repeat these last three steps for the throttle cable.

If either cable has stretched, or has become kinked or frayed, or tends to stick, the entire cable should be replaced.

19- Connect the fuel supply hose to the lower of the two fittings on the carburetor.

Carburetor Priming

The carburetor must be primed before starting the engine to save the battery from excessive cranking.

20- Place a shop towel under the fuel return hose fitting. Blow through the hose until fuel starts to drip from the fuel return fitting on the carburetor. The carburetor is now full of fuel and the engine is ready for cranking after the hose has been reconnected.

Mixture Screws with Limiter Caps

Since about 1989, the manufacturer has installed small black plastic limiter caps which snap over both the low speed and high speed adjustment screws. The purpose of these caps is to prevent any unauthorized tampering with the factory setting.

Each carburetor is individually adjusted at the factory using a flow meter. After the adjustment is completed, a limiter cap is installed over the head of the screw to prevent accidental tampering.

Each cap has a tab and the tab fits into a slot in the carburetor body at the base of the mixture screw. Therefore, the optimum position of each screw is with the tab down and indexed into the slot.

However, if the mixture screws are removed during service procedures, the low and high speed mixtures must be adjusted and both caps installed over the screws with the tabs on the caps indexed into slots in the carburetor body.

Idle Adjustment Screw

Identify the idle adjustment screw located on the throttle linkage just below the high speed and low speed mixture screws.

Move the craft to a test tank, to a body of water, or connect a flush attachment and hose to the engine. For maximum performance, the idle rpm should be adjusted under actual operating conditions. Connect a tachometer to the engine.

Start the engine and allow it to warm to operating temperature.

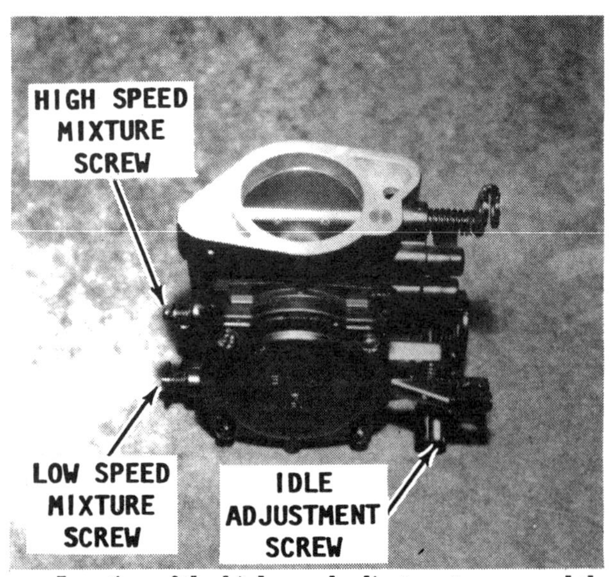

Location of the high speed adjustment screw and the low speed adjustment screw on a Keihin carburetor used on some 650 Series Models.

CAUTION

Water must circulate through the jet pump — to and from the engine, anytime the engine is operating. Circulating water will prevent overheating — which could cause damage to moving engine parts and possible engine seizure.

NEVER, AGAIN NEVER, operate the engine at high speed with a flush device attached. An engine operating at high speed with such a device attached, would **RUNAWAY** from lack of a load on the impeller shaft, causing extensive damage.

With the engine running, slowly rotate the idle adjustment screw until the engine idles at 1,150 to 1,350 rpm.

If the engine idle is set too low, the self-circling radius -- described in Chapter 1 -- increases and the craft may lose its self-circling feature.

If the engine idle is set too high, the craft may circle too fast for a displaced rider to climb aboard the craft.

Low and High Speed Mixture Adjustment

Each carburetor is individually adjusted at the factory using a flow meter and should not need any adjustment. Hovever, if replacement screws were installed the manufacturer suggests a starting point from which fine tuning on the water may be necessary.

Lightly seat both mixture screws. From this position back out the low speed screw 1-1/8 turns. Back out the high speed screw 5/8 of a turn.

High Altitude Operation

The idle system should never be changed when operating the craft at higher altitudes -- the low speed systems are not affected enough by changes in altitude to make an adjustment practical.

The idle adjustment just made is the optimum setting for sea level. When operating at higher elevations, the required leaner mixture is obtained by rotating the high speed screw the specified number of turns **CLOCKWISE** according to the following list.

3,300 ft (1,000 m) -- 1/8
6,600 ft (2,000 m) -- 1/4
9,900 ft (3,000 m) -- 3/8

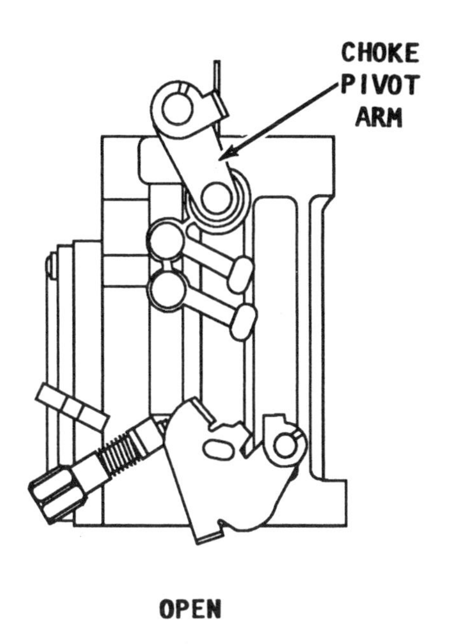

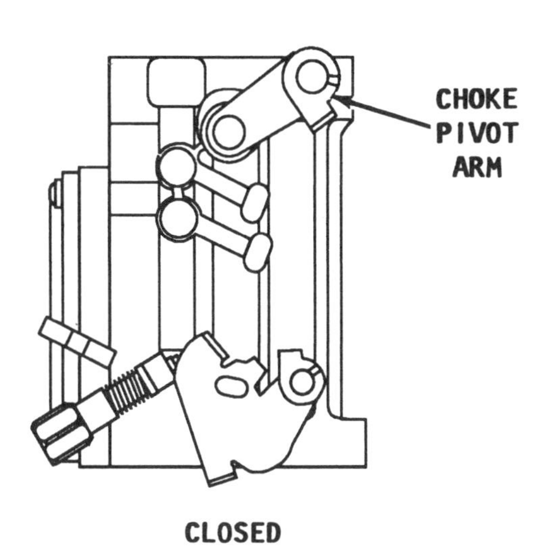

Line drawing to indicate the choke cable adjustment on the Model 650 Series with a CDK-34 carburetor. The choke is indicated open (top) and closed (bottom).

Choke Cable Adjustment

Rotate the choke knob to the full counterclockwise position. Observe the choke pivot arm on the side of the carburetor. The arm should be in the position shown in the accompanying illustration, with a minimum of slack in the choke cable, with the choke butterfly valve fully open.

If the choke needs adjustment: Rotate the choke knob to the full clockwise position. Loosen the two locknuts, one on each side of the cable support bracket. Set the choke pivot to the fully closed position, as shown. Tighten the two locknuts on the support bracket to hold this position, allowing a small amount of slack in the cable.

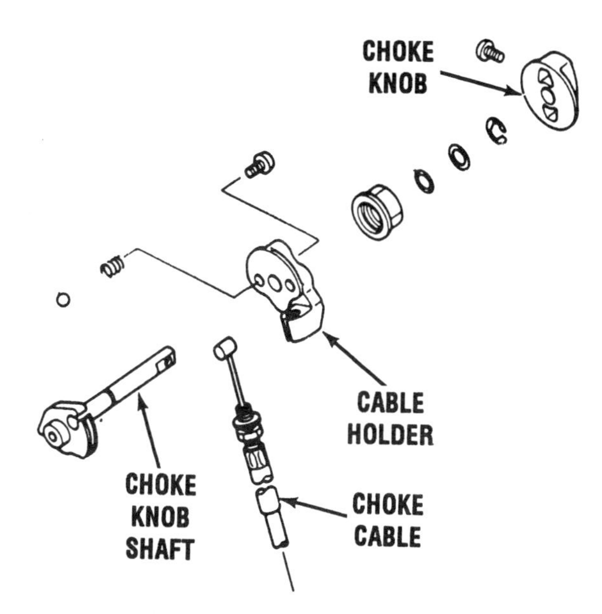

Exploded drawing of the choke knob and associated parts, which have been identified -- on a 650 Series Model.

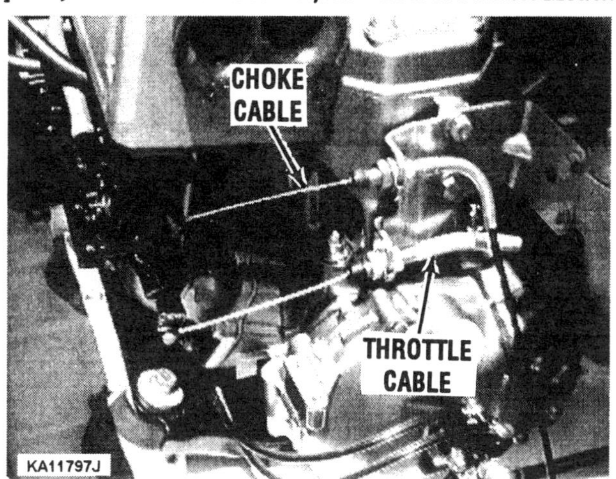

Typical choke and throttle cable / linkage assembly on a 750 Series Model engine. The choke in the above illustration is shown in the closed position.

If the choke cable has stretched, or has become kinked or frayed, or tends to stick, the entire cable must be replaced.

Throttle Cable Adjustment

The idle speed must be adjusted **BEFORE** adjusting the throttle lever free "play".

With the throttle lever released and the throttle closed, the idle adjustment screw should rest against the stopper arm of the throttle linkage, with a slight amount of slack in the throttle cable, as shown in the accompanying illustration.

With the throttle lever in the WOT position, the stopper arm of the throttle linkage should rest against the stop on the carburetor.

If the throttle cable needs adjustment: Loosen and rotate the two locknuts around

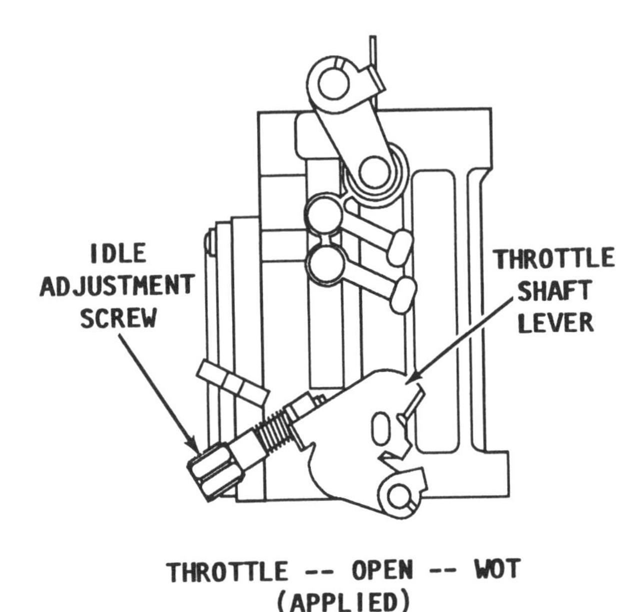

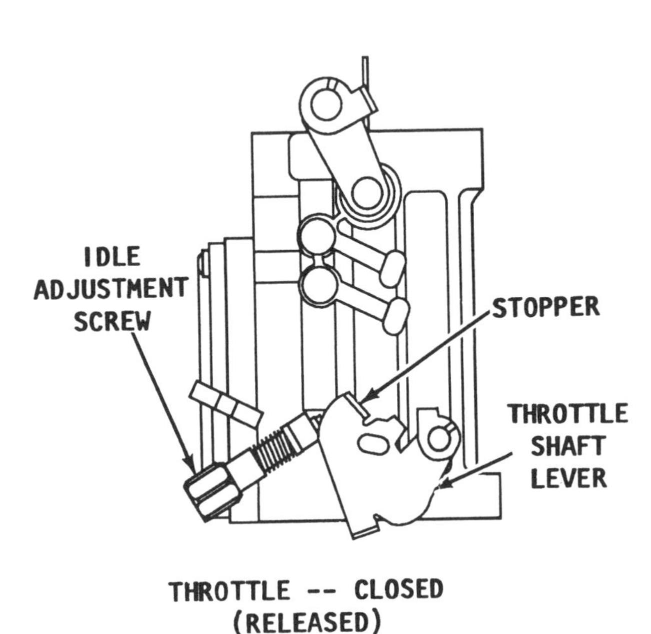

Line drawings to indicate the throttle cable position on the Model 650 Series with a CDK-34 carburetor. The throttle is indicated WOT -- wide open (top,) and closed (bottom). An exploded drawing of the throttle case is shown in the adjacent column.

the throttle cable at the cable support bracket until, with a minimum amount of slack in the cable, the idle adjustment screw rests against the throttle stopper with the throttle released. Tighten the two locknuts to hold this new adjustment.

If the throttle cable has stretched, or has become kinked or frayed, or tends to stick, the entire cable must be replaced.

Start the engine and check the completed work.

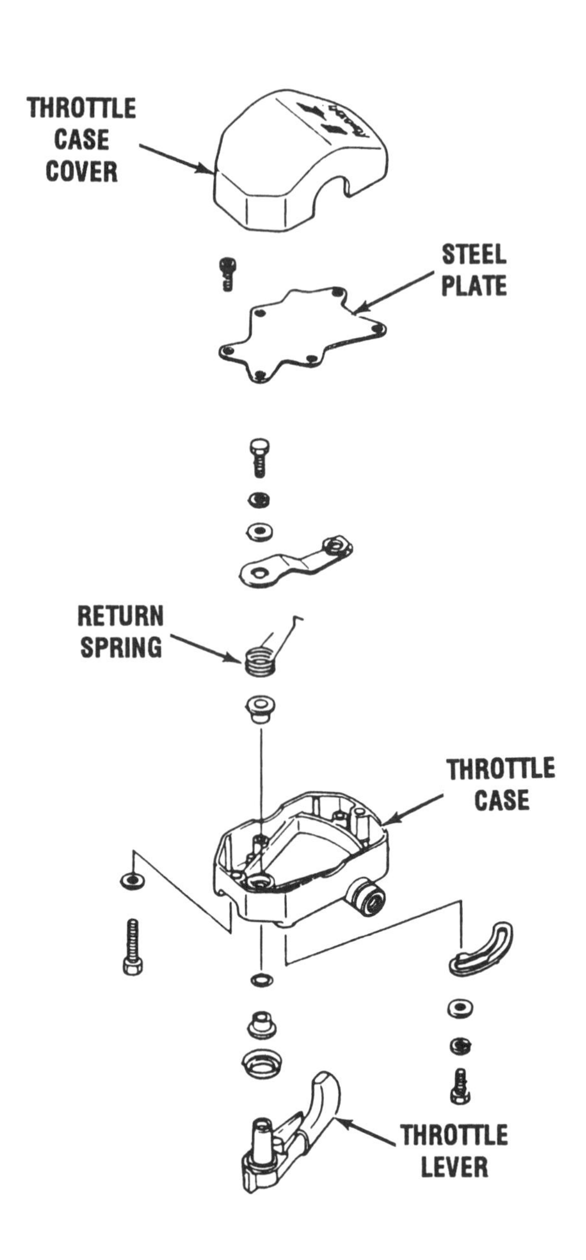

Exploded drawing of the throttle case on Model 650 Series watercraft. When assembling -- use a non-permanent locking agent on the bolts securing the two halves together and on the steel plate bolts. Apply a light coating of quality grease to the throttle lever.

CAUTION

Water must circulate through the jet pump -- to and from the engine, anytime the engine is operating. Circulating water will prevent overheating -- which could cause damage to moving engine parts and possible engine seizure.

NEVER, AGAIN NEVER, operate the engine at high speed with a flush device attached. An engine operating at high speed with such a device attached, would **RUNAWAY** from lack of a load on the impeller shaft, causing extensive damage.

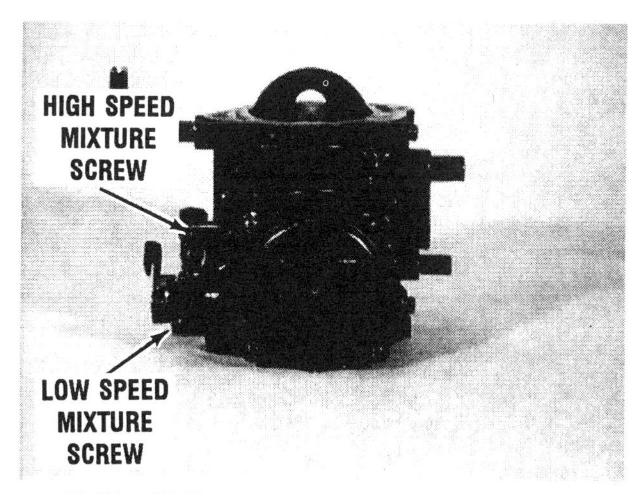

Keihin CDK-40 carburetor removed from the "rack" and ready for service work. Bear-in-mind, it is not necessary to remove the carburetor from the "rack" in order to perform a normal overhaul. The "front" side is shown in this illustration -- the "back" side contains the integral fuel pump. The center carburetor on a triple installation does not have the integral fuel pump.

6-7 SERVICE CDK-38 & 40 CARBURETOR WITH INTEGRAL FUEL PUMP

This section provides complete detailed procedures for removal, disassembly, cleaning and inspecting, assembling including bench adjustments, installation, and operating adjustments for the Keihin CDK-38 or CDK-40 carburetor.

The Keihin carburetors are identified by the manufacturer as CDK-38 with a 33mm venturi installed on the three cylinder Model 900 Series and 1100 Series engines; and CDK-40 with a 38mm venturi installed on Model 750 Series engines.

A "rack" of three CDK-40 Keihin carburetors being removed from the engine in preparation to an overhaul.

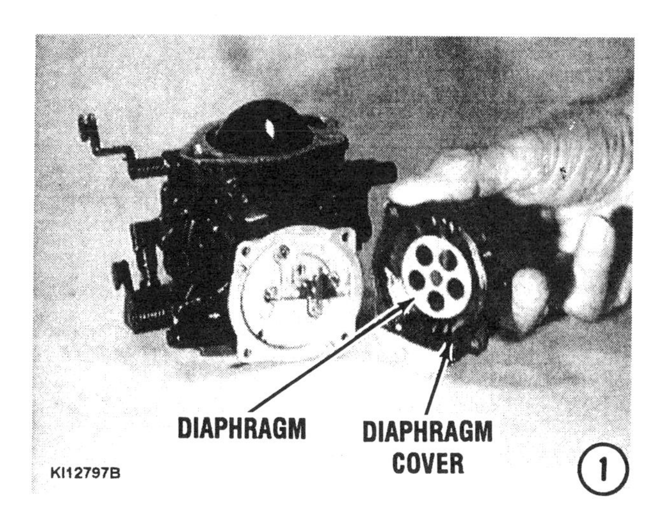

CARBURETOR DISASSEMBLING

SPECIAL WORDS

If working on a multiple carburetor assembly, servicing a carburetor may be completed while it remains mounted on the rack. The accompanying illustrations show the carburetor removed from the rack for photographic clarity.

If a carburetor IS removed from the rack, be sure to identify its position to ensure it is installed in the same location on the rack during assembling. If working on a three carburetor system, the middle -- No. 2 -- carburetor does not have an integral fuel pump.

SPECIAL DISCONNECT WORDS

Take time to identify and tag hoses and linkages **BEFORE** making the disconnect. Tagging these items now will facilitate speedy and correct connections during installation.

Carburetor "Front" Side

1- Remove the four screws securing the diaphragm cover to the carburetor and lift off

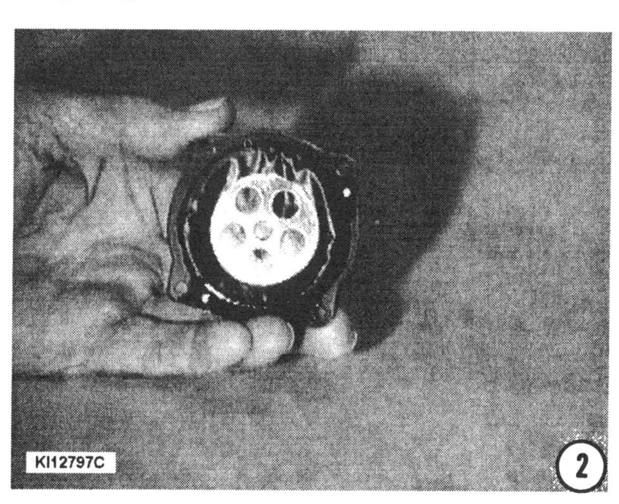

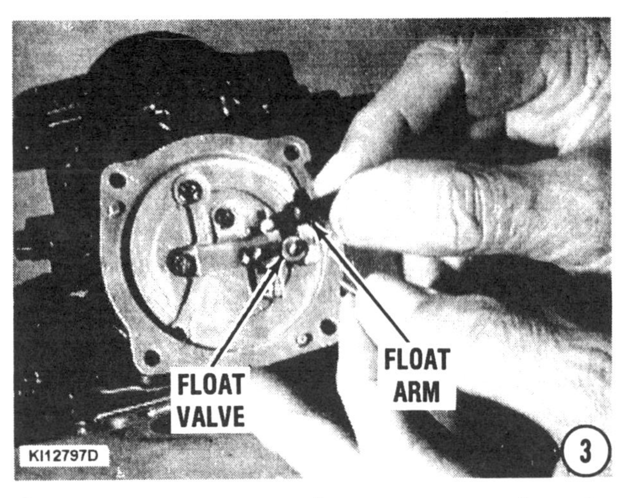

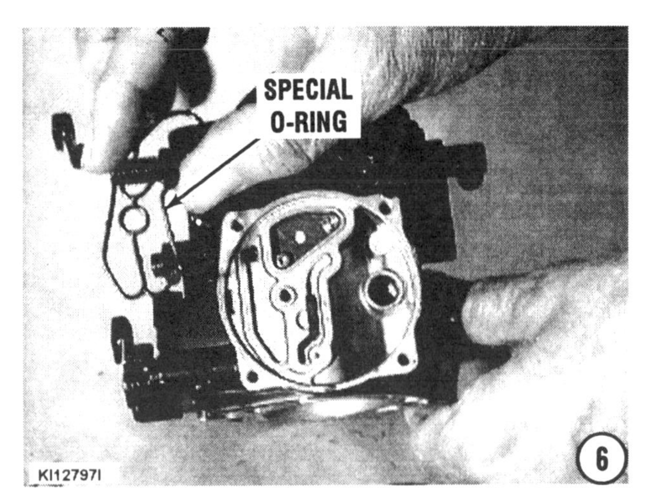

the cover. Remove the screw securing the diaphragm to the carburetor cover.

2- Inspect the diaphragm for any damage, i.e. pinholes or torn surface.

3- Remove the screw securing the float arm into its recess. Now, remove the float arm and float valve.

4- Remove the Phillips head screw securing the jet plate to the carburetor. Remove the jet plate. Inspect the jets to be sure the passageways are clear.

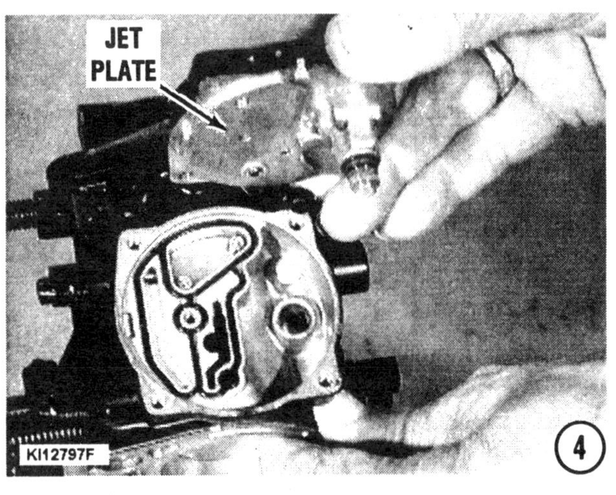

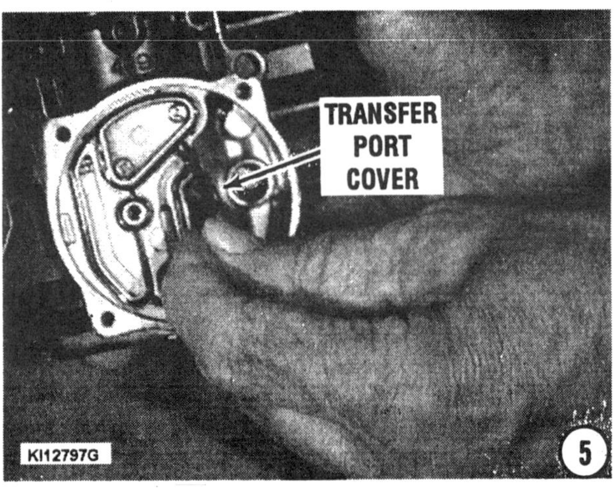

Inspect the fuel screen positioned below the fuel valve. Put the screen up to a light and inspect the screen for any blockage or damage.

5- Remove the transfer port cover. Verify the passageway is clear of debris.

6- Remove and inspect the special "spaghetti-type" O-ring.

7- Remove the two Phillips head screws securing the main chamber check valve assembly to the carburetor. Lift the check valve assembly -- including the metal plate, fiber reed valve, and nylon gasket free of the carburetor. Keep these items in proper order as an aid during assembling.

SPECIAL WORDS
ADJUSTING SCREWS

The low and high speed adjustment mixture screws must **NOT** be removed unless damage to the tapered end is suspected and the screws are to be replaced.

Under normal service, these two screws should remain undisturbed. Each Keihin carburetor is individually adjusted at the factory using a flow meter.

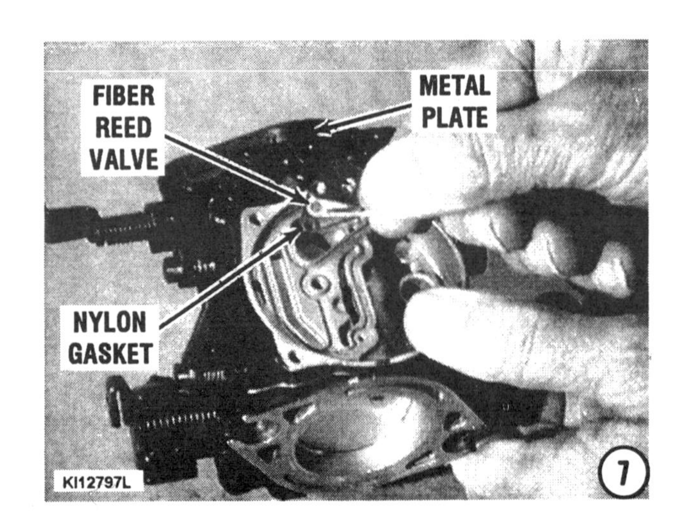

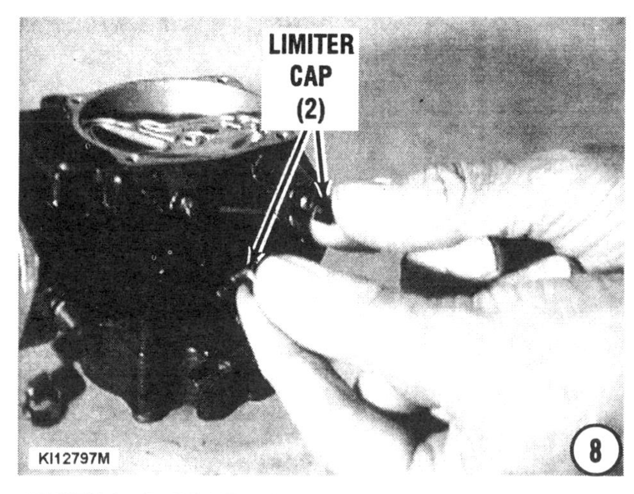

CRITICAL WORDS

Perform Step 8 **ONLY** if the high speed or low speed mixture cannot be adjusted properly. **ALSO** perform the step only if replacement screws have been obtained and are on hand. Failure to adjust the low speed or high speed mixture would suggest the tapered end of one -- but not necessarily both -- of the screws is damaged. In almost all cases, "damage" is caused by the screw being threaded in too tightly and the needle becoming "grooved", because it passed through the orifice. Once the needle is "grooved", proper mixture is not possible.

Each cap has a small flange which indexes into a slot in the carburetor to prevent the adjustment screws from being changed accidentally.

8- Pull off the limiter cap on the low speed mixture adjustment screw -- the screw next to the embossed **"L"**. Now, **CAREFULLY** count the number of turns required to seat the adjust-

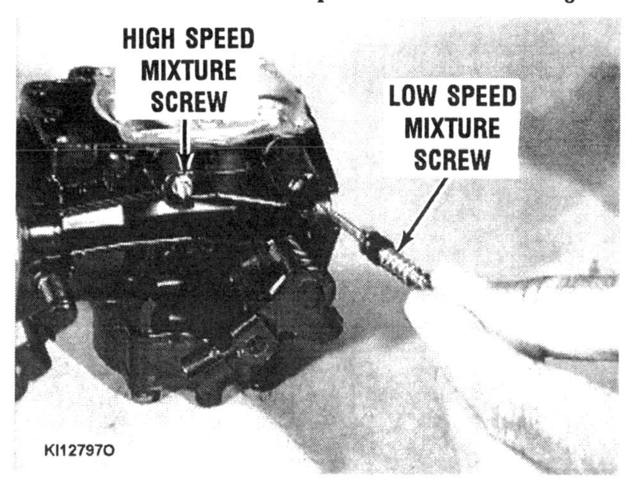

Before removing either adjustment screw, take time to count the number of turns required to seat the screw LIGHTLY, from its present position. Record the number which will be considered a "bench adjustment" during assembling.

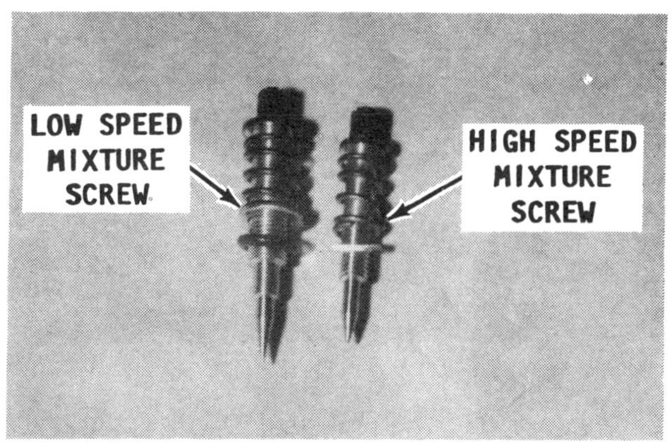

Comparison of the low speed and high speed mixture screws. Notice how the high speed screw is slightly shorter than the low speed screw.

ment screw **LIGHTLY** -- in quarter increments, and then back out the low speed adjustment screw. Pull off the limiter cap on the high speed mixture adjustment screw -- the screw next to the embossed **"H"** **CAREFULLY** count the number of turns in quarter turn increments required to seat the screw **LIGHTLY**, and then back out the high speed adjustment screw.

Take time to record the number of turns for each screw. Observe the difference in length and size between the high speed and low speed adjustment screws.

Carburetor "Back" Side (Pump Side)

Rotate the carburetor body around to gain access to the pump side of the carburetor.

9- Remove the screws securing the pump cover to the carburetor body. Lift off the pump cover.

10- Remove the diaphragm and special O-ring. Exercise care when "peeling off" the pump diaphragm to prevent damaging it.

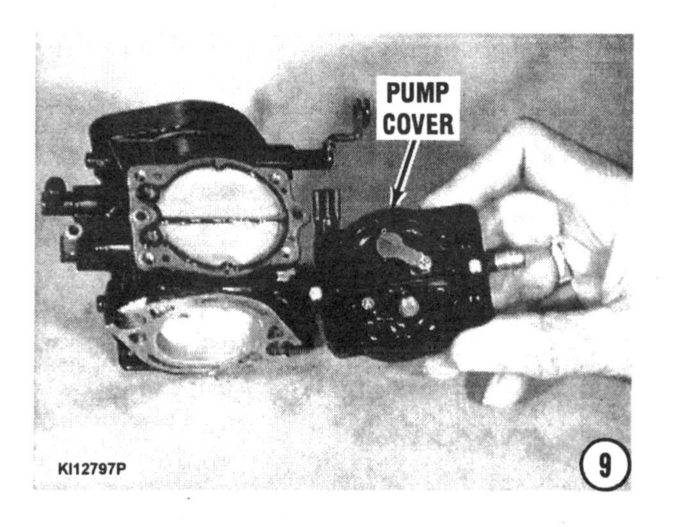

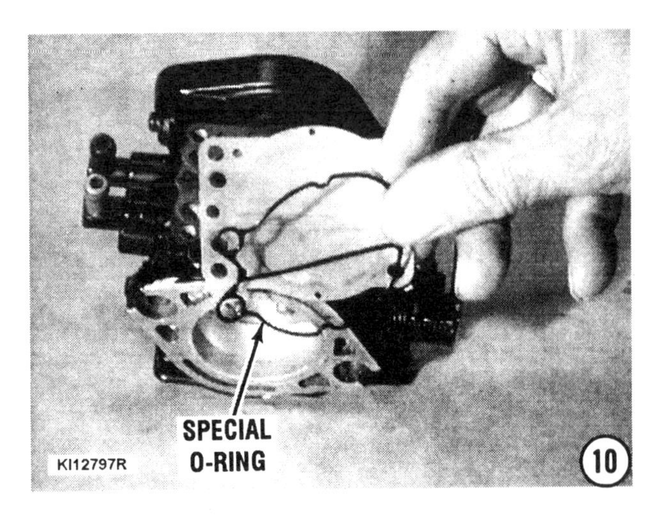

GOOD WORDS

Further disassembling of the fuel pump unit is not necessary, when performing a normal carburetor overhaul. Added to this fact is the requirement for special tools to disassemble the fuel pump.

CLEANING AND INSPECTING CDK-38 AND CDK-40

NEVER dip rubber parts, plastic parts, or diaphragms in carburetor cleaner. These parts should be cleaned **ONLY** in solvent, and then blown dry with compressed air.

Place all metal parts in a screen-type tray and dip them in carburetor cleaner until they appear completely clean, then blow them dry with compressed air. **Do NOT** immerse the carburetor body in a cleaner solution, because such action would surely destroy integral parts of the carburetor assembly.

Instead, use an aerosol can of Choke and Carburetor Cleaner to spray the assembly, as indicated in the accompanying illustration.

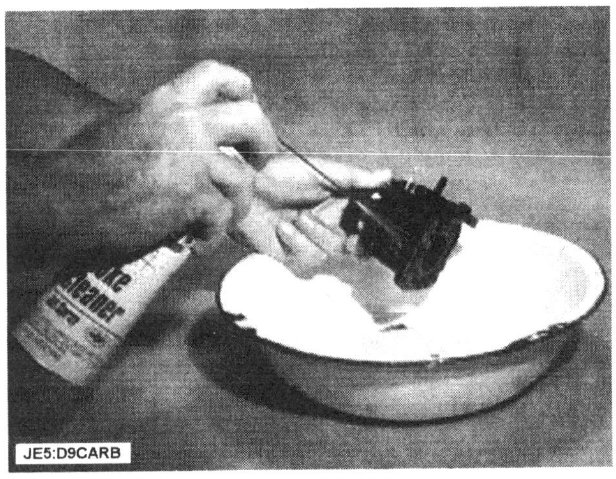

Using a Choke and Carburetor Cleaner on the carburetor body. ***NEVER*** *use a dip cleaner or hot tank solution, because such action will surely damage the sealing components in the carburetor body.*

Blow out all passages in the castings with compressed air. Check all parts and passages to be sure they are not clogged or contain any deposits. **NEVER** use a piece of wire or any type of pointed instrument to clean drilled passages or calibrated holes in a carburetor.

Move the throttle shaft back-and-forth to check for wear. If the shaft appears to be too loose, replace the complete throttle body, because individual replacement parts are **NOT** available.

CRITICAL AREAS

Inspect the condition of both the choke and throttle shafts. Most shaft designs incorporate a cutout portion to enable the butterfly valve to lay flat against the shaft. Two small holes are drilled in this cutout portion to accommodate the small screws securing the valve to the shaft.

The weakest point of the shaft is at the screw holes. Small cracks form at the holes due to metal fatigue. These cracks elongate the threaded holes and the small screws shake loose. Once loose, they are sucked into the carburetor, then the intake manifold, through the reed valves, and into the crankcase -- eventually finding their way into the combustion chamber. One of these small screws can cause very extensive and expensive damage to reed valves, pistons, rings, and cylinder walls. Therefore, check the cutout in the shaft **CLOSELY** for any evidence of a small crack.

Inspect the main body, airhorn, and venturi cluster gasket surfaces for cracks and burrs which might cause a leak. Check to be sure the float arm return spring has not been stretched. Check the float arm needle contacting surface and replace the diaphragm if this surface has a groove worn in it.

As previously mentioned, most of the parts which should be replaced during a carburetor overhaul are included in an overhaul kit available from the local marine dealer. One of these kits will contain a fuel inlet needle and seat. The needle should be replaced each time the carburetor is disassembled as a precaution against leakage.

Make a thorough inspection of the fuel chamber and fuel diaphragms for the tiniest pin hole. If one is discovered, the hole will only get bigger. Therefore, the diaphragm must be replaced in order to obtain full performance from the engine.

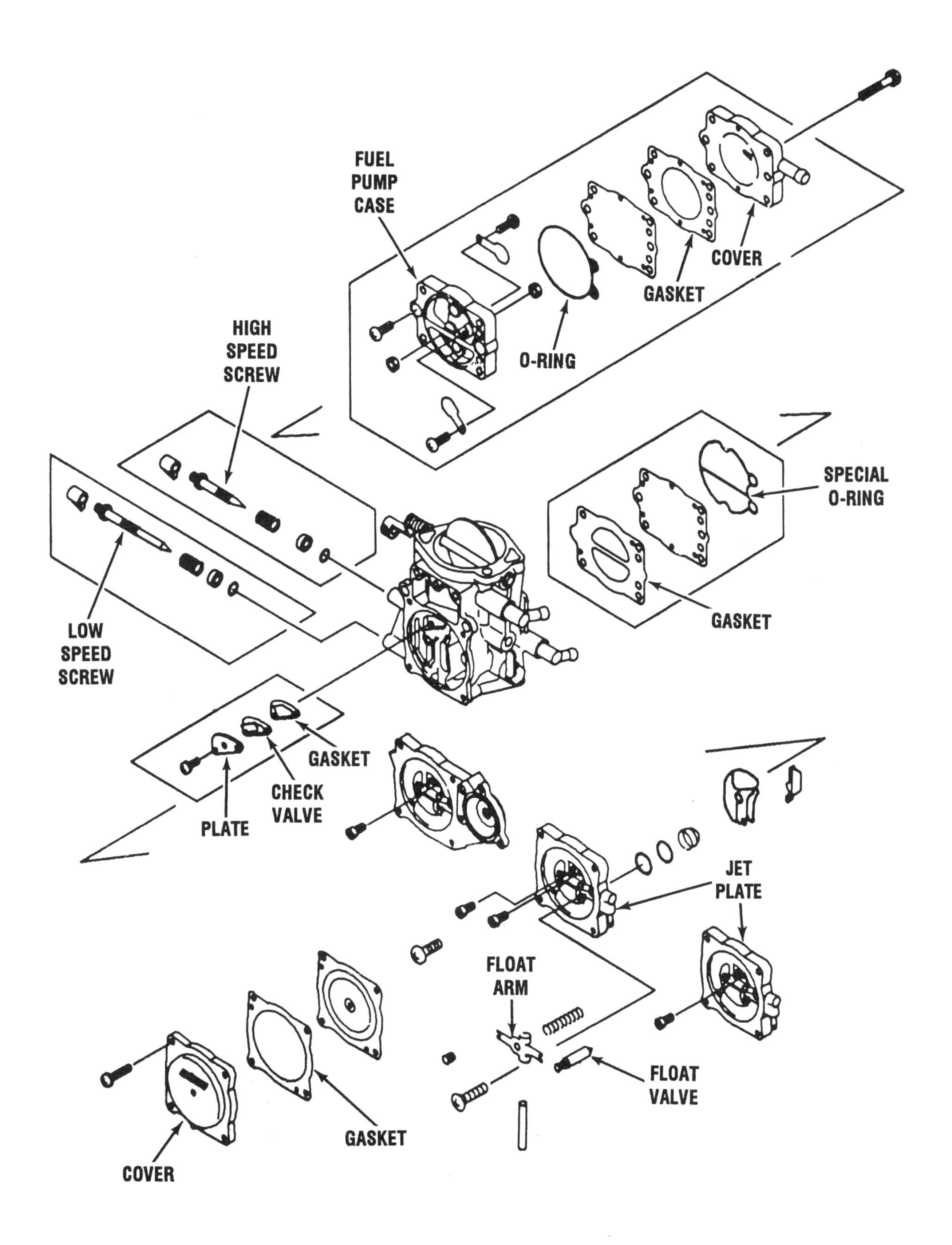

Exploded drawing of a Keihin CDK-40 carburetor with integral fuel pump on the back side. On a triple carburetor installation, the No. 2 (center) carburetor does not have the integral fuel pump.

ASSEMBLING CDK-38 AND CDK-40

The following procedures pickup the work after the cleaning and inspecting tasks have been completed, as outlined in the previous paragraphs and replacement parts are on hand.

Special Words on Mixture Screws

Observe the low and high speed mixture screws. These screws are **NOT** interchangeable. To identify the correct screw, place them side by side as shown in the accompanying illustration. The low speed screw is shorter than the high speed screw. Both screws are threaded into the carburetor at the base of the fuel pump. The low speed screw is installed next to the embossed letter **"L"**, and the high speed screw is installed next to the embossed letter **"H"**. Check to make sure both screws are equipped with a washer, a spring, another washer, and an O-ring.

1- Install the low speed screw into the opening embossed with the letter **"L"**. **CAREFULLY** thread the screw into the carburetor body until it is **LIGHTLY** seated. Now, back the screw out the same number of turns recorded during disassembling. If the number of turns out was not recorded, as instructed, back the low speed screw out 7/8 ±1/4 turn as a "bench adjustment" at this time. Install the safety cap over the screw with the flange on the cap indexed into the carburetor slot to prevent the screw from being rotated accidently.

CAREFULLY thread the high speed screw into the opening embossed with the letter **"H"**. until it is **LIGHTLY** seated, and then back it out the same number of turns recorded during disassembling. If the number of turns out was not recorded, as instructed, back the high speed screw out one full turn ±1/4 turn as a "bench adjustment" at this time. Install the safety cap over the adjustment screw, with the flange on the cap indexed into the slot in the carburetor.

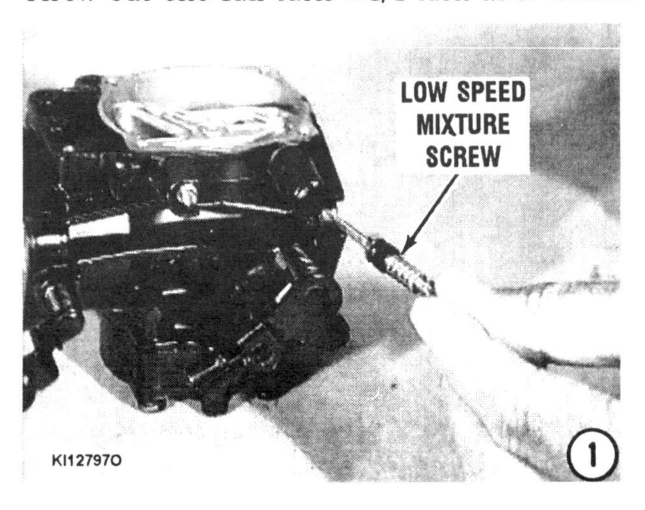

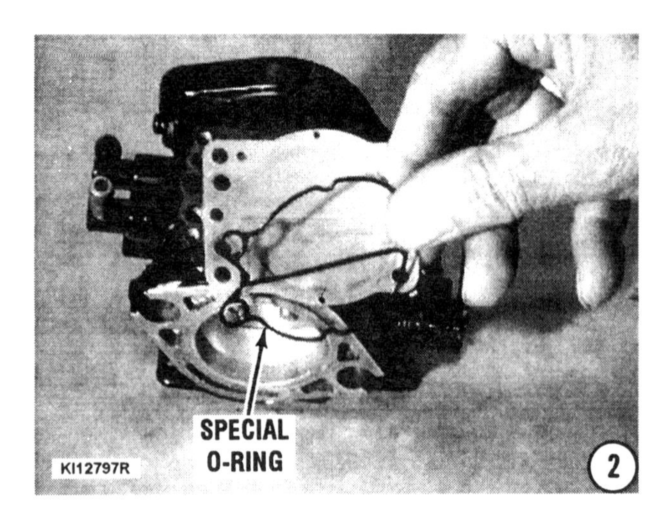

2- Install the pump diaphragm, followed by the special O-ring, into the pump chamber.

3- Position the pump cover over the carburetor body. Install and tighten the screws securely.

Rotate the carburetor body to gain access to the front side -- the cover with the "Keihin" name stamped into it.

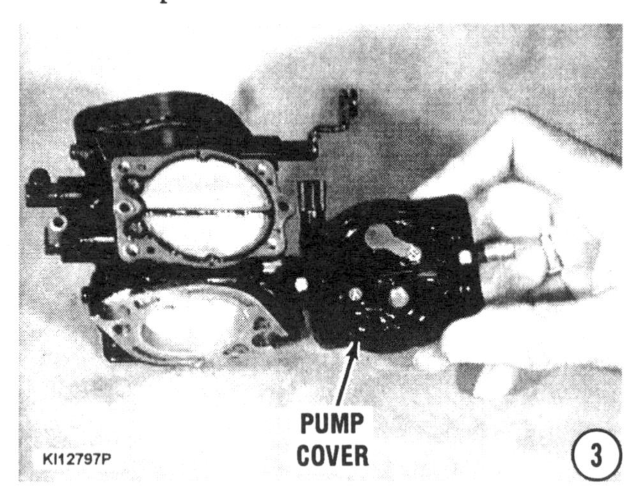

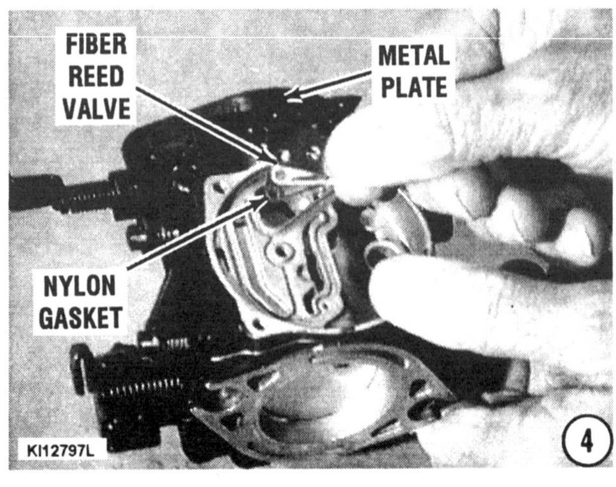

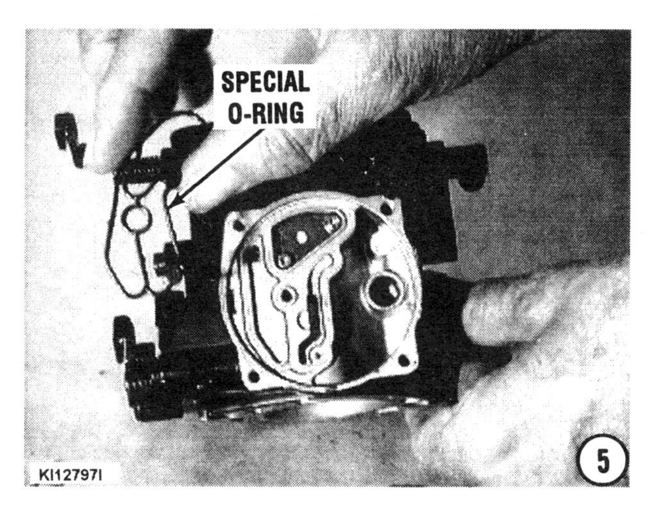

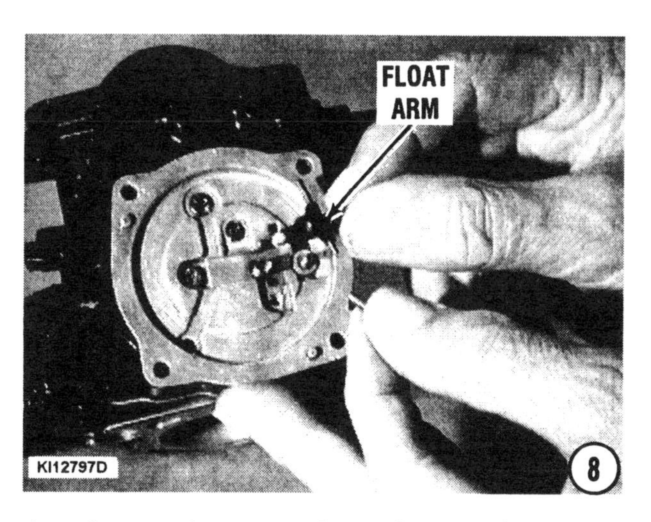

4- Obtain and install **IN SEQUENCE,** the main chamber check valve assembly as follows: First, the **NYLON GASKET** into the gasket recess, next, the **FIBER REED VALVE** into the recess, and finally the metal plate into the recess. Secure the check valve assembly in place with the attaching screws.

5- Insert the special **O**-ring into the groove.

6- Install the transfer port cover.

7- Insert the fuel screen into the jet plate below the fuel valve -- if it was removed. Position the jet plate over the carburetor body and secure it in place with the attaching screw.

8- Install the float arm into its recess and secure it with the screw.

9- Position the diaphragm in place on the inside of the diaphragm cover. Tighten the screw just "snug". Install the cover onto the carburetor body. Tighten the screws securely.

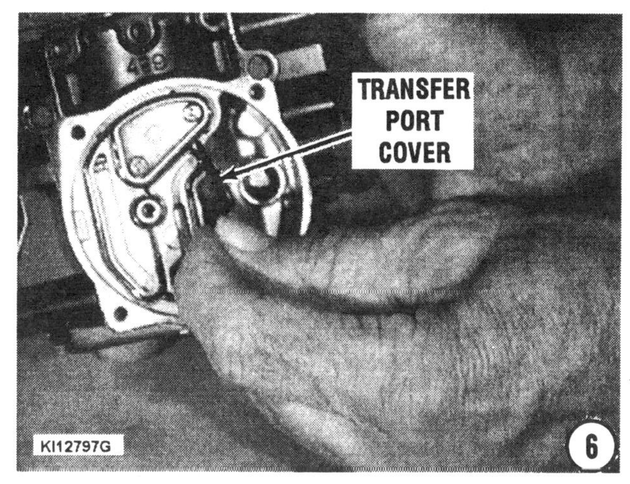

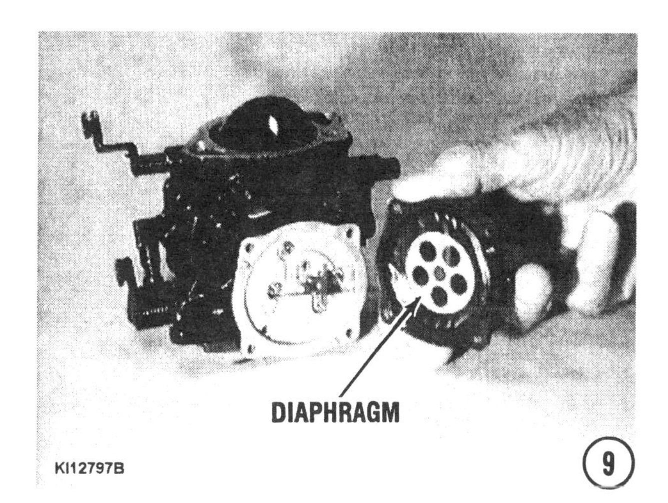

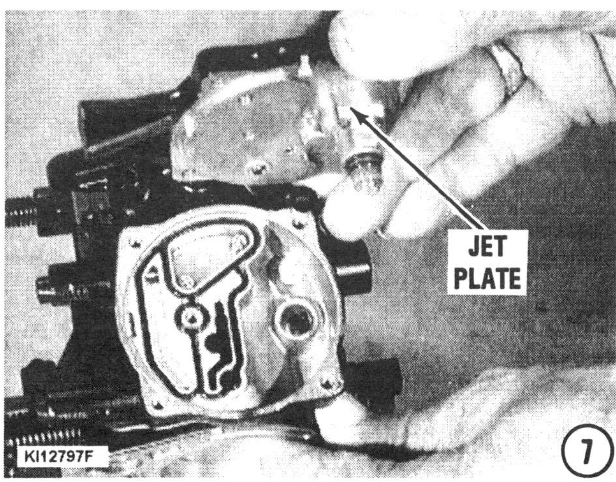

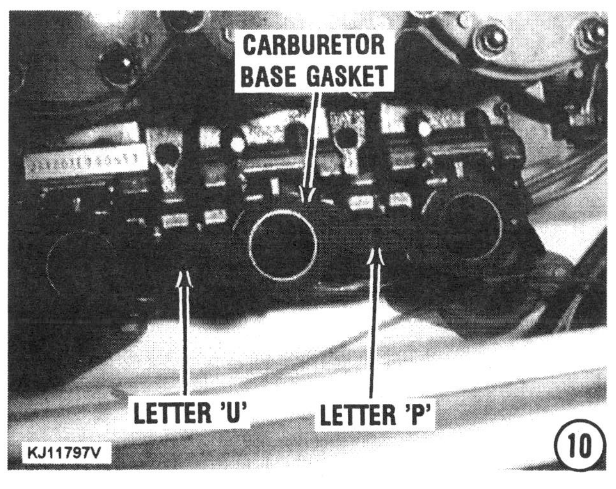

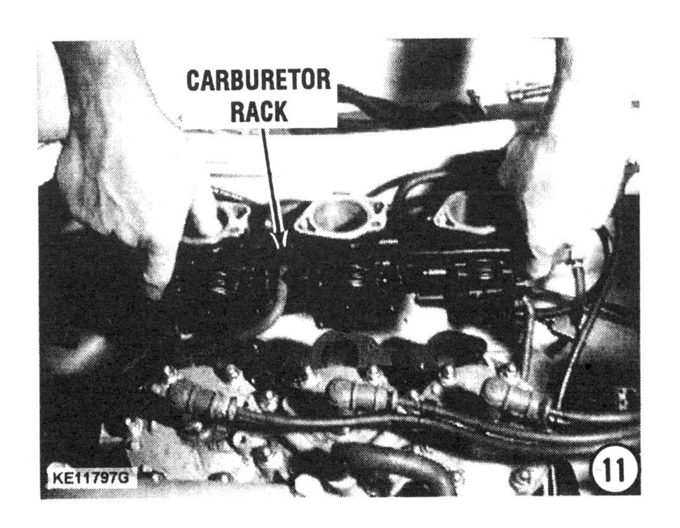

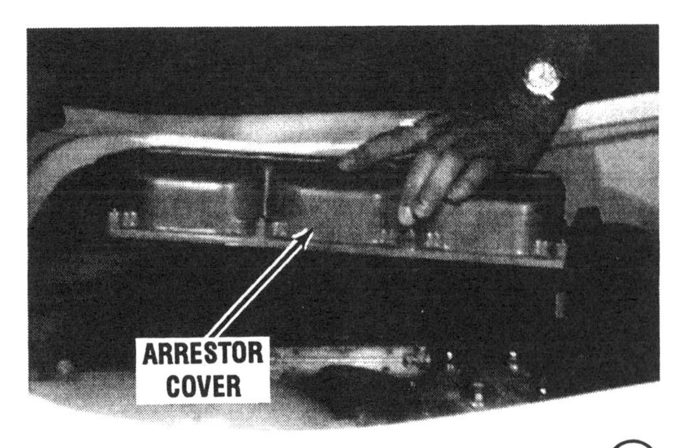

Carburetor Installation

If the carburetor of a multiple installation was removed from the rack, install the carburetor on the rack in the same position from which it was removed. Remember on a triple installation, the center carburetor does **NOT** have an integral fuel pump. Connect the hoses and linkage.

10- Place a new carburetor mounting gasket on the intake manifold, with the cutout word **"UP"** on the gasket facing up and readable.

11- Lower the carburetor rack onto the base gasket atop the intake manifold, with the bolt holes aligned.

12- Install the flame arrestor box onto the carburetor rack assembly. The rubber cylinders serve as gaskets between the carburetors and the flame arrestor box. Install and tighten the arrestor box/carburetor mounting bolts to a torque value of 69 in lb (7.8Nm). Position and seat each flame arrestor screen properly into the recessed area.

13- Guide the flame arrestor cover into position over the arrestor box. Install and tighten the bolts securing the cover to a torque value of 69 in lb (7.8Nm).

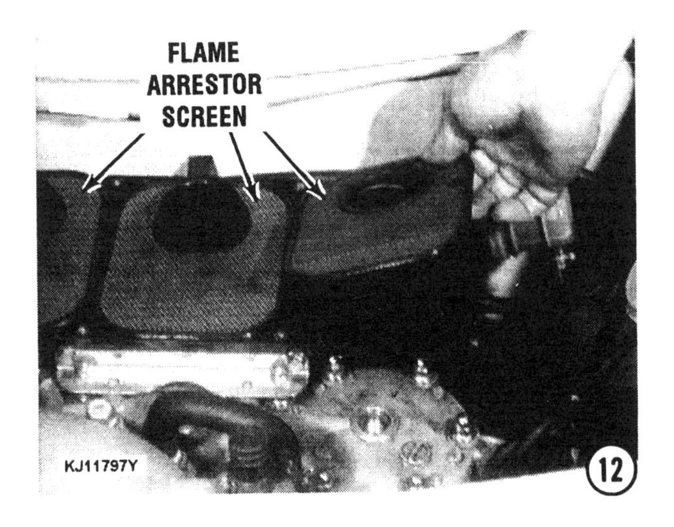

Final Connections

Connect the hoses and linkage items per the tags attached during the disassembling. If the tags were not attached as instructed, in most cases, the shape and length of the hoses and linkage will indicate where the connection is to be made.

Carburetor Priming

The carburetor must be primed before starting the engine to save the battery from excessive cranking.

Place a shop towel under the fuel return hose fitting. The hose is usually embossed with the word "RETURN". Blow through the hose until fuel starts to drip from the fuel return fitting on the carburetor. The carburetor is now full of fuel and the engine is ready for cranking after the hose has been reconnected.

Choke Cable Adjustment

Adjust the choke cable according to the procedures outlined in the previous section beginning on Page 6-23.

Rotate the choke knob to the full counterclockwise position. Observe the choke pivot arm on the side of the carburetor. The arm should be in the position shown in the accompanying illustration, with a minimum of slack in the choke cable, with the choke butterfly valve fully open.

If the choke needs adjustment: Rotate the choke knob to the full clockwise position. Loosen the two locknuts, one on each side of the cable support bracket. Set the choke pivot to the fully closed position, as shown. Tighten

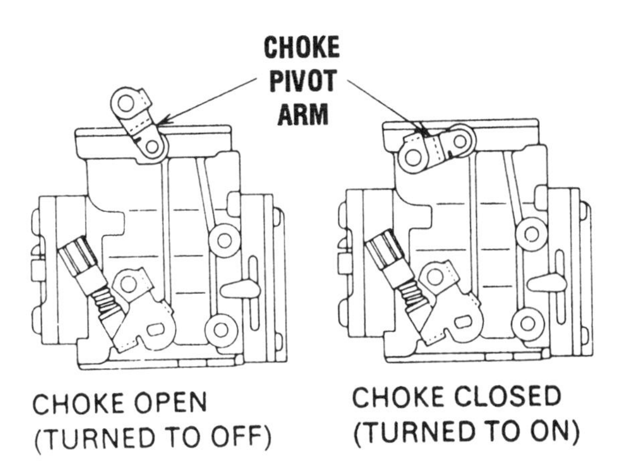

Simple line drawing to show the two positions of the choke -- choke open (left), and choke closed (right).

the two locknuts on the support bracket to hold this position, allowing a small amount of slack in the cable.

If the choke cable has stretched, or has become kinked or frayed, or tends to stick, the entire cable must be replaced.

Throttle Cable Adjustment

Adjust the throttle cable by following the procedures outlined in the previous section beginning on Page 6-23.

The idle speed must be adjusted **BEFORE** adjusting the throttle lever free "play".

With the throttle lever released and the throttle closed, the idle adjustment screw should rest against the stopper arm of the throttle linkage, with a slight amount of slack in the throttle cable.

With the throttle lever in the WOT position, the stopper arm of the throttle linkage should rest against the stop on the carburetor.

If the throttle cable needs adjustment: Loosen and rotate the two locknuts around the throttle cable at the cable support bracket until, with a minimum amount of slack in the cable, the idle adjustment screw rests against the throttle stopper with the throttle released. Tighten the two locknuts to hold this new adjustment.

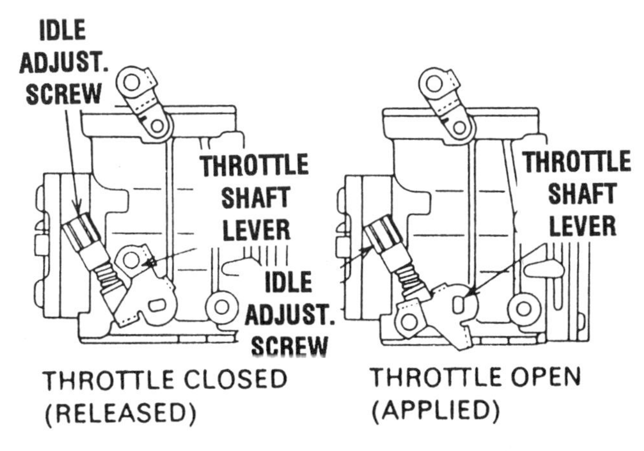

Line drawing to illustrate the throttle shaft lever in the closed position (left), and in the open position (right).

If the throttle cable has stretched, or has become kinked or frayed, or tends to stick, the entire cable must be replaced.

Start the engine and check the completed work.

CAUTION

Water must circulate through the jet pump -- to and from the engine, anytime the engine is operating. Circulating water will prevent overheating -- which could cause damage to moving engine parts and possible engine seizure.

NEVER, AGAIN NEVER, operate the engine at high speed with a flush device attached. An engine operating at high speed with such a device attached, would **RUNAWAY** from lack of a load on the impeller shaft, causing extensive damage.

6-8 REMOTE FUEL PUMP

This section presents complete procedures to service the remote fuel pump installed on a bulkhead in the engine compartment of the Model 650 Series engines with a Keihin CDK-34 carburetor. All other engines use a Keihin carburetor with an integral fuel pump. Detailed illustrated procedures to service the integral carburetor are presented in the previous section -- 6-7 Service CDK-38 & 40 carburetors.

THEORY OF OPERATION

The next few paragraphs briefly describe operation of the remote fuel pump used on personal watercraft covered in this manual. This description is followed by detailed procedures for testing the pressure, testing the volume, removing, servicing, and installing the fuel pump.

The fuel pump used is a diaphragm displacement type. The pump is secured to the bulkhead. A hose from the crankcase supplies crankcase impulses to the pump. After initial engine cranking to fill the pump and engine start, the pump supplies adequate fuel to the carburetor to meet engine demands under all speeds and conditions.

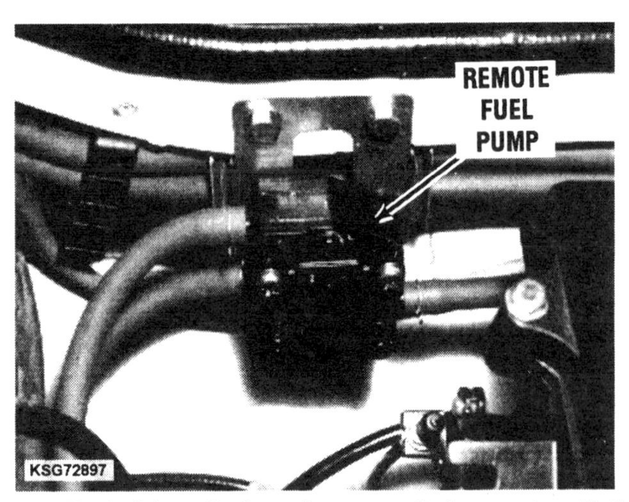

Typical installation of a remote fuel pump installed on the outboard bulkhead of a Model 650 watercraft.

The pump consists of two diaphragms, two similar check valves -- one for inlet (suction) and the other for outlet (discharge) -- a front and back cover, and a fuel pump body with a fitting for the impulse hose from the crankcase. The suction and compression created -- as the piston travels up and down in the cylinder -- causes the diaphragms to flex.

As the piston moves upward, the diaphragms will flex inward displacing volume on its opposite side to create suction. This suction will draw liquid fuel in through the inlet check valve.

When the piston moves downward, compression is created in the crankcase. This compression causes the pump diaphragms to flex in the opposite direction. This action causes the discharge check valve to lift off its seat. Fuel is then forced through the discharge valve into fuel supply hose and then on to the carburetor.

Problems with the fuel pump are limited to possible leaks in the flexible neoprene suction lines; a punctured diaphragm; air leaks between sections of the pump assembly, or possibly from the check valves becoming distorted and failing to seat properly.

The pump is activated by one cylinder. If this cylinder indicates a wet fouled condition, as evidenced by a wet fouled spark plug, be sure to check the fuel pump diaphragms for possible puncture or leakage.

Even the tiniest pin hole in a diaphragm will cause loss of engine performance.

If the pump is opened and disassembled, extreme care should be exercised when handling the internal components.

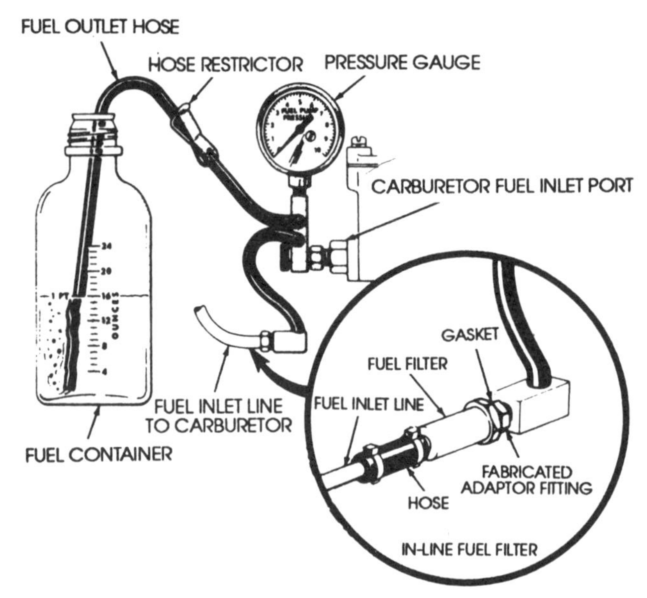

Test setup to check fuel pump pressure, as outlined in the text.

PUMP PRESSURE CHECK

Lack of an adequate fuel supply will cause the engine to operate lean, lose rpm, or cause piston scoring.

Check the fuel strainer on the end of the pickup in the fuel tank to be sure it is not too small or is not clogged. Be sure to check the in-line filter at the carburetor. If this screen becomes clogged, adequate fuel cannot pass through into the carburetor to meet engine demands.

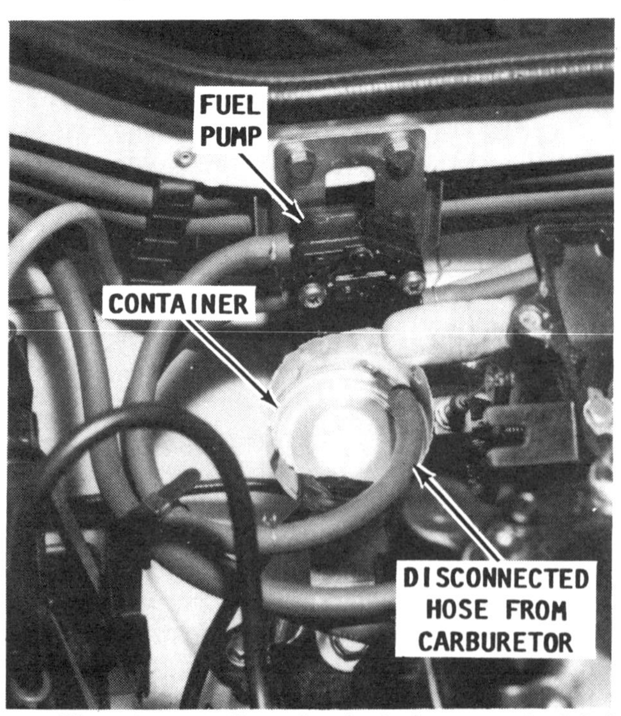

Close view to indicate the physical setup to check fuel pump volume output, as described in the text.

TO TEST

Install a fuel pressure gauge in the fuel line between the fuel pump and the carburetor.

Start the engine and allow it to warm to operating temperature and then check the fuel pressure.

CAUTION

Water must circulate through the jet pump -- to and from the engine, anytime the engine is operating. Circulating water will prevent overheating -- which could cause damage to moving engine parts and possible engine seizure.

NEVER, AGAIN NEVER, operate the engine at high speed with a flush device attached. An engine operating at high speed with such a device attached, would **RUNAWAY** from lack of a load on the impeller shaft, causing extensive damage.

Operate the enigne at full throttle and check the pressure reading. The gauge should indicate at least 2 psi.

PUMP VOLUME CHECK

First, These Words

The following test applies only to personal watercraft equipped with a remote type fuel pump separate from the carburetor.

On a Model 650 engine with remote fuel pump -- if air is allowed to enter the vacuum line between the intake manifold and the fuel pump, the pump will fail to operate properly.

CAUTION

Gasoline will be flowing in the engine area during this test. Therefore, guard against fire by grounding the high-tension leads to prevent possible sparking, which could easily ignite fuel or fuel vapors.

A good safety procedure is to ground both spark plug leads. Disconnect the fuel line at the carburetor. Place a suitable container over the end of the fuel line to catch discharged fuel. Crank the engine and observe the flow of fuel discharged into the container.

If no fuel is discharged from the hose, check the crankcase vacuum hose connections and the fuel hose for leaks. Another possible problem may be a break or obstruction in the fuel line.

If the fuel pump is operating properly, a healthy stream of fuel should pulse from the fuel hose.

Continue cranking the engine and catching fuel for about 15 pulses to determine if the amount of fuel decreases with each pulse or maintains a constant amount. A decrease in the discharge indicates a restriction in the line. If the fuel line is plugged, the fuel stream may stop.

If there is fuel in the fuel tank but no fuel is discharged while the engine is being cranked, the problem may be in one of five areas:

1- The line from the fuel pump to the carburetor may be plugged.

2- The fuel pump may be defective.

3- The line from the fuel tank to the fuel pump may be plugged or the line may be leaking air.

4- The in-line fuel filter may be plugged.

5- The vacuum line from the intake manifold to the fuel pump may be leaking air.

ROUGH ENGINE IDLE

If an engine does not idle smoothly, the most reasonable approach to the problem is to perform a tune-up to eliminate such areas as faulty spark plugs, excessive resistance in high tension leads, or timing out of adjustment.

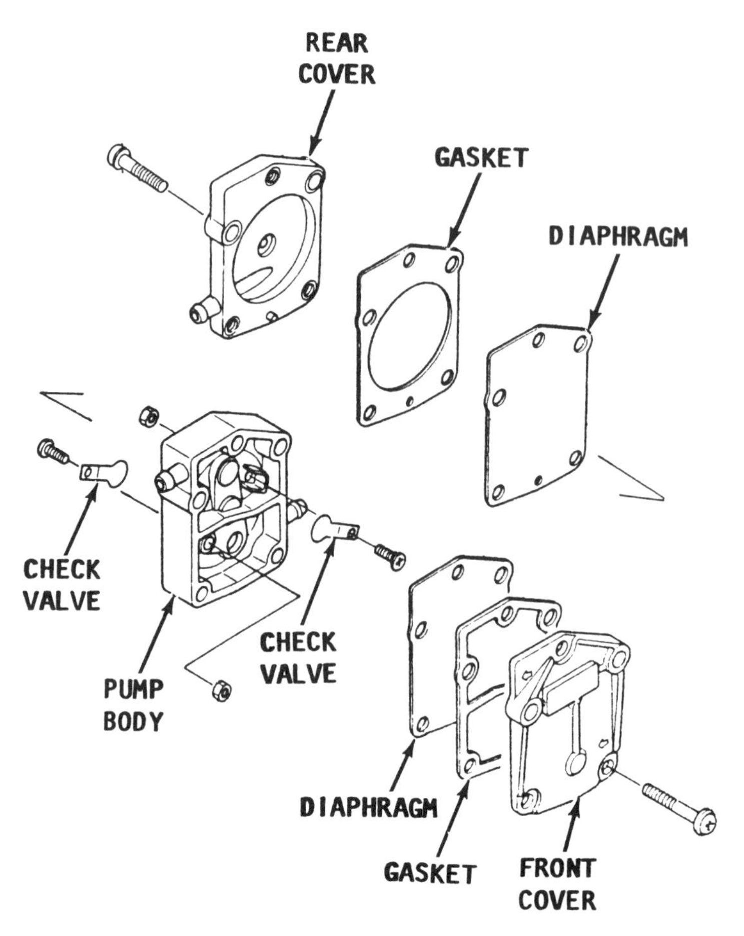

Exploded drawing of the remote fuel pump used on the Model 650 Series engine with the Keihin CDK-34 carburetor. Major parts have been identified.

Other problems contributing to rough engine idle may include:

An air leak in the intake manifold.
Uneven compression between cylinders.
Sticky or broken reeds, if so equipped.

SERVICING THE FUEL PUMP

The fuel pump should **NOT** be disassembled unless internal damage is suspected or the pump is leaking fuel. A fuel pump rebuilding kit must be on hand to ensure faulty parts will be replaced. Most fuel pump kits include new diaphragms, gaskets and check valves.

Disassembling and assembling should be performed on a clean work surface. Make every effort to prevent foreign material from entering the fuel pump or adhering to the diaphragms.

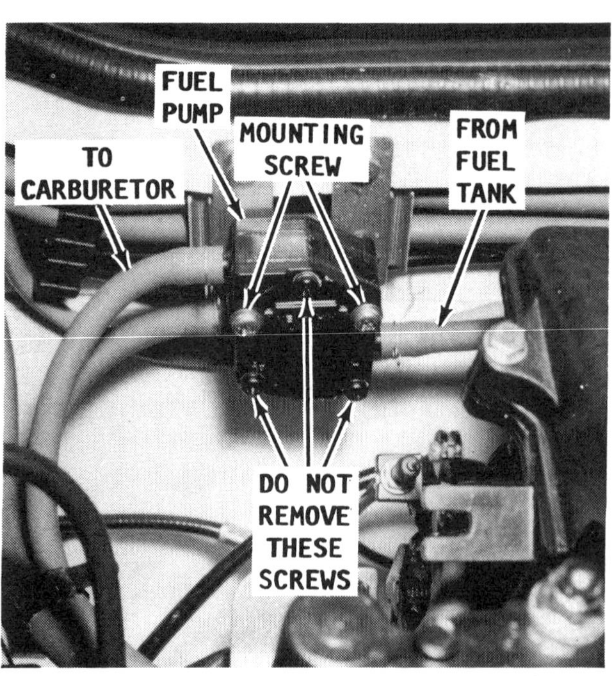

*Head-on view of the remote fuel pump clearly showing the **TWO** mounting screws and the three screws to be left in place -- as instructed in the text. Always handle internal fuel pump parts with EXTREME care.*

REMOVAL AND DISASSEMBLY

The pump consists of fragile diaphragms which are easily damaged. The pump should not be disassembled unless replacement diaphragms and gaskets are on hand.

Disconnect the vacuum hose, the inlet hose and the outlet hose from the fuel pump. Take time to plug the disconnected ends of all three hoses to prevent contaminants from entering the crankcase and fuel system. Observe the five screws on the fuel pump cover. Two of the screws are silver and the remaining three are black. Only the two silver screws are used to mount the pump to the bracket, the black screws hold the internal components of the the pump together. Therefore, remove **ONLY** the two silver screws.

Now, **CAREFULLY** separate the parts and keep them in **ORDER** as an assist in assembling. As parts are removed, take time to observe and remember how each check valve faces, because it must be installed in exactly the same manner, or the pump will not function.

Remove the pump and move it to a suitable clean work surface. Refer to the accompanying exploded drawing of the fuel pump as an aid to disassembling and assembling. Remove the three black screws securing the pump parts together.

CLEANING AND INSPECTING

Wash all parts thoroughly in solvent, and then blow them dry with compressed air. **USE CARE** when using compressed air on the check valves. **DO NOT** hold the nozzle too close because the check valve can be damaged from an excessive blast of air.

Inspect each part for wear and damage. Verify that the valve seats provide a flat contact area for the check valves. Tighten both check valve connections firmly as the valves are replaced.

The check valves should allow air to pass through in one direction, but not in the opposite direction.

Check the diaphragms for pin holes by holding it up to the light. If pin holes are detected or if the diaphragm is not pliable, it **MUST** be replaced.

ASSEMBLING

Proper operation of the fuel pump is essential for maximum engine performance. Therefore, always use **NEW** gaskets and diaphragms.

NEVER use any type of sealer on fuel pump gaskets.

TAKE CARE not to damage the very fragile flat surface of the check valve. Secure each check valve in place with a Phillips head screws. Tighten the screws securely.

Install one of the diaphragms and the gasket with the large circle cut out, over the rear pump body and then install the rear cover -- with the single crankcase pulse fitting.

Hold these parts together and turn the pump over. Install the second diaphragm, followed by the partitioned gasket and outer pump cover onto the front side of the pump body.

Check to be sure the holes for the screws are all aligned through the covers, diaphragms, and gaskets. If the diaphragms are not properly aligned, a tear would surely develop when the screws are installed.

Install the three black Phillips head screws through the various parts and tighten the screws securely.

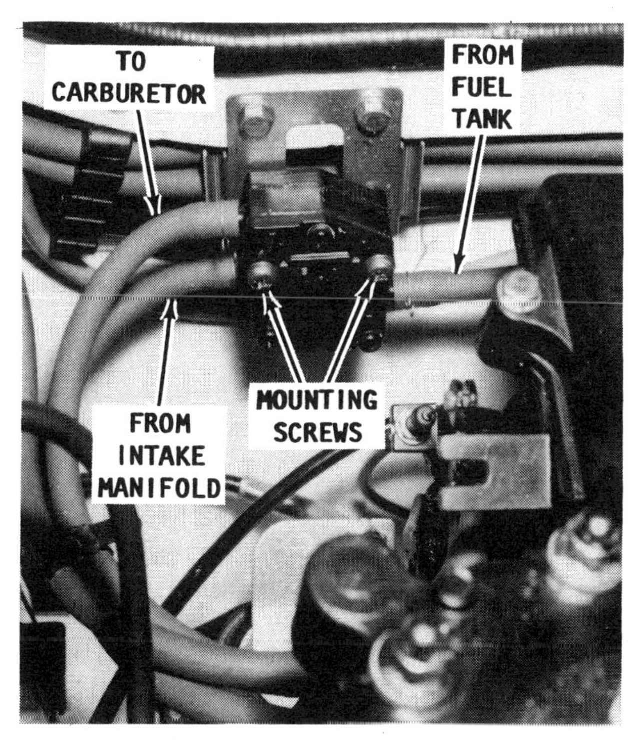

Hose identification for the remote fuel pump. In most cases, the length and shape of the hose will help identify.

Secure the pump to the bracket with the two silver Phillips head screws.

Connect the fuel line from the filter to the fitting on the lower right side. This fitting is embossed with an arrow pointing inward -- toward the center of the pump. Connect the fuel line from the carburetor to the upper left fitting -- the fitting embossed with an arrow facing away from the pump. Connect the crankcase pulse hose to the lower left fitting. This fitting is embossed with the "infinity" sign (∞). Secure the fuel and pulse hoses with wire type clamps.

Start the engine and check the completed work. Be sure to inspect all fuel connections for leaks.

CAUTION
Water must circulate through the jet pump -- to and from the engine, anytime the engine is operating. Circulating water will prevent overheating -- which could cause damage to moving engine parts and possible engine seizure.

NEVER, AGAIN NEVER, operate the engine at high speed with a flush device attached. An engine operating at high speed with such a device attached, would **RUNAWAY** from lack of a load on the impeller shaft, causing extensive damage.

6-9 OIL INJECTION

Oil injection is standard equipment on all series engines covered in this manual. This oil injection system replaces the age-old method of manually mixing oil with the fuel for lubrication of internal moving parts in a two-stroke engine.

Actually two different oil delivery methods are used on the models covered in this manual. One method delivers the oil directly into the intake manifold -- Model 650 series. The other method delivers oil to each carburetor where it is mixed with the fuel before entering the cylinder.

The rate of oil delivery is controlled by engine rpm because the oil pump is directly coupled to the forward end of the crankshaft. Therefore, as engine rpm increases the demand for more oil is met because the crankshaft is rotating faster -- driving the pump at a higher speed. The oil delivery rate is approximately 40:1 at low and idle speed.

Oil Mixture

Fill the oil tank to the top line embossed on the tank. Use only two-cycle NMMA engine oil (or an equivalent oil with a BIA certified rating TC-W). This oil is suitable for use through a temperature range of 14°F to 140°F (-10° to 60°C).

Break-In Period

In order for the engine to receive extra lubrication during the first 10 hours of break-in, the manufacturer recommends oil be added directly into the fuel tank to give a mixture of 25:1.

Professional mechanics, who own oil injected units, constantly add an extra ounce of oil per each gallon of fuel after the break-in period. This extra oil may prevent engine seizure in the event of an oil pump failure or clog in an oil delivery line.

Mechanics are quite cost conscious. They realize all too well how much cheaper replacing fouled spark plugs can be when weighed against an engine overhaul to replace expensive parts damaged from lack of adequate lubrication.

Location of the oil injection pump mounted on the flywheel cover on a Model 650 engine. The oil tank is to the left -- slightly out of focus.

Typical oil tank installation -- this one on a Model 900 Series. Major parts of the tank are shown in the line drawing in the adjacent column.

Use only outboard marine oil in the mixture, never automotive oils. Four-stroke automotive engine oil is not formulated to burn completely, only to lubricate. Therefore, automotive oil, if used in a two-stroke engine, will leave undesirable residue.

SYSTEM COMPONENTS

Principle components of the oil injection system include:

- Main oil tank.
- Oil gauge -- some models.
- Two one-way check valves.
- A breather valve.
- An oil injection pump.
- A network of hoses.

Oil Tank

The oil tank is mounted in different locations in the engine compartment -- depending on the series model being serviced -- with the filler easily accessible when the engine cover is removed. Capacity of the oil tank is as follows:

Model 650	2.9 qts (2.7L)
Model JH750	3.5 qts (3.3L)
Model JT750	2.9 qts (2.7L)
Model 900	3.5 qts (3.3L)
Model 1100	3.5 qts (3.3L)

Oil supply to the oil injection pump is gravity fed from the oil tank. Therefore, a one-way check valve is mounted on top of the tank. The other end of the valve is open to the atmosphere. The check valve allows air to enter the tank replacing oil used, but will not allow oil to flow out of the tank -- making the tank "spill-proof".

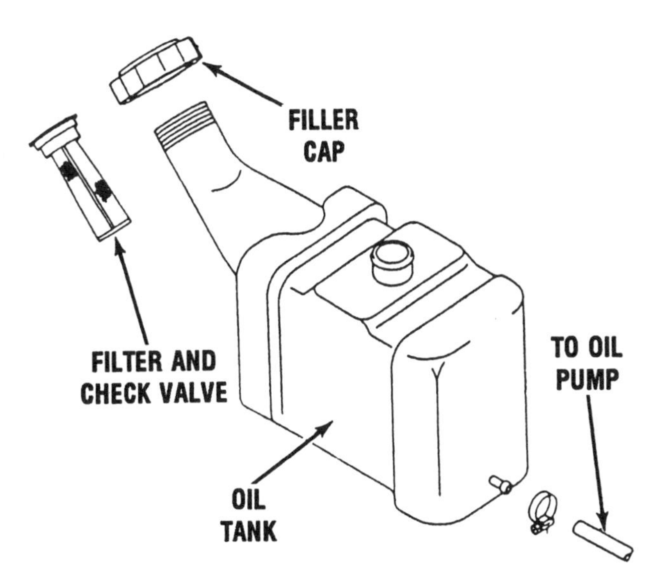

Simple line drawing to depict the oil tank, and associated parts.

Oil Injection Pump

The oil injection pump is a positive displacement type unit driven by a short shaft indexed into the crankshaft. With each revolution of the crankshaft the short shaft moves a plunger cam up and down, pumping oil through transparent hose/s to the intake manifold or carburetor/s, depending on the engine series.

System Inspection

Remove the oil tank filler cap. Inspect the condition of the seal beneath the cap, and replace the seal if it is worn or cracked.

Disconnect the ventilation hose from the fitting on the oil tank.

To test operation of the check valve:

First, pull off the two hoses and blow through the fitting centered on one face of

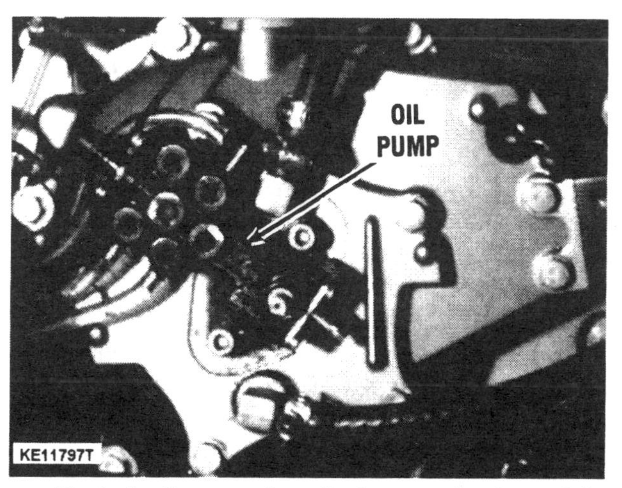

Typical oil pump installation on the face of the flywheel cover.

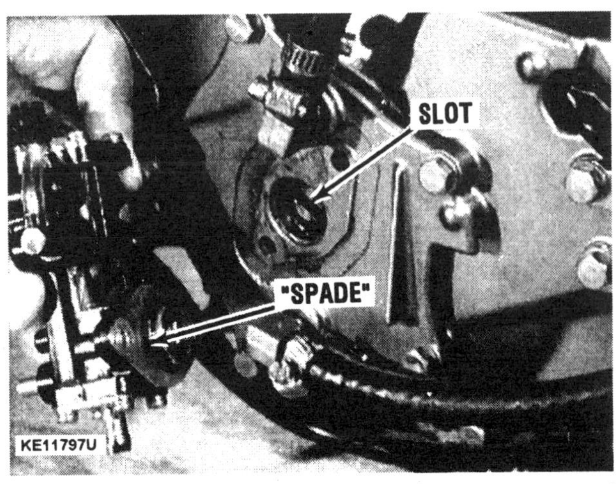

Oil pump removed showing the slot in the forward end of the crankshaft and the "spade" on the pump shaft. The "spade" indexes into the crankshaft slot.

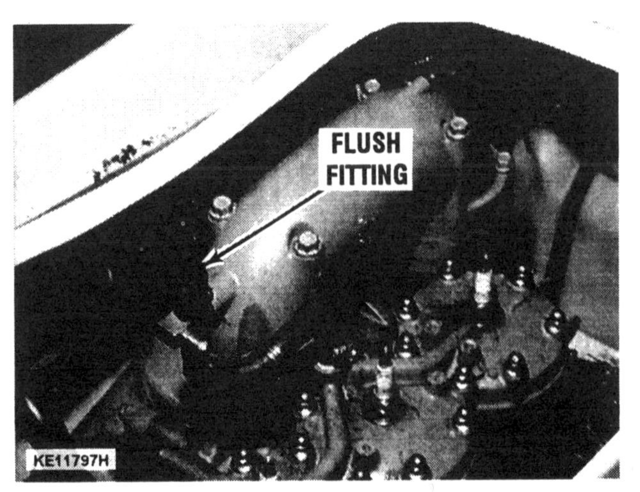

Typical location of the fitting for a garden hose attachment to provide circulating water to the engine when it is being operated during testing.

the valve, in the direction of the embossed arrow. The air should pass through the valve **ONLY** in this direction. Now, blow through the fitting which is off-center, on the other face of the valve. The valve should prevent air flow in this direction.

If the valve passes both tests, install the valve with the embossed arrows toward the tank. If the valve fails one or both tests, replace the valve.

If the arrows are missing, the fitting in the center of the valve face is the air "in" fitting -- air should pass through in only one direction. The off-center fitting should have the hose connected to the oil tank.

A faulty check valve will starve the engine of lubrication and quickly cause extensive and expensive damage.

Oil Pump Output Test

Prepare about a ten minute supply of 40:1 pre-mix using a good grade of two-stroke oil. "Jury-rig" a setup to provide this mixture to the fuel pump during this purging procedure. Check the level of oil in the oil tank and replenish as necessary.

Obtain a transparent measuring cylinder graduated in mL. If such a container is not available, use a measuring cup graduated in ounces. A conversion table on the first page of the Appendix may be used to convert ounces to milli-litres.

Start the engine and allow it to warm to operating temperature.

CAUTION

Water must circulate through the jet pump -- to and from the engine, anytime the engine is operating. Circulating water will prevent overheating -- which could cause damage to moving engine parts and possible engine seizure.

NEVER, AGAIN NEVER, operate the engine at high speed with a flush device attached. An engine operating at high speed with such a device attached, would **RUNAWAY** from lack of a load on the impeller shaft, causing extensive damage.

SPECIAL WORDS

During break-in or after long periods of storage, the oil/fuel and oil premix being fed through the carburetor will cause the engine to operate on a heavy mixture of oil.

Using a pair of needlenose pliers, squeeze the clamp on the transparent delivery hose. Push the clamp up the hose. Gently pull the hose from the fitting on the intake manifold and immediately hold it over the container.

Observe the flow of oil for four full minutes to verify the pump is functioning properly.

With the engine operating at 3000 rpm for 2 minutes, the pump will deliver 5.9-7.2mL for 650 Series and 750 Series engines; 5.2-6.3mL for 900 Series engines and 10.1-12.3mL for 1100 Series engines.

More Special Words

The oil/fuel mixture prepared earlier will provide adequate lubrication while the system is being tested.

A steady slow pulsing flow with no air bubbles may be expected. Reconnect the line. Shut down the engine and remove the flushing device.

Replenish the oil tank with two-stroke oil.

Troubleshooting

Unfortunately, most problems with the oil injection system are only discovered after it is too late. If either of the two hoses becomes clogged, kinked, or restricted, the oil delivery to the cylinders will be reduced.

Insufficient oil delivery to the engine will cause the oil level to drop slower than normal; the engine will overheat due to inadequate lubrication; and moving internal parts will wear more quickly or be severely damaged.

Excessive oil delivery to the cylinders will cause the oil level to drop at a much faster rate than normal. The engine will smoke -- especially at idle speed -- and the spark plugs will become fouled, causing the engine to misfire.

First Checks
Oil Delivery

If excessive or reduced oil delivery occurs, as described above, one or more of the following areas may require immediate attention:

Defective oil pump.
Clogged oil tank check valve.
Kinked oil hose.
Blocked oil hose.
Air in the system.
Leaking oil hose.
Improper idle fuel adjustment.
Poor quality of fuel.
Poor quality of two-stroke oil.

BAD NEWS

No overhaul kits or spare parts are available for the oil pump used on these units. Therefore, a defective oil pump must be replaced.

Systematically check each oil line and connection. A free flow of oil through an unrestricted line is **CRITICAL** for adequate engine lubrication.

CAUTION

Any time the oil tank hose is disconnected, the oil injection pump **MUST** be purged ("bled") of any trapped air. Failure to "bleed" the system could lead to engine seizure due to lack of adequate lubrication.

Instructions for "bleeding" -- purging -- air from the system follows.

Purging Air From Oil Injection System

The following procedures are to be performed any time the oil injection system has been opened (other than to add oil to the tank) and air has entered the system.

Prepare about a ten minute supply of 40:1 premix using two-stroke oil. "Jury-rig" a setup to provide this mixture to the fuel pump during this purging procedure. The oil/fuel mixture prepared will provide adequate lubrication while the system is being purged. Check the level of oil in the oil tank and replenish as necessary.

Start the engine and allow it to warm to operating temperature.

CAUTION

Water must circulate through the jet pump -- to and from the engine, anytime the engine is operating. Circulating water will prevent overheating -- which could cause damage to moving engine parts and possible engine seizure.

NEVER, AGAIN NEVER, operate the engine at high speed with a flush device at-

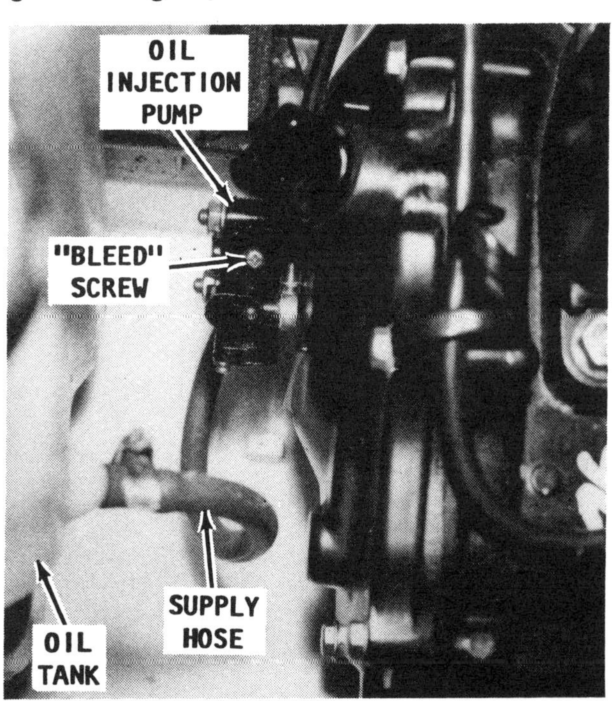

Location of the "bleed" screw on the oil injection pump on a Model 650 Series engine. Location of the "bleed" bolt on other model engines is shown at the top of the next page.

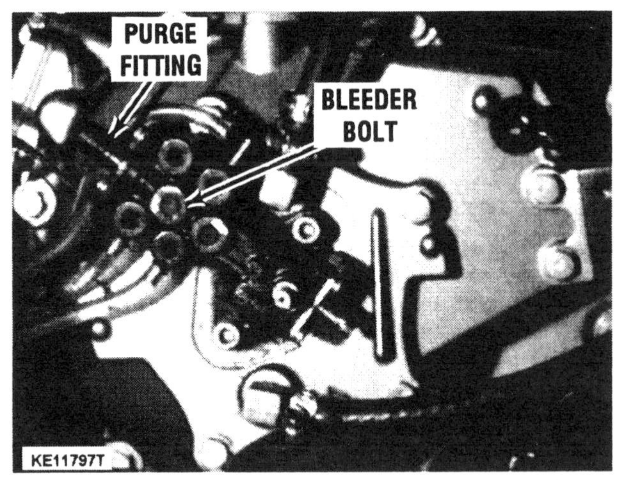

Close view of the oil pump to clearly show the air bleeder bolt. This bolt is only removed when it is necessary to "bleed" (purge), air from the system or from the pump and the line to the tank, as explained in the text.

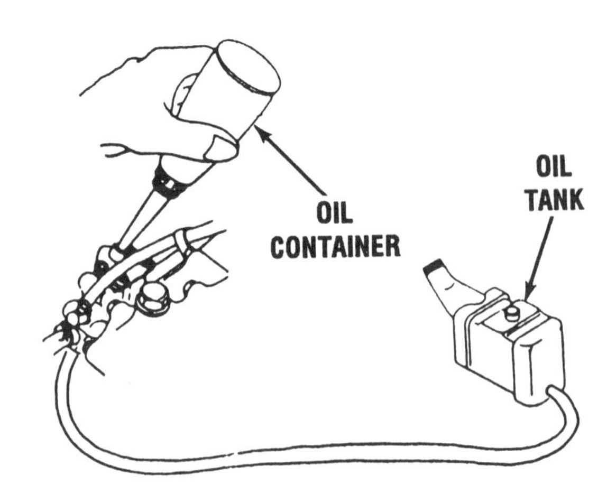

*Simple line drawing to depict the method of "bleeding" (purging), air from the oil pump and the line to the oil tank. The container **MUST** be squeezed in one continuous motion. The text explains why.*

tached. An engine operating at high speed with such a device attached, would **RUNAWAY** from lack of a load on the impeller shaft, causing extensive damage.

SPECIAL WORDS

Because the fuel and oil premix is being fed through the carburetor, in addition to oil from the oil injection system also being fed directly into the manifold, the engine will be operating on a heavy mixture of oil.

Obtain a small container and have a shop towel handy to catch any oil drips. Loosen the bleed screw on the oil pump a few turns and wait until a slow pulsing flow of oil emerges without any air bubbles. Observe the flow of oil for four full minutes to verify the pump is functioning properly and all air has been expelled from the system. Tighten the screw securely.

Replenish the oil tank with two-stroke oil.

Purging Air From The Oil Pump and Line To The Tank

Obtain a container which is soft enough to permit being squeezed by hand. Obtain some type of nozzle which may be inserted into the oil pump bleed fitting.

Fill the container with roughly 3.5 ounces of recommended two-stroke engine oil.

Remove the air bleeder bolt on the oil pump.

Loosen the oil tank cap slightly to permit air to escape.

Now, inject oil slowly through the oil pump bleed fitting by squeezing the container with one continuous motion. **DO NOT** stop halfway or at any time. This amount of oil is enough to fill the line from the oil pump to the oil tank and means air trapped within the pump and any air in the line will be expelled back into the tank.

CRITICAL WORDS

Squeezing the container intermittently could permit air to enter the oil line and defeat the purpose of attempting to purge air from the pump and the line. Any air left in the pump or the line could cause reduction in oil flow and possible damage to the engine through lack of adequate lubrication.

Install and tighten the air bleeder bolt. Tighten the oil tank filler cap.

7
IGNITION

7-1 INTRODUCTION AND CHAPTER COVERAGE

The less a marine engine is operated, the more care it needs. Allowing a marine engine to remain idle will do more harm than if it is used regularly. To maintain the engine in top shape and always ready for efficient operation at any time, the engine should be operated every 3 to 4 weeks throughout the year.

The carburetion and ignition principles of two-cycle engine operation **MUST** be understood in order to perform a proper tune-up on a marine engine.

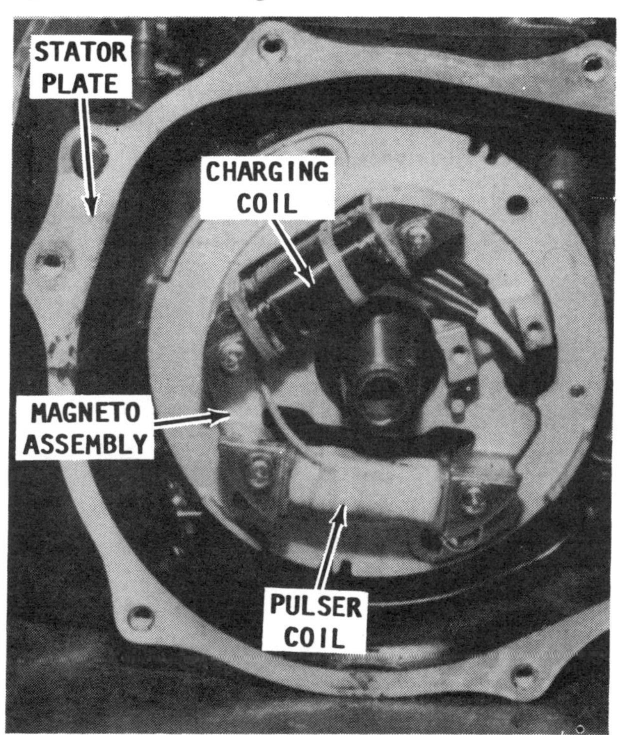

The ignition coil (pulser) coil and the charging coil on a Model 550 and 650 Series in mounted and well protected on the magneto assembly behind the flywheel.

If you have any doubts concerning your understanding of two-cycle engine operation, it would be best to study the operation theory section in the first portion of Chapter 8, before tackling any work on the ignition system.

A Capacitor Discharge Ignition (CDI) is used on all engine models covered in this manual.

7-2 SPARK PLUG EVALUATION

Removal

Remove the spark plug wires by pulling and twisting on only the molded cap. **NEVER** pull on the wire or the connection inside the cap may become separated or the boot damaged. Remove the spark plugs. **TAKE CARE** not to tilt the socket as you remove the plug or the insulator may be cracked.

Examine

Carefully examine all plugs to determine the firing conditions in the cylinder. If the side

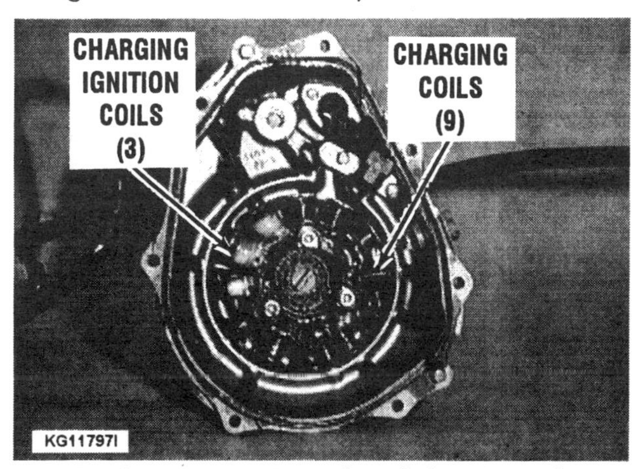

The charging ignition coils and charging coils on a Model 750, 900 and 1100 Series are mounted on the stator plate inside and protected by the flywheel cover.

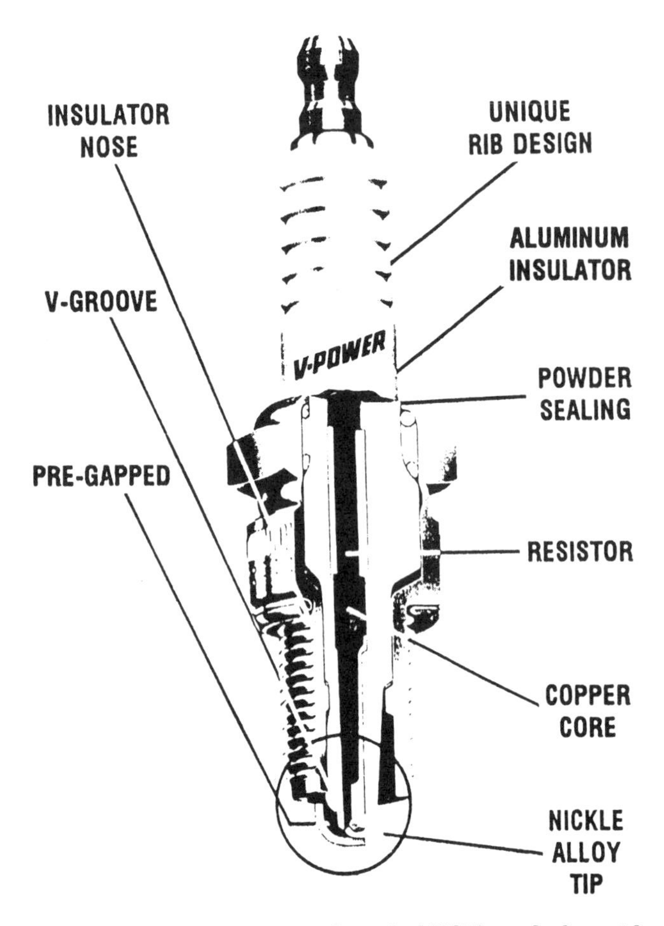

Cross-section drawing of a typical NGK spark plug with some of the features identified.

electrode is bent down onto the center electrode, the piston is traveling too far upward in the cylinder and striking the spark plug. Such damage indicates the piston pin or the rod bearing is worn excessively.

In most cases, an engine overhaul is required to correct the condition. To verify the cause of the problem, rotate the flywheel by hand. As the piston moves to the full up position, push on the piston crown with a screwdriver inserted through the spark plug hole, and at the same time rock the flywheel back-and-forth. If any play in the piston is detected, the engine must be rebuilt.

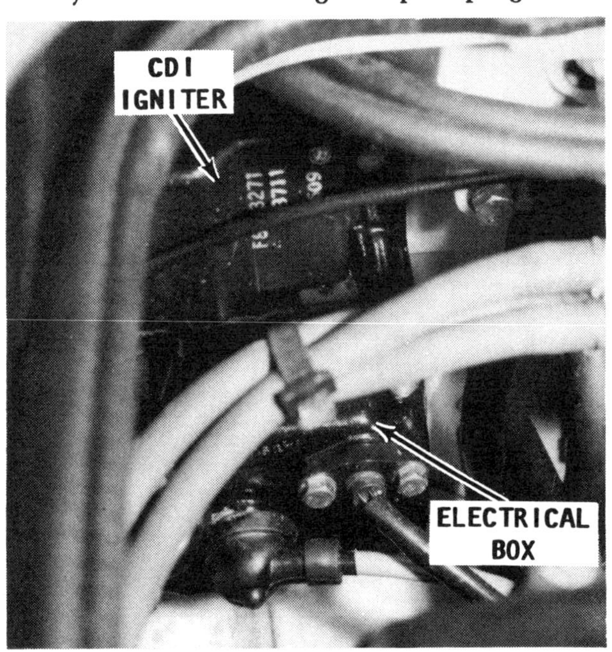

The combination CDI module and ignition coil (known as the CDI igniter), is mounted above the electrical box in the engine compartment on the Model 650 Series engine.

Correct Color

A proper firing plug should be dry and powdery. Hard deposits inside the shell indicate too much oil is being mixed with the fuel. The most important evidence is the light gray to tan color of the porcelain, which is an indication this plug has been running at the correct temperature. This means the plug is one with the correct heat range and also that the air/fuel mixture is correct.

Rich Mixture

A black, sooty condition on both spark plug shells and the porcelain is caused by an excessively rich air/fuel mixture, both at low and high speeds. The rich mixture lowers the combustion temperature so the spark plug does not run hot enough to burn off the deposits.

Deposits formed only on the shell is an indication the low-speed air/fuel mixture is too rich. At high speeds with the correct

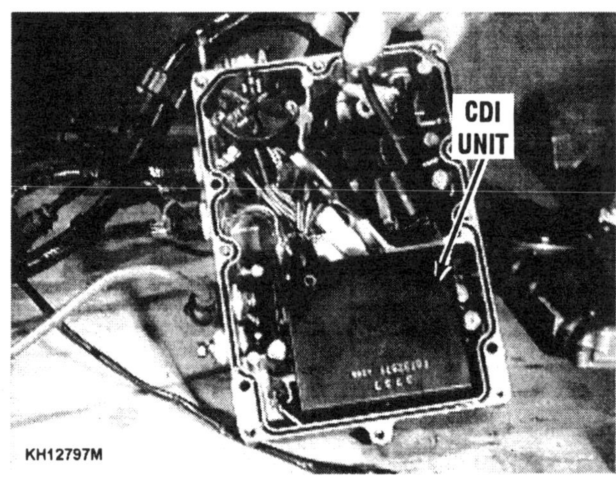

*The CDI unit on a Model **750**, **900** and **1100** Series engine is mounted inside and protected by the electrical box. The electrical box also houses the fuse holder, the cranking motor relay -- under the CDI unit -- the regulator/rectifier and other electrical components.*

mixture, the temperature in the combustion chamber is high enough to burn off the deposits on the insulator.

Too Cool

A dark insulator, with very few deposits, indicates the plug is running too cool. This condition can be caused by low compression or by using a spark plug of an incorrect heat range. If this condition shows on only one plug it is most usually caused by low compression in that cylinder. If both plugs have this appearance, then it is probably due to the plugs having a too-low heat range.

Fouled

A fouled spark plug may be caused by the wet oily deposits on the insulator shorting the high-tension current to ground inside the shell. The condition may also be caused by ignition problems which prevent a high-tension pulse from being delivered to the spark plug.

Carbon Deposits

Heavy carbon-like deposits are an indication of excessive oil in the fuel. This condition may be the result of worn piston rings or excessive ring end gap.

Overheating

A dead white or gray insulator, which is generally blistered, is an indication of overheating and pre-ignition. The electrode gap wear rate will be more than normal and in the case of pre-ignition, will actually cause the electrodes to melt. Overheating and pre-ignition are usually caused by overadvanced timing, detonation from using too-low an octane rating fuel, an excessively lean air/fuel mixture, or problems in the cooling system.

*Damaged spark plugs. Notice the broken electrode on the left plug. The missing part **MUST** be found and removed before returning the engine to service, to prevent serious damage to expensive internal parts.*

This spark plug is foul from operating with an overrich air/fuel mixture, possibly caused by an improper carburetor adjustment.

Electrode Wear

Electrode wear results in a wide gap and if the electrode becomes carbonized it will form a high-resistance path for the spark to jump across. Such a condition will cause the engine to misfire during acceleration. If both plugs are in this condition, it can cause an increase in fuel consumption and very poor performance at high-speed operation. The solution is to replace the spark plugs with a rating in the proper heat range and gapped to specification.

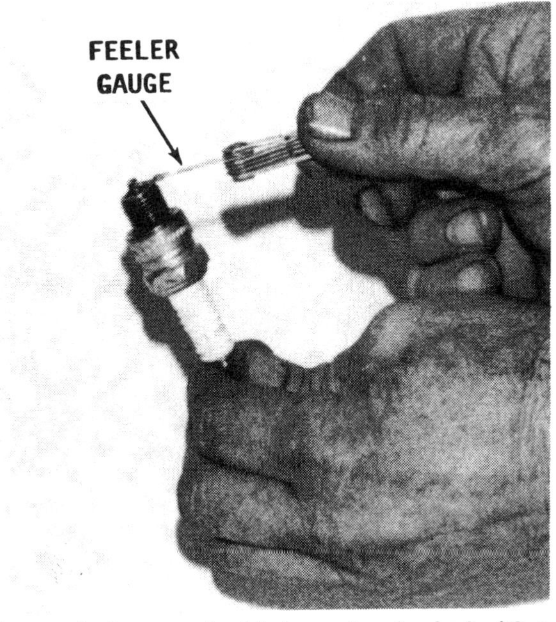

The spark plug gap should always be checked with a wire-type feeler gauge before installing new or used plugs.

Red rust-colored deposits on the entire firing end of a spark plug can be caused by water in the cylinder combustion chamber. This can be the first evidence of water entering the cylinders through the exhaust manifold because of an accumulation of scale. This condition **MUST** be corrected at the first opportunity. Refer to Chapter 8, Engine.

7-3 CDI (CAPACITOR DISCHARGE IGNITION) AND CHARGING SYSTEM

Model 550 and 650 Series

On the Model 550 and 650 Series engines about half of the ignition circuit is housed and well protected behind the flywheel. The remainder of the system components are mounted inside the engine compartment.

The charging coils are mounted next to the pulser coils behind the flywheel. Therefore, the charging system is presented in this chapter and in this section because once the flywheel is removed, the components for both systems are exposed.

Model 750, 900, and 1100 Series

On the engines listed in the heading, the stator is in front of the flywheel and protected by the flywheel cover. The remainder of the ignition components are housed and protected inside an electrical box.

The charging coils on the Model 750, 900, and 1100 series are mounted on the inside of the flywheel cover. This means it is not necessary to remove the flywheel in order to service these components.

The ignition coils on the Model 750, 900 and 1100 Series are mounted outside the electrical box.

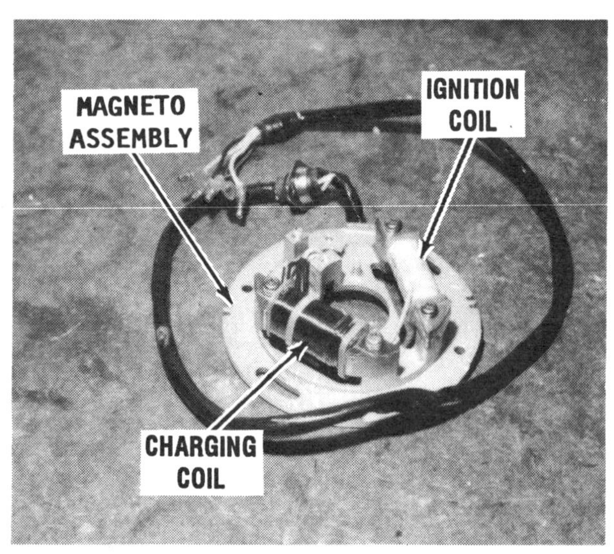

Ignition coil and charging coil installed on the magneto assembly and ready for installation on a Model ***650*** *Series engine.*

DESCRIPTION AND OPERATION IGNITION CIRCUIT

The following paragraphs describe -- in "layman's" language -- the CDI system and how it functions. Every effort has been made to keep "technical" terms to a minimum.

Basically, two major components comprise the CDI system -- namely an exciter coil and a "black box" CDI igniter. On the Model 550 and 650 Series engines, the exciter coil is housed behind the flywheel and the CDI igniter is mounted next to an electrical box mounted in the engine compartment.

The CDI igniter on the Model 750, 900 and 1100 Series engines is housed inside an electrical box mounted in the engine compartment.

The exciter coil is sometimes called a "trigger coil". Some series units have a pulser coil in the ignition circuit. The exciter coil and the trigger coil provide a source of AC current and are used to sequentially time electrical impulses.

As mentioned, the CDI igniter is a sealed "black box" type unit and is a combination CDI module and ignition on a Model 550 and 650 series engine.

The ignition module circuitry converts AC current to DC current. The module stores this DC current and releases it to the ignition coil automatically at the correct time.

The three charging ignition coils and nine charging coils are clearly visible, once the flywheel cover has been removed on a Model ***750, 900*** *or* ***1100*** *Series engine. Notice the flywheel is still in place.*

A series of diodes, an SCR gate, and a capacitor comprise the module circuitry. A diode is a solid state device designed to allow current flow in one direction but prevents flow in the opposite direction. A SCR gate is a solid state electronic switching device which permits voltage to flow only after it has been triggered by another source.

The ignition coil circuitry boosts the DC voltage instantly to approximately 20,000 volts. This high voltage is passed on to the spark plugs through the high tension leads.

Operation

Consider the engine operating and the flywheel rotating. At the point in time when the ignition timing marks align, an alternating coil is induced in the exciter coil. This voltage charges a capacitor in the CDI igniter circuit.

As the flywheel continues to rotate, another voltage is induced in the pulser coil, if so equipped. If the system does not have a pulser coil, the same voltage is induced in the exciter coil.

This voltage flows through a diode and on to the SCR gate. The SCR gate signals the capacitor to discharge the voltage to the primary coils inside the CDI igniter. The voltage in the primary windings of the ignition coil induces a high voltage in the secondary windings and causes a spark to jump to ground across the spark plug electrodes.

In this manner, a spark at the spark plug may be accurately timed by the marks on the flywheel relative to the magnets in the flywheel to provide as many as 100 sparks per second for an engine operating at about 6,000 rpm.

*The CDI igniter on a Model **550** Series engine is a combination of CDI module and ignition coil.*

SPECIAL TIMING WORDS

Only the Model 550 and 650 Series have adjustable timing. The figures for all other models are presented in this table for reference only. The timing is set at the factory and is non-adjustable. If an ignition timing problem is suspected, the "black box" must be replaced.

The stator on the Model **550** and **650** Series has slots opposite each other for the mounting bolts. These slots permit the stator to be rotated slightly in order to obtain the correct degree of timing.

Detailed ignition timing procedures for the Model 550 and 650 Series is presented in Section 7-5 of this chapter beginning on Page 7-14.

TROUBLESHOOTING IGNITION CDI SYSTEM

Always attempt to proceed with the troubleshooting in an orderly manner. The "shot in the dark" approach will only result in wasted time, incorrect diagnosis, replacement of unnecessary parts, and frustration.

Begin the ignition system troubleshooting with the spark plug and continue through the system until the source of trouble is located.

Spark Plugs

1- Check the plug wires to be sure they are properly connected. Check the entire length of the wires from the plugs to the magneto assembly on the stator plate. If the wires are to be removed from the spark plug, **ALWAYS** use a pulling and twisting motion as a precaution against damaging the connection.

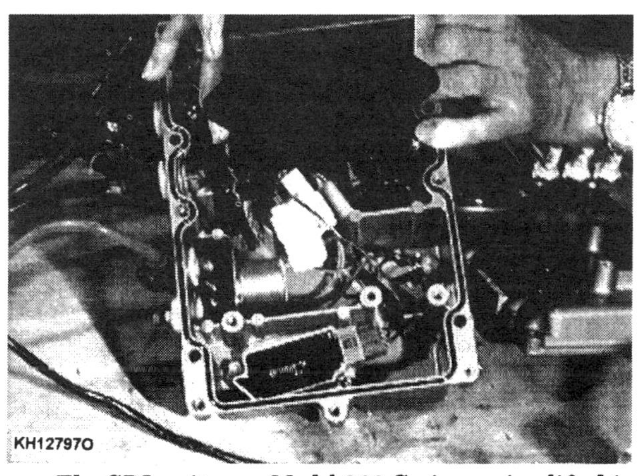

*The CDI unit on a Model **900** Series engine lifted in preparation to removing from the electrical box.*

Attempt to remove the spark plug by hand. This is a rough test to determine if the plug is tightened properly. The attempt to loosen the plug by hand should fail. The plug should be tight and require the proper socket size tool. Remove the spark plug and evaluate its condition as described in Section 7-2.

2- Use a spark tester and check for spark. If a spark tester is not available, hold the plug wire about 1/4" (6.4mm) from the engine. Crank the engine through a few revolutions using the cranking motor and check for spark. A strong spark over a wide gap must be observed when testing in this manner, because under compression a strong spark is necessary in order to ignite the air/fuel mixture in the cylinder. This means it is possible to think a strong spark is present, when in reality the spark will be too weak when the plug is installed. If there is no spark, or if the spark is weak,

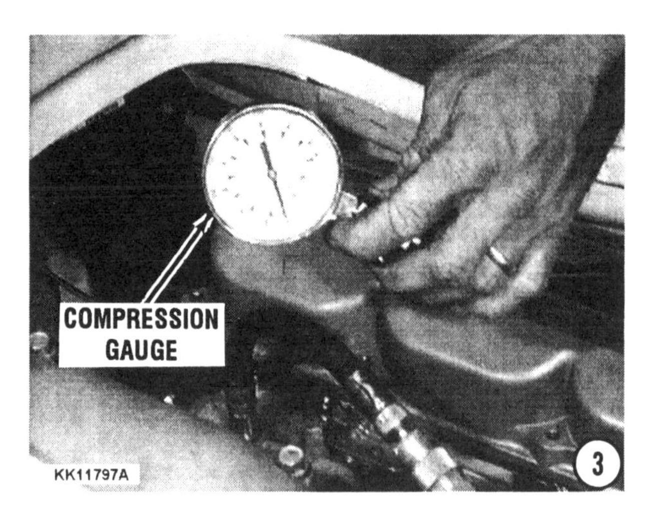

the trouble is most likely under the flywheel in the stator assembly.

Compression

3- Before spending too much time and money attempting to trace a problem to the ignition system, a compression check of the both cylinders should be made. If a cylinder does not have adequate compression, troubleshooting and attempted service of the ignition or fuel system will fail to give the desired results of satisfactory engine performance.

Remove the spark plug wire by pulling and twisting **ONLY** on the molded cap. **NEVER** pull on the wire because the connection inside the cap may be separated or the boot may be damaged. Remove the spark plug. Insert a compression gauge into the cylinder spark plug hole. Ground the spark plug leads to prevent damage to the

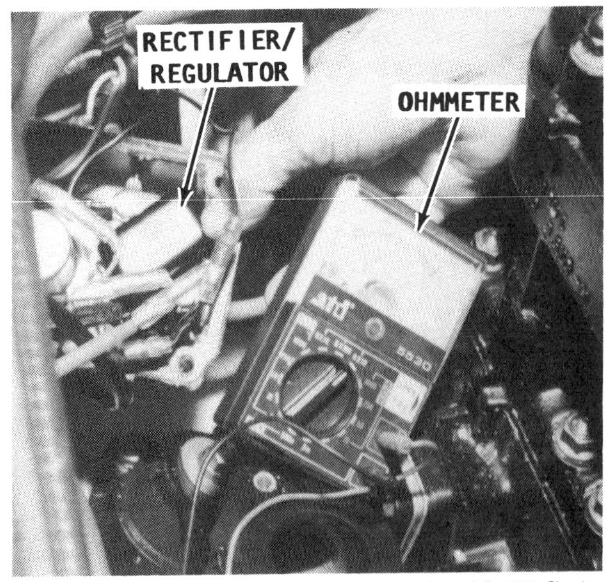

Testing the rectifier/regulator on a Model 650 Series engine after making a visual inspection first.

ignition coil. If a lead is not grounded, the coil will attempt to match the demand created by the spark trying to jump from the electrode to the nearest ground.

Crank a cold engine through several revolutions with the electric cranking motor. Note the compression reading.

The following table gives factory suggested acceptable pressures for the Model Series listed.

FACTORY ACCEPTABLE PRESSURES

MODEL/SERIES	PRESSURE RANGE
550SX	78 - 125 psi
650X-2	125 - 192 psi
750SX	121 - 187 psi
750ZXi	83 - 135 psi
750ST & STS	129 - 199 psi
750XiR	129 - 199 psi
900STX	83 - 135 psi
1100ZXi	95 - 151 psi

The differences of readings between cylinders is actually more important than the actual numerical number, but should not vary by more than 10%.

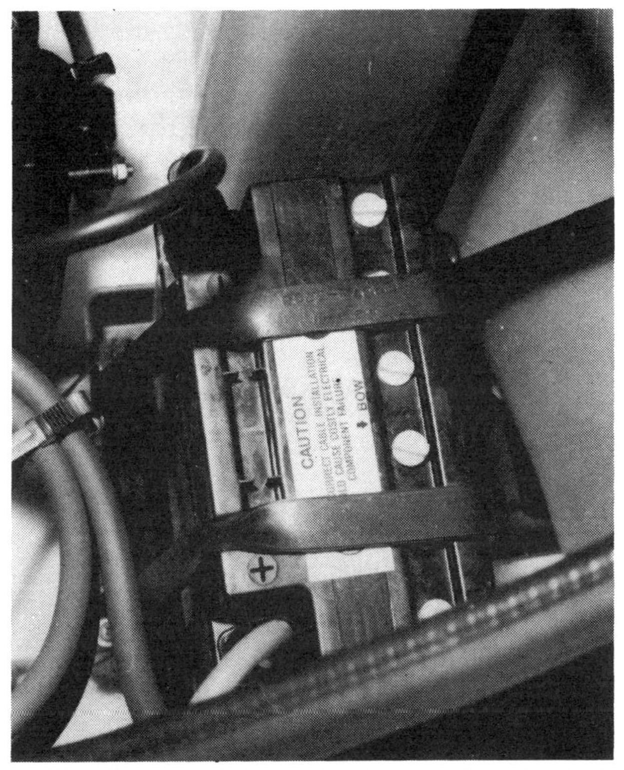

***NEVER** attempt to verify the charging circuit by operating the engine with the battery disconnected. Such action will damage the rectifier/regulator.*

7-4 TESTING IGNITION COMPONENTS

General Information

Due to the complex nature of the circuitry and high energy output pulses of microsecond duration, conventional testing devices such as a Volt/Ohm/Ammeter will not measure electrical output with the degree of accuracy required. A Kawasaki Ignition Tester, as used by the local Kawasaki dealer is an electrical device capable of measuring the peak energy output of capacitor discharge ignition units, as well as the exciter and pulser coils. This instrument was designed to troubleshoot the CDI ignition system. Unfortunately this tester may not be easily accessible. Therefore, this chapter includes only those tests which may be performed with instruments normally and easily available.

WORDS ON TESTERS

Resistance tests performed with a Volt/Ohmmeter are not as accurate or as conclusive as tests performed using a Kawasaki tester.

To replace a component on a Model 550 or 650 Series engine, the flywheel must be "pulled". See Section 8-3, Chapter 8, beginning on Page 8-9.

To replace components on other models covered in this manual, it is not necessary to "pull" the flywheel, only the flywheel cover.

WORDS FROM EXPERIENCE

During the tests, if a reading is slightly different from the specifications, but the engine still operates, then there is no real need to replace the affected component, until it actually fails. Bear in mind, in **MOST** cases electrical components are not returnable, once the item leaves the store. Therefore, make every attempt to avoid the **BUY** and **TRY** method of troubleshooting. Such a practice will lead only to wasted time, unneeded cash outlay, and frustration.

Intermittent or Multiple Problems

Many ignition problems occur only during engine operation -- when the component is subject to vibration, the engine is operating under a load, and/or there is a marked increase in temperature -- from the engine operating.

SPECIAL WORDS ON IGNITION AND CHARGING CIRCUIT TESTING

Most of the components in the ignition and charging circuits have leads with quick-disconnect fittings or harness plugs located inside the electrical box. Therefore, testing begins inside the electrical box.

Some components in these cirucits are housed inside the box -- others are housed inside the flywheel cover or behind the flywheel. If a component inside the electrical box is found to be defective, it may be replaced immediately -- usually through attaching hardware. If a component under the flywheel on a Model 550 or 650 Series engine -- the exciter, pulser, or charge coil -- is found to be defective, as mentioned earlier, the flywheel must be "pulled" in order to remove and replace the item. See Section 8-3, Chapter 8, beginning on Page 8-9 for instructions to "pull" and install the flywheel.

CRITICAL WORDS

Never attempt to verify the charging circuit by operating the engine with the battery disconnected. Such action would force current -- normally directed to charge the battery -- back through the rectifier and damage the diodes in the rectifier.

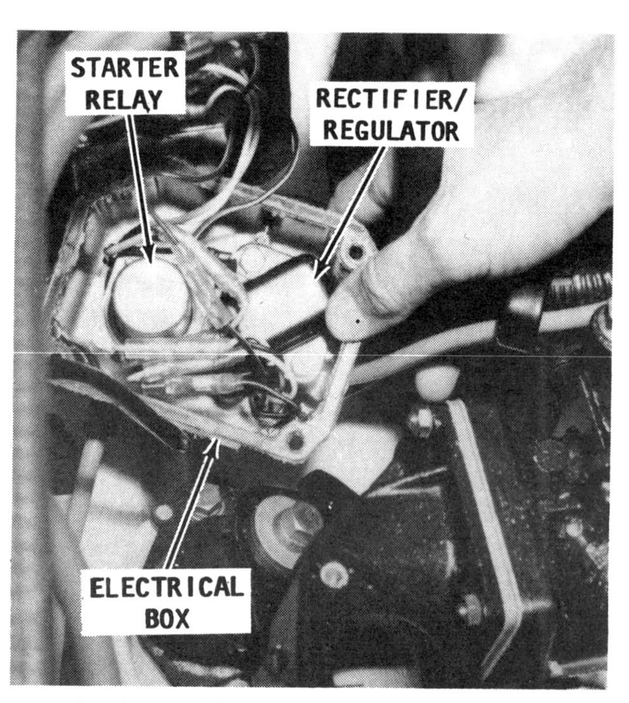

The electrical box in the engine compartment of a Model 650 Series being removed for testing or replacement of components, as described in the text.

Electrical Box Removal
Model 550 and 650 Series

Disconnect the battery ground cable from the battery. Remove the spark plug leads from the spark plugs. Remove the two bolts securing the connector cover to the electrical box. Pull off the cover. Disconnect the 4-pin connector and the four free leads at their quick-disconnect fittings.

Pull back the two "boots" over the large cables at the cranking motor relay. Disconnect both large cables from the relay terminals.

Remove the six bolts securing the electrical box to the engine compartment bulkhead.

Remove the two bolts securing the two halves of the box together, and then open the box.

Electrical Box Removal
All Models Except 550 and 650

1- Drain or siphon the oil from the oil tank. Loosen the clamp and pull the oil level sensor free of the oil tank. Remove the straps securing the oil tank in place. Disconnect the oil pump hose, and then move the oil tank forward, up and over the engine clear of the engine compartment.

Remove the high-tension leads at the spark plugs. Disconnect the positive (+) battery lead from the electrical box. Disconnect the cranking motor lead from the box. Disconnect the electrical box connector, the temperature sensor leads, the Start/Stop switch connector, and the main harness. Remove the four mounting bolts and lift the electrical box up and out of the engine compartment.

2- Once the electrical box has been removed and is on a suitable work surface, remove the oil tank mounting plate.

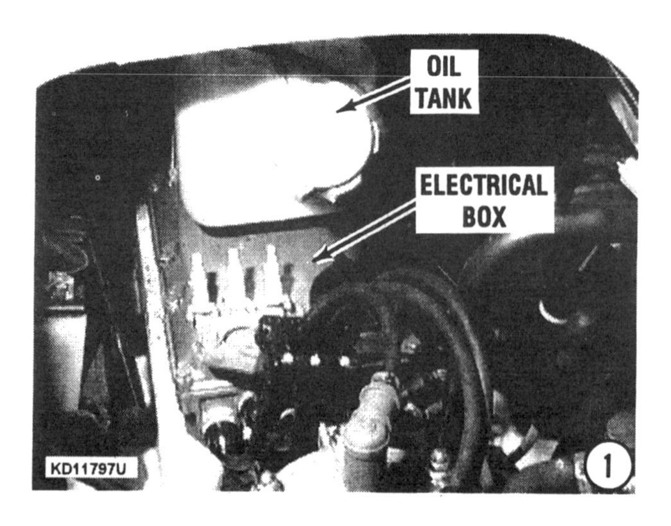

Electrical box removed from a Model 750 Series and on the work bench ready to be opened and components inside tested or replaced.

3- Remove the bolts securing the electrical box cover. Many of the components are now exposed.

4- Remove the hardware securing the CDI unit in place, and then lift the CDI unit to gain access to components housed under the unit.

Exciter Coil
Resistance Test
Model 550 Series

Identify the magneto assembly harness with a Red, Black, Gray and two light Green leads. Disconnect the two halves of the 6-prong plug -- one is empty. Select the Rx100 ohm scale on an ohmmeter.

Make contact with the Red meter lead to the Red male prong in the plug from the magneto assembly. Make contact with the Black meter lead to the Black male prong in the plug from the magneto assembly. The meter should register 112 - 168 ohms.

Pulser Coil
Resistance Test
Model 550 Series

Select the R x 10 ohm scale on the ohmmeter. Keep the Red meter lead on the Red male prong in the plug from the magneto assembly. Make contact with the Black meter lead to the Gray male prong in the plug from the magneto assembly. The meter should register 14.4 - 21.6 ohms.

Coil Results

If the reading is less than expected, a short probably exists in the coil and it must be replaced.

If the coil passes the resistance test, but the problem is still determined to be in the magneto assembly, the cause is most likely a loss of magnetism in the flywheel magnets. The flywheel must be replaced.

Exciter Coil
Resistance Test
Model 650 Only

Obtain an ohmmeter and set the meter to the 100 ohm scale. Make contact with one meter lead to the Black/Red from the exciter coil. Make contact with the other meter lead to the Black/Yellow lead from the exciter coil.

The meter should indicate 250 - 380 ohms. If the meter reading is less than the value just given, the exciter coil most likely has a short. No reading (infinity) indicates the coil has an open circuit. Low reading or no reading means the coil must be replaced.

If the coil is in satisfactory condition, the problem cause is probably a loss of magnetism in the flywheel. The flywheel must be replaced.

CDI Igniter Resistance Test
Model 550 and 650 Series

The secondary winding of the ignition coil portion of the CDI igniter can be checked for a shorted or broken winding using an ohmmeter. Select the R x 1k (1,000) ohm scale on the meter. Make contact with either meter lead to the spark terminal inside the "boot" of one spark plug high tension lead.

Make contact with the other meter lead to the other spark plug terminal in the other high tension lead.

550 Series

The meter should indicate 4.5k - 6.7k ohms.

650 Series

The meter should indicate 2.1k - 3.1k ohms.

If the meter reading is not as listed, replace the CDI igniter.

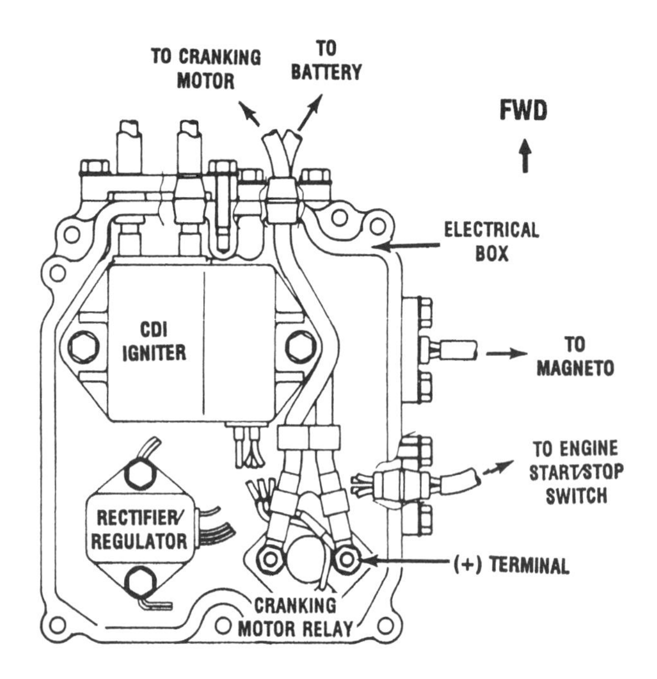

Line drawing to depict interior components of the electrical box for a Model 550 and 650 Series engine.

CDI Igniter Removal
Model 550 & 650 Series

First, remove and open the electrical box, as outlined in Ignition and Charge Circuit Testing, beginning on Page 7-8.

Next, disconnect the 6-pin connector inside the box. For each high-tension lead, back off the Black cap from the White threaded fitting on the side of the electrical box. Pull out the small Black grommet from inside the threaded fitting. The high-tension lead is an integral part of the CDI igniter.

Now, pull off the spark plug "boot" and remove the terminal from the end of the high-tension lead. Remove the two attaching screws and lift the igniter up and out of the electrical box, and at the same time feed the high-tension leads through the White threaded grommet caps in the side of the box.

CDI Igniter Installation
Model 550 and 650 Series

Apply a coating of non-permanent locking substance to the threads of all securing bolts and screws, just prior to installation.

Begin by passing the high-tension leads through the White threaded fittings on the side of the electrical box. Next, position the igniter over the two mounting holes and secure it in place with the two Phillips head screws.

For each high-tension lead: Apply a coating of multi-purpose water resistant lubricant to the small Black grommet, and then install it onto the end of the high-tension lead followed by the Black grommet cap.

Connect the leads between the CDI igniter and the magneto harness 6-prong connector matching color-to-color. Connect the leads from the start switch and the stop switch matching color-to-color.

Install the spark plug terminals onto the ends of the high-tension leads, followed by the spark plug "boot".

Install the electrical box.

Ignition Coil Winding
Resistance Test
Model 550 & 650

Secondary Winding

The ignition coil portion of the CDI igniter can be checked for a broken or badly shorted winding using a standard shop VOA meter, as

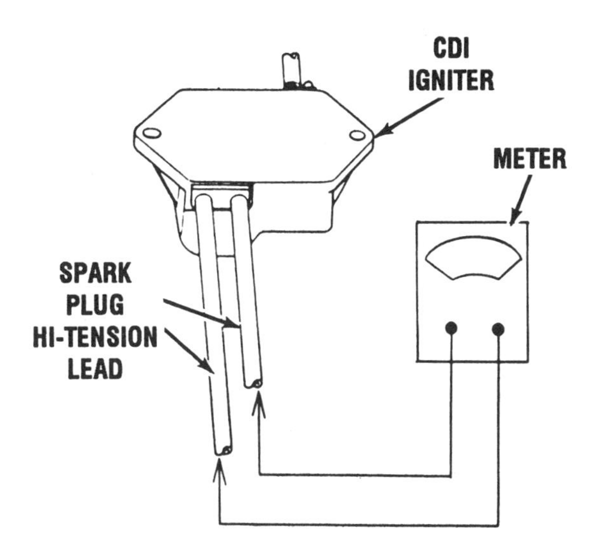

*Simplest line drawing possible to illustrate the meter hookup for testing the CDI Igniter Resistance secondary winding on a Model **550** or Model **650** Series engine.*

depicted in the accompanying line drawing. Such a shop hand tester is not able to detect layer shorts or shorts resulting from insulation breakdown under high voltage.

First, pull the spark plug cap off the high-tension lead.

Next, connect an VOA meter between the spark plug leads, as shown.

Set the meter to the x 1k ohm scale.

For the **550** Series -- the meter should indicate 4.5 - 6.7k ohms.

For the **650** Series -- the ohmeter should indicate 2.1 - 3.1k ohms.

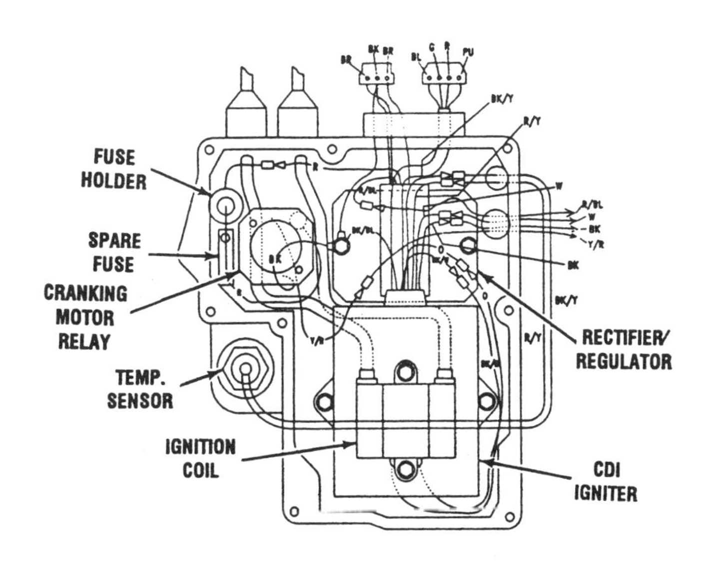

Line drawing to identify components inside the electrical box for a Model 750 SX engine installation.

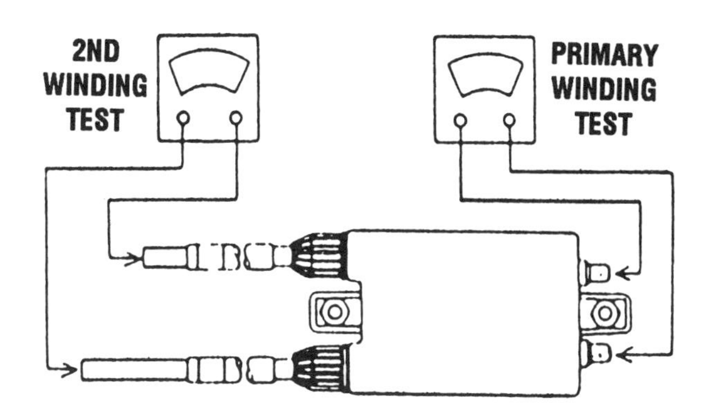

*Line drawing to depict the primary winding and secondary winding ignition coil resistance test hookup for the Model **750** Series.*

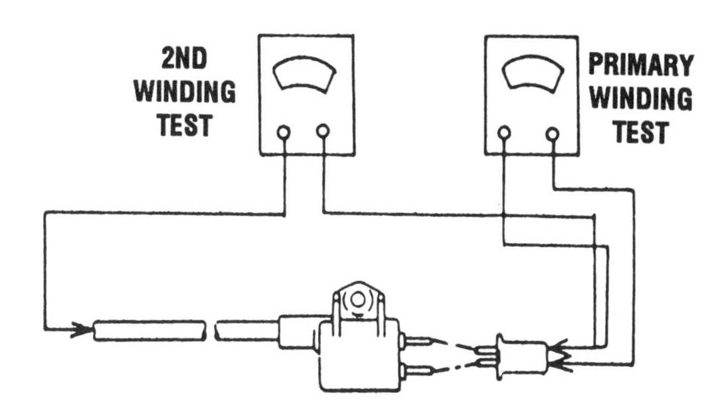

*Line drawing to depict the primary winding and secondary winding ignition coil resistance test hookup for the Model **900** and Model **1100** Series engines.*

Ignition Coil Winding Resistance Test Model 750, 900 & 1100 Series

The ignition coil portion of the CDI igniter can be checked for a broken or badly shorted winding using a standard shop VOA meter, as depicted in the accompanying line drawings. Such a shop hand tester is not able to detect layer shorts or shorts resulting from insulation breakdown under high voltage.

Primary Winding

Begin by disconnecting the primary leads from the terminals on the coil.

Next, connect the meter between the coil terminals.

Set the meter to the x 1 ohm scale. The meter should indicate resistance for the primary winding as follows:

Model **750** Series -- 0.08 - 0.010 ohm.
Model **900** Series -- 0.018 - 0.24 ohm.
Model **1100** Series -- 0.018 - 0.24 ohm.

Secondary Winding

Begin, by rotating the plug caps counterclockwise and removing the caps.

Next, connect the ohmmeter between the spark plug leads.

Set the meter to the x 1k ohm scale. The meter should indicate resistance for the secondary winding as follows:

Model **750** Series -- 3.5 - 4.7k ohms.
Model **900** Series -- 2.7 - 3.7k ohms.
Model **1100** Series -- 2.7 - 3.7k ohms.

GOOD WORDS

If the ignition coil windings passed the above tests, the coil is probably in good condition. Check the spark plug high-tension leads for visible damage and replace as necessary. **HOWEVER,** if the ignition system still fails to perform properly, and all other components have been checked as satisfactory, replace the coil.

Coil removal and installation procedures are outlined in the following paragraphs.

CDI Igniter Removal
Model 650 Series

Begin by opening the electrical box as outlined earlier in this chapter -- Page 7-8. Disconnect the three or four igniter leads at their quick-disconnect fittings. Remove the two Phillips head screws securing the igniter to the electrical box.

Disconnect the two high-tension leads from the spark plugs and feed the grommet, with the small leads attached, through the hole in the electrical box.

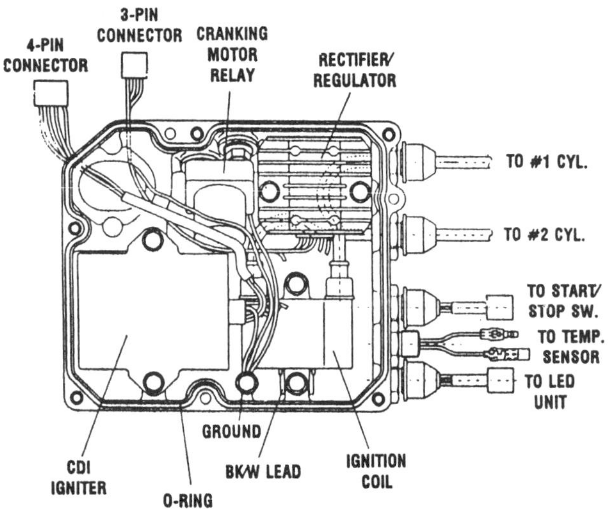

Line drawing to identify components inside the electrical box for a Model 750XiR Series engine. Major ignition items have been identified.

CDI Igniter Installation
Model 650 Series

Apply a coating on non-permanent locking substance to the threads of all securing bolts and screws, just prior to installation.

Insert the grommet, with the CDI leads attached, into the recess in the side of the electrical box. Secure the igniter to the side of the electrical box with the two Phillip head screws. If the unit being serviced has a ground lead, make the connection.

Connect the three or four igniter leads - matching color-to-color. If difficultly is encountered with the lead connections, refer to the wiring diagrams in the Appendix (using a magnifying glass, if necessary), for the unit being serviced. Connect the two high-tension leads to the spark plugs.

Install the electrical box. Tighten the attaching bolts to a torque value of 69 in. lbs (7.8 Nm).

CDI Igniter Removal
Model 750, 900 & 1100

Remove and open the electrical box, as outlined earlier in this chapter -- Page 7-8.

Remove the CDI igniter through the attaching hardware. Separate the connectors and the unit is free.

CDI Igniter Installation
Model 750, 900 & 1100

Apply a non-permanent locking agent to the mounting bolts. Bring the connectors together, and then install and secure the CDI unit with the attaching hardware. Tighten the bolts to a torque value of 69 in. lbs (7.8Nm).

Pickup Coil Resistance
Model 750, 900 & 1100

Remove the electrical box connector. Separate the 6-pin connector.

Obtain an ohmmeter and set it to the x 100 scale range. Make contact with one lead of the meter to the Green pickup coil terminal and make contact with the other meter lead to the Blue coil terminal.

The meter should indicate 396 - 594 ohms.

If the meter indicates a higher reading, the coil has an open lead and **MUST** be replaced. If the meter reading is much less than the minimum figure given, the coil has a short and must be replaced.

TEXT CONTINUES ON PAGE 7-14

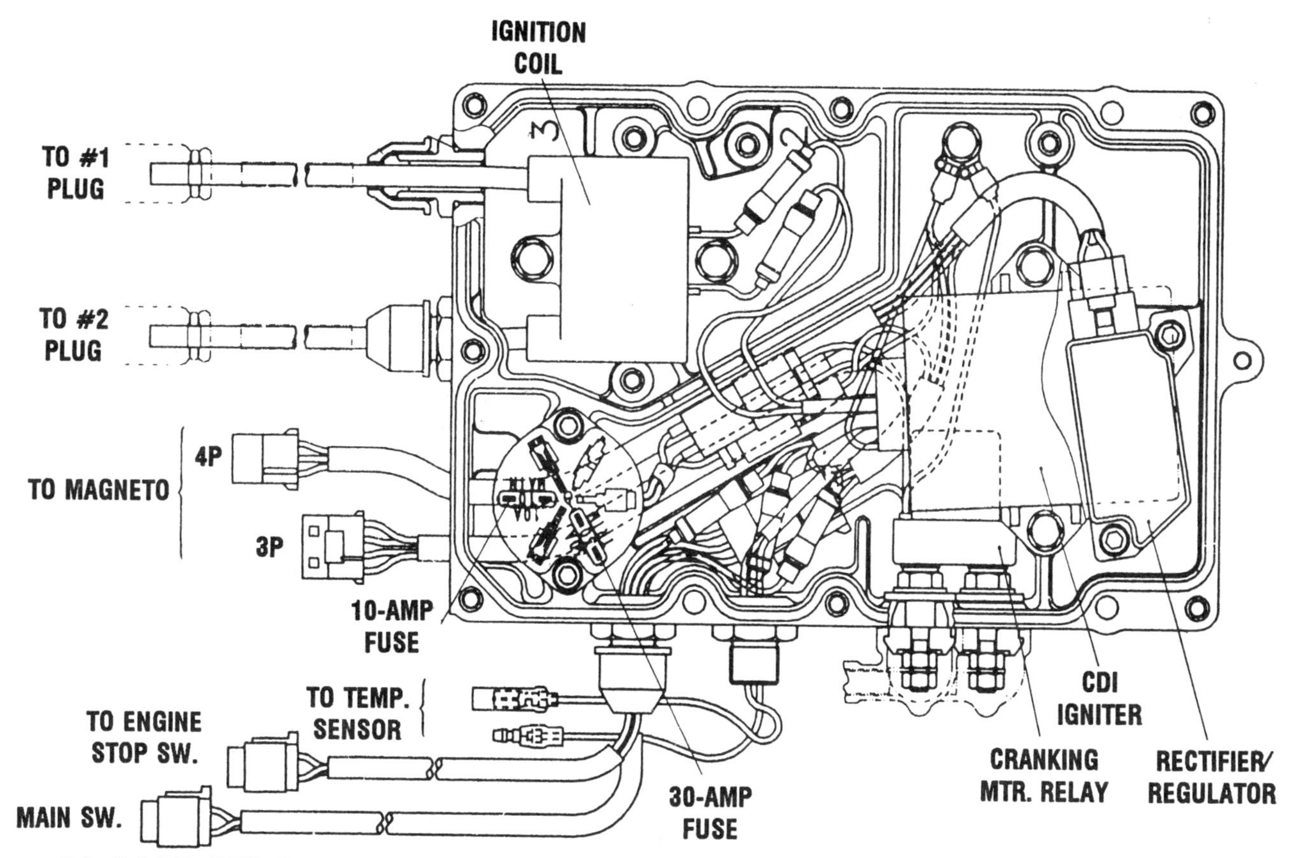

*Model **750ZXi** Detailed line drawing to establish position of components housed inside the electrical box. Major ignition items are identified.*

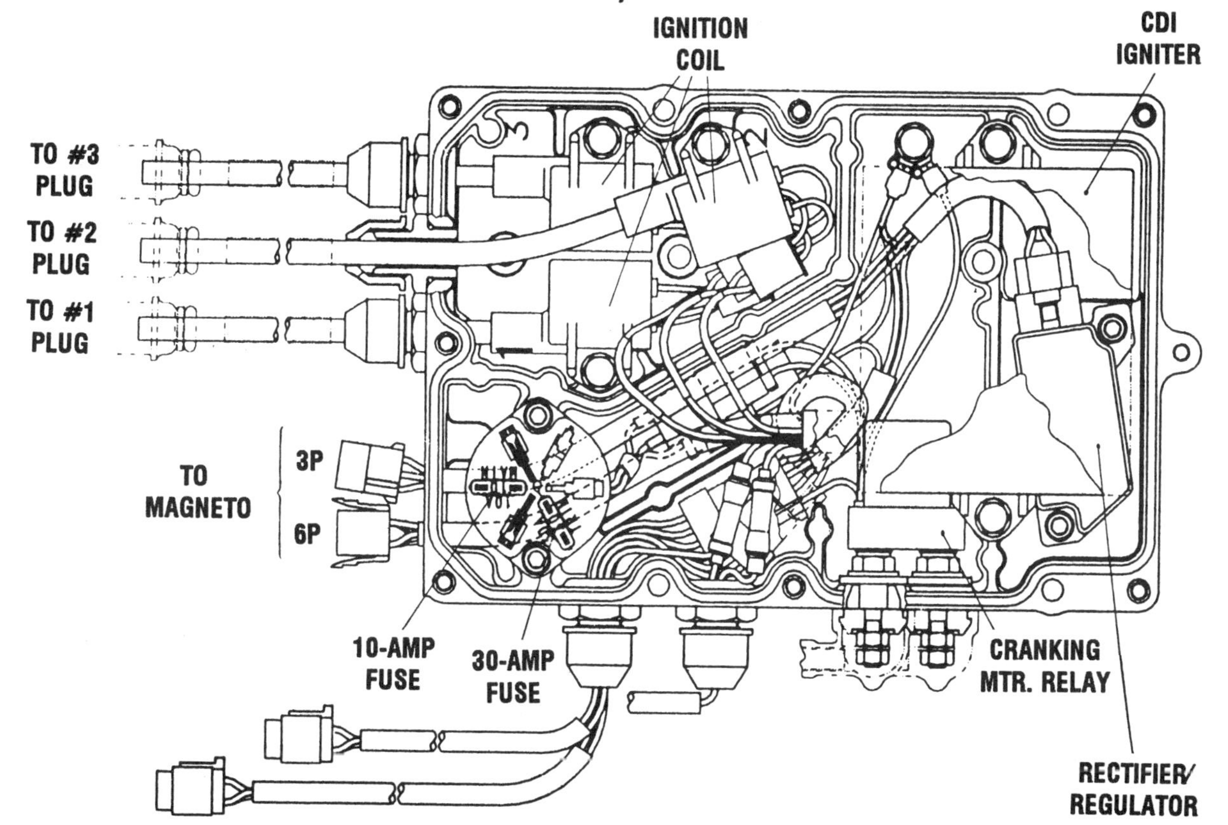

*Model **900ZXi** Detailed line drawing to establish position of components housed inside the electrical box. Major ignition items are identified.*

7-5 IGNITION TIMING ADJUSTMENTS

All Kawasaki personal watercraft covered in this manual are equipped with an automatic advance type CDI system. Ignition timing is accomplished electronically and signal output increases proportionately to engine speed increase -- thereby advancing the timing. Such a system has no moving or sliding parts. Therefore, periodic adjustments are not necessary.

The timing is set correctly at the factory and should never need adjustment **UNLESS** the magneto assembly is misaligned or an ignition component -- such as the magneto assembly -- has been replaced.

To adjust the timing on the Model **550** and the Model **650** Series engines, the fuel tank must first be removed.

The oil injection system need not be disturbed on the Model **650** Series to adjust the timing.

TEST TANK

The craft must be moved to a test tank or to a body of water with lines securing attached to prevent the craft from moving, because the engine will be operated at high speed.

Fuel Tank Removal
Model 550 and 650 Series

Begin by loosening the filler cap on the fuel tank slightly to release any vapor pressure built up inside the tank.

If the level of fuel in the tank is above the inlet neck fitting, siphon some of the fuel from the tank to avoid spillage when the tank is removed.

Next, disconnect the oil tank vent hose from the fuel tank vent hose at the tank fitting. Lift out the fuel outlet assembly. Loosen the large hose clamp around the inlet neck and ease the hose from the fitting on the tank. Now, unhook the two rubber straps securing the tank in the hull. Lift the tank up and out of the craft, and then set it down close by. Reconnect the fuel outlet and retainer nut to the tank.

Preliminary Timing Tasks
Model 550 Series

Remove the fuel tank, as outlined in the previous paragraphs.

Remove the flywheel cover. Remove and **SAVE** the sealing ring.

Preliminary Timing Tasks
Model 650 Series

First, remove the fuel tank, as outlined in the previous paragraphs above, under "Fuel Tank Removal". Remove the magneto breather plug on the flywheel cover. There is no need to disturb the oil injection system on the Model **650** Series for this timing procedure.

Determining TDC (Top Dead Center)
Model 550 and 650 Series

Remove the forward (No. 1) spark plug. Install a dial indicator into the spark plug opening and determine TDC of the piston.

GOOD WORDS

If a dial indicator is not available, use a pencil, straw, or equivalent, resting on top of the piston dome. Determine the highest point to which the inserted object rises when the crankshaft is rotated.

Timing Marks
Model 550 and 650

Locate the timing mark -- may be a line, a dot, a letter, or even a projection embossed onto the flywheel. If necessary apply a dab of "Whiteout" or other white substance to the mark to make it more visible.

With the piston at TDC -- make an aligning mark on a stationary surface near the timing mark using the "Whiteout". This stationary mark may be on the rim of the stator plate on

*The timing marks on the magneto assembly **MUST** align with the marks on the stator plate during installation.*

a Model **550** Series, or the mark can be on the edge of the flywheel cover on the Model **650** Series engine. These two marks will be compared during the timing procedure. Therefore, make them as close together as practical -- extend the embossed mark, if necessary.

Connect a timing light to the battery and hook the pickup around the forward spark plug high-tension lead. Check to be sure the direction of the embossed arrow on the timing light pickup is **TOWARD** the spark plug.

Connect a tachometer to the engine.

SPECIAL WORDS ON TACHOMETERS

The manufacturer recommends the use of a Kawasaki Electronic Tachometer. Instructions for connecting the tachometer leads are provided with the meter.

HOWEVER, on a two-stroke engine with CDI type ignition, the tachometer leads are usually connected across the charging coil leads. As a general rule, the two leads of the same color in the magneto plate harness lead at the electrical box, are the leads from the charging coil.

In a test tank or with the craft in a body of water well secured to prevent the craft from moving -- start the engine and allow it to warm to operating temperature.

CAUTION

Water must circulate through the jet pump -- to and from the engine, anytime the engine is operating. Circulating water will prevent overheating -- which could cause damage to moving parts and possible engine seizure. Therefore, do not operate the engine for more than 15-seconds without circulating water. NEVER advance the throttle to high rpm with no load on the engine.

NEVER, AGAIN NEVER operate the engine at high speed with a flush device attached. An engine operating at high speed with such a device attached, could **RUNAWAY** from lack of a load on the impeller shaft, causing extensive and expensive damage.

DYNAMIC TIMING CHECK -- MODEL 550 & 650

After the engine has reached normal operating temperature -- with the timing light and the tachometer connected -- advance the throttle lever until the engine has reached 6,000 rpm.

Aim the timing light at the two timing marks. If the mark on the flywheel aligns with the stationary mark, the timing is set correctly. However, if the mark on the flywheel is not aligned with the stationary mark -- a timing adjustment is required. Shut down the engine and perform the following procedure.

FIRST, THESE CRITICAL WORDS

The two most important items in establishing and maintaining perfect timing on a two-stroke engine are the magneto assembly securing screws and the crankshaft woodruff key.

Magneto Assembly Securing Screws

The magneto assembly is mounted on top of the stator plate and has two elongated slots for the securing screws to pass through and permit rotating the assembly slightly to achieve proper timing. If the magneto assembly is allowed to rotate just a small amount because the securing screws have worked loose, the ignition timing may be off as much as 15°. The symptoms of this misalignment are loss of power and difficult start-up.

Crankshaft Woodruff Key

The same symptoms apply to a sheared Woodruff key allowing the flywheel - actually the flywheel magnets -- to become out of position on the crankshaft.

During wave jumping or using a wet ramp, when the craft clears the water, the engine rpm increases dramatically -- approaching the limit set by the rpm limiter in an unloaded condition.

Once the craft returns to the water surface, the sudden load on the impeller is transferred along the driveline including the crankshaft. Speed of the rotating pump shaft and the crankshaft is suddenly reduced while the flywheel attempts to continue rotating rapidly due to its mass.

This condition, especially when repeated over a period of time, may shear the Woodruff and the flywheel becomes repositioned on the crankshaft.

A very small misalignment will result in a change in ignition timing. The craft operator will notice a drastic loss of power. A worse condition would be the flywheel shifting to the degree engine start would not be possible.

Adjusting Ignition Timing

If servicing a Model **550** or **650** unit, remove the flywheel cover with the oil injection pump still attached -- Model **650** only.

Observe the two cutout slots in the forward face of the flywheel. Rotate the flywheel until the two magneto assembly securing screws align with the cutouts.

Loosen, but **DO NOT REMOVE** these two screws.

WORDS FROM EXPERIENCE

If this is the first work performed since the unit left the factory, these screws are almost impossible to loosen "cold". While attempting to rotate the screwdriver **COUNTERCLOCKWISE**, tap the end of the screwdriver with a hammer to jar the screw loose.

Once the screws have been loosened slightly, very **CAREFULLY** -- working through the flywheel cutouts -- rotate the magneto assembly either clockwise or counterclockwise, as necessary to correct the timing. When the two marks align, and this process is "trial and error", tighten the two securing screws very tightly.

If these screws should happen to vibrate loose, they could jam between the rotating flywheel magnets and the coils on the magneto assembly -- causing very serious and expensive damage.

If the series being serviced does not have cutouts in the face of the flywheel allowing access to the magneto assembly screws, the flywheel must be removed. To remove the flywheel, see Section 8-3 in Chapter 8, beginning on Page 8-9.

IGNITION TIMING
(Only Necessary If Component Replaced)

SPECIAL TIMING WORDS

As mentioned several times earlier, only the Model **550** and **650** Series have adjustable timing. The figures for all other models are presented in the following table for reference only. The timing is set at the factory and is non-adjustable. If an ignition timing problem is suspected, the "black box" must be replaced.

The stator and the flywheel on the Model **550** and **650** Series have a slot opposite each other for the mounting bolts. These slots permit the stator to be rotated slightly in order to obtain the correct degree of timing.

Model 550

Ign. Timing	21° BTDC @ 6,000 rpm
Ign. Coil 2nd Winding	4.5 - 6.7k ohms
Exciter Coil (Mtr. Bk to Ground Mtr. Red to Bk)	112 - 168 ohms
Pulser Coil (Mtr. Bk to Ground Mtr. to Gray	14.4 - 21.6 ohms

Model 650

Ign. Timing	17° BTDC @ 6,000 rpm
Ign. Coil 2nd Winding	2.1 - 3.1k ohms
Exciter Coil Resistance	250 - 380 ohms

Model 750SX

Ign. Timing	13° BTDC @ 1,250 rpm 16° BTDC @ 2,500 rpm
Ign. Coil Primary Winding	0.08 - 0.1 ohm
Ign. Coil 2nd Winding	3.5 - 4.7k ohms

Model 750 XiR

Ign. Timing	13° BTDC @ 1,250 rpm 16° BTDC @ 2,500 rpm
Ign. Coil Primary Winding	0.08 - 0.1 ohm
Ign. Coil 2nd Winding	3.5 - 4.7k ohms
Pickup Coil	396 - 594 ohms
Pickup Coil gap (rotor & core)	0.8 - 1.0 mm

Model 750 High Performance

Ign. Timing	13° BTDC @ 1,250 rpm 20.2° BTDC @3,000 rpm
Ign. Coil Primary Winding	0.08 - 0.10 ohm
Ign. Coil 2nd Winding	3.5 - 4.7 k ohms

Pickup Coil 396 - 594 ohms
Pickup Coil Gap
(rotor & coil) 0.8 - 1.0 mm

Model 900 High Performance
Ign. Timing 13° BTDC @ 1,250 rpm
20.2° BTDC @3,000 rpm
Ign. Coil Primary
Winding 0.18 - 0.24 ohm
Ign. Coil 2nd
Winding 2.7 - 3.7 ohms

Model 1100 High Performance
Ign. Timing 17° BTDC @ 1,250 rpm
27° BTDC @ 3,000 rpm
Ign. Coil Primary
Winding 0.18 - 0.24 Ohm
Ign. Coil 2nd
Winding 2.7 - 3.7 ohms
Pickup Coil 396 - 594 ohms
Pickup Coil Gap
(rotor & core) 0.8-1.0 mm

Closing Tasks
Model 550 Series

If the flywheel was removed to adjust the position of the magneto assembly, install the flywheel and the flywheel cover -- per the instructions in Chapter 8, beginning on Page 8-34.

Disconnect the fuel outlet and retainer nut from the fuel tank. Lower the tank into the craft and set it in position on the two tank mats. Hook the two rubber restraining straps around the tank securing the tank in the hull.

Install the inlet neck onto the tank fitting, and then secure it in place with a large hose clamp. Install the fuel outlet assembly. Make a final inspection of all fuel fittings and check valves.

Closing Tasks
Model 650 Series

If the flywheel was removed to adjust the position of the magneto assembly, install the flywheel, the flywheel cover, and the oil pump. Detailed illustrated flywheel installation procedures are presented in Chapter 8, beginning on Page 8-62. The oil pump should have been left undisturbed on the flywheel cover.

Install the magneto breather plug into the flywheel cover.

Disconnect the fuel outlet and retainer nut from the tank. Lower the tank into the craft and set it in position on the two tank mats. Hook the two rubber restraining straps around the tank securing the tank in the hull.

Install the inlet neck onto the tank fitting, and then secure it in place with a large hose clamp. Install the fuel outlet assembly. Make a final inspection of all fuel fittings and check valves. If the oil pump was disturbed, purge the pump and system per the procedures outlined in Section 6-9, beginning on Page 6-38.

7-6 CHARGING CIRCUIT

The charging coils are installed on the magneto. Therefore, the charging circuit section has been included in this chapter -- Ignition.

DESCRIPTION AND OPERATION

The charging circuit consists of the following items:

Permanent magnets -- attached to the inside perimeter of the flywheel.
Charging coils -- installed on the magneto assembly.
A rectifier/regulator -- located either inside or attached to the electrical box.
An external battery
Necessary wiring -- connecting all together.

The **NEGATIVE** (minus) side of the rectifier/regulator is always grounded. The **POSITIVE** side is always connected to the battery.

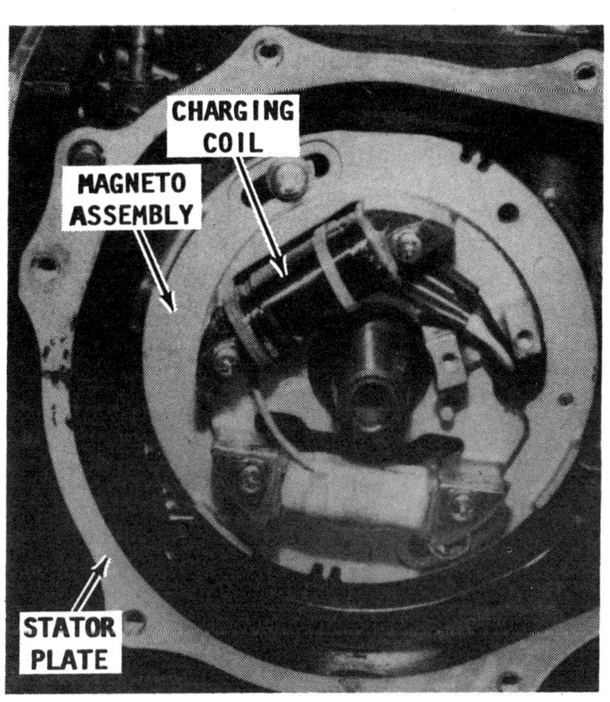

Charging coil on a Model 550 or Model 650 mounted on the stator plate behind the flywheel where it is well protected.

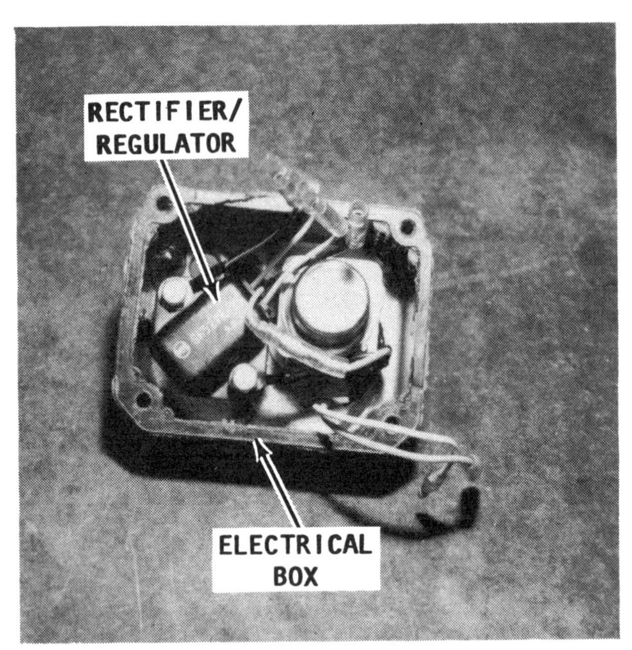

*Rectifier/regulator installed in the electrical box on a Model **550** or **650** Series where it is well protected.*

The negative terminal on the battery must be connected to a good ground on the engine.

As the flywheel magnets rotate, voltage is induced in the charging coil mounted on the magneto assembly under the flywheel. The alternating current passes through a series of four diodes inside the rectifier/regulator and is transformed to DC current.

The rectifier/regulator allows a maximum of 14 volts to pass on to the battery. Any excess voltage is passed to an SCR gate and back through the charging coil. In this manner, overcharging the battery is avoided. The life of components in the charging circuit is greatly extended due to the absence of power surges.

The rectifier/regulator is actually a sealed "black box" type unit. The regulator circuit contains four diodes. If any one of the diodes becomes defective, the complete rectifier/regulator must be replaced.

The alternating current generated in the charging coil windings passes to the rectifier/regulator. The rectifier portion changes the alternating current (AC) to direct current (DC) to charge the 12-volt battery. A second function of the rectifier is to smooth the pulses of current to form a steady DC current flow.

The regulator portion of the rectifier/regulator limits the current produced and sends any excess current back through the charging coil windings.

Electronic components in the charging circuit, including the battery, will last much longer if subjected to a low steady voltage. A sudden voltage surge could overheat and damage charging circuit components.

On a Model **550** and **650** Series engine, the magneto assembly, including the charging coil is located behind and protected by the flywheel.

On a Model **750, 900** and **1100** Series engine, the magneto is housed inside the flywheel cover. Therefore, to service charging components on these model engines it is not necessary to "pull" the flywheel.

Because these components are so well protected by the flywheel or the flywheel cover, the magneto assembly, including the charging coil/s, seldom causes problems in the charging circuit. When problems do occur, they can usually be traced to the rectifier/regulator or to the battery.

If either the charging coil/s or the rectifier/regulator fails troubleshooting tests, the

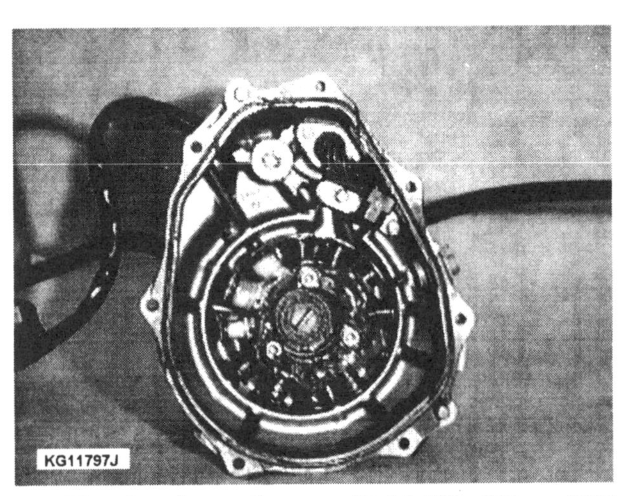

*The charging coils on a Model **750, 900** or **1100** Series engine are installed on the stator plate inside the flywheel cover. Therefore the flywheel does **NOT** have to be "pulled" in order to test or replace the charging coils.*

*Location of the rectifier/regulator inside the electrical box on a Model **900** and **1100** Series. Exact location of the rectifier/regulator inside the box may vary slightly for the Model **750** Series.*

defective unit cannot be repaired, it **MUST** be replaced.

TROUBLESHOOTING CHARGING SYSTEM

As mentioned earlier, the charging coil on a Model **550** and **650** series engines is mounted behind and protected by the flywheel. On the Model **750**, **900** and **1100** Series engines the charging coils are mounted on the stator inside the flywheel cover where they are well protected.

The charging coil/s seldom cause problems in the charging circuit. Most problems can be traced to the rectifier/regulator, the battery, or to the connecting wiring.

CAUTION

The following three conditions will cause damage to the rectifier/regulator.

a- Battery cables connected in reverse. **However**, the Model **750**, **900** and **1100** have a fuse installed to prevent damage to the rectifier/regulator in the event the battery is not connected properly.

b- Operating the engine with one or both battery cables disconnected.

c- A loose or broken connection in the charging circuit.

The voltage generated by the charging coil/s is intended to charge the battery. Anytime the circuit is incomplete, the voltage has actually no where to go. As a result, the voltage will actually reverse itself. Such reversed voltage will damage the internal structure of a diode in the circuit. The diode will then have a scorched or burned appearance.

Perform a visual inspection of the following areas before making any resistance tests on charging circuit components.

a- Check to be sure the battery cables are connected to the correct terminals.

b- Check condition of the battery and recharge, if necessary.

c- Inspect the wiring and the connections between the charging coil, the rectifier/regulator, and the battery. Replace any damaged or deteriorated cable insulation. Pay particular attention for loose, corroded, or disconnected terminals or connections.

TESTING CHARGING COIL OUTPUT

The procedure for testing the charging coil/s are very similar, but the color code differs for the model engines covered in this manual. Therefore, the following "Plan Ahead", "Safety Words" and the "Caution" are valid for all models. Separate actual testing procedures with the correct color codes are presented.

PLAN AHEAD

Each of the leads which are disconnected will be tested with a voltmeter connected across one lead and a suitable ground connection, and then the other lead and a suitable ground connection with the engine operating.

Therefore, position the leads to make the test quickly and efficiently and decide on the ground connection to be used. Have a pencil and paper handy to record both readings.

SAFETY WORDS

The manufacturer warns this test **MUST NOT** be performed with the craft in a body of water due to the possibility of electric shock to the individual performing the test. In light of this warning, this test **MUST** be performed in a test tank with the person performing the test on dry land.

Therefore, move the craft to a test tank or secure it in a body of water next to a float, as in a regular boat "slip". Secure the craft fore and aft to prevent movement during the following tests.

CAUTION

Water must circulate through the jet pump -- to and from the engine, anytime the engine is operating. Circulating water will prevent overheating -- which could cause damage to moving parts and possible engine seizure. Therefore, do not operate the engine for more than 15-seconds without circulating water. NEVER advance the throttle to high rpm with no load on the engine.

Model 550 Series Charging Coil Output Test

Obtain a voltmeter. Remove the electrical box connector.

Disconnect the 6-pin connector.

Start the engine and make the following tests with the engine operating at approximately 6,000 rpm.

Model 550 Identification of connector pin openings mentioned in the text for making a charging coil output test.

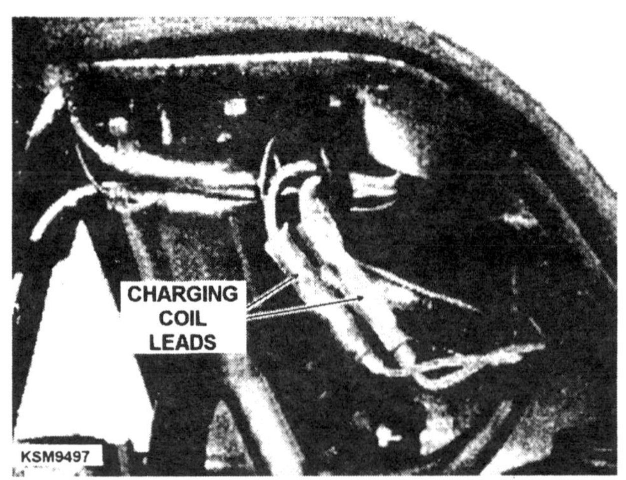

Model 650 Identification of the Yellow charging coil connectors to be separated in preparation to making charging coil output test, described in the text.

Move the meter to the 25 VAC setting, and then make contact with the Black meter lead to the Black pin opening in the connector, as indicated in the accompanying illustration. Make contact with the Red meter lead to one of the Light Green pin openings in the connector. Record the meter reading.

With the engine still operating, make contact with the Black meter lead again to the Black connector pin opening. Make contact with the Red meter lead to the other Light Green pin opening. Record the meter reading. Shut down the engine.

If the voltage is satisfactory -- 12 - 15 VAC -- check the regulator. If the charging coil output voltage is low, check the charging coil resistance with a ohmmeter as follows:

Set the meter to the R x 1 scale. Make contact the Black meter lead to the Black connector pin opening. Make contact with the Red meter lead to one of the Light Green pin openings. Meter should indicate a resistance of 1.2 - 1.8 ohms.

Make contact with the Black meter lead again to the Black pin opening and the Red meter lead to the other Light Green pin opening. The meter should again indicate 1.2 - 1.8 ohms.

Now, make contact with the Black meter lead to one of the Light Green pin openings and the Red meter lead to the other Light Green pin opening. The meter should indicate 2.4 - 3.6 ohms.

BAD NEWS

If the resistance of the coil is satisfactory, but the voltage check indicates the charging system to be defective, the permanent magnets in the flywheel have most likely become weak. If such is the case, the flywheel must be replaced.

Model 650 Series
Charging Coil Output Test

READ AND OBEY

Take time to read and follow the "Plan Ahead", "Safety Words" and the "Caution" presented in the previous paragraphs.

Obtain a Voltmeter and select the 250 VAC setting. Open the electrical box. Remove the magneto leads from the electrical box.

Disconnect the charging coil leads -- the Yellow leads.

Start the engine and advance the throttle to about 3,000 rpm. With the engine operating, make contact with the positive (Red) meter lead to one of the Yellow leads -- from the magneto. Make contact with the Black meter lead to the other Yellow lead.

The meter should indicate 38 VAC.

If the voltage output is satisfactory, the regulator portion of the rectifier/regulator is defective.

Charging Coil Output Test
Model 750 and 900 series

READ AND OBEY

Take time to read and follow the "Plan Ahead", "Safety Words" and the "Caution" presented in the previous paragraphs, on Page 7-19.

Obtain a voltmeter and select the scale to register 50 volts.

Remove the electrical box connector. Disconnect the 3-pin connector.

Temporarily connect the magneto leads, but not the charging coil Brown leads.

Start the engine and advance the throttle to 6,000 rpm. Make contact with the positive (Red) meter lead to one of the Brown charging coil leads. Make contact with the negative (Black) meter lead to the other charging coil Brown lead. The meter should indicate 50 volts as a standard value.

If the charging coil output is satisfactory, check the regulator.

If the charging coil output is low, check the charging coil resistance according to the following procedures.

Charging Coil Resistance Test Model 750 & 900

Remove the electrical box connector. Disconnect the 3-pin connector. Temporarily connect the magneto leads, but not the charging coil Brown leads.

Obtain an ohmmeter and select the R x 1 ohm scale. Make contact with the positive (Red) meter lead to one of the Brown charging coil leads. Make contact with the negative (Black) meter lead to the other Brown charging coil lead.

The meter should indicate 0.7 - 1.1 ohms.

If the resistance in the charging coil is satisfactory, but the voltage check indicates the charging system to be defective, the permanent magnets in the flywheel have become weak. The flywheel should be replaced.

Exciter Coil Resistance Test Model 750 Series

Obtain an ohmmeter and select the R x 100 scale.

Make contact with one meter lead to the purple coil lead. Make contact with the other meter lead to the Red coil lead.

The meter should indicate 265.6 - 398.4 ohms.

Exciter Coil Resistance Test Model 900 Series

Obtain an ohmmeter and select the R x 100 scale. Make contact with one meter lead to the Purple coil lead. Make contact with the other meter lead to the Red coil lead.

The meter should indicate 348.8 - 523.2 ohms.

Now, select the 1 x 10 meter scale. Make contact with one meter lead to the Yellow coil lead. Make contact with the other meter lead to the Black coil lead.

The meter should indicate 21.6 - 32.4 ohms.

If the meter readings are not satisfactory, replace the coil.

Charging Coil Output Test Model 1100 Series

READ AND OBEY

Take time to read and follow the "Plan Ahead", "Safety Words" and the "Caution" presented in the previous paragraphs, on Page 7-19.

Obtain a voltmeter and select the scale to register 50 volts.

Remove the electrical box connector. Disconnect the 3-pin connector.

Temporarily connect the magneto leads, but not the charging coil Brown leads.

Start the engine and advance the throttle to 6,000 rpm. Make contact with the positive (Red) meter lead to one of the Brown charging coil leads. Make contact with the negative (Black) meter lead to the other charging coil Brown lead. The meter should indicate 50 volts as a standard value.

If the charging coil output is satisfactory, check the regulator.

If the charging coil output is low, check the charging coil resistance according to the following procedures.

Charging Coil Resistance Test Model 1100

Remove the electrical box connector. Disconnect the 3-pin connector. Temporarily connect the magneto leads, but not the charging coil Brown leads.

Obtain an ohmmeter and select the R x 1 ohm scale. Make contact with the positive (Red) meter lead to one of the Brown charging coil leads. Make contact with the negative (Black)

*The two magneto assembly securing screws on a Model 550 or Model 650 Series engine **MUST** be tight to prevent the magneto plate from rotating during engine operation. The slightest movement of the plate will disturb the ignition timing.*

meter lead to the other Brown charging coil lead.

The meter should indicate 0.7 - 1.1 ohms.

If the resistance in the charging coil is satisfactory, but the voltage check indicates the charging system to be defective, the permanent magnets in the flywheel have become weak. The flywheel should be replaced.

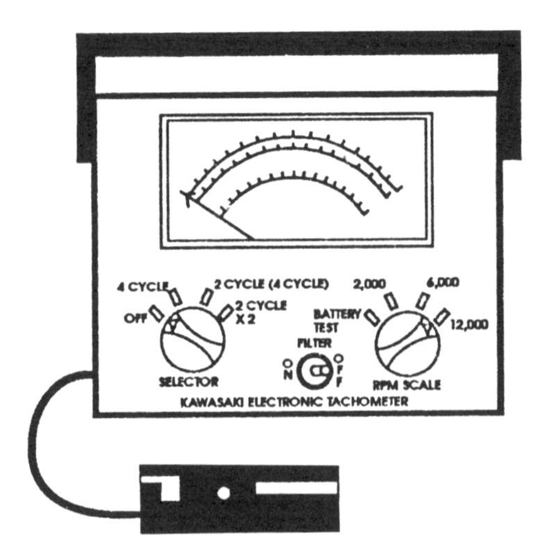

*Maximum engine performance can only be obtained through proper timing and tuning using a tachometer. The text on page 7-15 gives some good words on the use of tachometers with a two-stroke engine. Actually some tachometers **CANNOT** be used with a two-stroke unit.*

The day's fun on a PWC is greatly enhanced when the fuel and ignition systems are properly adjusted and functioning efficiently.

8
ENGINE

8-1 INTRODUCTION AND CHAPTER ORGANIZATION

The following Kawasaki Series are covered in this chapter from 1992 through 1998.

JS550 Series	1992-1994
JF650 Series	1992-1996
JL650 Series	1992-1995
JS650 Series	1992-1993
JS750 Series	1992-1996
JH750 Series	1992 & On
JT750 Series	1994 & On
JH900 Series	1995 & On
JT900 Series	1997 & On
JH1100 Series	1996 & On
JT1100 Series	1997 & On

The working sections of this chapter are divided into seven main sections as follows:

8-2 Description and Operation -- Two-Cycle Engine
8-3 Removal and Disassembling -- Two-Cylinder Engine
8-4 Assembling and Installation -- Two-Cycle Engine
8-5 Removal and Disassembling -- Three-Cycle Engine
8-6 Assembling and Installation -- Three-Cycle Engine
8-7 Cleaning and Inspecting -- All Engines
8-8 Sealants, Adhesives, Lubricants, and Fuel Stabilizers

The carburetion and ignition principles of two-cycle engine operation **MUST** be understood in order to perform proper service work on the marine engines covered in this manual.

Repair Procedures

Service and repair procedures will vary slightly between individual models, but the basic instructions are almost identical. Any variations between models are clearly indicated. Special tools may be called out in certain instances. These tools may be purchased from the local dealer. In certain instances, an alternate tool may be obtained or fabricated.

Torque Values

All torque values must be met when they are specified. Torque values for various parts of each engine are given in the text.

A torque wrench is essential to correctly assemble the engine. **NEVER** attempt to assemble an engine without a torque wrench. Attaching bolts **MUST** be tightened to the required torque value in two progressive sequences, following the specified tightening order. On the first sequence, tighten the bolts to 1/2 the torque value. On the second sequence, tighten the bolts to the full torque value.

Other Engine Components

Service procedures for the carburetor, fuel pumps, cranking motor and other components are given in separate Chapters. See the Table of Contents.

Reed Valve Service

The reeds on Kawasaki two-cylinder, two-stroke engines are contained in an externally mounted reed block. Therefore, the engine need not be disassembled in order to replace a broken reed.

550 Series Only

Due to the unique design of this engine, reeds are not necessary.

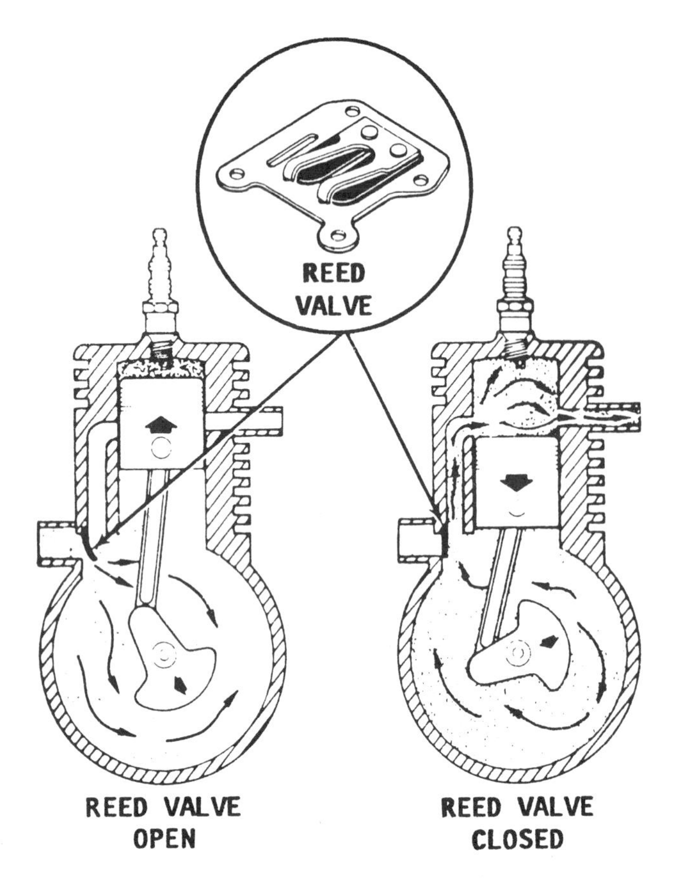

Reed valves are used to control the flow of air/fuel into the crankcase and eventually into the cylinder. As the piston moves upward in the cylinder, the resulting suction in the crankcase overcomes the spring tension of the reed. The reed is pulled free from its seat and the air/fuel mixture is drawn into the crankcase.

Parts **MUST** be cleaned and thoroughly inspected before they are assembled, installed, or adjusted. Use proper lubricants, or their equivalent, whenever they are recommended.

8-2 TWO-CYLCE ENGINE DESCRIPTION AND OPERATION

Intake/Exhaust

Two-cycle engines utilize an arrangement of port openings to admit fuel to the combustion chamber and to purge the exhaust gases after burning has been completed. The ports are located in a precise pattern in order for them to be opened and closed at an exact moment by the piston as it moves up and down in the cylinder. The exhaust port is located slightly higher than the fuel intake port. This arrangement opens the exhaust port first as the piston starts downward and therefore, the exhaust phase begins a fraction of a second before the intake phase.

Lubrication

A two-cycle engine is lubricated by mixing oil with the fuel. Therefore, various parts are lubricated as the fuel mixture passes through the crankcase and the cylinder. On the 300 Series and the 650 Series engines, an oil injection system delivers oil directly into the intake manifold. On the 440 Series and 550 Series engines, oil is mixed with the gasoline in the fuel tank in the age old manner.

Physical Laws

The two-cycle engine is able to function because of two very simple physical laws.

One: Gases will flow from an area of high pressure to an area of lower pressure. A tire blowout is an example of this principle. The high-pressure air escapes rapidly if the tube is punctured.

Two: If a gas is compressed into a smaller area, the pressure increases, and if a gas expands into a larger area, the pressure is decreased.

If these two laws are kept in mind, the operation of the two-cycle engine will be more easily understood.

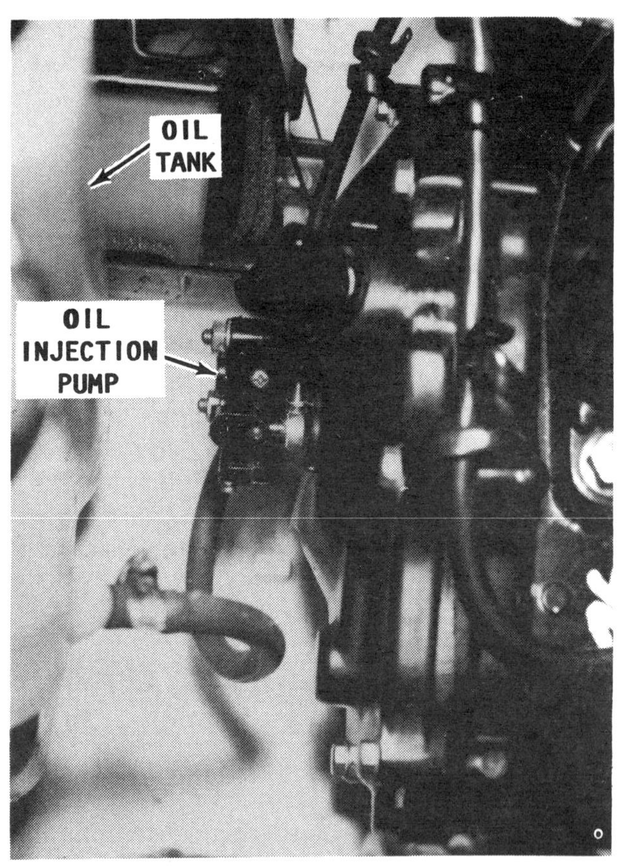

All engines covered in this manual, with the exception of the Model 550 Series, are equipped with an oil injection system. The Model 550 still uses an oil and gas premix.

ACTUAL OPERATION TWO-CYCLE ENGINE

Beginning with the piston approaching top dead center on the compression stroke: The intake and exhaust ports are closed by the piston; the reed valve is open; the spark plug fires; the compressed air/fuel mixture is ignited; and the power stroke begins. The reed valve was open because as the piston moved upward, the crankcase volume increased, which reduced the crankcase pressure to less than the outside atmosphere.

As the piston moves downward on the power stroke, the combustion chamber is filled with burning gases. As the exhaust port is uncovered, the gases, which are under great pressure, escape rapidly through the exhaust ports. The piston continues its downward movement. Pressure within the crankcase increases, closing the reed valves against their seats. The crankcase then becomes a sealed chamber. The air/fuel mixture is compressed ready for delivery to the combustion chamber. As the piston continues to move downward, the intake port is uncovered. A fresh air/fuel mixture rushes through the intake port into the combustion chamber striking the top of the piston where it is deflected along the cylinder wall. The reed valve remains closed until the piston moves upward again.

When the piston begins to move upward on the compression stroke, the reed valve opens because the crankcase volume has been increased, reducing crankcase pressure to less than the outside atmosphere. The intake and exhaust ports are closed and the fresh fuel charge is compressed inside the combustion chamber.

Pressure in the crankcase decreases as the piston moves upward and a fresh charge of air flows through the carburetor picking up fuel. As the piston approaches top dead center, the spark plug ignites the air/fuel mixture, the power stroke begins and one full cycle has been completed.

TIMING -- TWO-STROKE ENGINES

The exact time of spark plug firing depends on engine speed. At low speed the spark is retarded, fires later than when the piston is at or beyond top dead center. Engine timing is built into the unit at the factory. At high speed, the spark is advanced, fires earlier than when the piston is at top dead center.

Summary

More than one phase of the cycle occurs simultaneously during operation of a two-cycle engine. On the downward stroke, power occurs above the piston while the ports are closed. When the ports open, exhaust begins and intake follows. Below the piston, fresh air/fuel mixture is compressed in the crankcase.

On the upward stroke, exhaust and intake continue as long as the ports are open. Compression begins when the ports are closed and continues until the spark plug ignites the air/fuel mixture. Below the piston, a fresh air/fuel mixture is drawn into the crankcase ready to be compressed during the next cycle.

8-3 SERVICE -- 2-CYLINDER ENGINES

ADVICE

Before commencing any work on the engine an understanding of two-cycle engine operation will be most helpful. Therefore, it would be well worth the time to study the principles of two-cycle engines, as outlined briefly in the previous section -- 8-2. A Polaroid, or equivalent instant-type camera is an extremely useful item, providing the means of accurately recording the arrangement of parts and wire connections **BEFORE** the disassembly work begins. Such a record is invaluable during assembling.

Under normal condtions, timing adjustments are not required on the engines covered in this manual. However, if an adjustment is necessary, for any reason, the flywheel must be "pulled" to gain access to the magneto assembly and stator plate, as shown on this Model 650.

The reed block on a Model 650 Series engine can be serviced while the engine remains in place in the engine compartment.

IMPORTANT WORDS

The following series of illustrations were taken of a 650 Series engine. The 550 Series engine differs slightly in appearance, but has the same basic components. Where differences occur, such as bolt tightening patterns, the differences will be clearly identified.

REED BLOCK WORDS

If the only work to be performed on the engine is servicing the reed block assembly, remove the flame arrestor, carburetors, and perform Disassembling Steps 9, 10, and 11, beginning on Page 8-16.

Clearance between the reed block and the reed valve is 0.2mm maximum.

After the reed block and intake manifold assemblies are installed, install the carburetor and the flame arrestor.

Preliminary Tasks
Engine Overhaul

Disconnect the negative and positive leads from the battery. Disconnect the throttle and choke cables from the carburetor. Disconnect the fuel hoses at the fuel tank. Disconnect the stator harness lead at the electrical box. Remove the coupling cover.

Remove the carburetor.

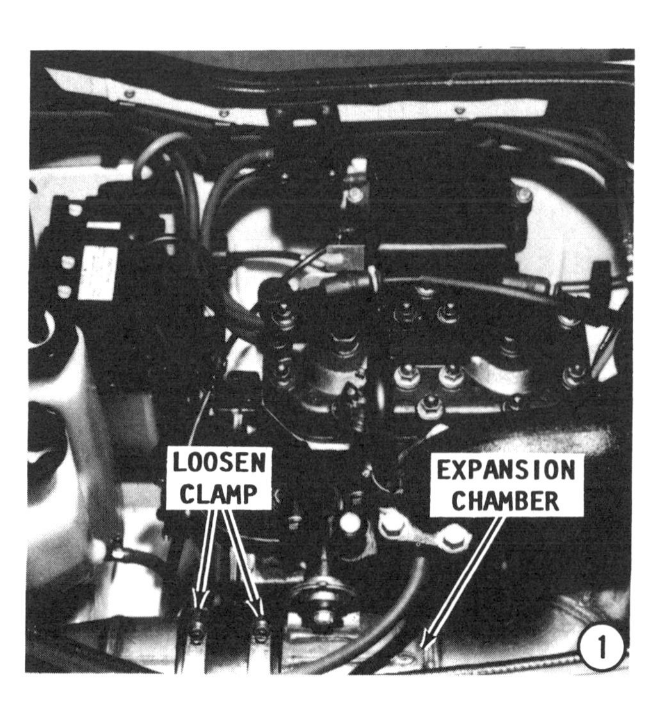

NOW, THESE WORDS

As mentioned earlier, the following procedures and supporting illustrations were developed while servicing a 650 Series engine. Since some difference may occur between engine models, any differences between the 650 Series and other two-cylinder series engines will be clearly identified.

REMOVAL

1- Loosen, but do not remove the two large hose clamps at the forward end of the expansion chamber.

2- Release the pliable metal clamps securing the electrical leads to the engine block.

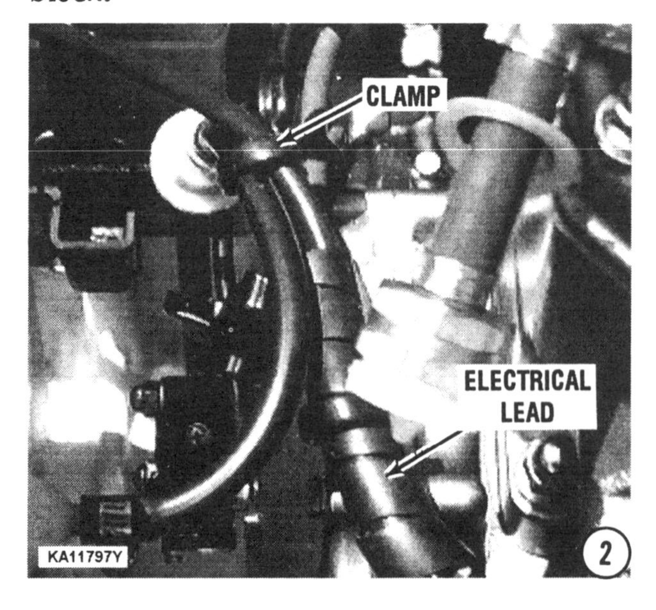

3- Disconnect the pulse hose from the fitting on the intake manifold.

Model 650 Series With Oil Injection

4- Pinch the oil supply hose, and then disconnect the hose from the oil injection pump. Insert a screw into the disconnected hose end to prevent oil loss and a mess in the bilge.

Disconnect the oil delivery hose from the fitting on the intake manifold at the base of the carburetor.

5- Remove the two nuts and washers securing the oil injection pump to the fly-wheel cover. Pull the pump free of the studs.

All Two-Cylinder Engines -- This Section

6- Release the rubber "bungee" type straps securing the fuel tank to the tank tray. Remove the fuel tank cap, and set it to one side.

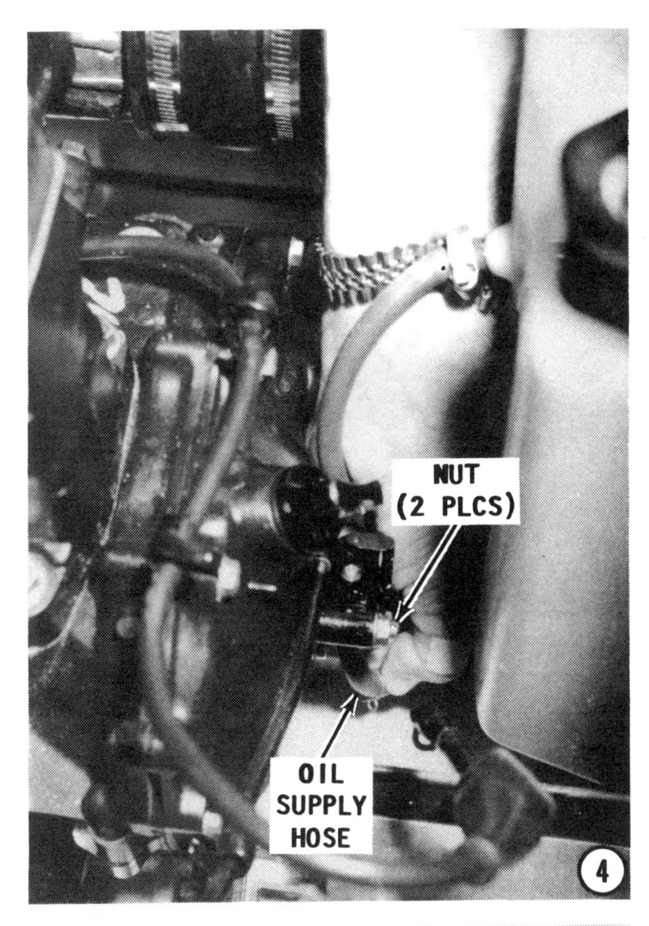

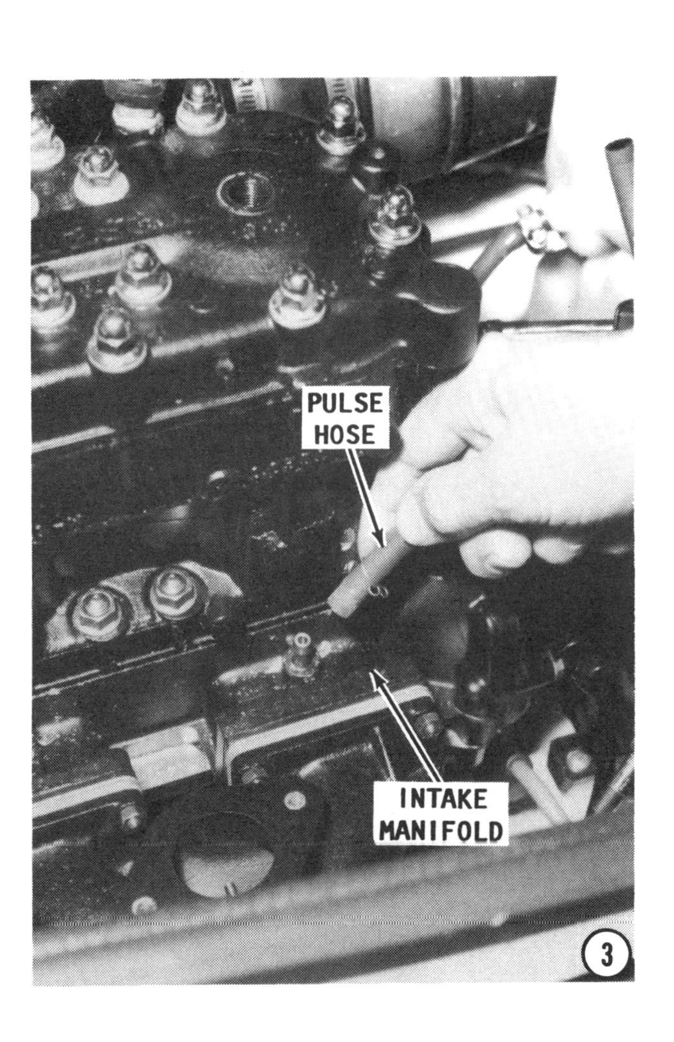

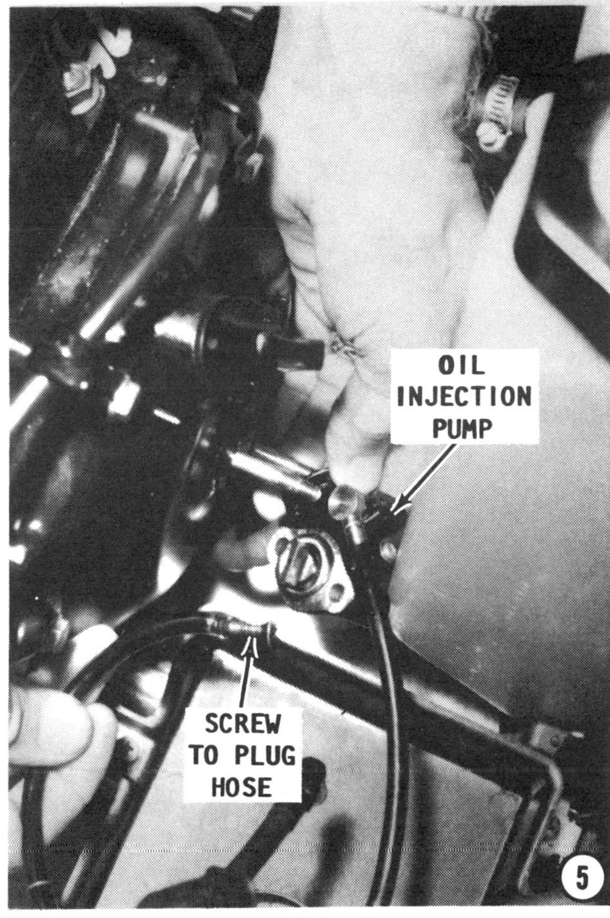

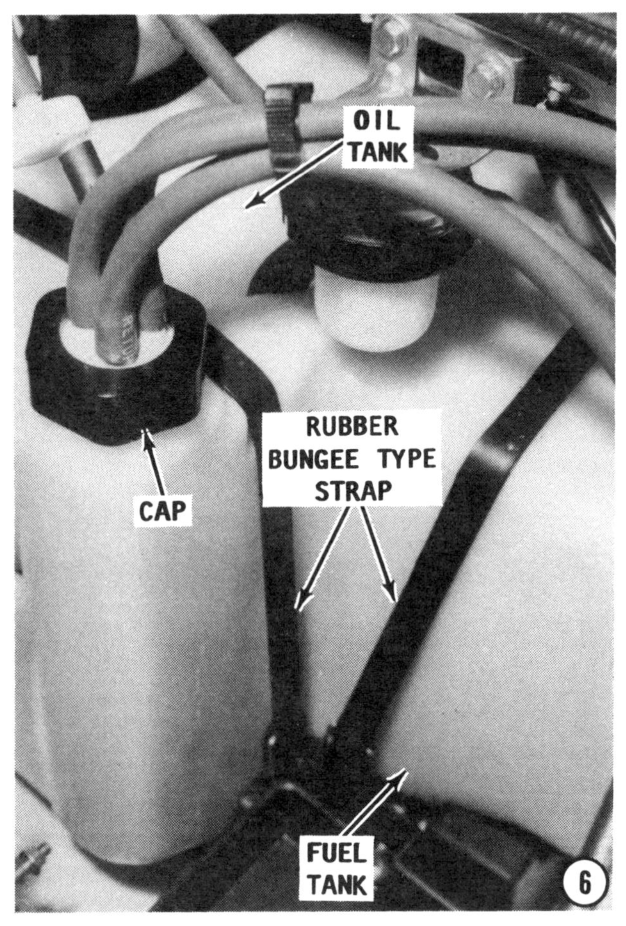

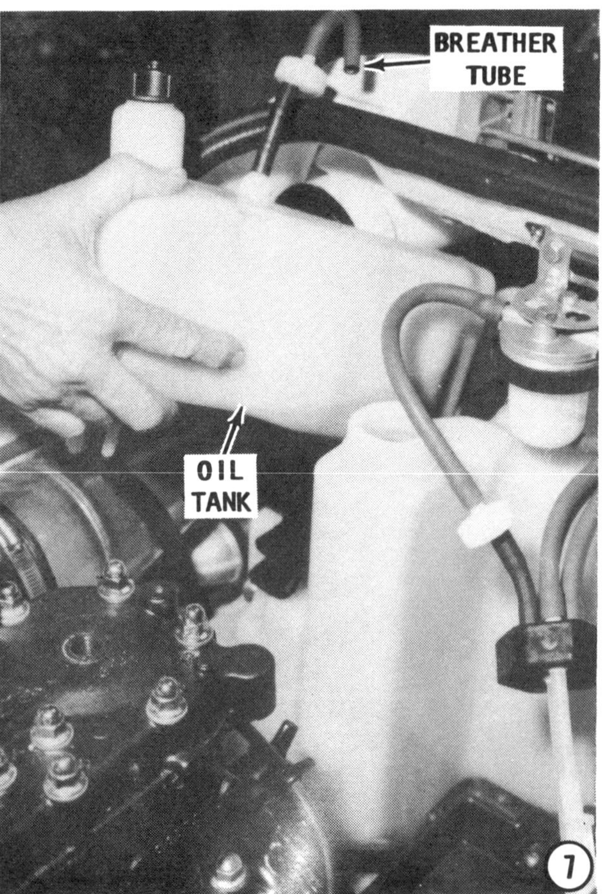

Oil & Fuel Tank Removal

7- Lift the oil tank, with the breasther tube attached up over the engine and clear of the engine compartment.

8- Reach behind the fuel tank and loosen both filler tube clamps around the large filler hose. Work the hose free of the tank fitting. Now, lift the tank up and around the forward end of the engine, then up and free of the engine compartment.

SPECIAL WORDS

The exhaust system is now exposed. Therefore, any service work required to this system, including the water box muffler and exhaust pipe may now be performed.

9- Now, reach under the exhaust manifold and disconnect the cooling water supply hose from the fitting on the manifold.

10- Disconnect the bypass hose from the fitting on the exhaust elbow.

11- Remove the large single bolt securing the expansion chamber to the exhaust manifold. Remove the bolts and washers securing the exhaust elbow to the manifold.

Disconnect the water valve cable from the exhaust elbow flange.

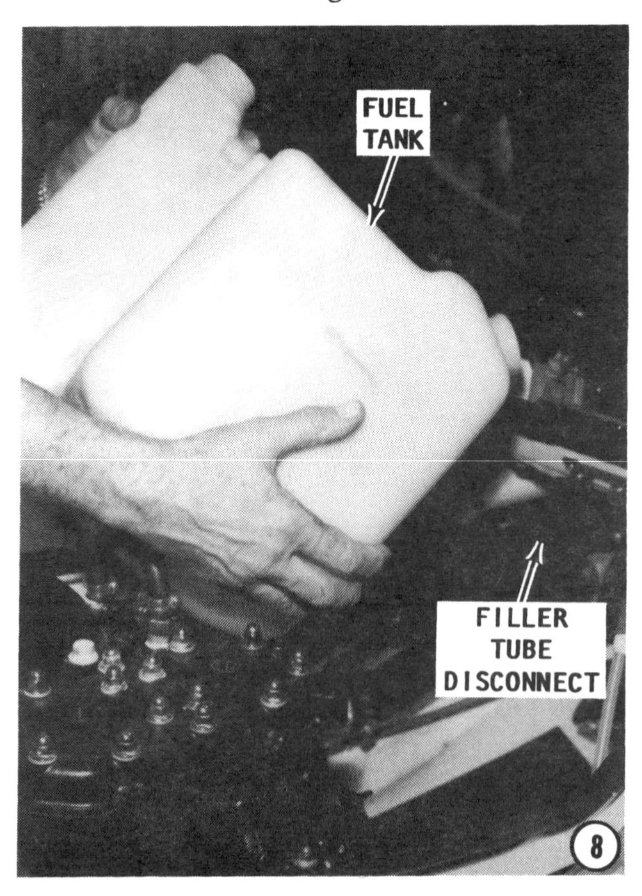

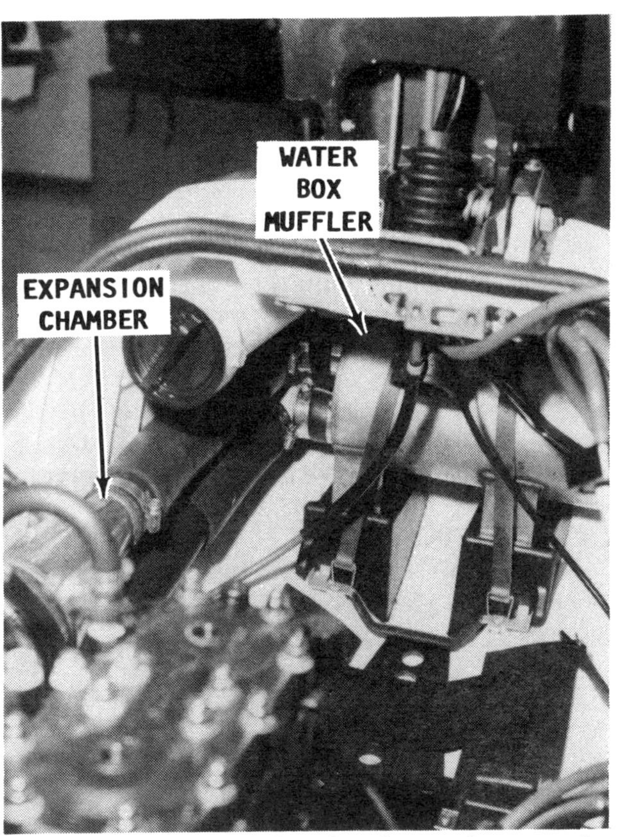

The fuel tank and the oil tank must be removed to gain access for service work on the exhaust system on a Model 650 engine. The fuel tank on a Model 550 must be removed to gain access to the water box muffler.

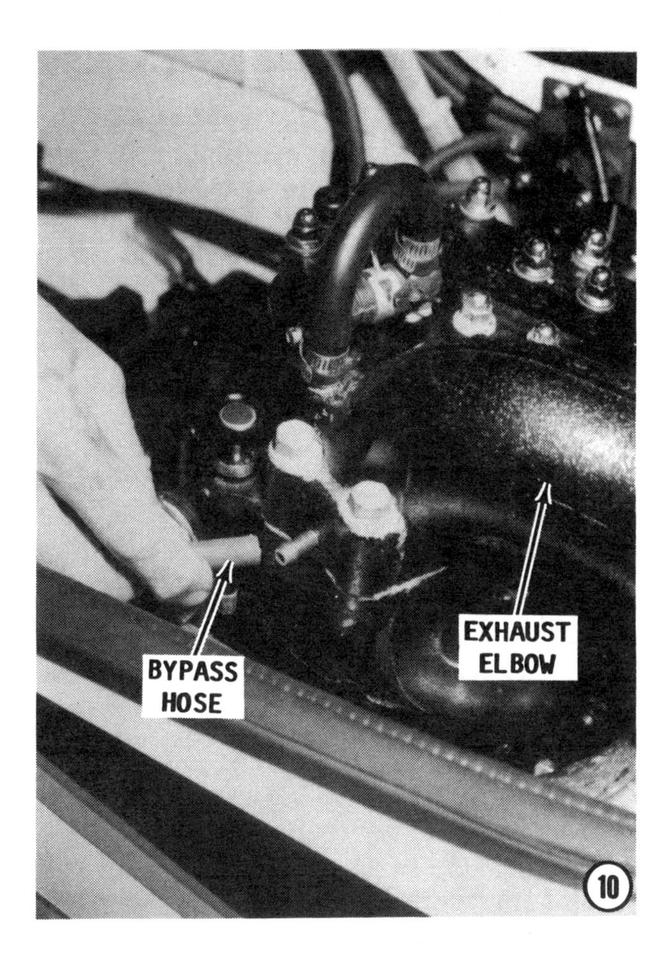

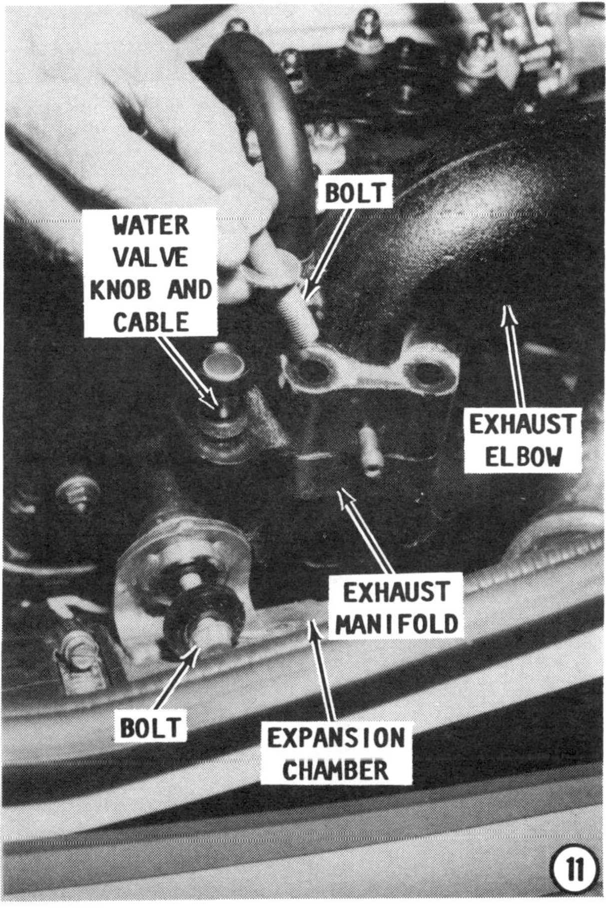

12- Remove the cooling hose from the fitting on the exhaust elbow. Work the exhaust elbow to break the seal between the elbow and the exhaust manifold. The restraining clamps were loosened in Step 1. Move the exhaust elbow to one side to provide clearance for engine removal. Remove the gasket from the surface of the manifold.

VERY IMPORTANT SHIM PACK WORDS

Make a note of the number of pieces of shim material used at each of the four engine mounts -- between the engine bed bracket and the rubber dampers. Keep the four stacks of shim material separate to ensure they will be installed in their original locations. The engine **MUST** be aligned with the impeller shaft to avoid vibration and coupler wear. The amount of shim material at each of the four engine mounts, necessary for accurate alignment, is determined at the factory or dealer/installer. The **SAME** amount must be placed under the mounts to maintain alignment. Therefore, it is **ESSENTIAL** to identify the amount and location of each stack of shim material.

13- Loosen the four engine mounting bolts. Two adults should have no difficulty in lifting this engine. Have an assistant lift the engine slightly, and then slide the shim material out from under each of the four engine brackets. Keep the shim material in groups and identified. After the shim material has been removed for all four mounts, remove the engine mounting bolts.

14- Move the engine forward to disengage the rubber coupler from the jet pump coupler. Lift the engine and watch for any hose or electrical leads still attached to the engine. Transport the engine to a suitable work surface. Remove the rubber coupler.

CRITICAL WORDS

Again, **SAVE** the shim material used between the engine bed bracket and the rubber dampers. Keep the shim material together in groups and identify each group to **ENSURE** proper engine alignment during installation.

Engine Preparation

The following instructions pickup the work after the preliminary tasks listed above have been completed and the engine is on a suitable work surface.

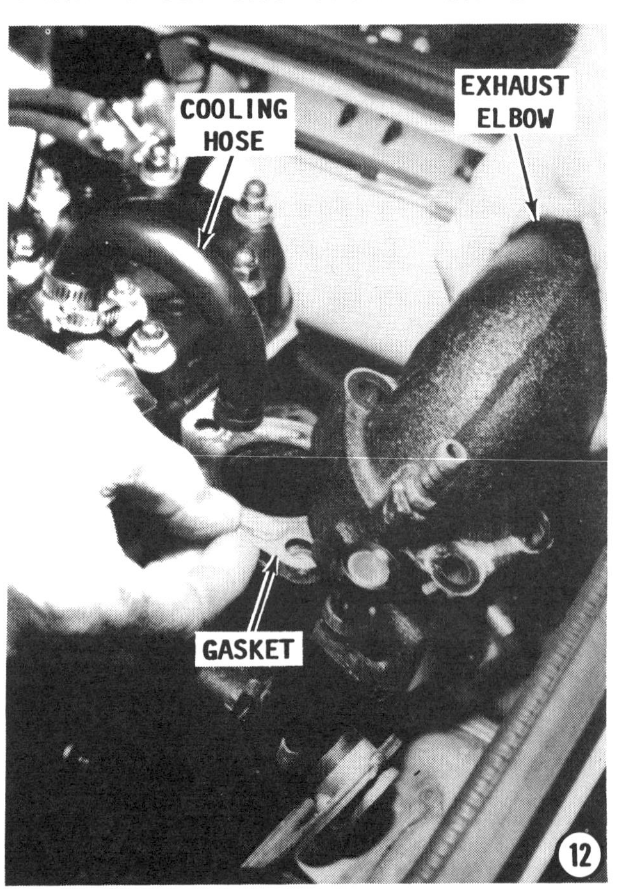

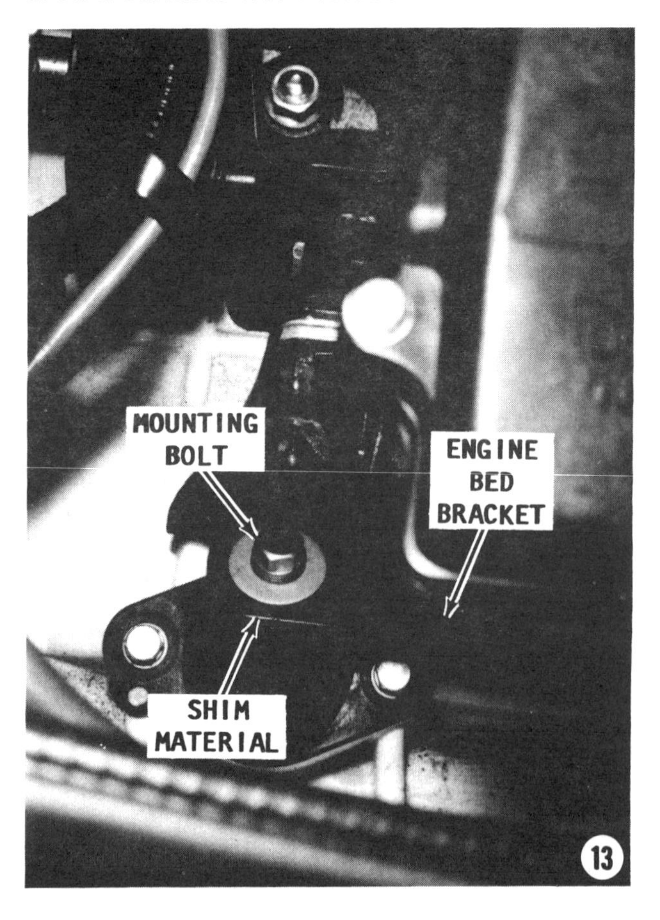

FIRST, THESE WORDS

Procedural steps are given to remove and disassemble virtually all items of the engine. However, as the work moves along, if certain items, i.e. bearings, bushings, seals, etc. are found to be fit for further service, simply skip the disassembly steps involved. Proceed with the required tasks to disassemble the necessary components.

COMPLETE DETAILED procedures to clean and service virtually all parts of the engine are presented in Section 8-7, beginning on Page 8-67.

DISASSEMBLING TWO-CYLINDER

SPECIAL WORDS ON SPECIAL TOOLS

The following list of special tools are required in order to perform some of the necessary work on the engine. Some tasks become extremely difficult, if not impossible, without the special tools or some type of equivalent. Manufacturers' part numbers are given. In specific steps of the procedures -- whenever possible -- a special effort has been made to give some type of equivalent or homemade tool which may be substituted. However, the cost of a special tool may be offset by money saved if an expensive part is not damaged.

550 Series

Piston Pin Puller	P/N 57001-910
Magneto Puller	P/N 57001-259
Flywheel Holder	P/N 57001-1313
Coupler Holder	P/N T57001-276

650 Series

Piston Pin Puller	P/N 57001-910
Magneto Puller	P/N 57001-259
Flywheel Holder	P/N 57001-1156
Coupler Holder	P/N 57001-1230

750 Series

Piston Pin Puller	P/N 57001-910
Flywheel Holder	P/N 57001-1313
Coupler Holder	P/N 57001-1230

900 & 1100 Series

Piston Pin Puller	P/N 57001-910
Flywheel Holder	P/N 57001-1313
Coupler Holder	P/N 57001-1230

SPECIAL WORDS ON REAR MAIN OIL SEALS

If the rear main oil seals require replacement, the metal coupler must first be removed. This metal coupler is removed while the special tool is in place to remove the flywheel nut.

If the rear main oil seal does not require service, the metal coupler can remain in place. Therefore, skip the instructions under "Special Words" following Step 3, and continue with Step 4.

If the rear main oil seals are to be replaced, the metal coupler must be removed while the flywheel is still in place. There is virtually no way on this green, green earth to remove the coupler without the flywheel installed.

If the rear main oil seals require replacement and the metal coupler must be removed, perform the special instructions under "Special Words" following Step 3, and then continue with the work as outlined in Step 4.

Once the flywheel is removed, it is impossible to remove the metal coupler because the crankshaft cannot be prevented from rotating.

GOOD CRANKING MOTOR WORDS

Procedures to remove the cranking motor from a 550 Series engine differ from removal of the cranking motor from a 650 Series or 750 Series engine.

Therefore, the procedures have been separated into separate steps for the 550 Series and the 650 and 750 Series. As clearly noted, the 550 Series engines use shim material at the two forward cranking motor mounting bolts.

Cranking Motor Removal Model 550 Series

1- Remove the two rear cranking motor mounting bolts. Remove and **SAVE** any shim material installed between the rear mounting bracket and the crankcase. Remove the two front cranking motor mounting bolts. Slide the cranking motor aft -- toward the rear of the engine. Disconnect the large Red cable from the terminal on the motor.

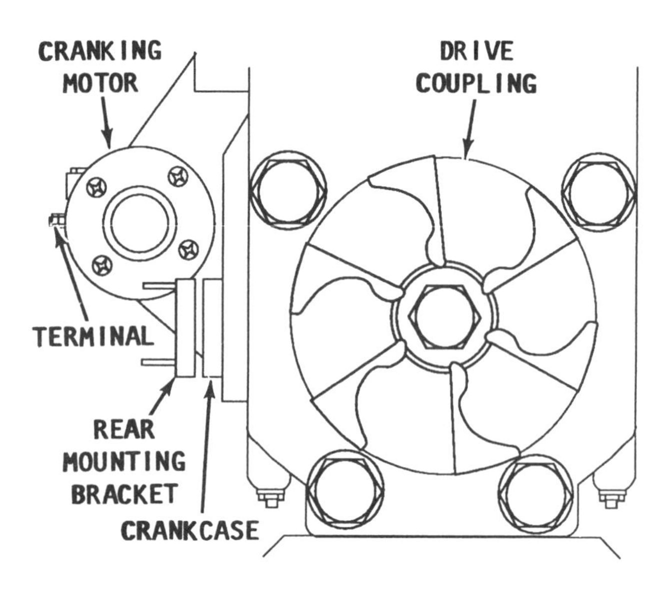

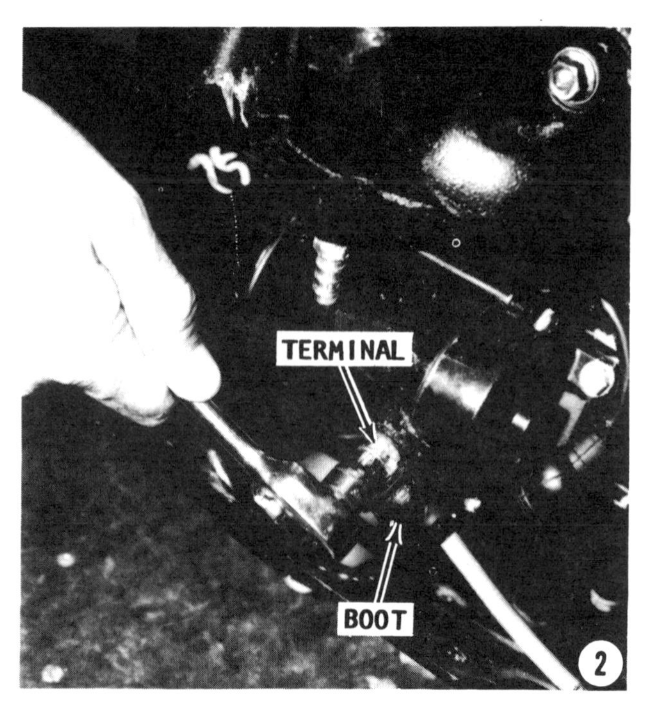

Cranking Motor Removal All Engines Except 550

2- Push back the protective "boot" from the terminal on the cranking motor and remove the large Red cable.

3- Remove the two mounting bolts securing the cranking motor to the crankcase. On some Model 650 Series engines, the lower mounting bolt also secures the battery ground cable to the crankcase.

4- Move the cranking motor aft and the armature shaft will slide free of the reduction gear inside the flywheel cover.

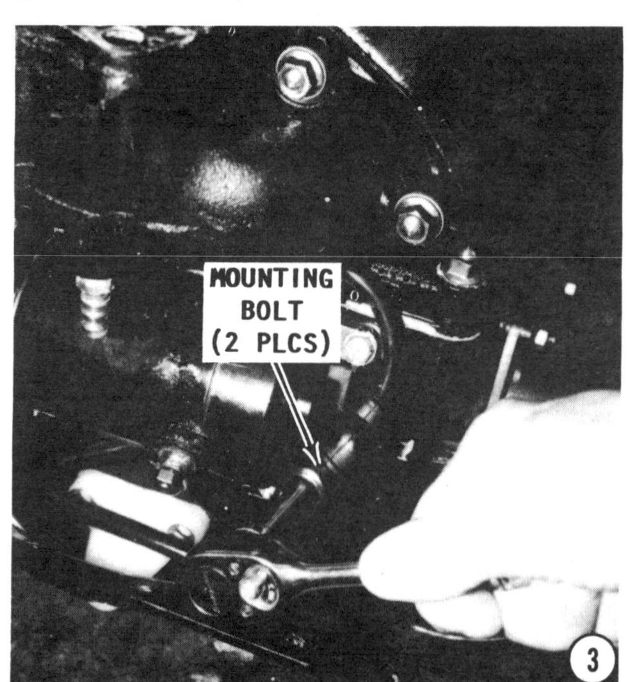

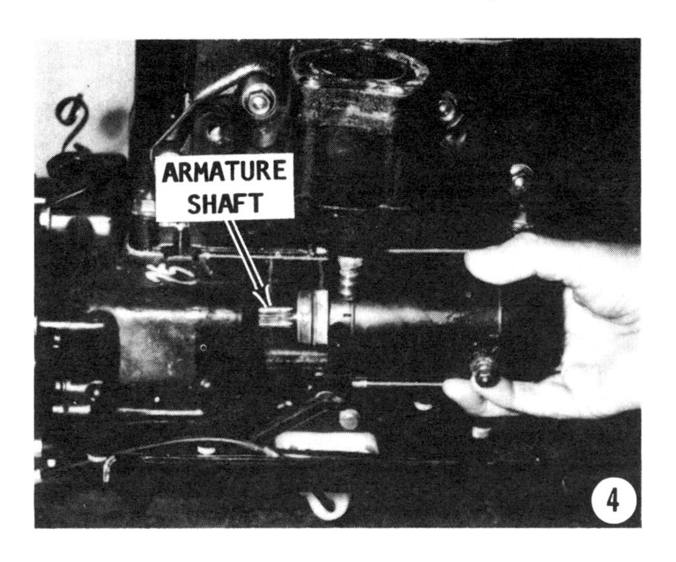

"PULLING" THE FLYWHEEL

1- Remove the spark plugs.

NOTE

If the flywheel cover has not been removed, perform Step 2 and Step 3. If the cover has been removed, proceed directly to "Special Words", just before Step 4.

2- Remove the bolts securing the flywheel cover to the stator plate.

3- Remove the flywheel cover from the stator plate. If the cover is stuck to the plate, obtain a soft head mallet and tap around the circumference of the cover to

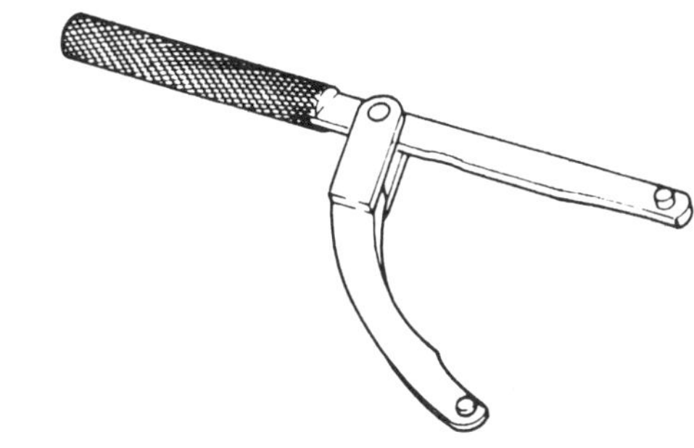

Quick rough line drawing to illustrate the flywheel holding tool for the Model 550 and 650 Series engines. Loosening the flywheel nut is not an easy task and this special tool is necessary.

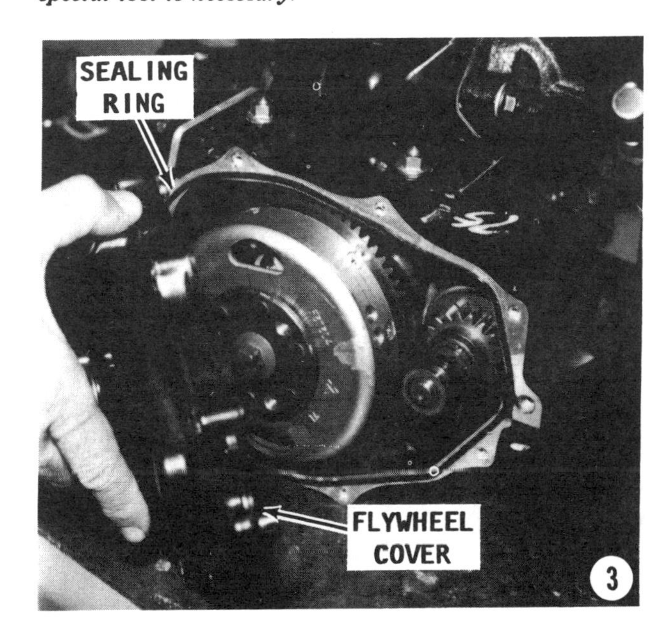

jar it free. Remove and discard the sealing ring, only if a replacement is available.

SPECIAL WORDS

If the two rear main oil seals are to be removed, install the flywheel holder to the flywheel and insert the coupler holder into the metal coupler. Prevent the flywheel from rotating and unthread the metal coupler from the rear of the crankshaft.

4- If servicing a 550 Series engine, bend back the lockwasher tabs behind the flywheel nut. Now, using the flywheel holder tool previously listed for the engine being serviced, insert the two indexing pins on the ends of the arms through the holes in the flywheel. Prevent the flywheel from rotating with the special tool. An assistant can help by holding the flywheel in place while loosening the flywheel fastener. Loosen, then remove the flywheel nut and washer -- 650 Series; the tabbed lockwasher -- 550 Series; flanged bolt -- 750 Series.

5- Install the flywheel puller onto the flywheel, and then using the center bolt, take a strain on the puller with the proper size wrench. Now, continue to tighten the center bolt and at the same time, **SHOCK** the crankshaft with a gentle to moderate tap with a hammer on the end of the center bolt. This shock will assist in "breaking" the flywheel loose from the crankshaft.

DO NOT "pound" on the center bolt. The force of such action would be transmitted to the seals and bearings along the crankshaft and cause damage. **NEVER** strike the flywheel because this type of shock would surely cause the flywheel magnets to loose their magnetism.

6- Pull the flywheel free of the crankshaft. Remove and save the Woodruff key from the crankshaft keyway.

SPECIAL WORDS
WOODRUFF KEYS

Inspect the condition of the Woodruff key and its keyway in the crankshaft. If the key shows any signs of shearing, the cause is probably from jumping waves and the effect will be a loss of power. When the craft is clear of the water, the engine RPM approaches the limit set by the RPM limiter in an unloaded condition.

When the craft returns to the water surface, the sudden load on the impeller is transferred along the driveline. Speed of the rotating pump shaft and the crankshaft is suddenly reduced while the flywheel

continues to spin rapidly due to its mass. This condition, especially when repeated over a period of time, often shears the Woodruff key and the flywheel becomes repositioned on the crankshaft.

A very small misalignment will result in a change in ignition timing. The craft operator will notice a drastic loss of power. A worse condition would be the flywheel shifting to the degree engine start would not be possible.

650 Series Only

Lift out the reduction gear assembly.

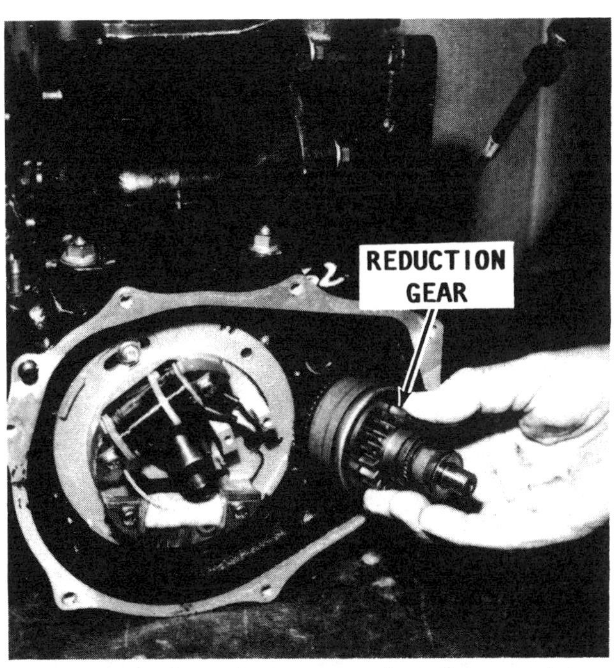

Removing the reduction gear from the armature shaft of the cranking motor on a 650 Series engine.

Magneto Assembly Removal

7- The securing screws pass through elongated slots in the magneto assembly. The assembly could be installed as much as 15° off its original position -- drastically affecting the timing characteristics of the engine. Therefore, at the elongated slot in the magneto assembly, make a set of aligned punch marks -- one on the magneto assembly -- the other on the stator plate. These marks will serve as an aid during installation to **ENSURE** correct positioning of the magneto assembly on the stator plate.

8- Remove the two screws securing the magneto assembly to the stator plate.

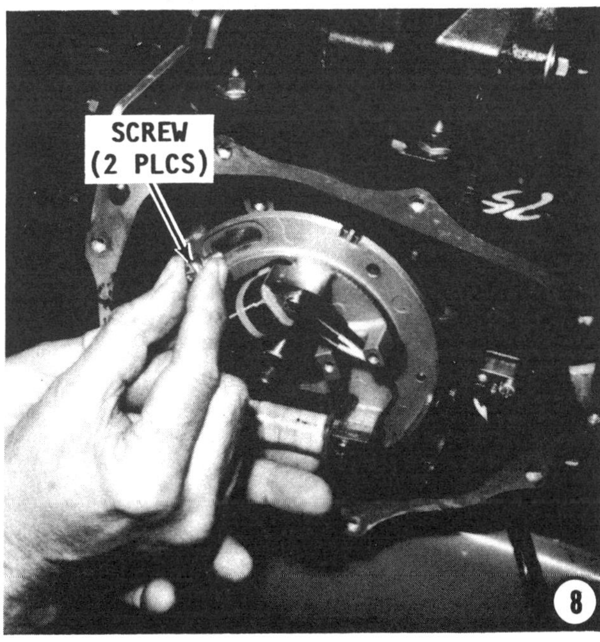

9- Remove the two small screws securing the magneto assembly harness to the grommet in the stator plate. Feed the harness and connector end through the opening in the stator plate and remove the magneto assembly from the crankshaft.

BLOCK DISASSEMBLING

Water Valve
650 Series and 750 Series

1- Remove the two bolts securing the water drain valve to the block. Pull the valve free of the block.

2- Separate the control valve housing from the reed valve cover. Remove the two

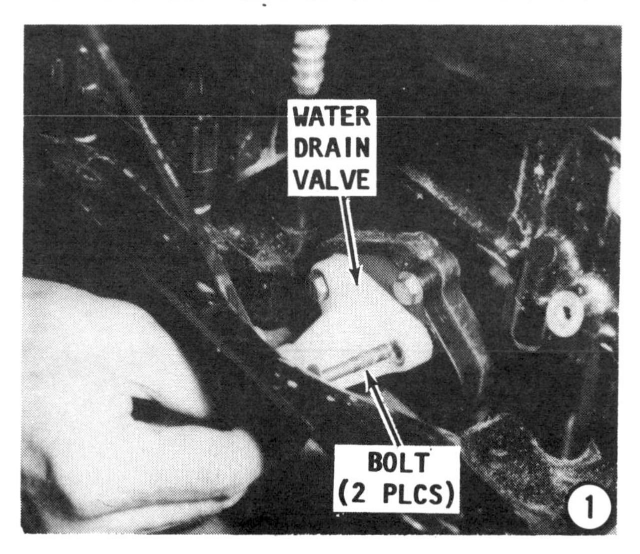

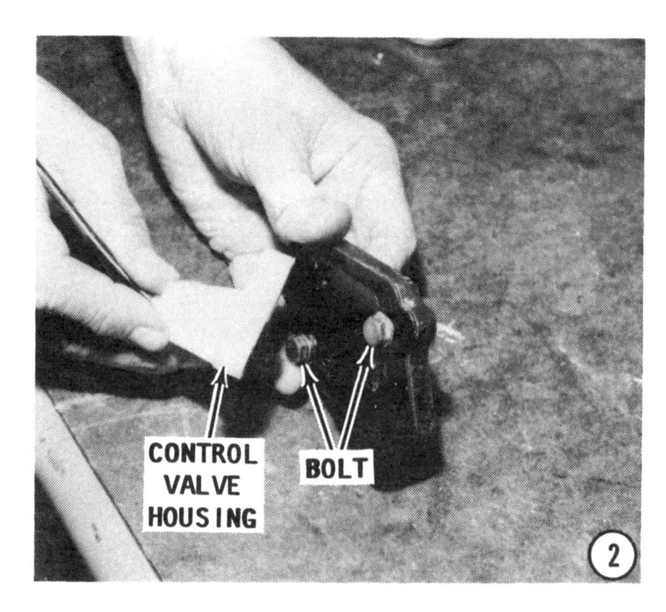

bolts securing the cover to the reed valve assembly.

3- Separate the cover from the reed valve assembly.

Cylinder Head Removal
All Series Engines

4- Remove the nuts and washers securing the cylinder head to the block. Pry the head loose from the block. **TAKE CARE** not to damage the smooth machined mating surfaces of the block or the head. If the head is stuck to the block, it may be necessary to tap the head with a soft head mallet to "jar" the head loose.

5- After the head has been "jarred" loose, keep the head as level as possible and lift it **STRAIGHT** up and off the studs.

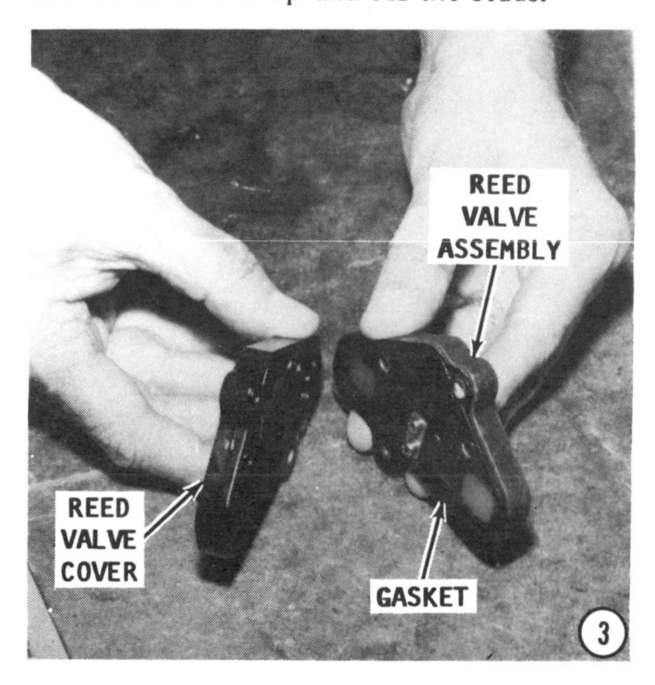

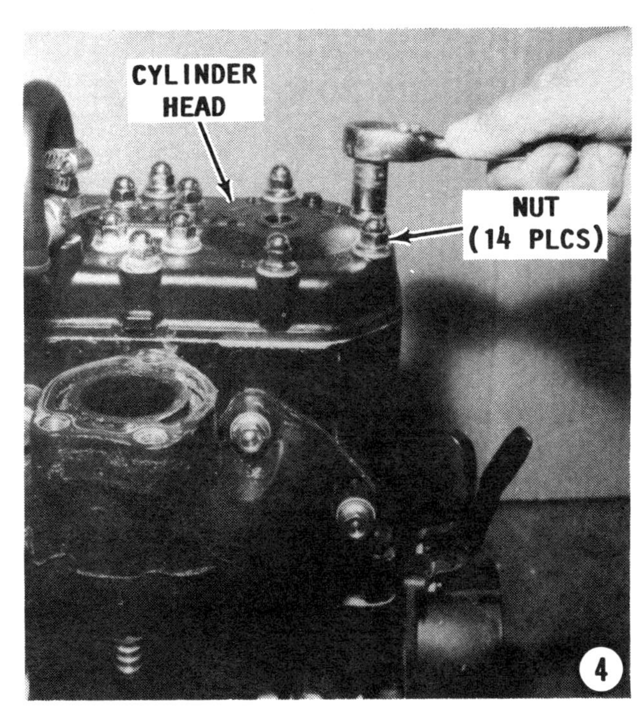

6- Remove and discard the head gasket. The head gasket can **NEVER** be used a second time.

Exhaust Manifold Removal

7- Remove the four nuts/bolts and washers securing the exhaust manifold to the block. "Jar" the manifold loose from the block.

8- Slide the exhaust manifold free of the studs. Remove all traces of gasket material from the two mating surfaces.

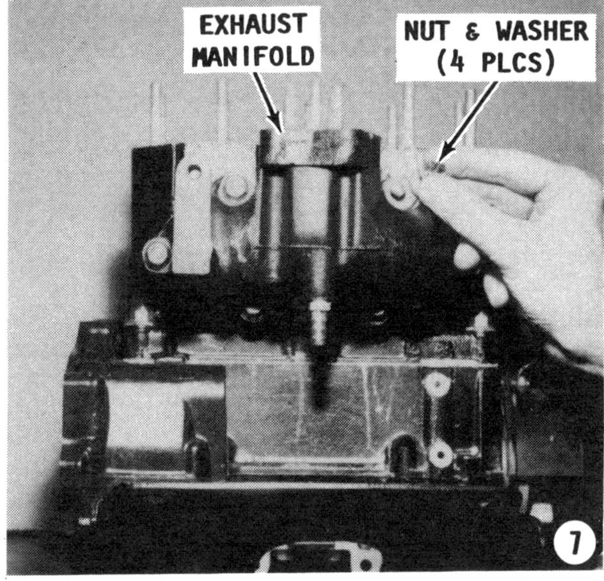

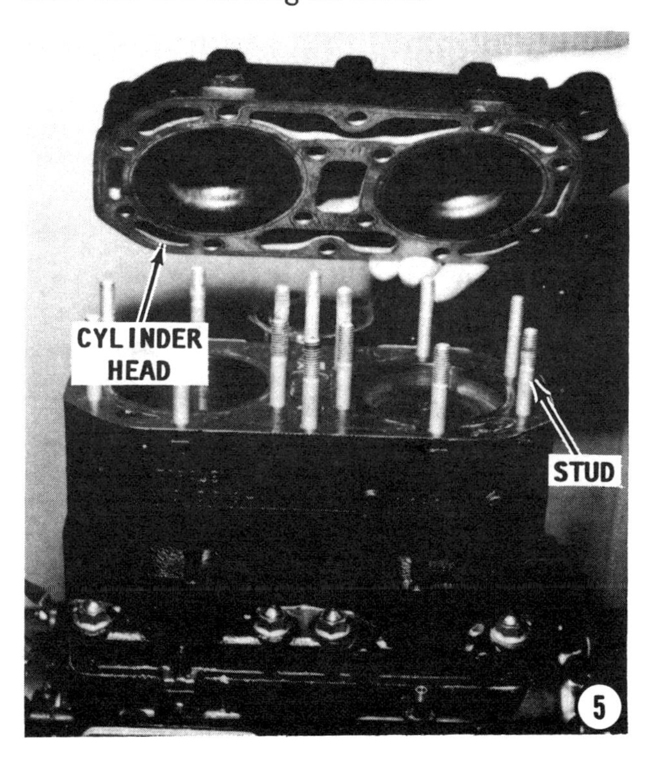

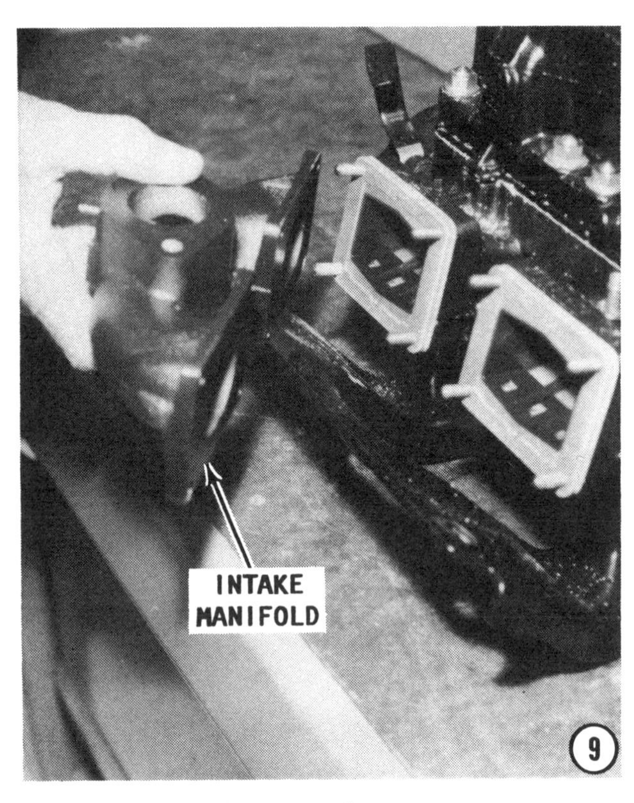

Intake Manifold Removal
All Series

9- Remove the nuts and washers securing the intake manifold to the block. Lift the manifold free of the block.

CRITICAL WORDS
REED BLOCK ASSEMBLIES
650 SERIES & 750 SERIES ONLY

Take great care in the next step when handling reed valve assemblies. The reed valve housing is coated with a special rubber substance. **TAKE CARE** not to damage this

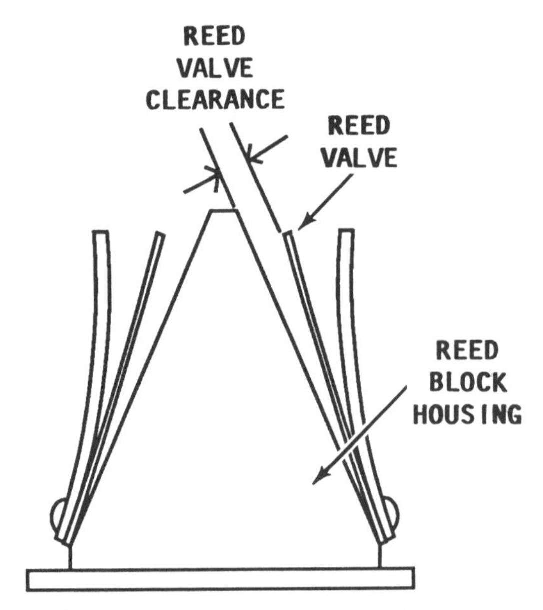

The maximum reed valve clearance for the 300 Series and the 650 Series is 0.007" (0.2mm).

coating. Once the assembly is removed, keep it away from sunlight, moisture, dust, and dirt. Sunlight can deteriorate valve seat rubber seals. Moisture can easily rust stoppers overnight. Dust and dirt -- especially sand or other gritty material -- can break reed petals if caught between stoppers and reed petals.

Make special arrangements to store the reed valves to keep them isolated from the elements while further work is being performed on the engine.

10- Lift out the two reed valve assemblies. Remove and discard the gaskets installed on both sides of the reed block housing -- four gaskets in all.

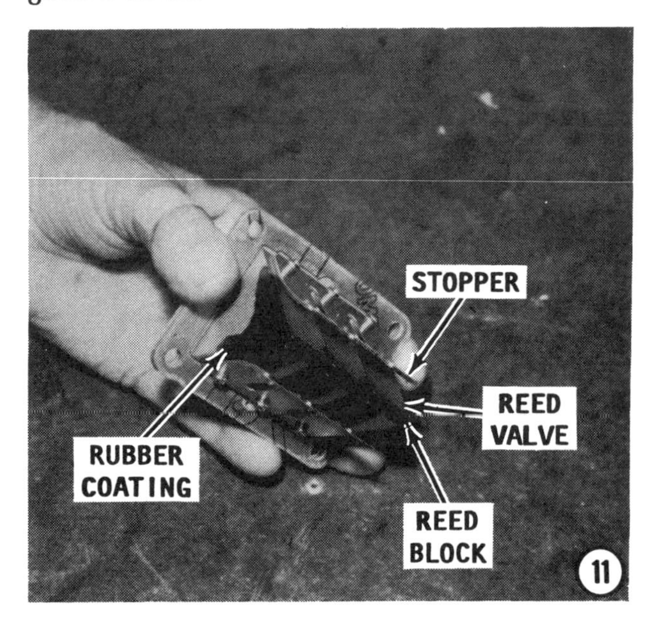

11- Measure the reed valve warpage with a feeler gauge. If the distance between the reed block housing and the reed valve is more than 0.2mm, the reed valve assembly **MUST** be replaced. Reed valve assemblies should be readily available from an authorized Kawasaki dealer in most areas.

READ AND BELIEVE

The reeds must never be turned over in an attempt to correct a problem. Such action would cause the reed to flex in the opposite direction and the reed would break in a very short time.

Block Disassembling

12- If servicing a Model 650 or Model 750 Series engine, remove the nuts securing the cylinder block to the upper crankcase. For the Model 550 Series, the block is secured between the head and the upper crankcase with the head nuts.

13- Keep the cylinder block as level as possible and lift it **STRAIGHT** up and clear of the studs. Two locating pins are indexed on one side of the Model 650 and Model 750 block and upper crankcase. These pins may remain with the cylinder block as it is lifted or the pins may remain with the lower crankcase. **TAKE CARE** not to lose these pins.

14- Remove and discard the base gasket.

15- Scribe a mark on the inside of the piston skirt to identify the No. 1 and No. 2 piston **BEFORE** the piston is removed from the connecting rod, as described in the next step.

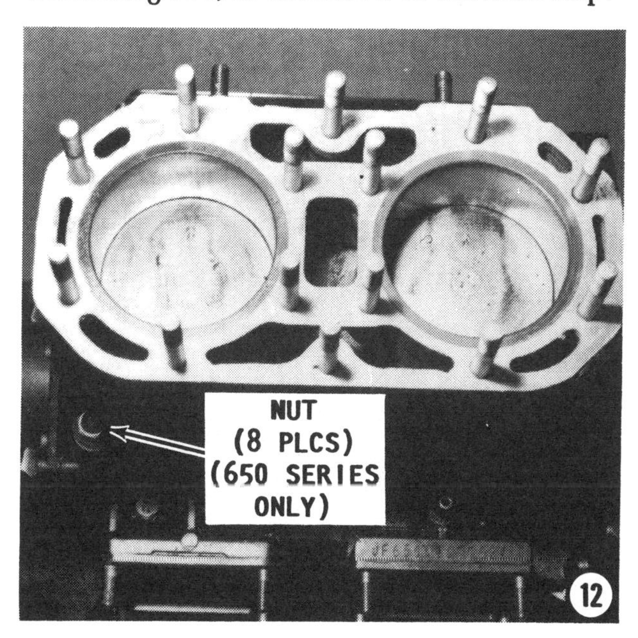

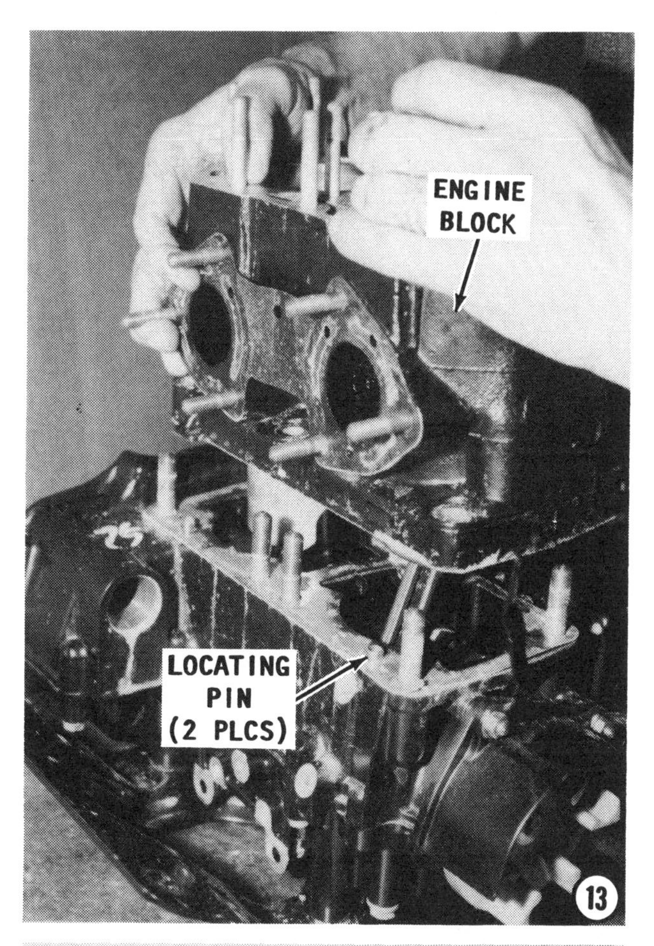

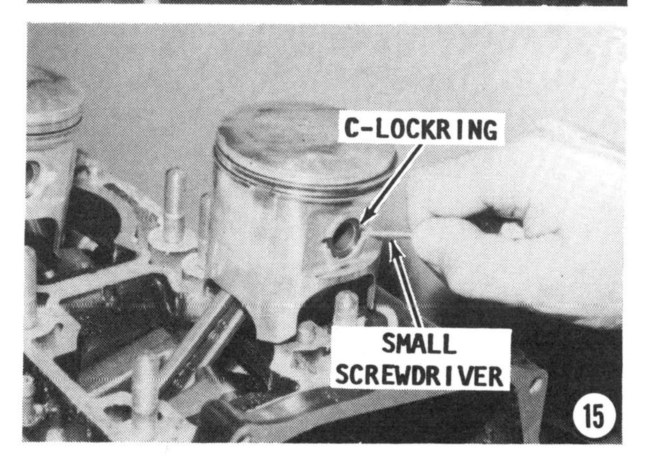

*C-lockrings are under tremendous tension during removal and installation, presenting a potential hazard. As an eye protection measure, safety glasses or a shield should **ALWAYS** be worn during work with such snap rings. Warn others in the area such work is in progress.*

SAFETY WORDS

The piston pin C-lockrings are made of spring steel and may pop out of the groove with considerable force. Therefore, warn other people in the area and **WEAR** eye protection glasses while removing the piston pin lockrings.

Remove the C-lockring from both ends of the piston pin using a screwdriver inserted into the small groove next to the piston pin opening. Discard the C-lockrings. These rings stretch during removal and must **NEVER** be used a second time.

16- Push the piston pin out free of the piston. If the pin is tight and refuses to budge, use the piston pin puller assembly, listed at the beginning of this chapter.

17- Remove the caged needle bearings from the small end of the connecting rod.

ADVICE
ALL SERIES ENGINES

New needle bearings should be installed in the connecting rods, even though they may appear to be in serviceable condition. New bearings will ensure lasting service after the overhaul work is completed. If it is necessary to install the used bearings, keep them separate and identified to **ENSURE** they will be installed onto the same connecting rod from which they were removed.

CRITICAL WORDS

The rod is an integral part of the crankshaft. The two are manufactured together and **CANNOT** be separated.

Slowly rotate the bearing races. If rough spots are felt, a new crankshaft must be purchased. The bearings are installed at the factory and are not replaceable.

GOOD WORDS

Good shop practice dictates to replace the rings during an engine overhaul. However, if the rings are to be used again, expand them **ONLY** enough to clear the piston and the grooves, because used rings are brittle and break very easily.

18- Gently spread the top piston ring enough to pry it out and up over the top of the piston. No special tool is required to remove the piston rings. Remove the lower ring in a similar manner. These rings are **EXTREMELY** brittle and have to be handled with care if they are intended for further service.

Be sure to keep all parts for each piston together (i.e. piston rings) for corresponding piston identification and installation, if more engine service work is required. In other words, invert the piston over and carefully place the matching C-lockrings, piston pin, rings, caged needle bearing and any applicable shim material, inside the piston for proper pairing of each piston assembly. Taking the time now to keep these parts together will **ENSURE** the correct parts will be installed with the proper cylinder during installation.

Engine Bed Bracket Removal

19- Turn the engine upside down on the work surface. Remove the four bolts securing the engine bed bracket to the lower crankcase.

Crankcase Separation

20- Remove the seven 6mm bolts and eight 8mm bolts securing the two halves of the crankcase together. Two prying tabs are cast into the crankcase to assist in separating the two halves. **ALWAYS** pry **ONLY** at these tab locations. Prying elsewhere could very likely damage an aluminum sealing surface.

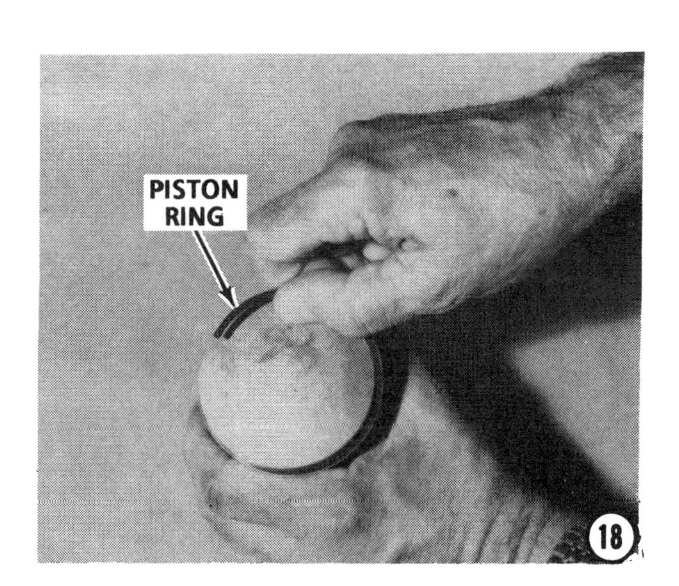

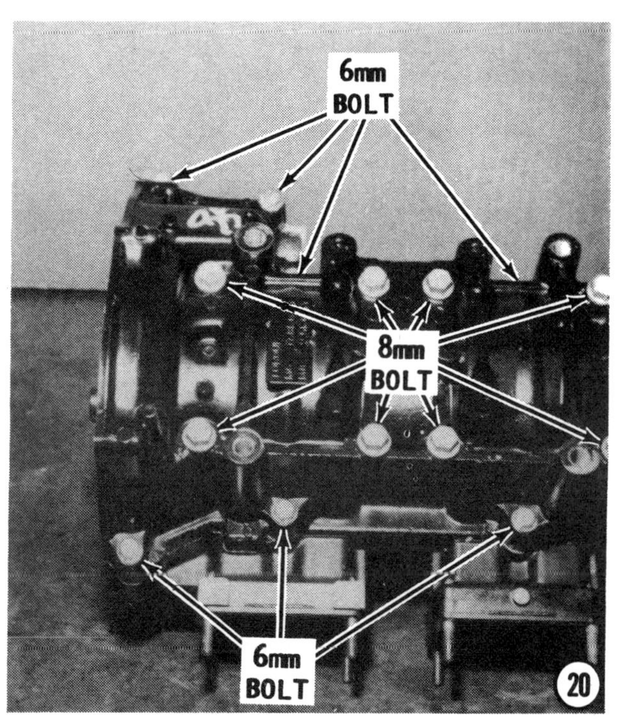

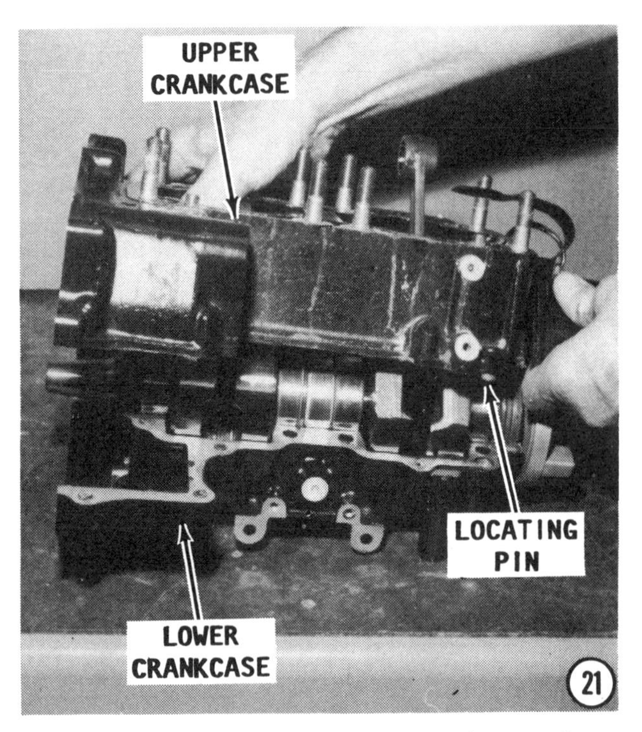

21- Separate the lower crankcase from the upper crankcase. **TAKE CARE** not to lose the two locating pins indexed diagonally across from each other. These pins may remain in either half when the crankcase is separated. Tap the tapered end of the crankshaft with a soft head mallet to "jar" it free of the crankcase.

Crankshaft Disassembling

22- Pull the forward oil seal free of the tapered end of the crankshaft. If the metal coupler was removed from the threaded end of the crankshaft, remove the two rear main oil seals.

CLEANING AND INSPECTING

COMPLETE and **DETAILED** instructions for cleaning and inspecting **ALL** parts of the engine, including the block, will be found in Section 8-7, beginning on Page 8-67.

8-4 ASSEMBLING AND INSTALLATION TWO-CYLINDER ENGINES

Detailed procedures are given to assemble virtually all parts of the engine, followed by instructions to install the engine.

Again, these procedures are identified for two-cylinder engines. Another section within this chapter is devoted to the disassembly, cleaning and inspecting, and assembly of three-cylinder engines, beginning on Page 8-45.

If certain parts were not removed or disassembled because the part was found to be fit for further service, simply skip the particular step involved and continue with the required tasks to return the engine to operating condition.

The following instructions, beginning on page 8-26 (after the exploded drawings), pickup the work after all parts of the engine have been thoroughly cleaned and inspected according to the procedures outlined in Section 8-7.

The Cleaning and Inspecting section should have revealed any parts unfit for further service.

ADVICE

Before commencing the assembling tasks, check to be sure replacement parts have been obtained and are on hand.

EXPLODED DRAWINGS

The exploded drawings of two-cylinder engines on Pages 8-21 through 8-25 will prove to be helpful during the assembling work.

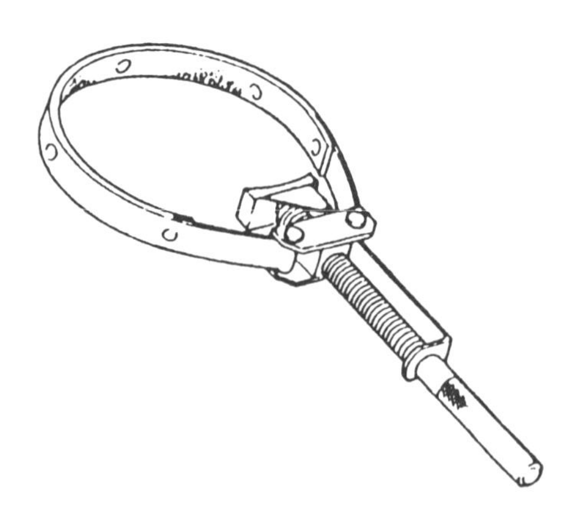

Quick line drawing of the flywheel holding tool used when installing or removing the flywheel on a Model 750 Series engine. This tool is necessary in order to tighten the flywheel nut to the required torque value.

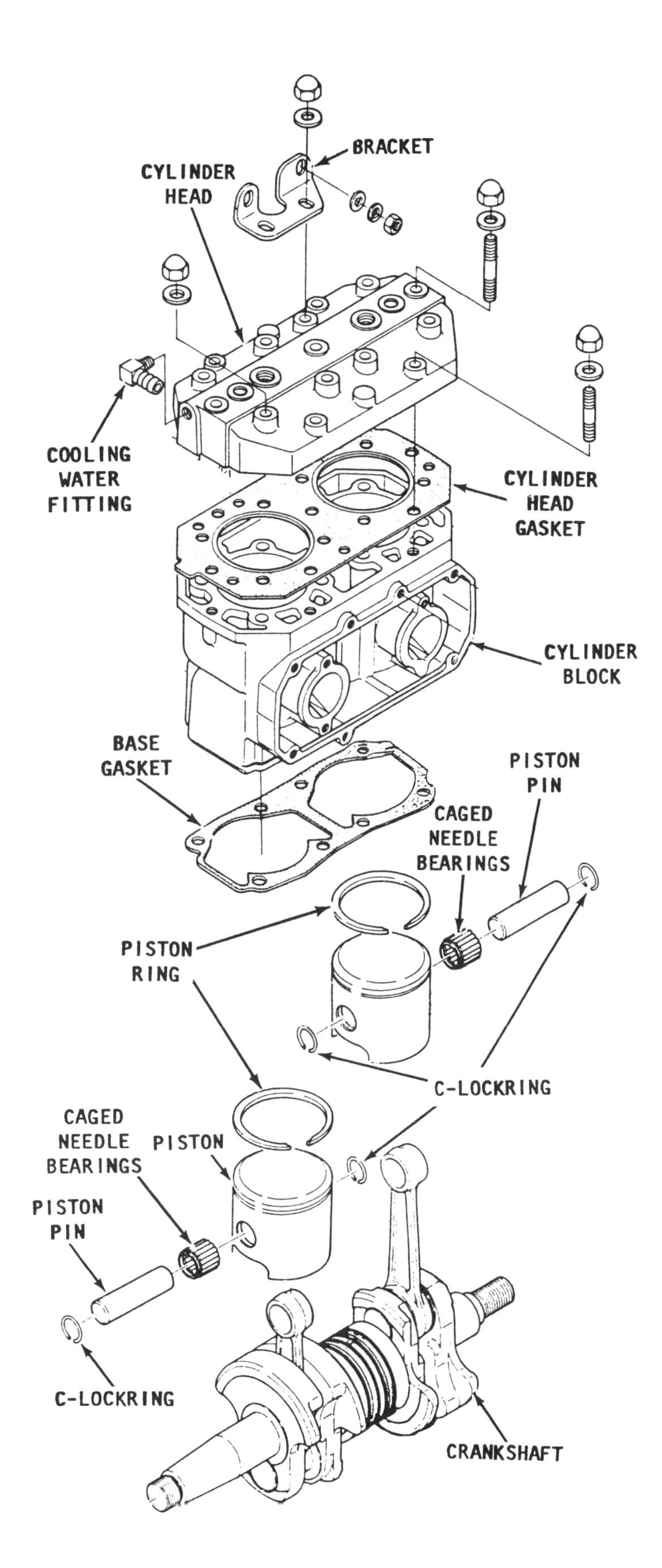

Exploded drawing of the upper half of the Model **550** *Series crankcase, with major parts identified.*

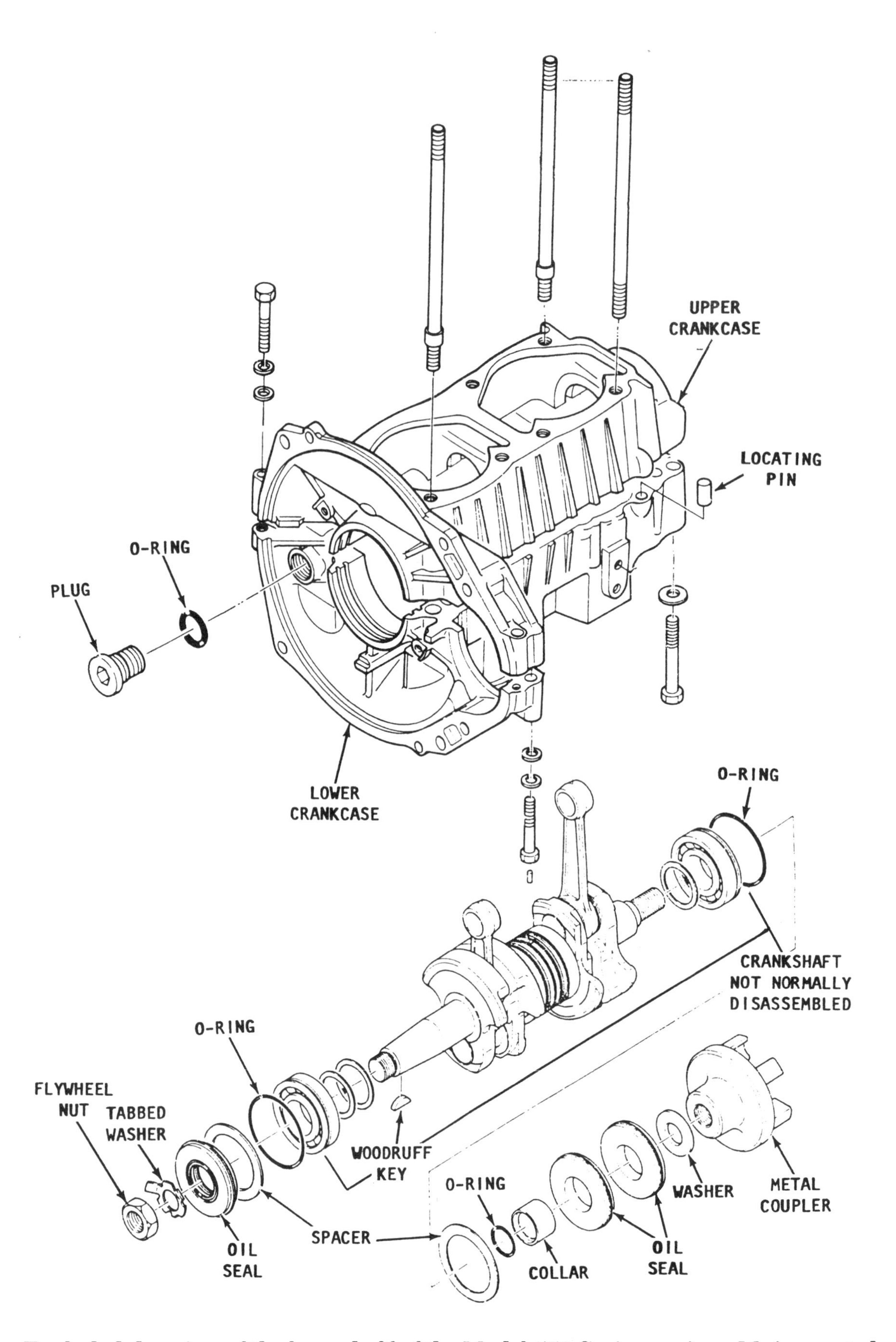

Exploded drawing of the lower half of the Model 550 Series engine. Major parts have been identified.

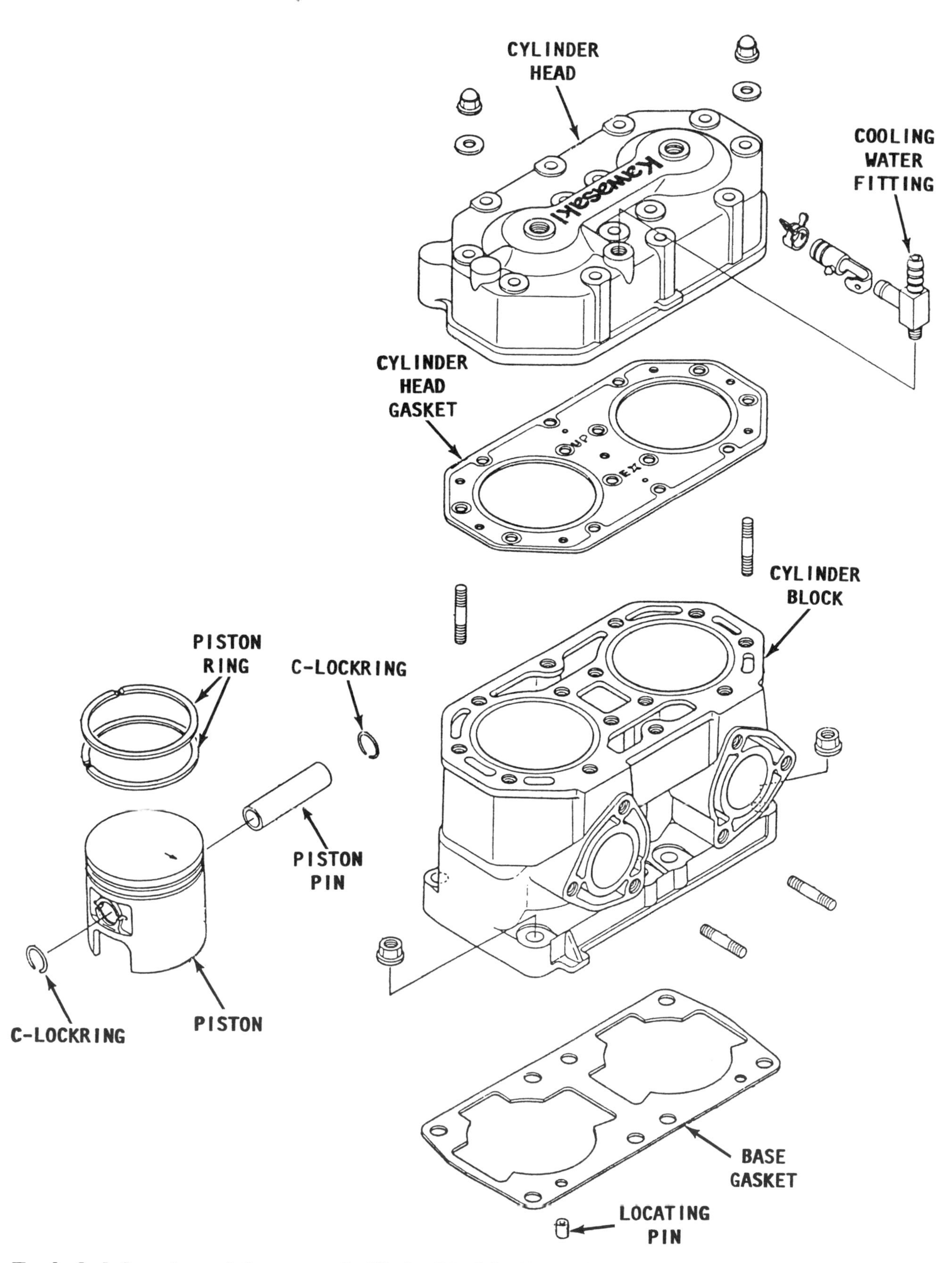

Exploded drawing of the upper half of a Model **650** *and the Model* **750** *Series engine with major parts identified. The lower crankcase half is shown on the following page.*

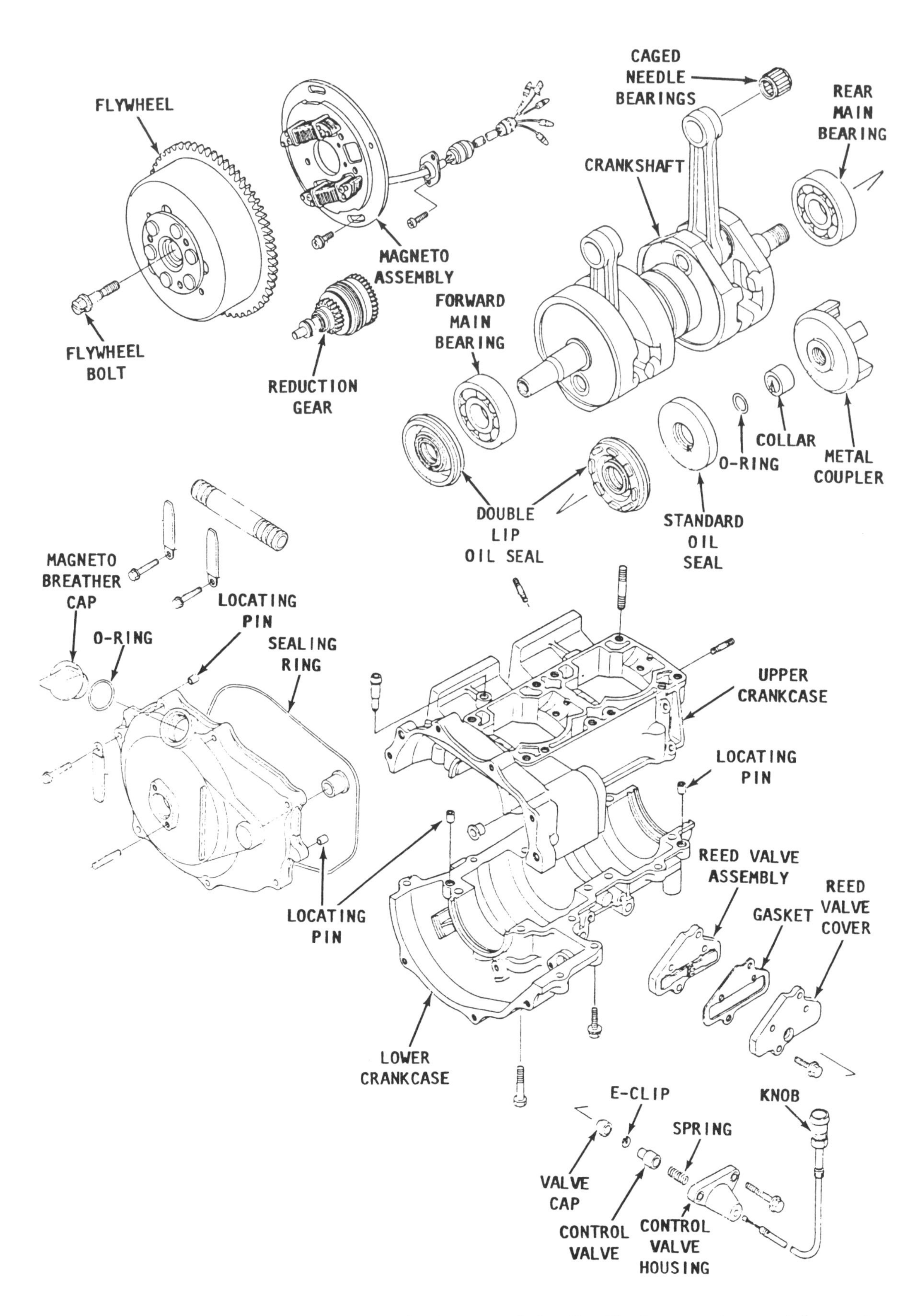

*Exploded drawing of the lower half of a Model **650** Series engine, with major parts identified. The upper crankcase half is shown on the previous page.*

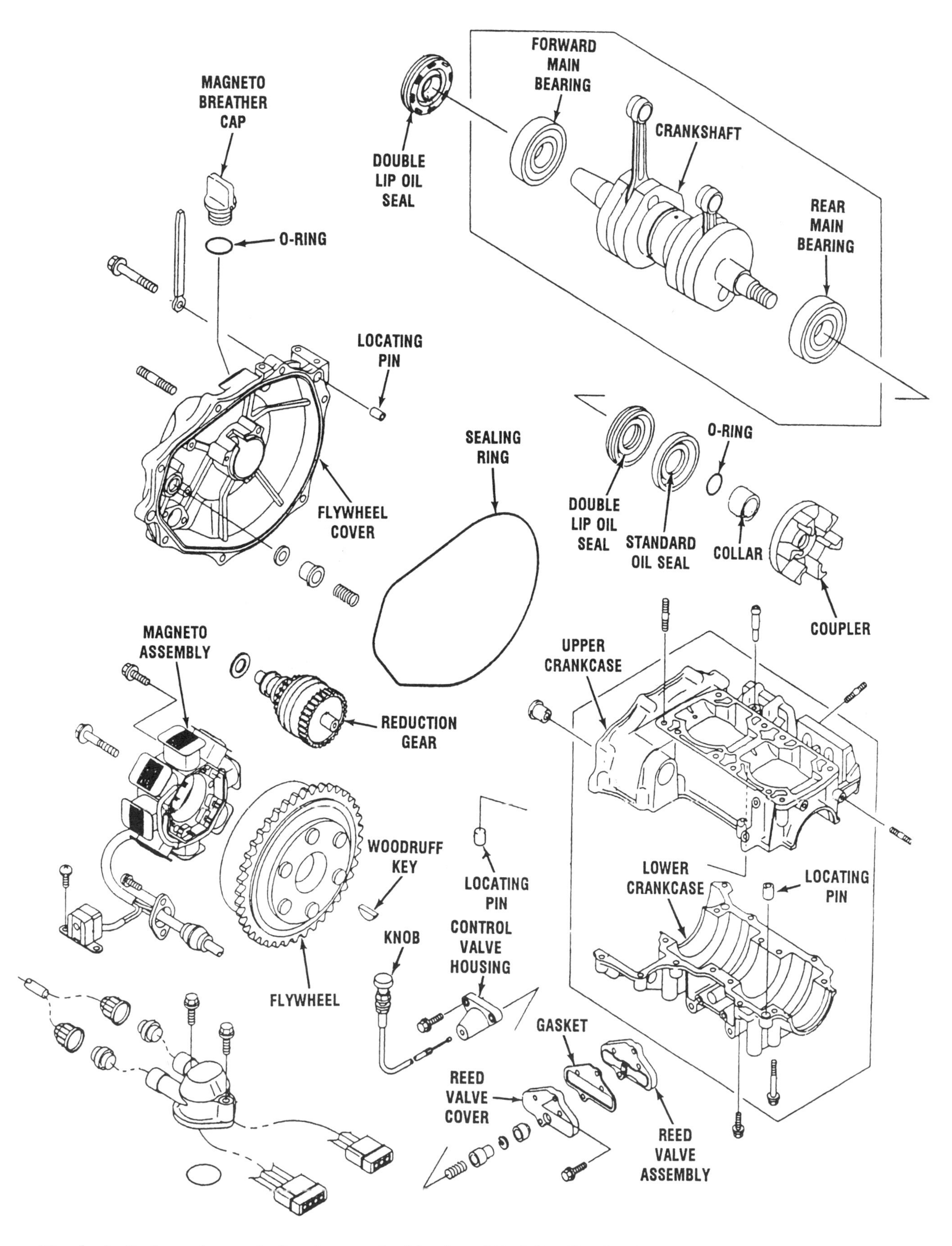

Exploded drawing of the lower half of a Model ***750*** *Series engine, with major parts identified. The upper crankcase half is shown on the previous page.*

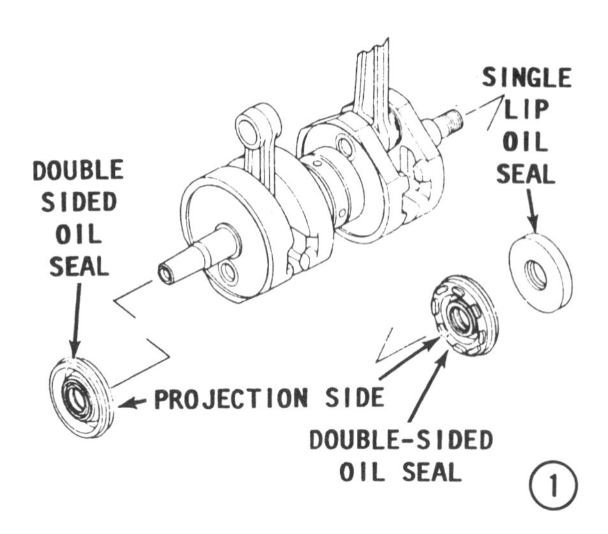

Applying Liquid Gasket Compound to the mating surface of a three-cylinder lower crankcase. The same material and technique is valid for a two-cylinder engine.

ASSEMBLING LOWER HALF

1- Pack the lips of the three oil seals with Shell Alvania EP1, or equivalent, lubricant. Install the double sided seals --one onto each end of the crankshaft, with the projection side facing **TOWARD** the crankshaft journals. Install the single lip seal onto the threaded end of the crankshaft with the seal lip facing **AFT.**

2- Apply a coating of Liquid Gasket compound to the mating surface of the lower crankcase -- looping around the bolt and locating pin holes.

3- Seat the crankshaft assembly into the lower crankcase half and fit the labyrinth seal circlip into the groove in the crankcase. Check to be sure the locating pins are in place on one of the crankcase halves.

4- Lower the upper crankcase half over the crankshaft -- guiding the connecting rods through the two openings. Bring the two crankcase halves together with the locating pins indexed into the matching holes in the other half.

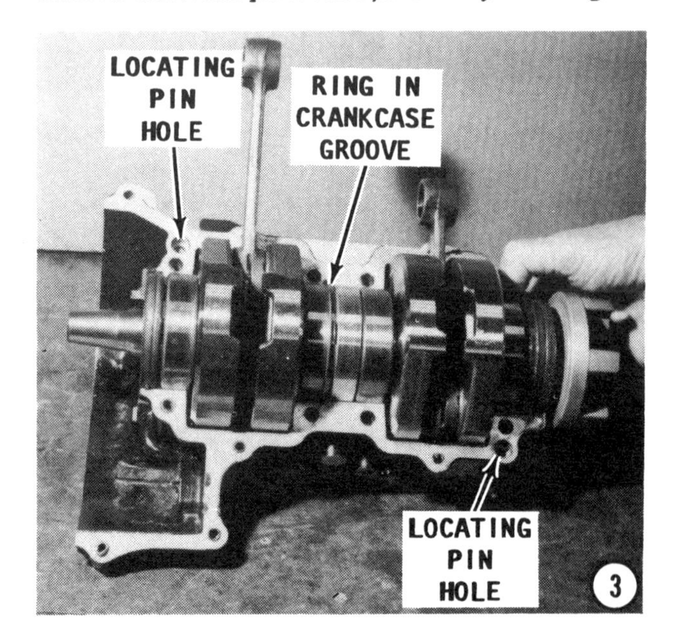

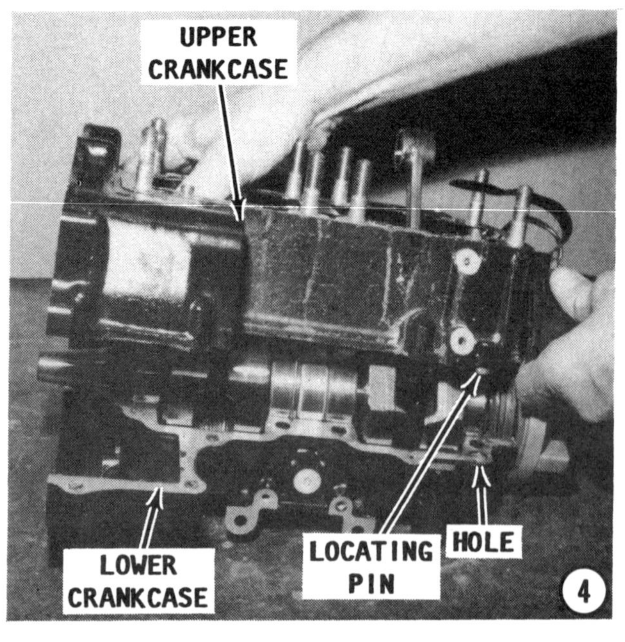

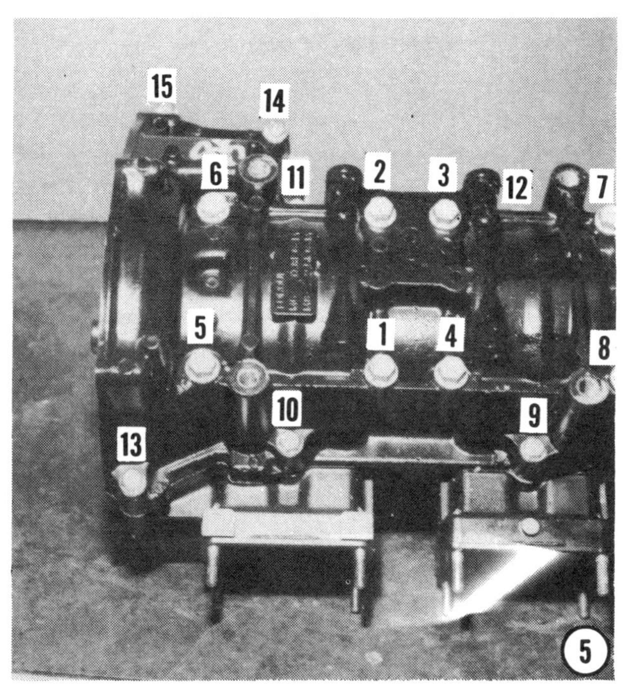

5- Apply a coating of non-permanent locking compound to the threads of the crankcase bolts.

CRITICAL BOLT WORDS

Refer to the accompanying illustrations for this step. Step 5 illustration applies to the 650 Series and 750 Series engines. The line drawing captioned illustration applies to the 550 Series engines.

For all engines, bring the bolts to "just snug". Then, tighten the bolts in the sequence shown to 1/2 torque value. Once the bolts are tightened to the first increment, repeat the sequence and bring the bolts to full torque.

For the 550 Series -- bolt numbers 1 thru 10 are the large bolts with flat washers. These bolts are to be tightened to a torque value of 16 ft lb (22Nm). Bolts numbered 11 thru 13 are smaller with flat washers **AND** lockwashers. These bolts are to be tightened to only 52 in lb (6Nm).

For the 650 Series -- bolt numbers 1 thru 8 are 8mm bolts and are to be tightened to a torque value of 18 ft lb (25mm). Bolts numbered 9 thru 15 are 6mm bolts and are to be tightened to only 69 in lb (7.8Nm). These fifteen bolts do not have flat washers or lockwashers.

For the 750 Series -- bolt numbers 1 thru 8 are 8mm bolts and are to be tightened to a torque value of 22 ft lb (29Nm). Bolts numbered 9 thru 15 are 6mm bolts and are to be tightened to only 69 in lb (7.8Nm). These fifteen bolts do not have flat washers or lockwashers.

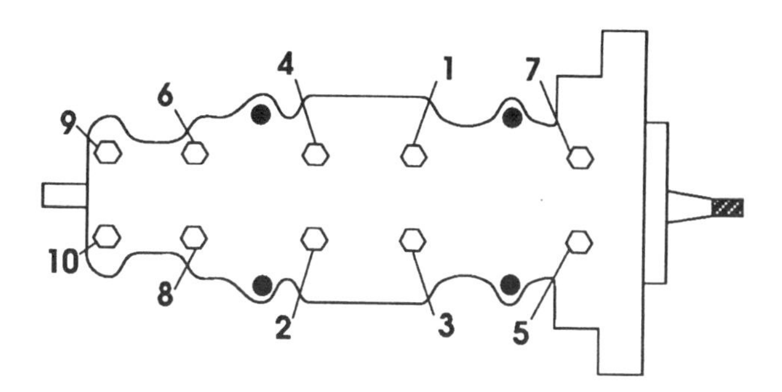

Crankcase tightening sequence for the Model 550 and Model 650 Series engines.

Pay attention -- to ensure the correct bolts are threaded into the proper openings.

Now, rotate the crankshaft by hand to be sure the crankshaft does not bind.

BAD NEWS

If binding is felt, it will be necessary to remove the crankcase and reseat the crankshaft. It will also be necessary to check the positioning of the labyrinth seal circlips and the bearing locating pins. If binding is still a problem after the crankcase has been installed a second time, the cause might very well be a broken piston ring.

Engine Bed Bracket Installation

6- With the engine inverted, position the engine bed bracket over the crankcase. For a 550 Series engine, the cutaway corners of the bracket **MUST** face **FORWARD**. For a 650 or 750 Series engine, the notches at the corners of the engine bed bracket, **MUST** be positioned as shown in the accompanying illustration.

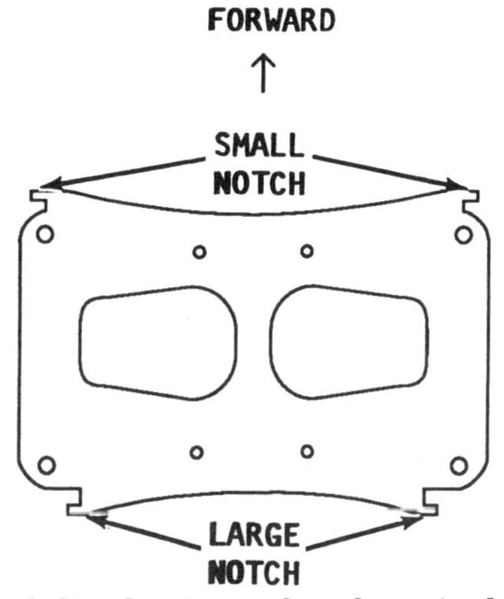

Simple line drawing to show the engine bed bracket with the large notches and the small notches, mentioned in the text.

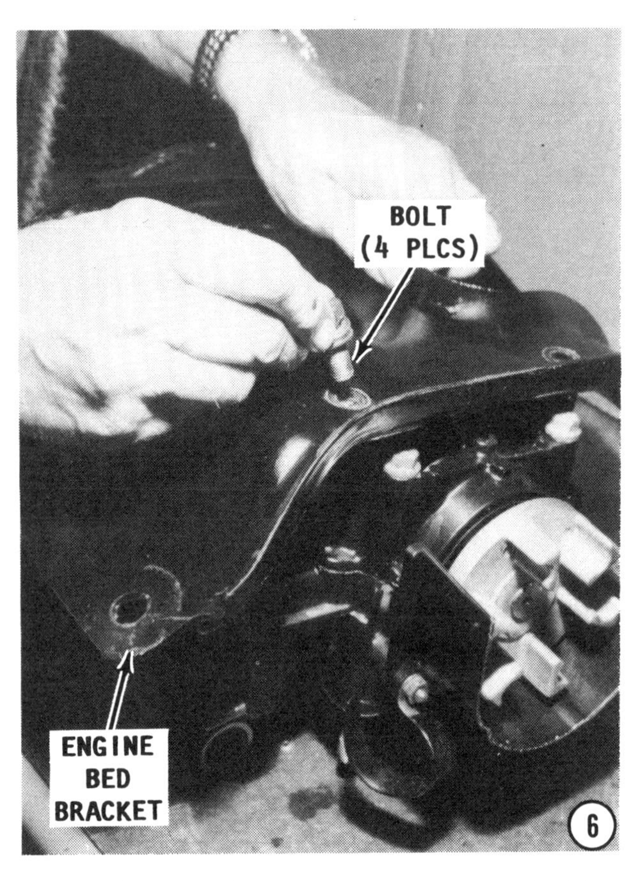

Apply a coating of non-permanent locking compound to the threads of the mounting bolts. Install and tighten the bolts to the proper torque value listed below for the appropriate engine series.

550 Series	35 ft lb (48Nm)
650 Series	27 ft lb (36Nm)
750 Series	27 ft lb (36Nm)

The 550 Series bolts have flat washers and lockwashers. Washers are not used on the 650 Series and 750 Series, they use flanged bolts.

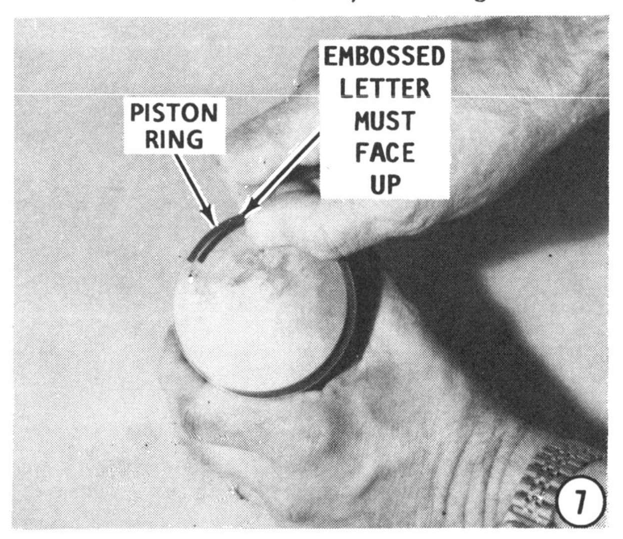

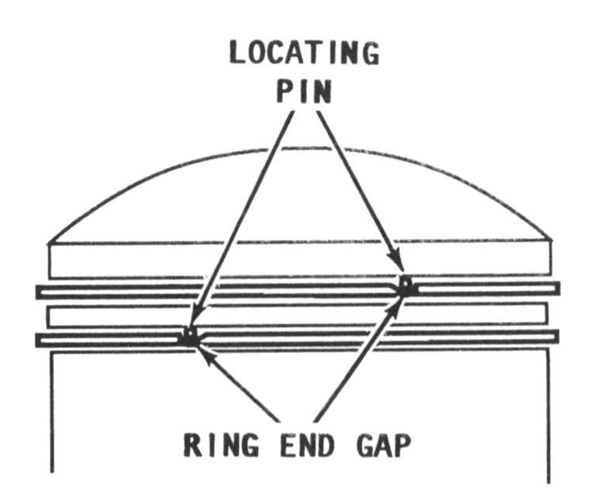

Align the ring end gap over the locating pin in each piston groove to prevent the ring from rotating in the groove. If the ring was free to rotate, an end might very well catch on the port opening and break.

Piston Installation
All Series Engines

7- Observe the embossed letter on each piston ring. The ring **MUST** be installed with this letter facing **UP.**

Install a new set of piston rings onto the piston/s. No special tool is necessary to install the piston rings. **HOWEVER,** take care to spread the ring only enough to clear the top of the piston. The rings are **EXTREMELY** brittle and will snap if spread beyond their limit. Align the ring end gap over the locating pin in each piston groove.

The purpose of the pin is to prevent the ring from rotating in the ring groove. The ring end gap must travel up and down a portion of the cylinder wall where there is no port opening. If the ring was free to rotate, an end might catch on the port opening and break the ring.

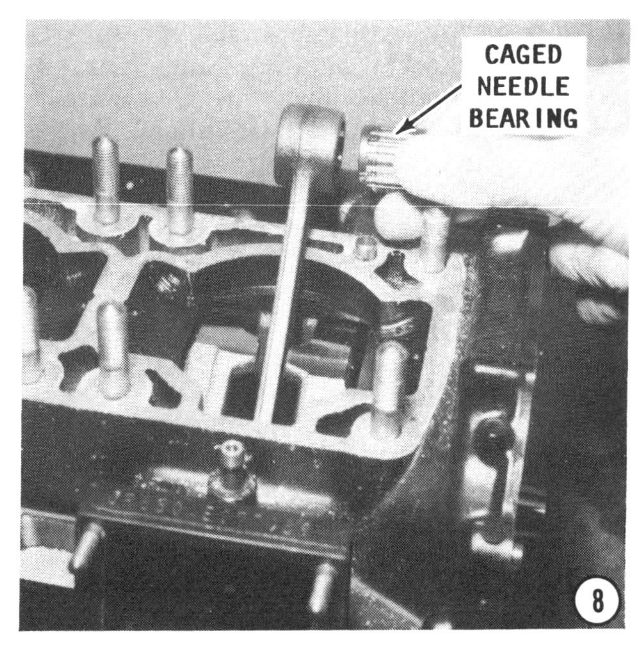

8- Dip the caged needle bearings into a container of engine oil, and then insert the cage into the small end of the connecting rod.

CRITICAL WORDS

An arrow is embossed on each piston crown. The direction of this arrow **MUST** be **TOWARD** the exhaust manifold -- port side.

9- Position the piston over the connecting rod with the arrow embossed on the crown facing **TOWARD** the exhaust manifold -- port side. Insert the piston pin into the piston and caged needle bearings. Center the pin in the piston.

SAFETY WORDS

The piston pin C-lockrings are made of spring steel and may slip out of the groove with considerable force. Therefore, warn other persons in the area and **WEAR** eye protection glasses while installing the piston pin lockrings in the next step.

10- Install a C-lockring at each end of the piston pin. The lockrings may be worked in by hand. Position the end gap away from the cutout on the piston opening, as shown in the accompanying line drawing.

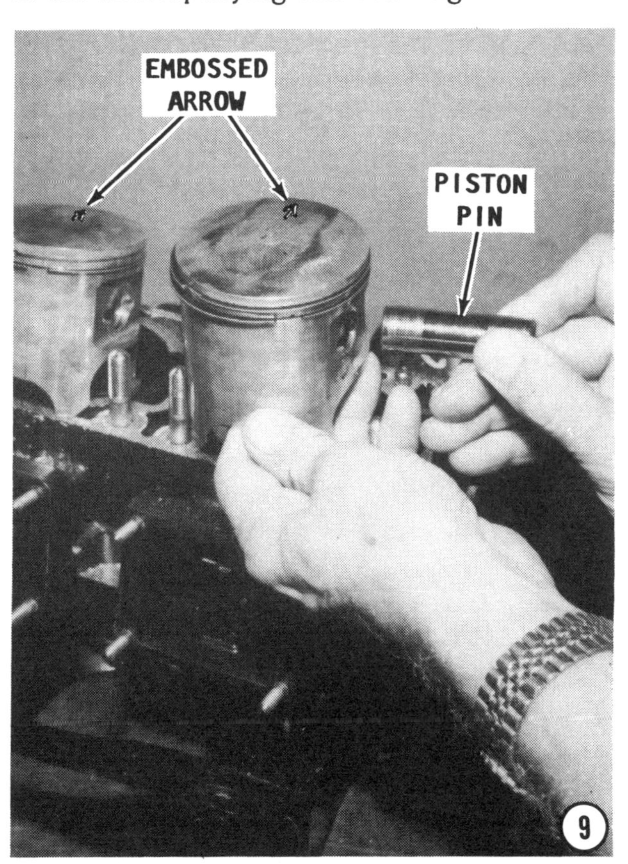

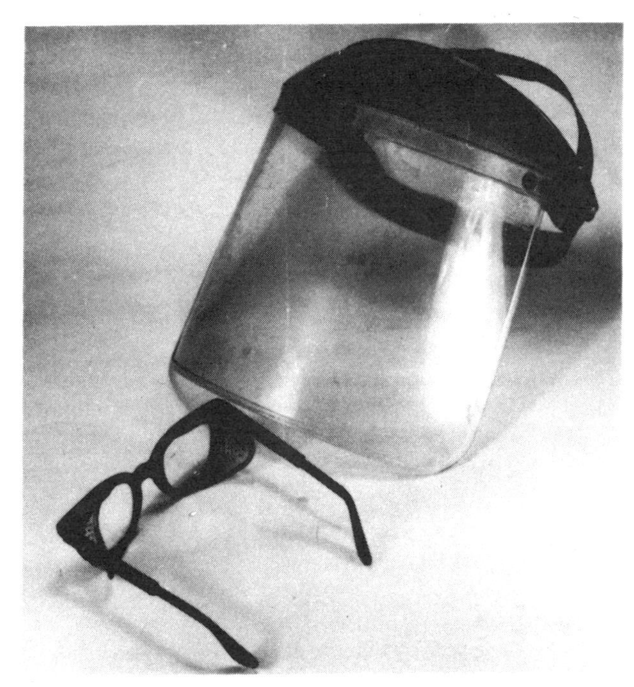

C-lockrings are under tremendous tension during removal and installation, presenting a potential hazard. As an eye protection measure, safety glasses or a shield should ***ALWAYS*** *be worn during work with such snap rings. Warn others in the area such work is in progress.*

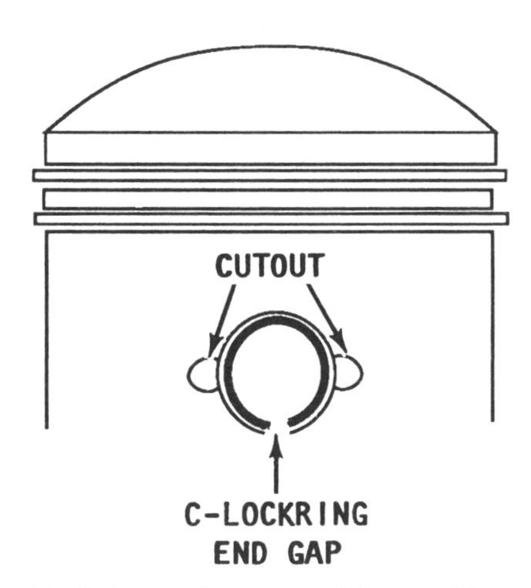

The C-lockring end gap must be positioned away from the two cutouts in the piston, as shown.

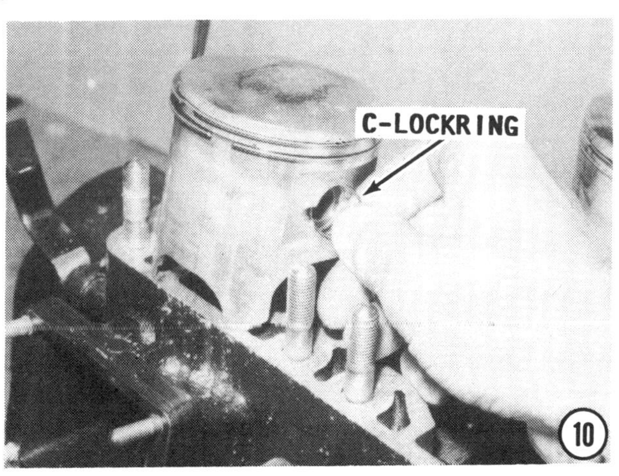

Block Installation

11- Install the base gasket over the crankcase studs. The manufacturer does not call for any type material be applied to either side of this gasket.

12- Coat the upper portion of each piston and the rings with 2-stroke engine oil.

Hold both pistons at right angles to the crankshaft. **SLOWLY** lower the cylinder block down over the long studs and onto the piston crown. Squeeze the rings inward around the locating pin and gently push the piston/s into the bore/s.

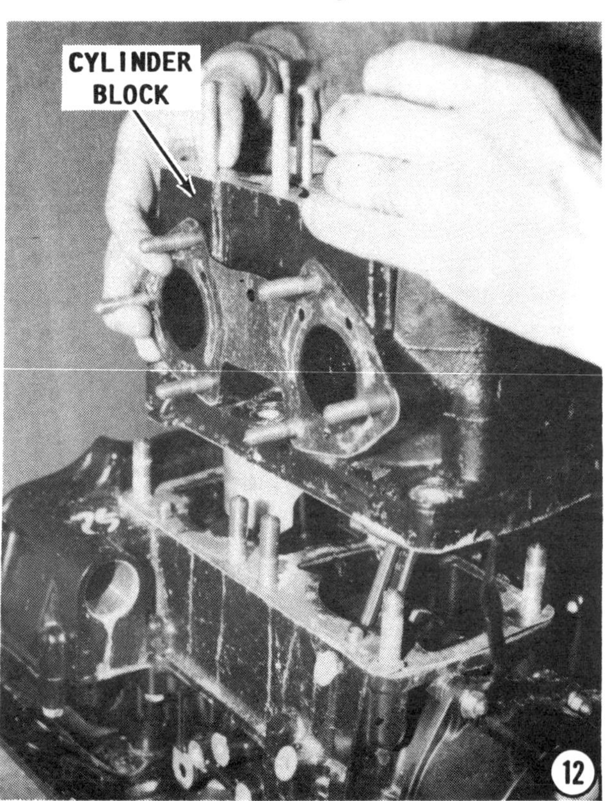

CRITICAL WORDS

If difficulty is experienced in fitting the piston/s into the cylinder/s, **DO NOT** force the piston. Such action might result in a broken piston ring. Make sure the ring end gap is aligned with the locating pin.

If more help is needed, obtain a large hose clamp and install the clamp around the piston to compress the rings. Try again -- it can be done.

13- Make a final check of the piston installation. The arrow **MUST** face **TOWARD** the exhaust side of the engine.

If servicing a 650 Series engine, install the nuts securing the cylinder block to the upper crankcase. Cross-tighten the nuts to a torque value of 25 ft lb (34Nm). For all other series engines, the block is secured between the head and the upper crankcase with the head nuts.

Reed Block Installation
650 Series Only

14- Observe the small hole in the long side of the reed block housing and the cutouts in the gaskets. Position a gasket on each side of the reed block housing. Slide the reed block housing over the mounting studs, with the hole at the bottom, as shown.

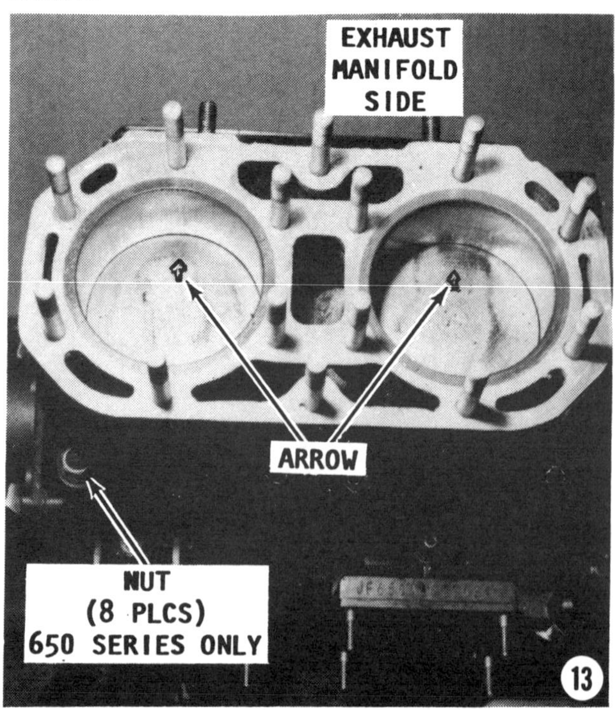

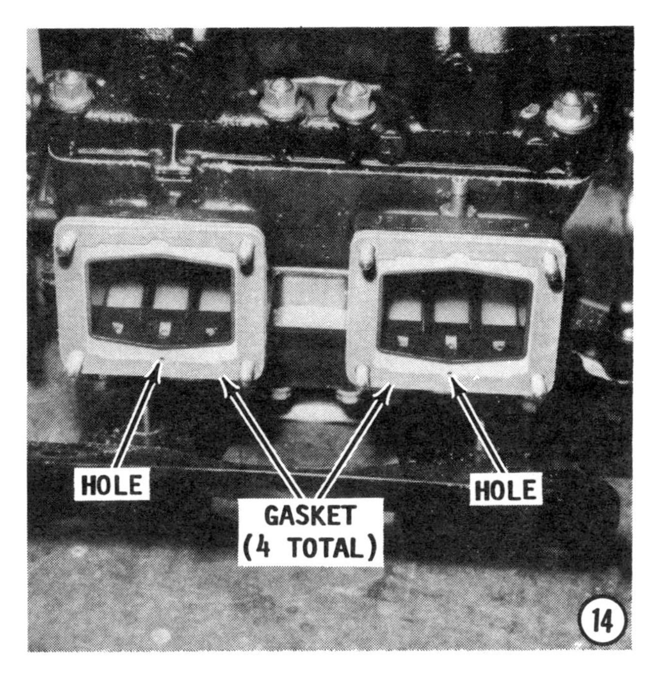

All Series Engines

15- Install the intake manifold. Tighten 550 and 650 Series mounting nuts to a torque value of 69 in lb (7.8Nm), and 750 mounting nuts to 87 in lb (9.8Nm).

Exhaust Manifold Installation
All Series Engines

16- Position a new exhaust manifold gasket onto the block.

17- Install the exhaust manifold to the engine block. Apply a coating of Loctite Lock N' Seal to the threads of the mounting studs of the 650 Series or 750 Series. or bolts. The 550 Series engines have exhaust manifold mounting bolts. Tighten the stud nuts and or bolts to the following torque values: 550 Series bolts -- 52 in lb (6Nm); 650 Series nuts -- 18 ft lb (25Nm); 750 Series nuts -- 13.5 ft lb (19Nm).

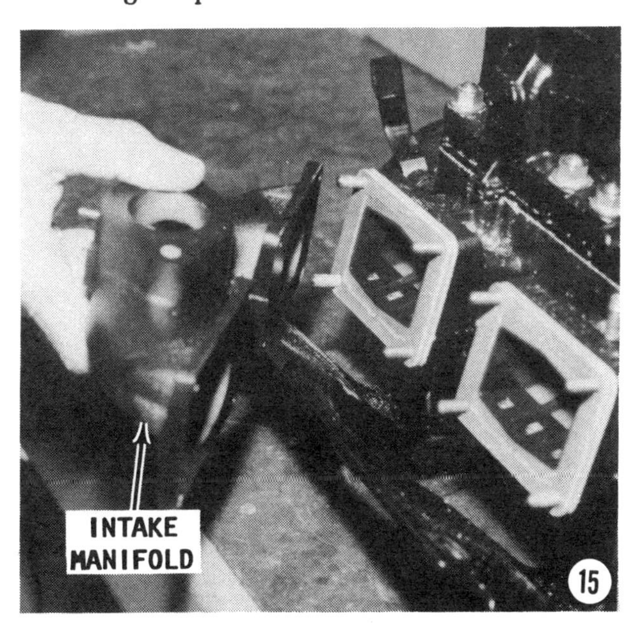

Cylinder Head Installation

18- Observe the words **"UP"** and **"EX"** embossed on a new cylinder head gasket. The gasket **MUST** be installed with these words facing **UP** and with the **"EX"** toward the exhaust -- port side -- of the engine. Install the gasket as described.

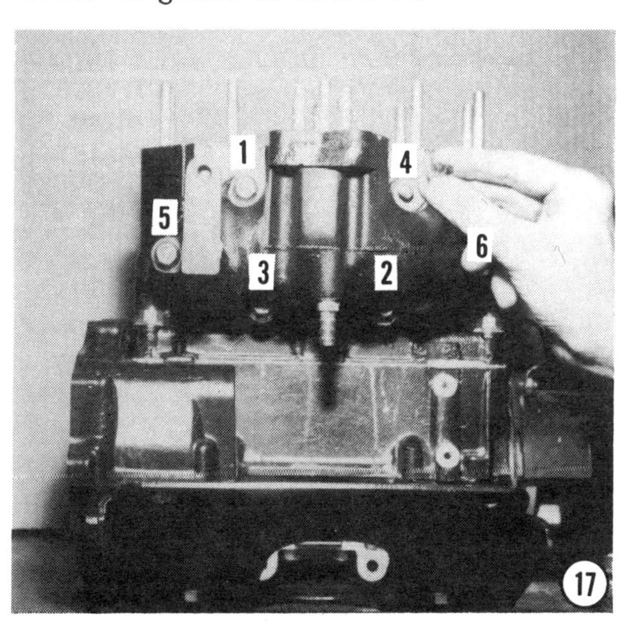

SPECIAL WORDS

The manufacturer recommends **NO** sealing agent be used on either side of the head gasket.

SEALING SURFACE WORDS
CRITICAL READING

Because of the high temperatures and pressures the sealing surfaces of the cylinder head and the block are the most prone to water leaks. No sealing agent is recommended **BECAUSE** it is almost impossible to apply an even coat of sealer. An even coat would be essential to ensure a air/water tight seal.

Some head gaskets are supplied with a "tacky" coating on both surfaces applied at the time of manufacture. This "tacky" substance will provide an even coating all around. Therefore, no further sealing agent is required.

HOWEVER, if a slight water leak should be noticed following completed assembly work and engine start up, **DO NOT** attempt to stop the leak by tightening the head bolts beyond the recommended torque value. Such action will only aggravate the problem and most likely distort the head.

FURTHERMORE, tightening the bolts, which are case hardened aluminum, may force the bolt beyond its elastic limit and cause the bolt to **FRACTURE. BAD NEWS,** very **BAD NEWS** indeed. A fractured bolt must usually be drilled out and the hole retapped to accommodate an oversize bolt, etc. Avoid such a situation.

Probable causes and remedies of a new head gasket leaking are:

a- Sealing surfaces not thoroughly cleaned of old gasket material. Disassemble and remove **ALL** traces of old gasket.

b- Damage to the machined surface of the head or the block. The remedy for this damage is the same as for the next case **"c"**.

c- Permanently distorted head or block. Spray a light **EVEN** coat of any type metallic spray paint on both sides of a new head gasket. Use only metallic paint -- any color will do. Regular spray paint does not have the particle content required to provide the extra sealing properties this procedure requires.

Assemble the block and head with the gasket while the paint is still **TACKY**. Install the head bolts and tighten them in the recommended sequence and to the proper torque value -- **NO MORE!**

Allow the paint to set for at least **24** hours before starting the engine. Consider this procedure as a temporary "band aid"

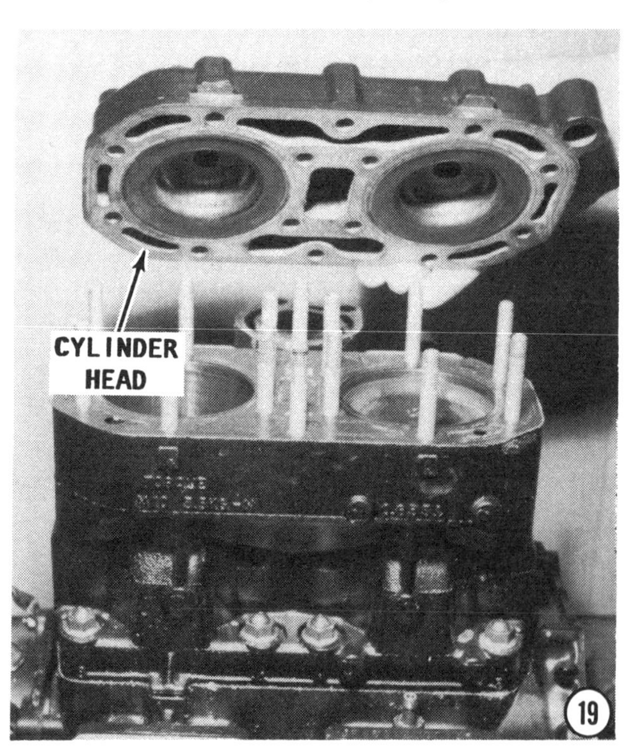

Cylinder head tightening sequence for the engines covered in this manual are almost always embossed on the head, as shown here. If for some reason they cannot be deciphered, the sequences are provided in the adjacent column.

type solution until a new head may be purchased or other permanent measures can be performed.

Under normal circumstances, if procedures have been followed to the letter, the head gasket will not leak.

19- Lower the head down over the mounting studs and onto the engine block.

20- Install the washers and nuts onto the studs. Tighten the nuts to the following

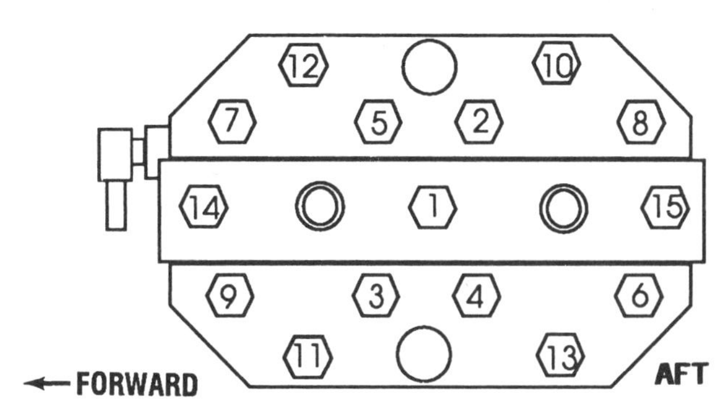

Cylinder head bolt tightening sequence for the Model 550 Series engines.

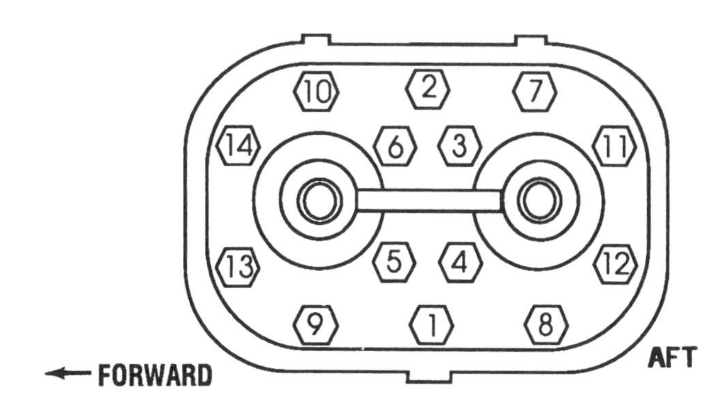

Cylinder head bolt tightening sequence for the Model 650 Series engines.

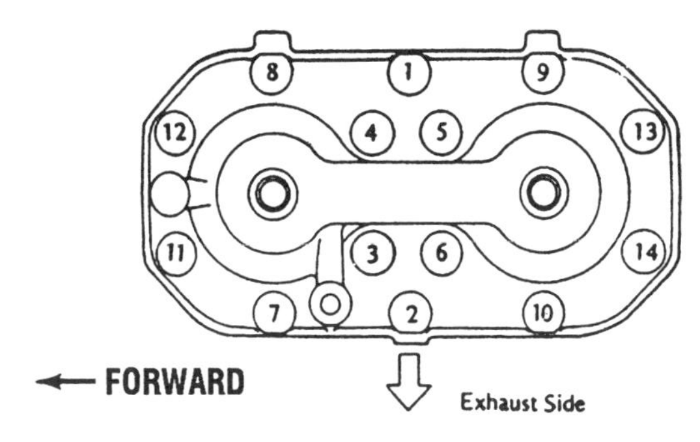

Cylinder head bolt tightening sequence for the Model 750 Series engines.

Head Bolt Torque Values

550 Series	16 ft lb (22Nm)
650 Series	22 ft lb (29Nm)
750 Series	22 ft lb (29Nm)

Water Drain Valve
650 Series and 750 Series

21- Position the reed valve cover over the reed valve assembly. Install and tighten the two securing bolts.

22- Install the control valve housing over the reed valve cover, with the valve

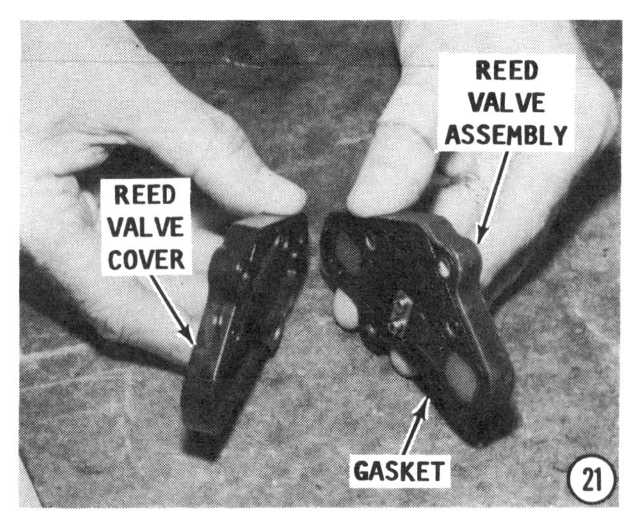

cap indexed into the drain hole. Lock the water drain valve. Pull the water drain valve out, and then tighten the locknut. Apply a coating of non-permanent locking compound to the threads of the two mounting bolts. Loosen the locknut slightly until the knob returns by spring tension.

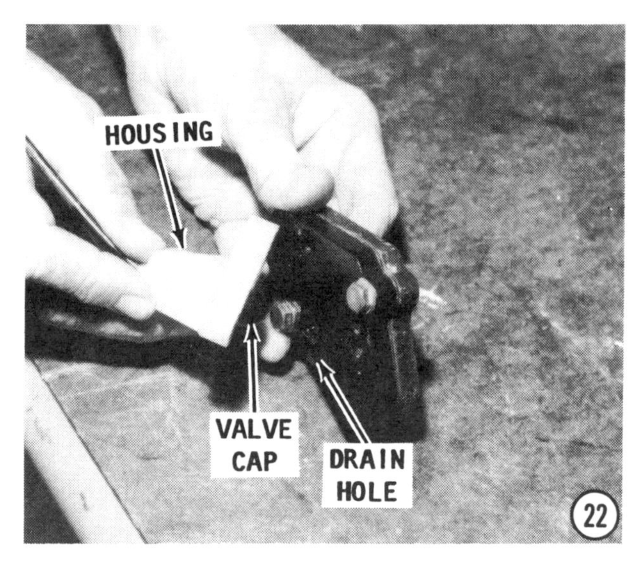

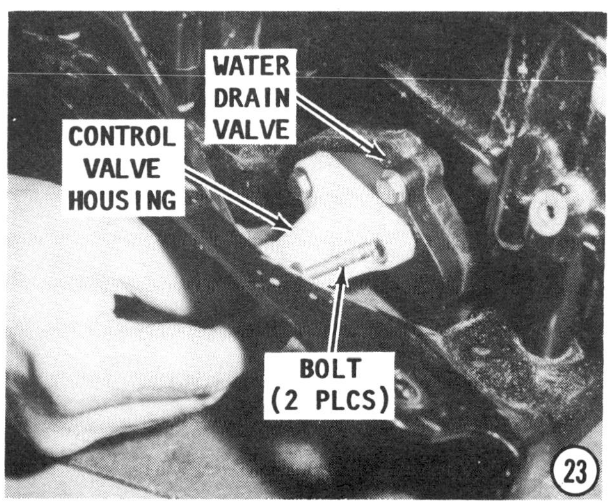

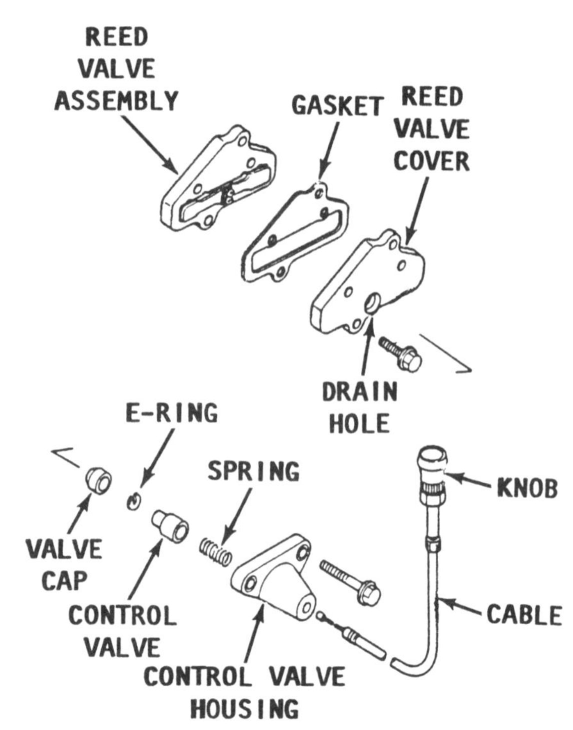

Exploded drawing of the water drain valve installed only on the 650 Series engines. Major parts are identified.

23- Install the water drain valve to the block and tighten the mounting bolts securely.

Flywheel Installation

24- Apply a coating of multi-purpose water resistant lubricant to the grommet in place on the stator harness. Feed the

harness connector through the opening in the stator plate and position the magneto assembly over the end of the crankshaft. Push the grommet into the opening and secure it in place with the two attaching screws. Tighten both screws securely.

Timing Marks

Identify the two marks made on the magneto assembly and the stator plate prior to removal. These marks serve to correctly position the magneto plate on the stator plate. The securing screws pass through elongated slots in the magneto assembly. Therefore, the assembly could be installed as much as 15° to its original position, drastically affecting the timing characteristics of the engine. If a **NEW** magneto assembly is being installed, match the new assembly to the old assembly and make a new mark as close to the original location as possible. The engine must be timed according to the procedures outlined in Section 7-7, of Chapter 7.

If **NO** timing marks were made prior to removal of the magneto assembly, as instructed in Disassembling, closely inspect the elongated slots in the magneto assembly and attempt to determine the original positioning of the screws. If no identifiable mark can be found, center the magneto assembly in the slot. This position may or may not be close to the correct position to permit the engine to start. Detailed timing procedure is presented in Section 7-7, of Chapter 7.

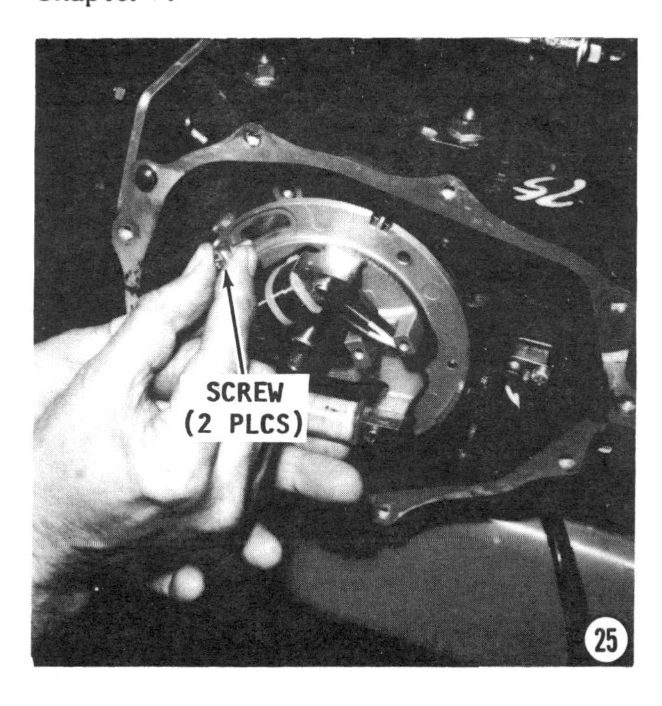

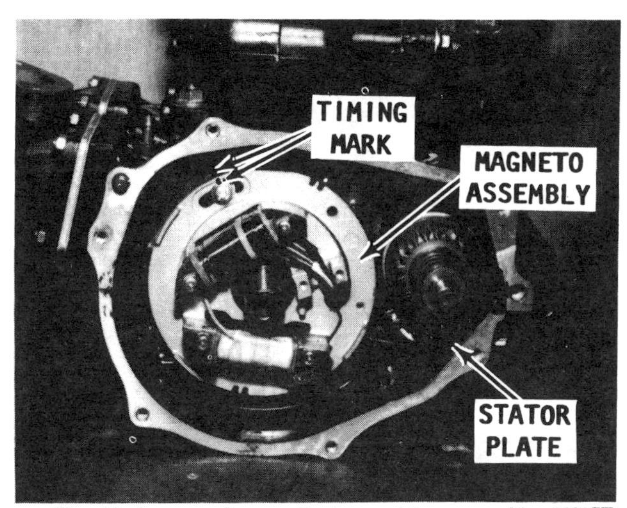

*The timing marks on the magneto assembly **MUST** align with the marks on the stator plate during installation.*

25- Apply a coating of Loctite #572, or equivalent, to the threads of the two magneto securing screws. Align the mark on the magneto assembly with the mark on the stator plate. Install and tighten the two screws securing the assembly to the stator plate **WITHOUT** disturbing the two aligned marks.

Flywheel Installation

FIRST, THESE WORDS

If the engine is to be timed, do not hesitate to install the flywheel using the Woodruff key and tightening the flywheel nut/bolt to the required torque value. The flywheel has cutouts affording access to the

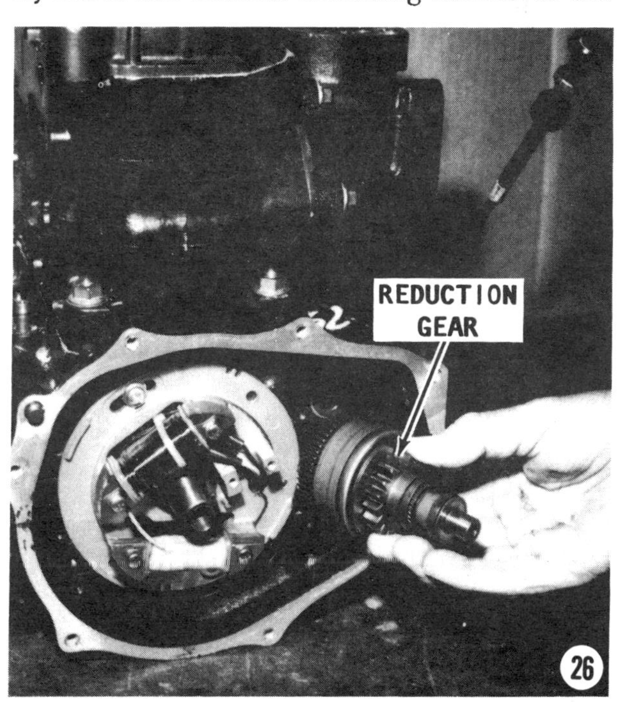

two magneto assembly securing screws. Therefore, the engine can be properly timed with the flywheel in place and secured.

650 Series and 750 Series

26- Slide the reduction gear over the splined armature shaft of the cranking motor. Apply a coating of multi-purpose water resistant lubricant to the reduction gear.

All Series

27- Apply just a "dab" of thick lubricant to the curved surface of the Woodruff key to hold it in place while the flywheel is being installed. Press the Woodruff key into place in the crankshaft recess. Clean any excess lubricant from the shaft to prevent the flywheel from "walking" during engine operation.

Check the flywheel magnets to ensure they are free of any metal particles. Double check the taper in the flywheel hub and the taper in the crankshaft to verify they are clean and contain no oil.

Now, slide the flywheel onto the crankshaft with the keyway in the flywheel aligned with the Woodruff key in place on the crankshaft. Rotate the flywheel **COUNTERCLOCKWISE** and at the same time check to be sure the flywheel does not contact any part of the magneto assembly or any wiring.

28- Thread the flywheel retaining device onto the crankshaft.

550 Series -- tabbed lockwasher and nut.
650 Series -- regular washer and nut.
750 Series -- flanged bolt

Use the special flywheel holding tool, or a spanner wrench, or the home made tool to prevent the flywheel from rotating while the flywheel nut/bolt is tightened to the following torque value for the series being serviced.

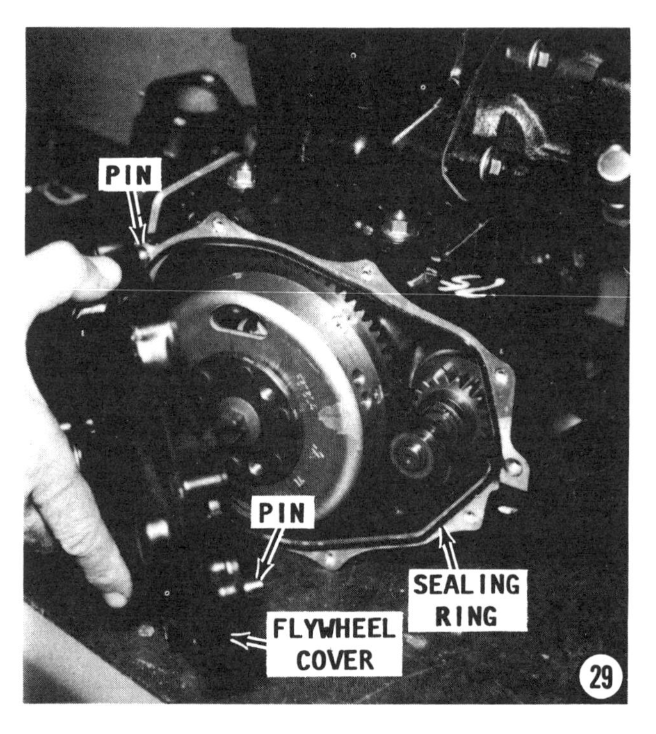

Flywheel Nut Torque Value

550 Series -- 115 ft lb (110Nm)
650 Series -- 72 ft lb (98Nm)
750 Series -- 94 ft lb (125Nm)

If servicing a Model 550 Series engine, use a hammer and punch to pry up the tabs on the tabbed lockwasher against the matching faces of the flywheel nut. If no tabs align, back off the nut slightly until the tabs align with a nut face.

29- Position a **NEW** sealing ring over the rim of the stator plate. Apply a coating of Silicone lubricant on the sealing ring of the flywheel cover. This process will help prevent the sealing ring from "jumping out" of the groove while positioning the flywheel cover.

Install the flywheel cover over the stator plate with the two pins -- one in the cover and the other in the plate -- indexed into their respective holes.

30- Apply a coating of Loctite Lock N' Seal to the threads of the attaching bolts, and then tighten the bolts in the sequence shown to a torque value of 12 ft lbs (16Nm).

31- Install and tighten the spark plugs to a torque value of 20 ft lb (16Nm). Secure the spark plug high tension leads to the spark

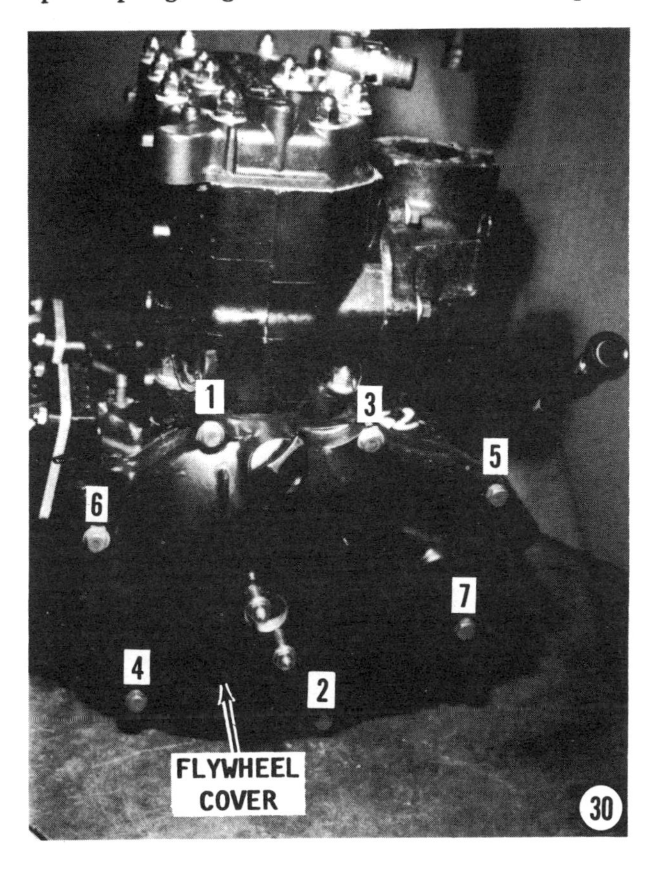

plugs. Work the rubber "boot" down over the spark plug. The rubber "boot" gives good protection to the spark plug against water or even moisture.

Cranking Motor Installation
550 Series Only

32- Apply a coating of Loctite SuperFlex to the O-ring around the forward end cap. Apply engine oil to the pinion gear. Position the motor from rear of the engine forward against the flywheel housing. Apply a coating of Loctite Lock N' Seal to the threads of all four mounting bolts. Install the two forward mounting bolts and tighten them to a torque value of 12 ft lb (16Nm).

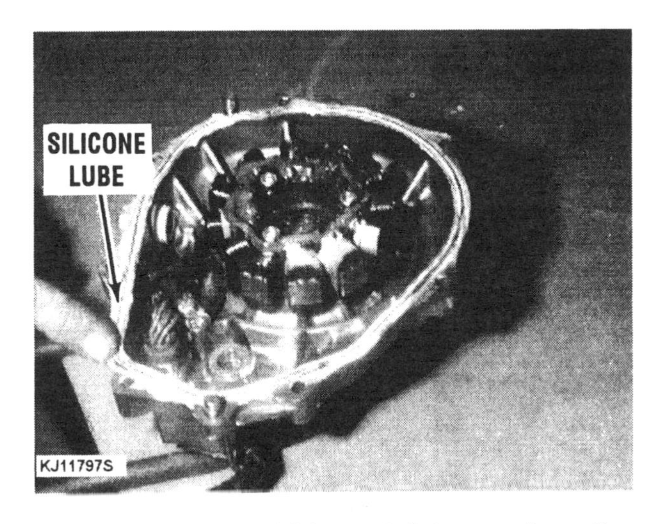

Applying a coat of Silicone Lubricant to the sealing ring on the flywheel cover of a Model 900 Series engine. The same material and technique is used on the flywheel cover of a two-cylinder engine, as mentioned in the text.

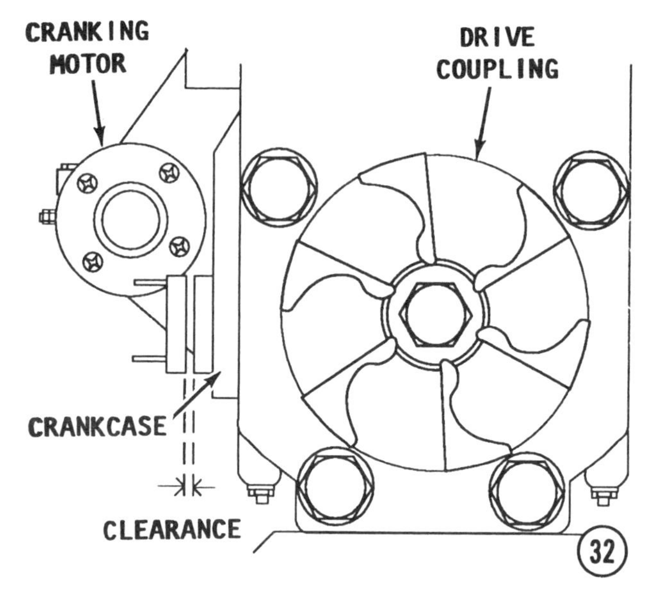

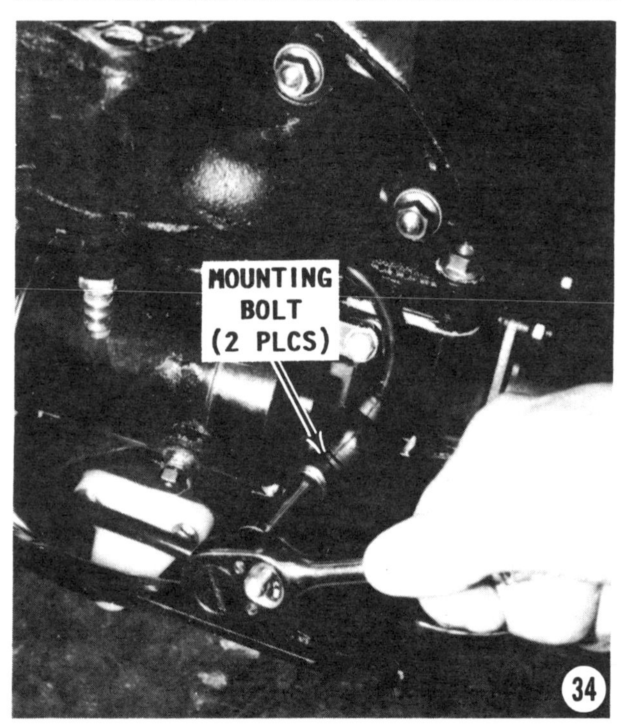

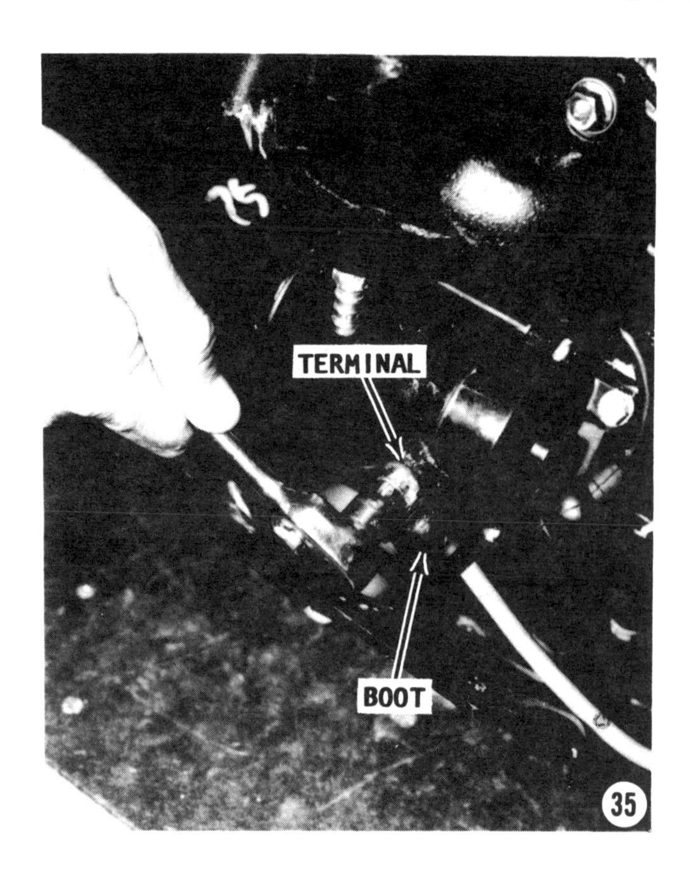

SHIM MATERIAL WORDS

Before installing the two aft mounting bolts, measure the distance between the rear mounting bracket and the crankcase. Fill this space with the original pieces of shim material, as saved during removal.

Shim material is available in thicknesses of 0.4, 0.6, and 0.8mm. Install the two aft mounting bolts and tighten them to a torque value of 54 in lbs (6Nm).

Cranking Motor Installation
650 Series and 750 Series

33- Clean the cranking motor lugs and the contact surface of the crankcase in the area where the motor is grounded. Apply a coating of engine oil to the large oil ring on the forward end cap. Position the motor against the crankcase and at the same time, insert the end of the armature shaft into the reduction gear inside the flywheel cover.

34- If the battery ground cable was originally connected to the crankcase with the lower mounting bolt, place the eye of the cable over the bolt and apply a non-permanent locking compound on both mounting bolts. Install and tighten the bolts securely.

35- Secure the large Red cable to the terminal on the motor frame and pull the boot over the connection.

ENGINE INSTALLATION

1- Insert the rubber coupler into the metal coupler on the jet pump driveshaft. Enlist the aid of an assistant and move the engine into the craft slightly forward of the engine mounts. Lower the engine slowly onto the rubber dampers and at the same time move the engine aft to engage the engine coupler with the rubber coupler on the jet pump driveshaft.

2- Apply a coating of Loctite #242 to the four engine mounting bolts. Temporarily thread the bolts -- just a few turns -- into the fixed engine rubber dampers.

SPECIAL WORDS

During disassembling, the instructions called for the shim material under each engine mount to be saved and identified to assist in engine alignment. If these groups of shim material are installed back in their original location, engine alignment should require very little adjustment.

Install the identified groups of shim material back into their original locations from which they were removed -- between the fixed engine rubber dampers and the engine bed bracket.

Engine Alignment

3- Several cross-section line drawings -- identified as **"A"** thru **"E"**, and an exterior view **"F"**, -- are included with the alignment procedures to indicate where measurements are to be taken; when proper alignment has been achieved; and when an adjustment is required.

Begin by placing a short straightedge across the two metal couplers, as indicated in illustration **"A"**. Attempt to insert a

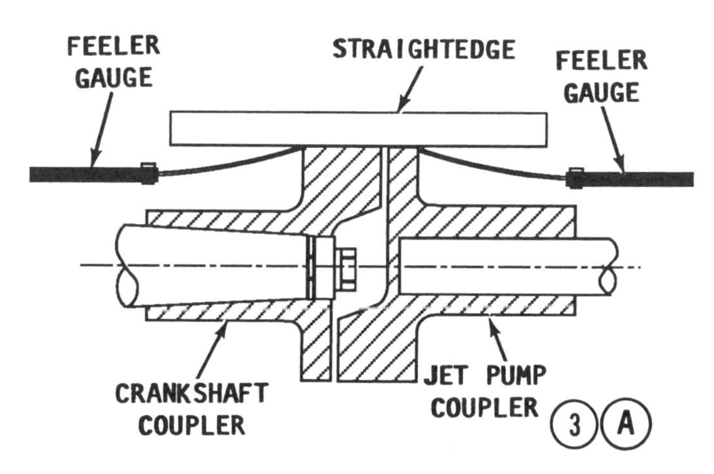

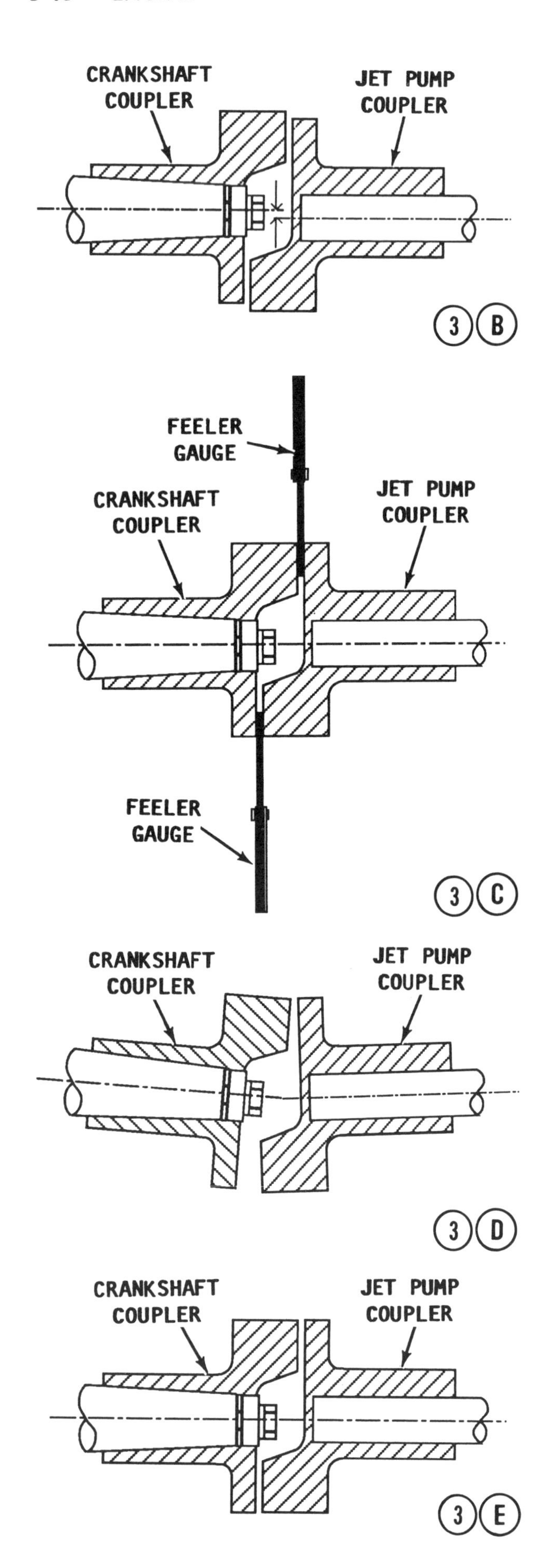

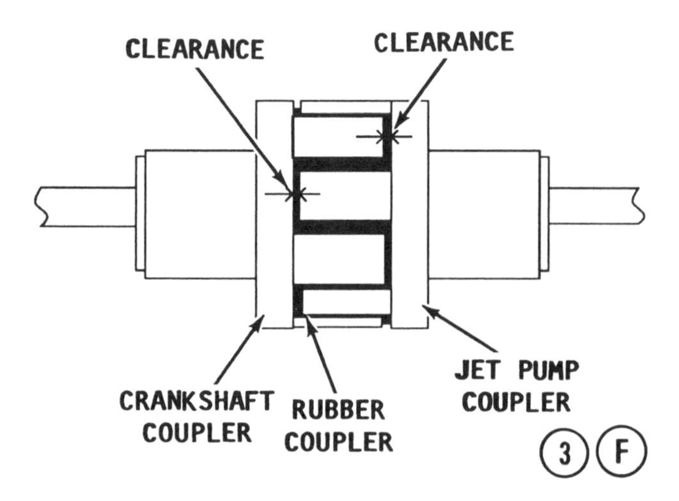

feeler gauge between the straightedge and the coupler on both sides. If the clearance exceeds 0.012" (0.3mm), an adjustment must be made to all four engine mounts to raise or lower the crankshaft. An extreme case is depicted in illustration **"B"**.

Next, insert a feeler gauge between the flanges of both couplers, as indicated in illustration **"C"**. Measure the clearance -- top and bottom --180° opposite. If the difference between the two measurements is more than 0.024" (0.6mm), the shafts are misaligned -- not on the same axis, as indicated in illustration **"D"**. Therefore, shim material must be added or removed from the two forward engine mounts to correct the condition.

Illustration **"E"** depicts proper alignment of the engine crankshaft and the jet pump driveshaft.

Coupler Wear

To determine if the rubber coupler is worn or has hardened, insert a feeler gauge between the rubber coupler and the engine metal coupler, as indicated in illustration **"F"**. Make a note of the clearance. Insert the feeler gauge between the rubber coupler and the jet pump metal coupler. Again make a note of the clearance. Now, add the two clearances. If the total exceeds 0.019" (0.5mm), the rubber coupler **MUST** be replaced.

Once proper alignment has been achieved, tighten the engine bed mounting bolts to the following torque value.

550 Series	27 ft lb (37Nm)
650 Series	16 ft lb (22Nm)
750 Series	27 ft lb (37Nm)

4- Position a new gasket on the exhaust manifold, and then ease the exhaust elbow into place.

5- Install and tighten the securing bolts to the following torque value.

Exhaust Elbow Bolt Torque Value

550 Series	12 ft lb (16Nm)
650 Series	36 ft lb (49Nm)
750 Series	14.5 ft lb (20Nm)

Install the water valve cable to the bracket on the exhaust elbow flange. Install the bolt securing the expansion chamber to the exhaust manifold. Tighten the bolt to a torque value of 14.5 ft lb (20Nm).

6- Install the bypass hose onto the fitting on the exhaust elbow.

Fuel Tank Installation

7- Check to be sure the mat is in position in the hull.

8- Lower the fuel tank into place and push the filler tube onto the fuel tank neck. Secure the tube to the neck with two hose clamps.

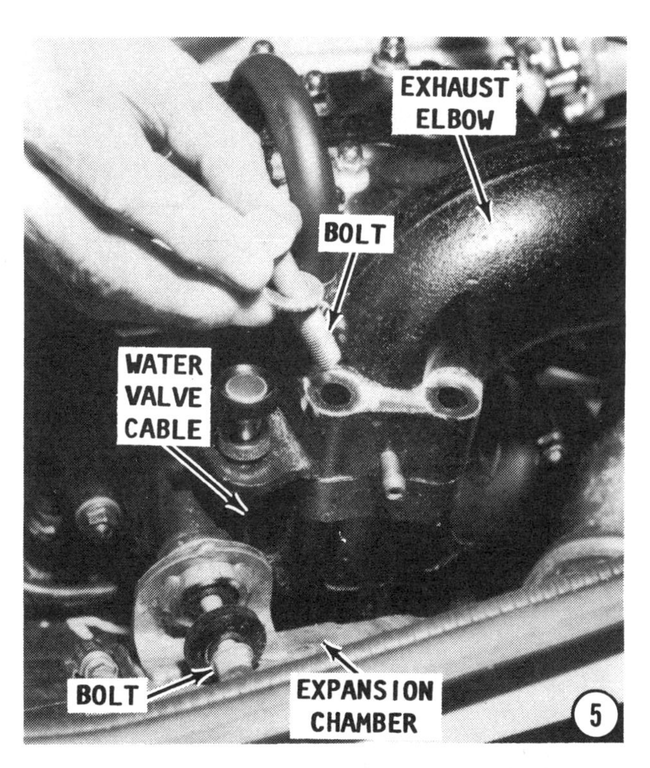

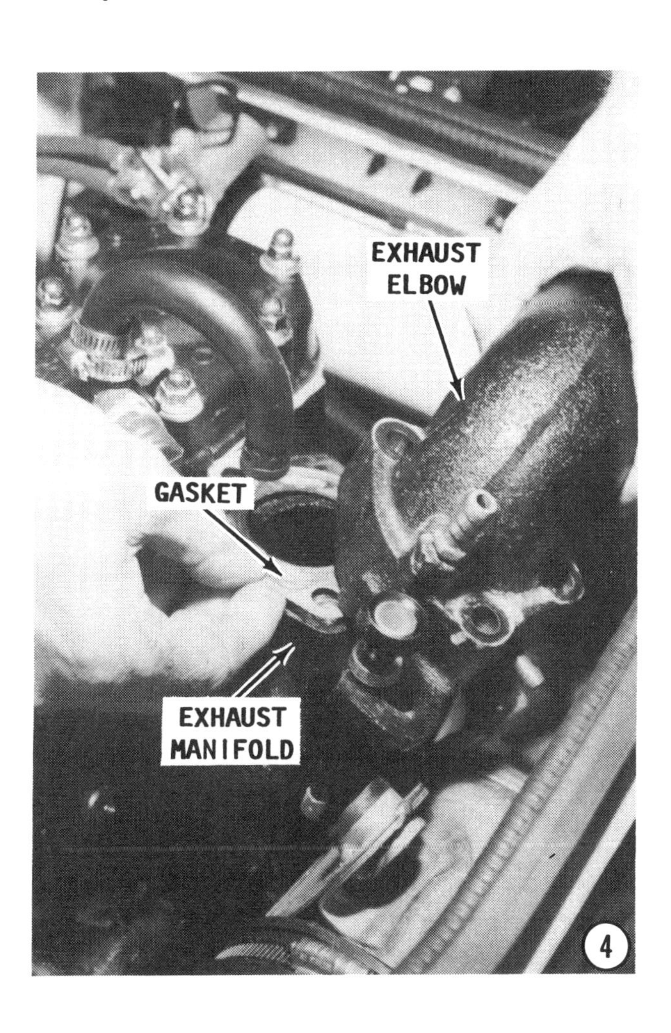

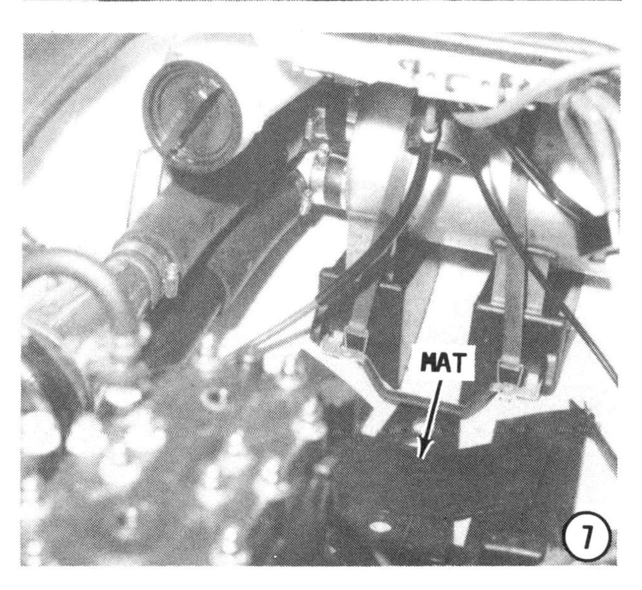

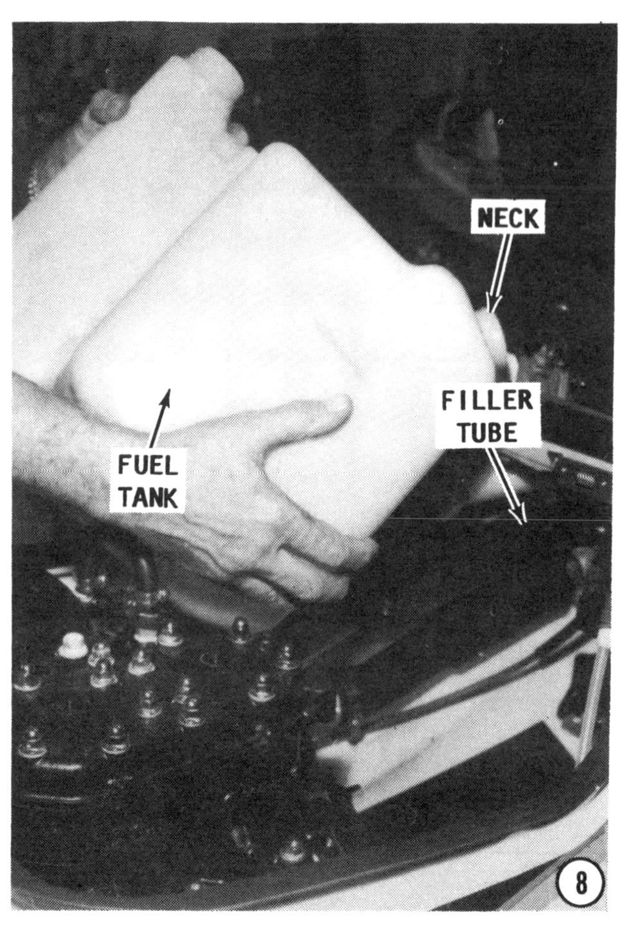

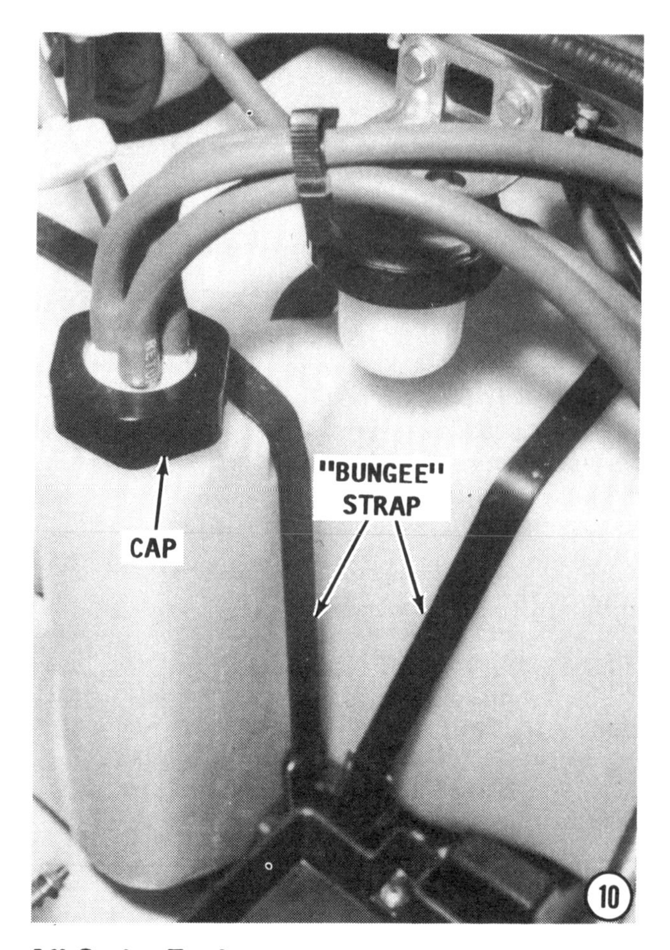

Oil Injection Tank Installation
650 Series and 750 Series

9- Move the oil injection tank, with the breather hose attached into the bracket. Check to be sure the rubber cushions surround the base of the tank.

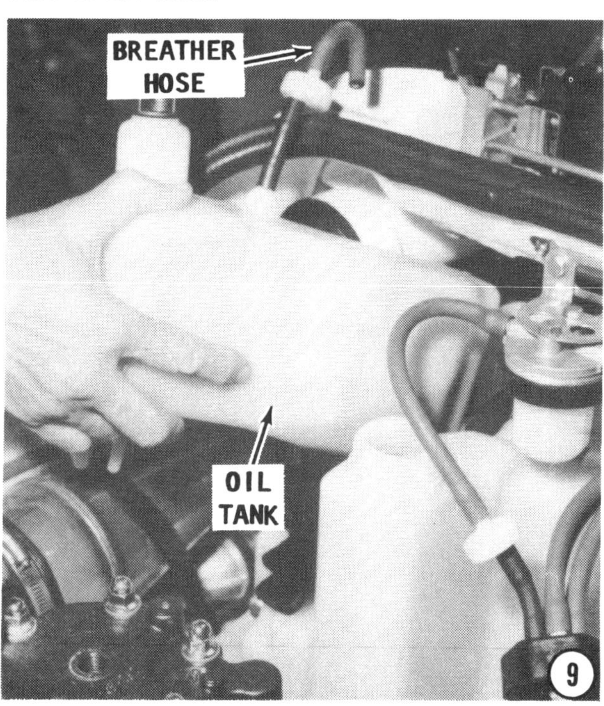

All Series Engines

10- Install the fuel tank filler cap. Secure the fuel tank and the oil tank -- if equipped -- with the two rubber "bungee" type restraining straps.

Oil Injection Pump Installation
650 Series and 750 Series

11- Slide the oil pump onto the two mounting studs with the oil pump shaft indexing into the crankshaft slot.

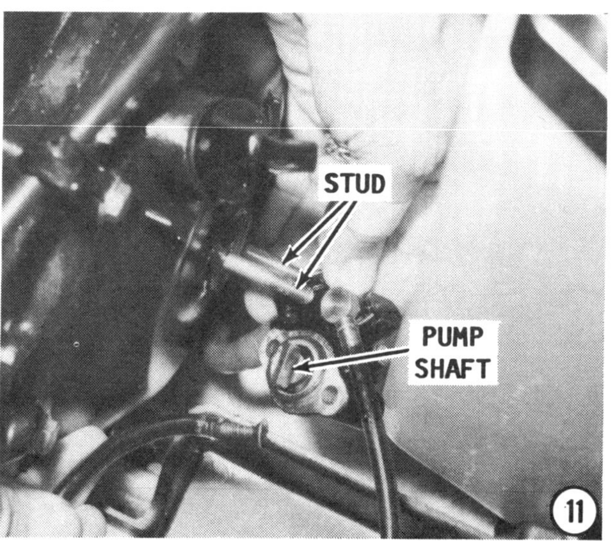

12- Secure the pump to the flywheel cover with the two nuts and washers. Tighten the nuts securely. Install the oil delivery hose to the fitting on the intake manifold.

Remove the screw from the end of the oil supply hose. Connect the hose to the inlet fitting on the pump. After all work has been completed and the engine is started, the oil injection **MUST** be "bled" -- purged of air -- according to the procedures outlined in Section 6-10, beginning on Page 6-49.

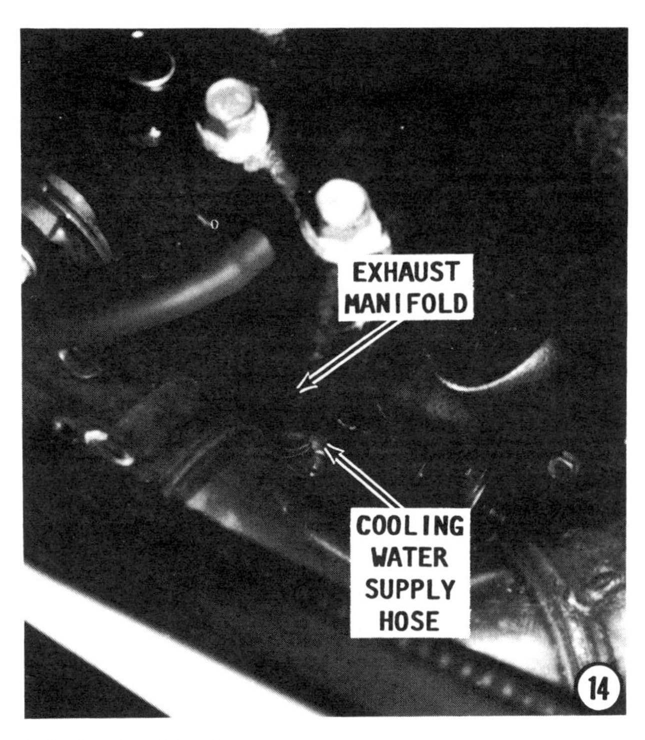

All Series Engines

13- Install the fuel pump pulse hose to the fitting on the intake manifold.

14- Secure the cooling water supply hose to the fitting on the underneath side of the exhaust manifold.

15- Secure the electrical leads to the engine block with the pliable metal clamps.

16- Tighten the hose clamps at the forward end of the expansion chamber. Tighten any and all other connections which may have been loosened.

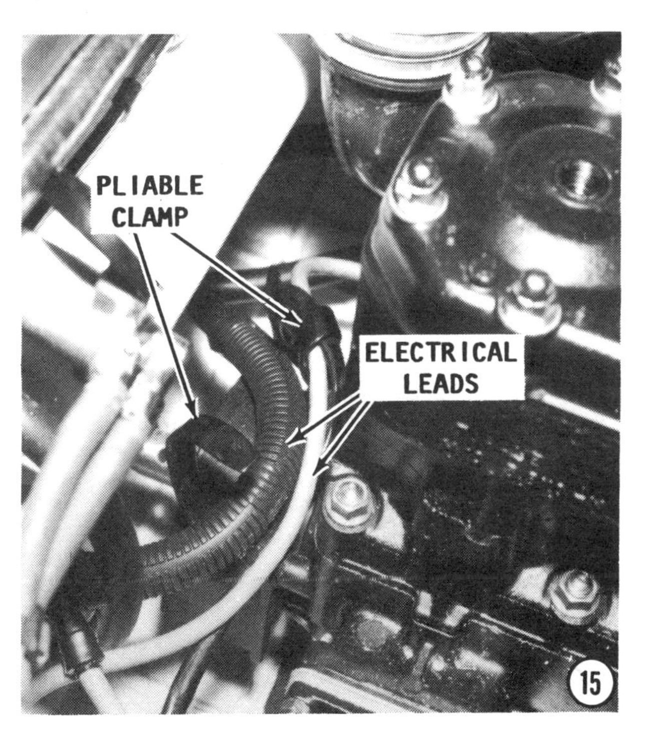

CLOSING TASKS

Position the cover over the coupler. Coat the threads of the attaching bolts with Loctite #242. Install and tighten the bolts securely.

All 2-Cylinder Engines

Install the carburetor/s associated hoses and throttle and choke cables -- see Chapter 6. Install the electrical box -- see Chapter 7.

Connect the stator harness lead at the electrical box. The oil injection system **MUST** be "bled" while operating the engine on a premix, as directed in Chapter 6, Page 6-41. Connect the positive and negative cables to the battery.

CRITICAL COOLING WATER WORDS

Move the craft to a body of water or connect a garden hose to the engine cooling water supply fitting or flush fitting on the cylinder head.

If a garden hose is used, start the engine and allow the rpm's to stabilize at idle speed **FOR JUST A FEW SECONDS,** and then turn the water on.

Adjust the water flow until a small trickle is discharged from the bypass outlet on the port side of the hull.

When the engine is to be shut down, turn the water off **FIRST** -- raise the aft portion of the hull -- **WHILE THE ENGINE IS OPERATING AT IDLE** -- "rev" the engine just a **COUPLE** times to clear water from the exhaust system -- and then shut it down.

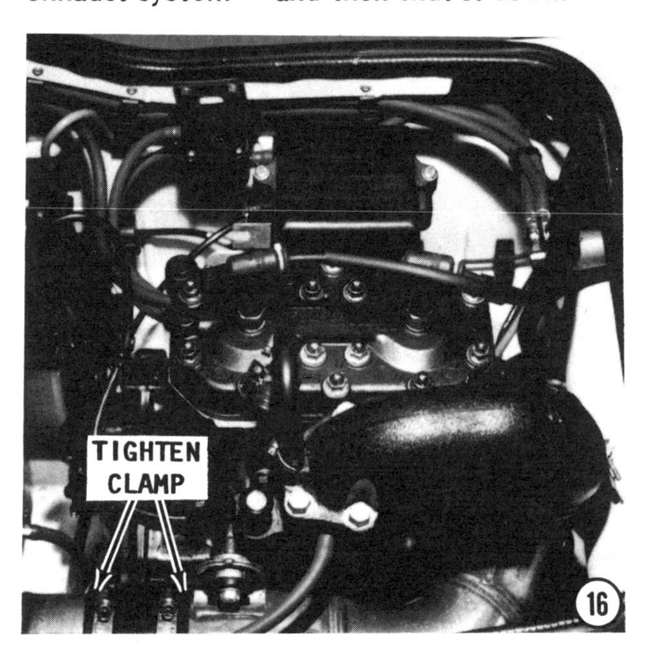

NEVER allow the engine to operate without cooling water for more than 15 seconds.

CAUTION

Water must circulate through the jet pump -- to and from the engine, anytime the engine is operating. Circulating water will prevent overheating -- which could cause damage to moving engine parts and possible engine seizure.

Attempt to start and run the engine without the engine cover in place. This will provide the opportunity to check for water or fuel leaks.

After the engine is operating properly, install the engine cover. Follow the break-in procedures with a load on the jet pump.

Break-in Procedures

As soon as the engine starts, **CHECK** to be sure a fine stream of water is discharged from the exhaust relief hole.

DO NOT operate the engine at full throttle except for **VERY** short periods, until after 10 hours of operation as follows:

a- Operate at 1/2 throttle, approximately 2500 to 3500 rpm, for 2 hours.

b- Operate at any speed after 2 hours **BUT NOT** at sustained full throttle until another 8 hours of operation.

c- Mix gasoline and oil during the break-in period, total of 10 hours, at a ratio of 25:1.

d- While the engine is operating during the initial period, check the fuel, exhaust, and water systems for leaks.

CLEANING AND INSPECTING -- ALL ENGINES

The success of the overhual work is largely dependent on how well the cleaning and inspecting tasks are completed. If some parts are not thoroughly cleaned, or if an unsatisfactory unit is allowed to be returned to service through negligent inspection, the time and expense involved in the overhaul work will not be justified with peak engine performance and long operating life.

Therefore, the procedures presented in Section 8-7, beginning on Page 8-67 should be followed closely and work performed with patience and attention to detail.

8-5 SERVICE THREE-CYLINDER ENGINES

This section provides complete detailed illustration procedures to remove, disassemble, assemble, and install a Kawasaki manufactured "out the door" PWC 3-cylinder engine from 1992 to and including 1998.

Very few references have been made to "high performance" units, although the **factory** high performance Model 750 Series engine is covered.

ADVICE

A Polaroid, or equivalent instant-type camera is an extremely useful item, providing the means of accurately recording the arrangement of parts and wire connections **BEFORE** the disassembly work begins. Such a record is invaluable during assembly.

REED BLOCK WORDS

If the only work to be performed on the engine is servicing the reed block assemblies, remove the flame arrestor, carburetors, and follow the Disassembling Steps 9, 10, and 11, beginning on Page 8-46.

Clearance between the reed block and the reed valve is 0.2mm maximum.

After the reed block and intake manifold assemblies are installed, install the carburetor and the flame arrestor.

Engine Removal

1- Disconnect the negative and positive leads from the battery. Loosen the fuel filler cap slightly to permit any pressure build-up in the fuel system to escape. Remove the bolts securing the flame arrestor to the carburetors. Lift the flame arrestor free. Disconnect the throttle and choke cables from carburetor #1 -- the forward carburetor.

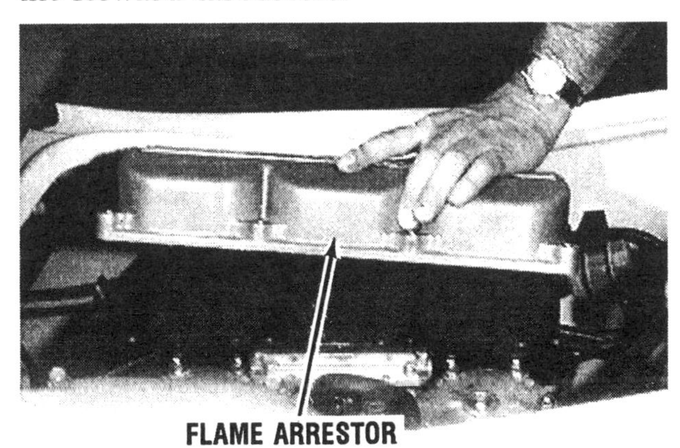

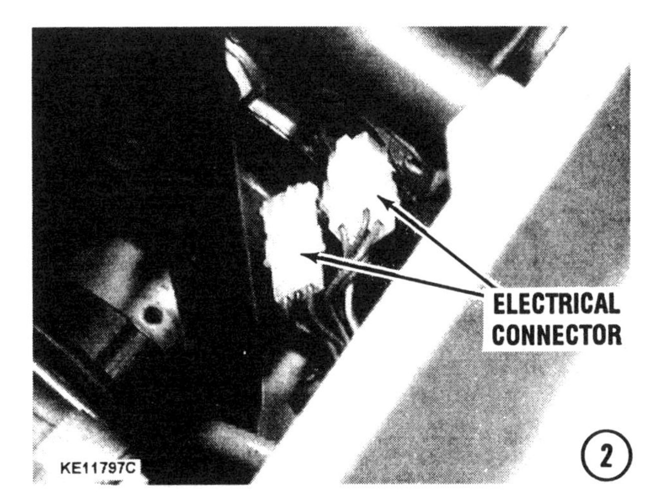

Cut the zip ties securing the fuel hose and pulse hoses. Remove and pinch off the fuel hoses with an appropriate clamp tool or, insert a screw or golf tee in the disconnected end of the hose to prevent loss of fuel.

2- Remove the two bolts securing the stator harness lead to the electrical box. Pull the cover, exposing two pin connectors. Unfasten these connectors.

3- Disconnect the three oil lines from the oil pump. It is a good idea to push the oil hose

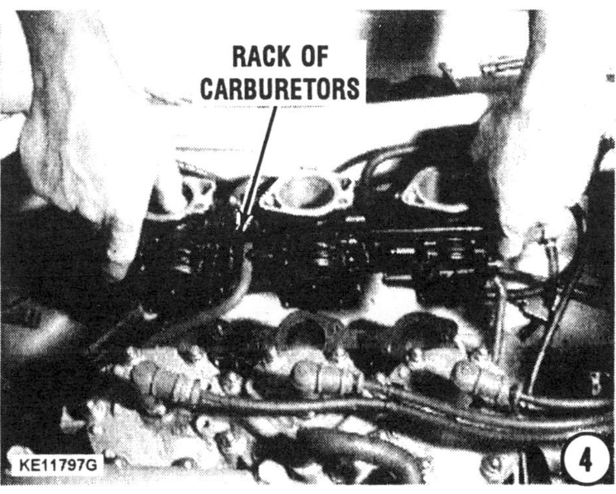

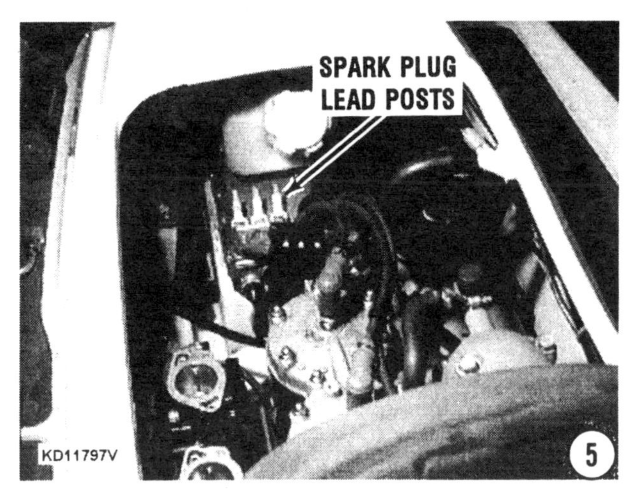

spring clips down the hose, to prevent them from accidentally finding their way into the intake manifold after the carburetors have been removed.

4- Lift the carburetor rack assembly off the intake manifold and set it aside ready for disassembly and overhaul work, per the detailed instruction in Chapter 6.

5- Remove the spark plug high tension leads and place them on the posts, as shown in the illustration.

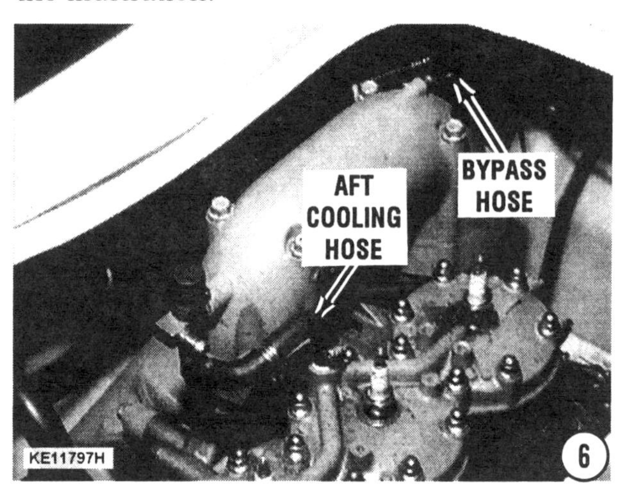

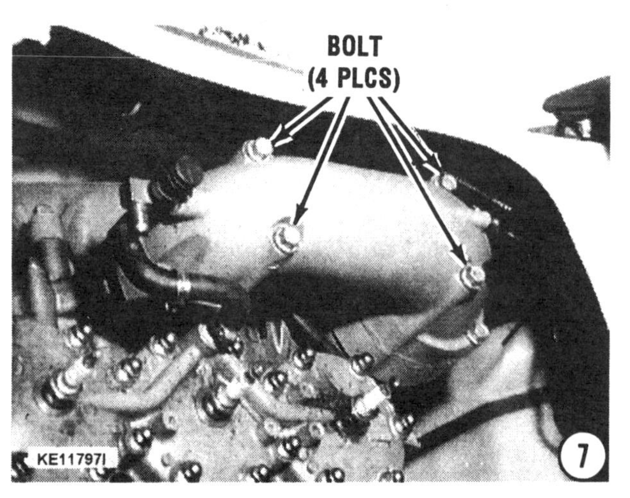

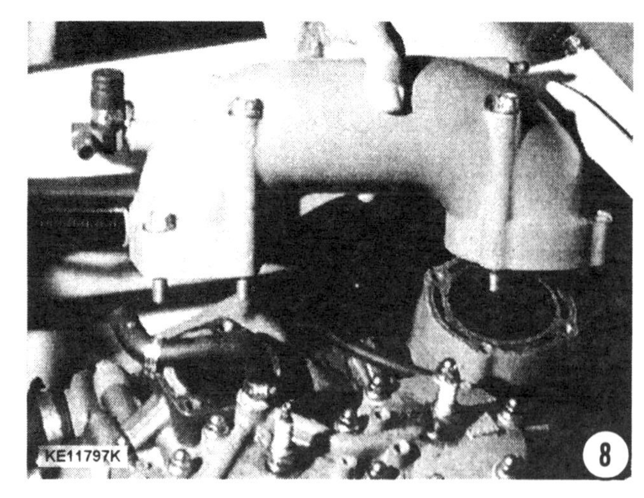

Exhaust Manifold Removal

6- Disconnect the cooling hose from the aft end of the exhaust elbow. Remove the smaller bypass hose, located at the top front end of the exhaust elbow.

7- Remove the four bolts securing the exhaust elbow to the exhaust manifold and expansion chamber. Take note of the longer bolts and their proper location.

8- Lift the exhaust elbow off the exhaust manifold and expansion chamber.

9- Disconnect the magneto cooling hose from the top center of the expansion chamber. This hose provides cooling to the flywheel cover housing. (The accompanying drawing shows the exhaust elbow still in place.)

10. Disconnect the tube from the aft end of the expansion chamber. Disconnect the cooling hose located on the underside of the expansion chamber -- just forward of the connecting tube, and then remove the hose from the expansion chamber.

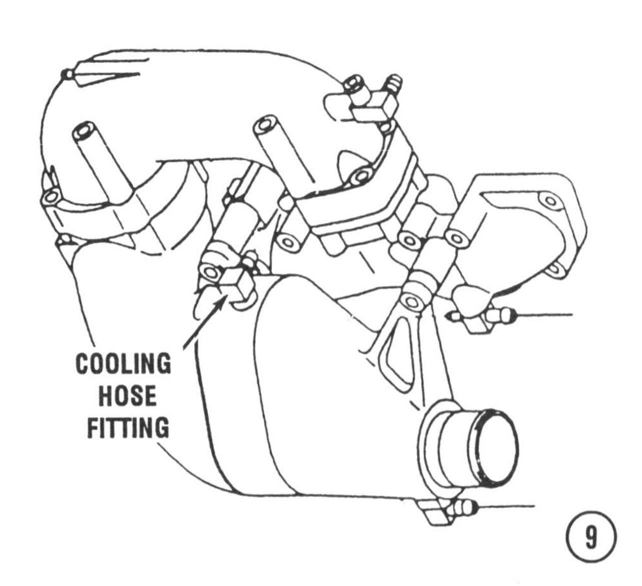

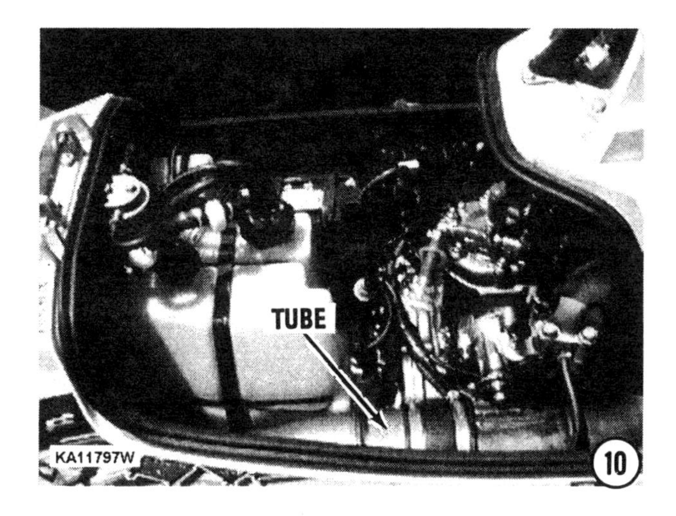

11- Remove the two bolts securing the expansion chamber to the exhaust manifold. Now, remove the expansion chamber from the engine compartment by sliding it forward and moving it around the forward end of the engine. Once the chamber is in front of the engine, the chamber may be lifted up and clear of the compartment.

Disconnect the main cooling hose, which is located at the bottom of the exhaust manifold. Wrap all hoses still connected to the engine assembly around the engine, in preparation to lifting the engine out of the hull.

12- Remove the four motor mounting bolts.

13- The engine is now ready to be lifted out of the watercraft. One adult can lift the engine by first straddling the engine compartment, as shown. Now, lift and slide the engine forward enough to disengage the rubber coupler from the jet pump coupler. Lift the engine up and watch for any hose or electrical leads still attached to the engine. Move the engine to a suitable work surface. Remove the rubber coupler.

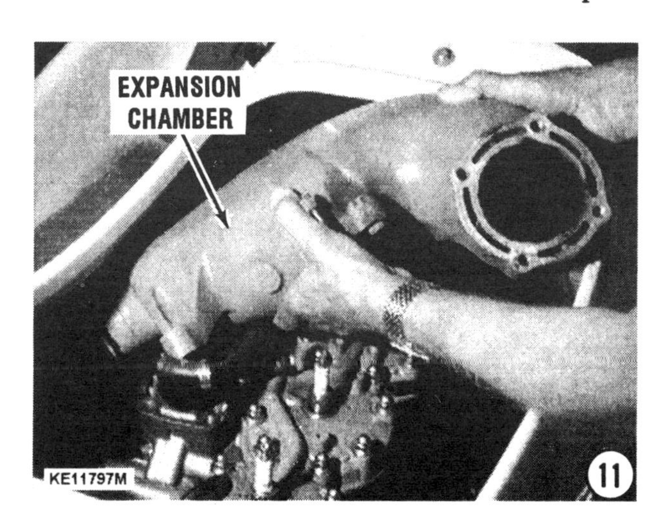

ENGINE DISASSEMBLING

The following instructions pickup the work after the preliminary tasks listed above have been completed and the engine is on a suitable work surface.

FIRST, THESE WORDS

Procedural steps are given to remove and disassemble virtually all items of the engine. However, as the work moves along, if certain items, i.e. bearings, bushings, seals, etc. are found to be fit for further service, simply skip the disassembly steps involved. Proceed with the required tasks to disassemble the necessary components.

COMPLETE and **DETAILED ILLUSTRATED** procedures to clean and service virtually all parts of the engine are outlined in Section 8-7, beginning on Page 8-67.

SPECIAL WORDS ON SPECIAL TOOLS

The following list of special tools are required in order to perform some of the necessary work on the engine. Some tasks become

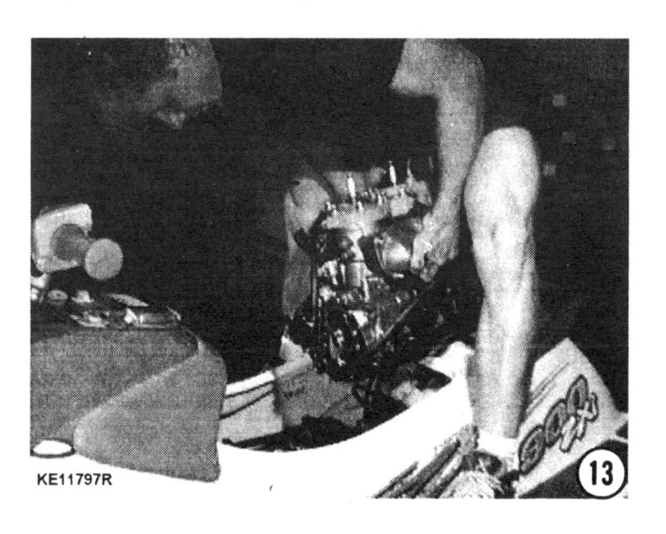

extremely difficult, if not impossible, without the special tools or some type of equivalent. The manufactures' part numbers are given. In specific steps of the procedures -- whenever possible -- a special effort has been make to give some type of equivalent or homemade tool which may be substituted. However, the cost of a special tool may be offset by the money saved if an expensive part is not damaged.

900 and 1100 Series

Piston Pin Puller	P/N 57001-910
Flywheel Holder	P/N 57001-1313
Coupler Holder	P/N 57001-1230
Flywheel Puller	P/N 57001-1223

Cranking Motor Removal
900 and 1100 Series

1- Push back the protective boot from the power terminal and the ground terminal on the cranking motor and remove the cables.

2- Remove the two mounting bolts securing the cranking motor to the crankcase. Move the cranking motor aft and the armature shaft will slide free of the reduction gear inside the flywheel cover.

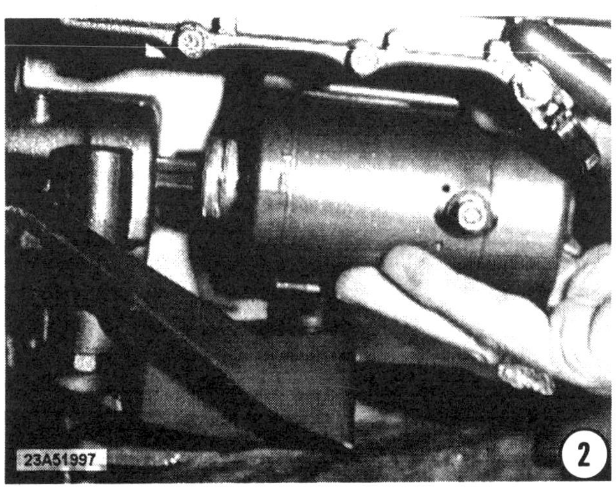

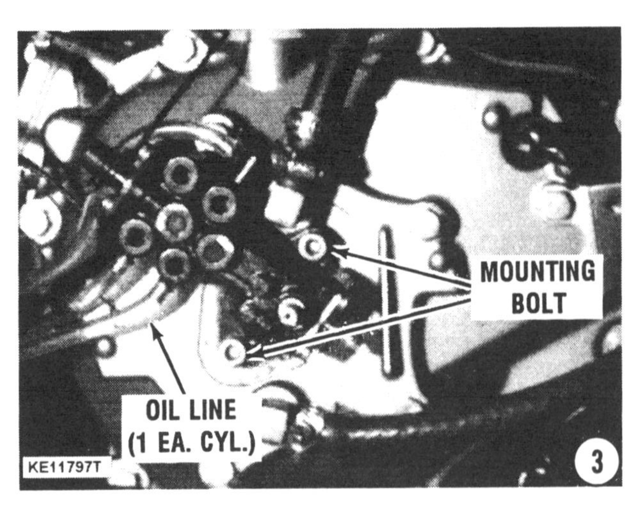

"Pulling" The Flywheel

3- Remove the oil lines from the straps secured to the flywheel cover. Disconnect the bottom cooling hose from the oil pump which leads to the intake manifold. Remove the electrical and hose clamp bolts.

HELPFUL NOTE

Placing a mark of "White-out" on these bolt heads and their locations will assist in the assembly of the engine later.

Remove the two allen head bolts securing the oil pump to the flywheel cover.

4- Pull the oil pump free of the flywheel cover. For detailed instructions to service the oil pump, including "bleeding" air from the system, see Chapter 6.

5- Remove the ten mounting bolts securing the flywheel cover to the crankcase. See Chapter 7 to service the ignition system housed inside the flywheel cover. Refer to Chapter 9 to service the electrical system.

6- Remove the flywheel cover from the crankcase. If the cover is stuck in place, obtain a soft head mallet and tap around the circumference of the cover to jar it free. Remove and

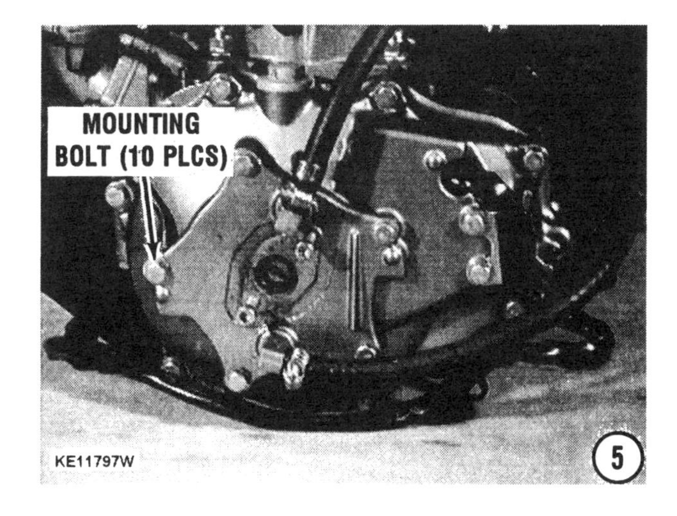

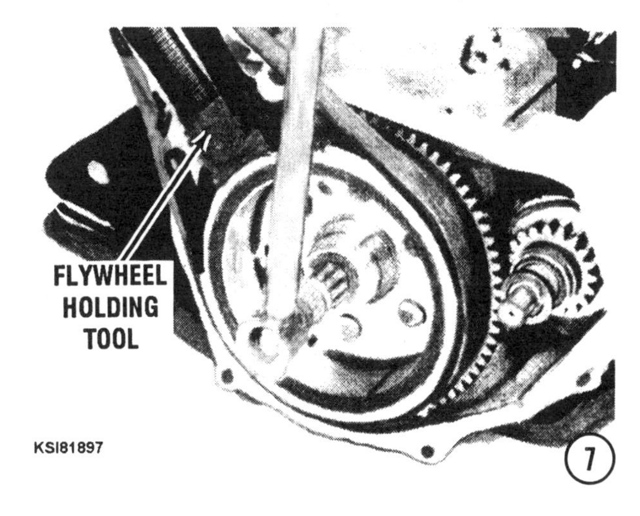

discard the sealing ring, only if a replacement is on hand.

7- Obtain special Flywheel Holder Tool P/N 57001-1313. Install the flywheel holder to the flywheel to prevent the flywheel from rotating. For this step, disregard the wrench on the flywheel nut.

8- Obtain special Coupler Holder Tool P/N 57001-1230. Insert the coupler holder tool into the metal coupler at the rear of the engine. Back out the metal coupler from the aft end of the crankshaft. Removal of the coupler is not an easy task. **HOWEVER,** use care to prevent possible damage to the coupler.

9- Leave the special flywheel holding tool in place on the flywheel. Using an extra long handled wrench, back off the special "flanged" bolt from the crankshaft. Again, this is **NOT** an easy task, but using the "shock" treatment, the bolt may "break" loose.

10- Obtain special Flywheel Puller Tool P/N 57001-1223. Secure the main part of the tool to the flywheel through the two attaching bolts, as shown. Using the center bolt, take a strain on the puller with the proper sized wrench. Con-

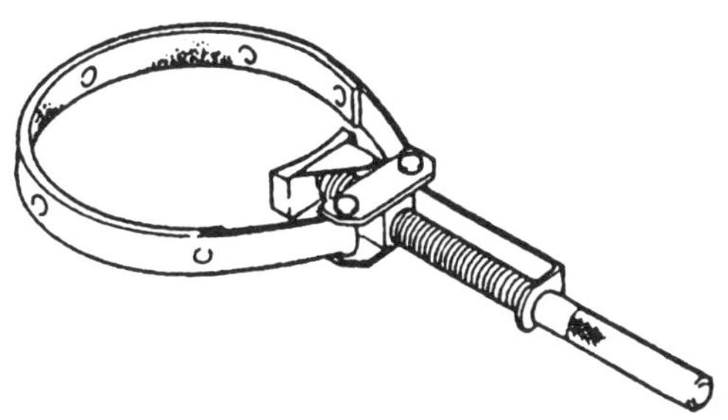

"Rough" line drawing of the flywheel holding tool used when removing/installing the flywheel nut, to prevent damage to the flywheel teeth. This tool is almost essential, because removing the nut is not an easy task.

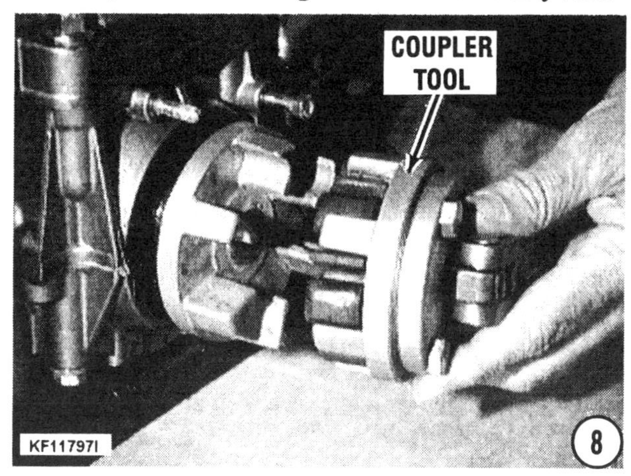

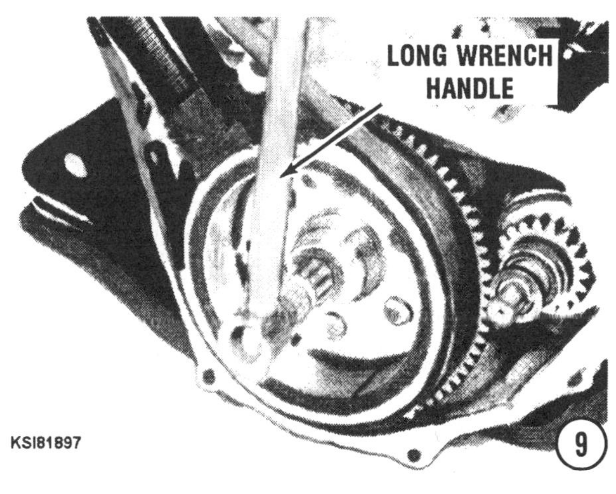

tinue to tighten the center bolt and at the same time, **SHOCK** the crankshaft with a gentle to moderate tap with a hammer on the end of the center bolt. This shock will assist in "breaking" the flywheel loose from the crankshaft.

DO NOT strike hard on the center bolt. The force of such action would be transmitted to the seals and bearings along the crankshaft and cause damage. **NEVER** make any kind of direct blow to the flywheel, because such a shock would surely cause the flywheel magnets to loose their magnetism.

11- Pull the flywheel free of the crankshaft. Remove and save the Woodruff key from the crankshaft keyway.

SPECIAL WORDS WOODRUFF KEYS

Inspect the condition of the Woodruff key and its keyway in the crankshaft. If the key shows any signs of shearing, the cause is probably from jumping waves and the effect will be a loss of power. When the craft is clear of the water, the engine RPM approaches the limit set by the RPM limiter in an unloaded condition.

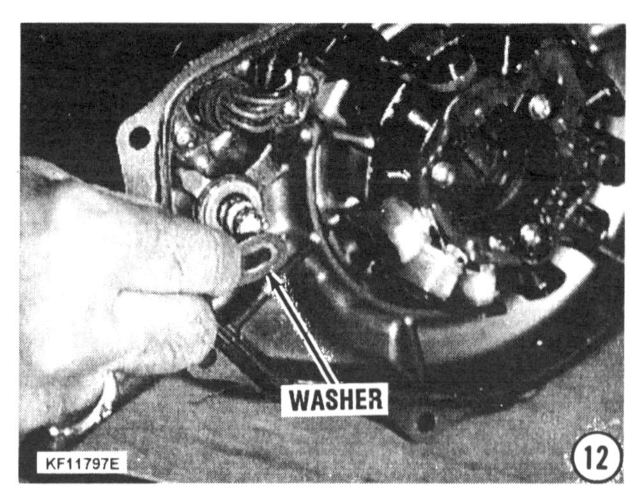

When the craft returns to the water surface, the sudden load on the impeller is transferred along the drive line. Speed of the rotating pump shaft and the crankshaft is suddenly reduced while the flywheel continues to spin rapidly due to its mass. This condition, especially when repeated over a period of time, often shears the woodruff key and the flywheel becomes repositioned on the crankshaft.

A very small misalignment will result in a change in ignition timing. The craft operator will notice a drastic loss of power. A worse condition would be the flywheel shifting to the degree engine start would not be possible.

12- Remove and **SAVE** the washer and spring in the flywheel cover. These two items are used on the forward portion of the reduction gear assembly driveshaft.

13- Pull the reduction gear assembly free of the engine crankcase.

BLOCK DISASSEMBLING

Cylinder Head Removal

14- Remove the nuts and washers securing the head to the block. Pry the head loose from

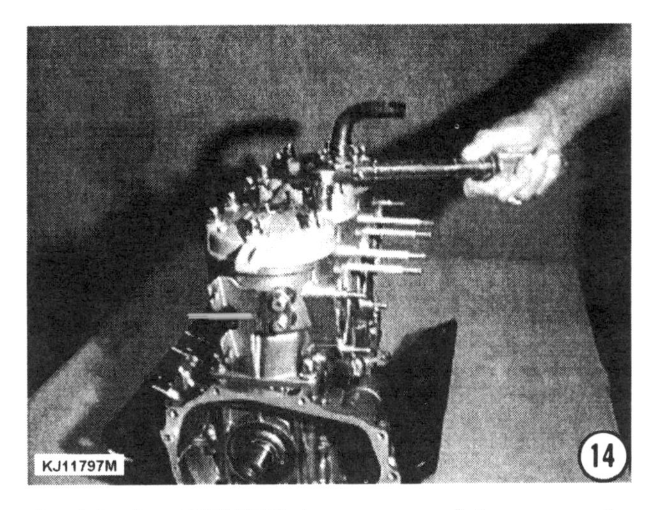

the block. **NEVER** insert a tool between the head and the block, because such action would surely mar the machined surface of either the head or the block or both.

If the head fails to come loose from the block, it may be necessary to tap the head with a soft head mallet to "jar" it loose.

15- After the head has been "jarred" loose, keep the head as level as possible and lift it **STRAIGHT** up and off the studs.

16- Remove and discard the head gasket. The head gasket can **NEVER** be used a second time.

Exhaust Manifold Removal

17- Remove the nuts securing the exhaust manifold to the block. "Jar" the manifold loose from the block with "moderate" blows with a soft head mallet. Lift the exhaust manifold free of the studs. Remove all traces of gasket material from the two mating surfaces.

Intake Manifold Removal

18- Remove the nuts securing the intake manifold. Lift the manifold free of the block.

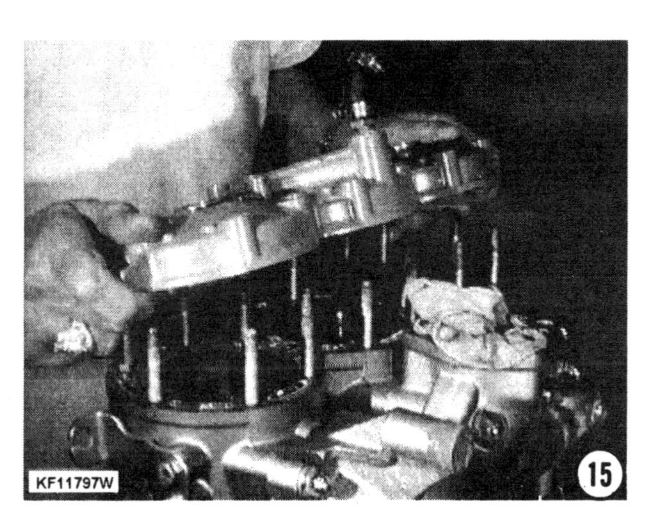

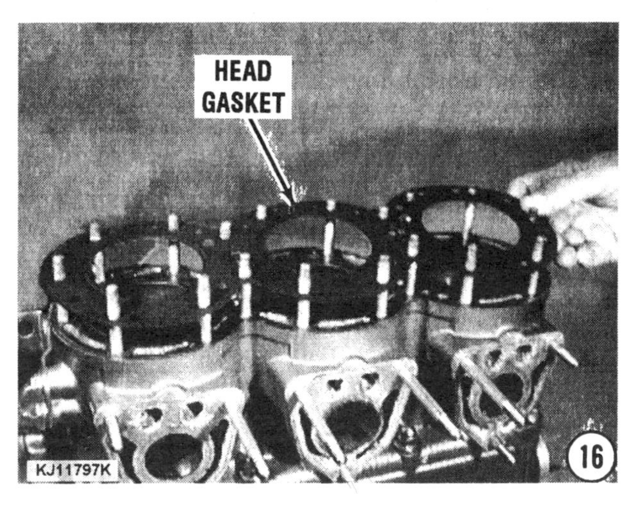

CRITICAL WORDS ON REED BLOCK ASSEMBLIES

Take care in the next step when handling reed valve assemblies. The reed valve housing is coated with a special rubber substance. Work slowly to prevent damage to this coating. Once the assembly is removed, keep it away from sunlight, moisture, dust and dirt. Sunlight can deteriorate valve seat rubber seals. Moisture can easily rust stoppers overnight. Dust and

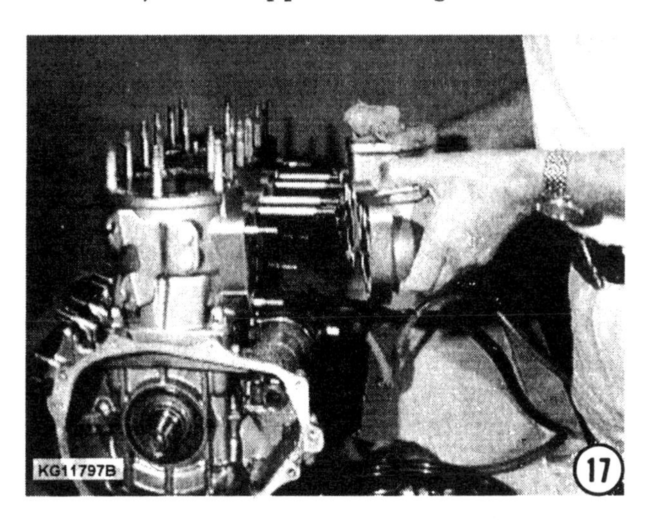

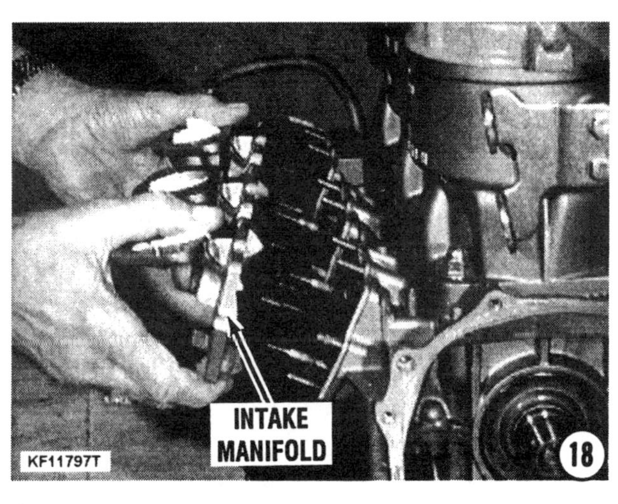

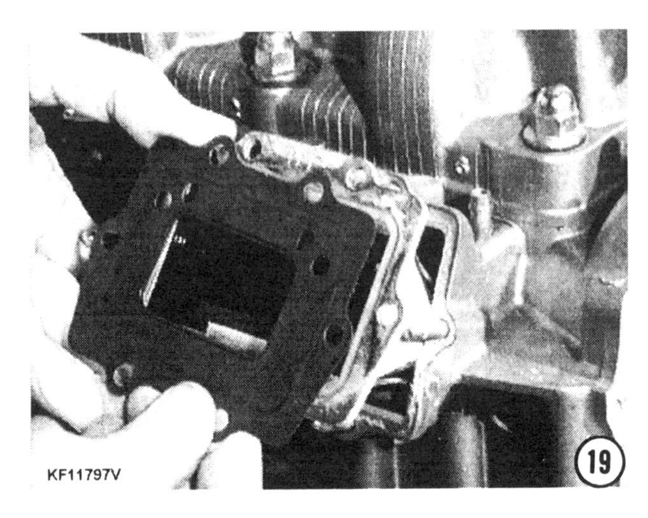
KF11797V

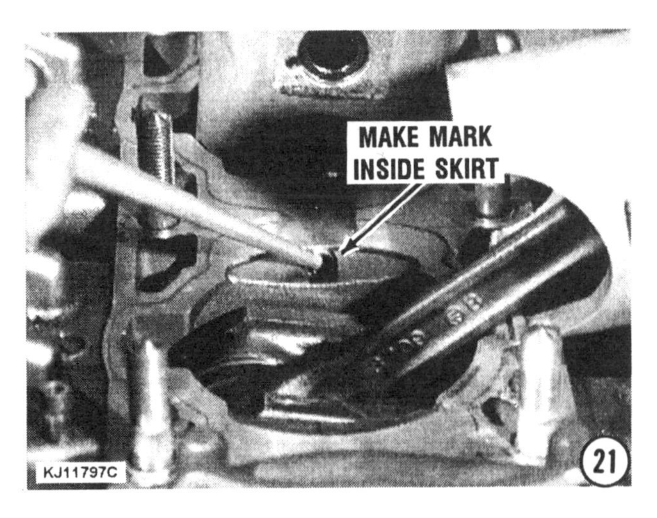

KJ11797C

dirt -- especially sand or other gritty material -- can break reed petals if caught between stoppers and reed petals.

Make special arrangements to store the reed valves to keep them isolated from the elements while further work is being performed on the engine.

19- Lift out the three reed valve assemblies. Remove and discard the gaskets installed on both sides of the reed block housing -- six gaskets in all.

READ AND BELIEVE

The reeds must never be turned over in an attempt to correct a problem. Such action would cause the reed to flex in the opposite direction and the reed would break in a very short time.

Block Disassembling

20- Remove the nuts securing the cylinder block to the upper crankcase.

Keep the cylinder block as level as possible and **SLOWLY** lift it **STRAIGHT** up and clear of the studs. If possible, have an assistant prevent the rods from striking the edge of the cylinders as the cylinder block is raised.

Two locating pins are indexed on one side of the head and the upper crankcase. These pins may remain with the cylinder block as it is lifted or the pins may remain with the lower crankcase. **TAKE CARE** not to lose these pins. Remove and discard the base gasket.

21- Scribe a mark on the inside surface of each piston skirt to identify the piston location **BEFORE** removal from the connecting rod, to **ENSURE** it will be installed back onto the same rod and into the same cylinder from which it was removed.

SAFETY WORDS

The piston pin C-lockrings are made of spring steel and may pop out of the groove with considerable force. Therefore, warn other people in the area and **WEAR** eye protection glasses while removing the piston pin lockrings.

22- Remove the C-lockring from both ends of the piston pin using a screwdriver inserted into the small groove next to the piston pin

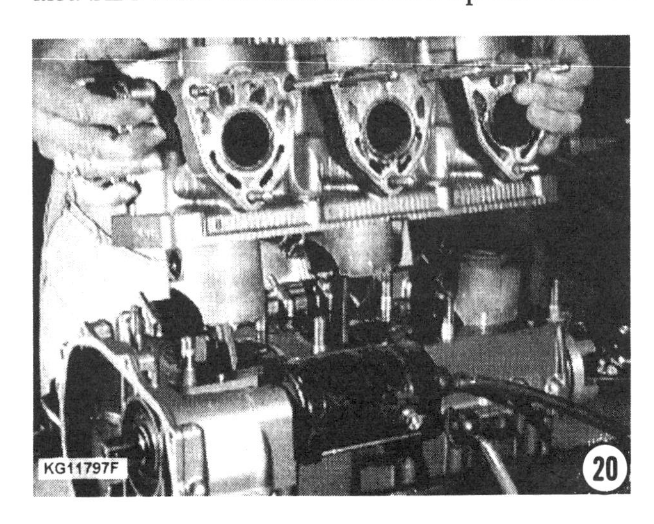
KG11797F

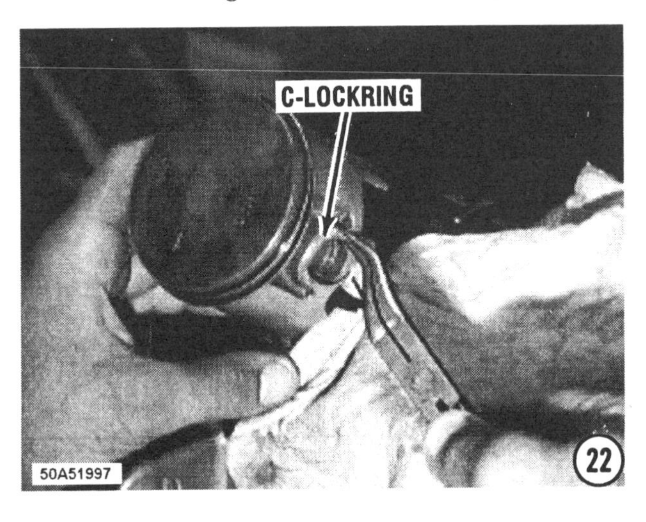

50A51997

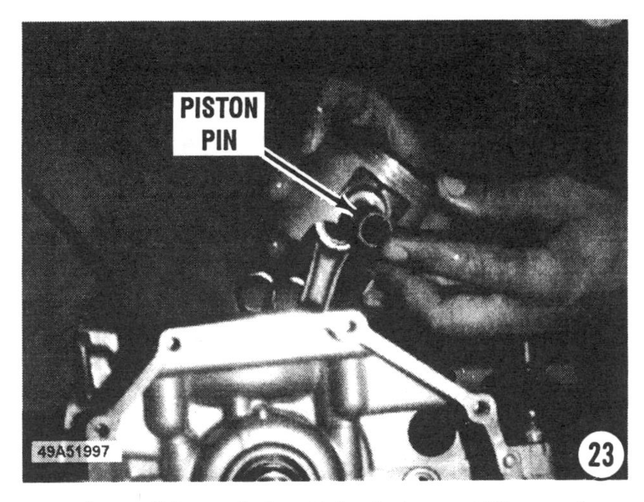

opening. Discard the C-lockrings. These rings stretch during removal and must **NEVER** be used a second time.

23- Push the piston pin out -- free of the piston. If the pin is tight and refuses to budge, use a very blunt punch and light taps with a hammer to start it moving.

24- Remove the caged needle bearings from the small end of the connecting rod. Associated parts of the connecting rod and piston are shown in the accompanying illustration.

ADVICE

New needle bearings should be installed in the connecting rods, even though they may appear to be in serviceable condition. New bearings will ensure lasting service after the overhaul work is completed. If it is necessary to install the used bearings, keep them separate and identified to **ENSURE** they will be installed onto the same connecting rod from which they were removed.

CRITICAL WORDS

The rod is an integral part of the crankshaft. The two are manufactured together and **CANNOT** be separated.

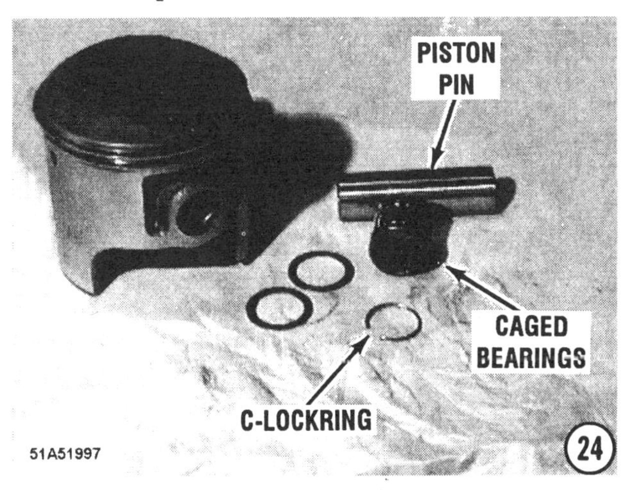

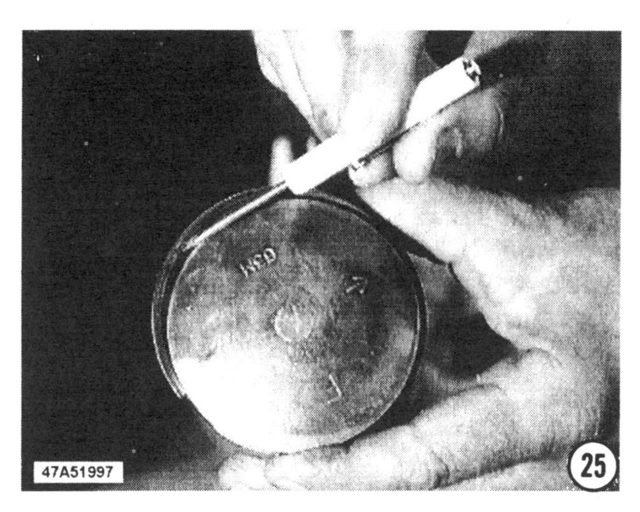

Slowly rotate the bearing races. If rough spots are felt, a new crankshaft must be purchased. The bearings are installed at the factory and are not replaceable.

GOOD WORDS

Good shop practice dictates to replace the rings during an engine overhaul However, if the rings are to be used again, expand them **ONLY** enough to clear the piston and the grooves, because used rings are brittle and break very easily.

25- Gently spread the top piston ring enough to pry it out and up over the top of the piston. No special tool is required to remove the piston rings. Remove the lower ring in a similar manner. These rings are **EXTREMELY** brittle and have to be handled with care if they are intended for further service.

Be sure to keep all parts for each piston together (i.e. piston rings and bearings) for corresponding piston identification and installation if more engine servicing is required. Turn the piston over and carefully place the matching C-lockrings, piston pin, rings, caged needle

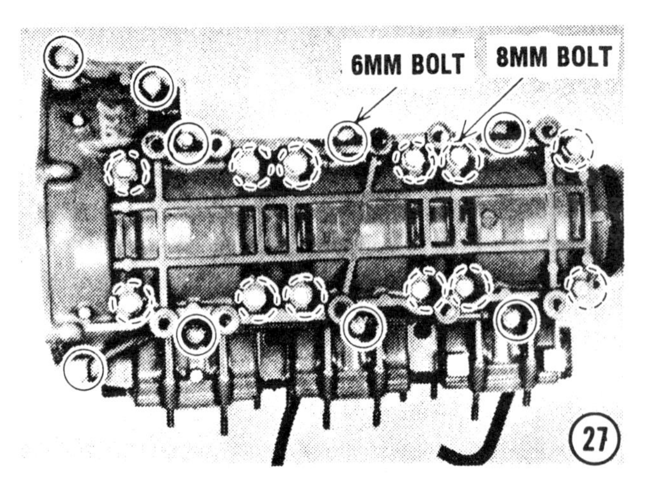

bearings inside the piston for proper matching during assembling.

CRITICAL ROD WORDS

The connecting rods are an integral part of the crankshaft and cannot be separated. Therefore handle the crankshaft with extra care during any movement, including cleaning and inspecting.

Engine Bed Bracket Removal

26- Stand the engine upright with the flywheel end on the work surface, as shown. Remove the four bolts securing the engine bed bracket to the lower crankcase.

Crankcase Separation

27- First, remove the 6mm bolts and then the 8mm bolts securing the two halves of the crankcase together. Two prying tabs are cast into the crankcase to assist in separating the two halves. **ALWAYS** pry **ONLY** at these two tab locations. Prying elsewhere could very likely damage an aluminum sealing surface.

Separate the lower crankcase from the upper crankcase. **TAKE CARE** not to lose the two locating pins indexed diagonally across from each other. These pins may remain in either half when the crankcase is separated.

28- Tap the tapered end of the crankshaft with a soft head mallet to "jar" it free of the crankcase. Lift the crankshaft with rods attached up and free of the upper crankcase half.

Crankshaft Disassembling

Pull the forward oil seal free of the tapered end of the crankshaft. If the metal coupler was removed from the threaded end of the crankshaft, remove the two rear main oil seals.

8-6 ASSEMBLING AND INSTALLATION THREE-CYLINDER ENGINES

Detailed procedures are given to assemble virtually all parts of the engine, followed by instructions to install the engine. Again, these procedures are identified for three-cylinder engines.

If certain parts were not removed or disassembled because the part was found to be fit for further service, simply skip the particular step involved and continue with the required tasks to return the engine to operating condition.

The following instructions pickup the work after all parts of the engine have been thoroughly cleaned and inspected according to the procedures outlined in Section 8-7.

The Cleaning and Inspecting section should have revealed any parts unfit for further service.

ADVICE

Before commencing the assembling tasks, check to be sure replacement parts have been obtained and are on hand.

EXPLODED DRAWINGS

The exploded drawings of the three-cylinder engines on Pages 8-55 and 8-56 will prove to be helpful during the assembling work.

Assembling Lower Half Three-Cylinder Engine

1- Pack the lips of the oil seals with Shell Alvania EP1, or equivalent, lubricant. Install the double lip seal -- onto the forward -- flywheel end -- of the crankshaft with "knobs" on the seal facing **INWARD -- TOWARD** the bearings. Install the other two seals -- back-to-back -- on the coupler end of the crankshaft.

Text Continues on Page 8-57

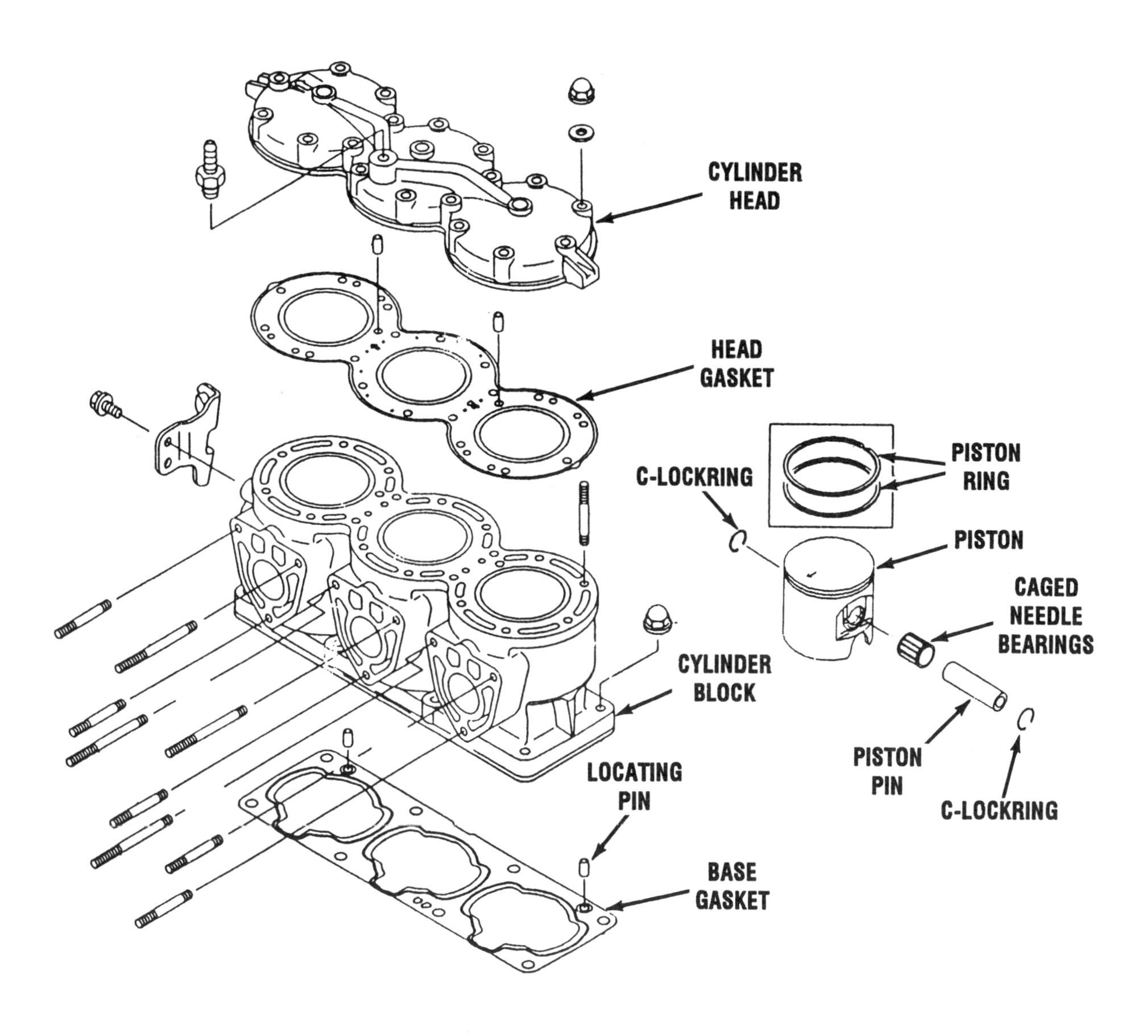

*Exploded drawing of the Model **900** and Model **1100** Series upper crankcase half, with major parts identified.*

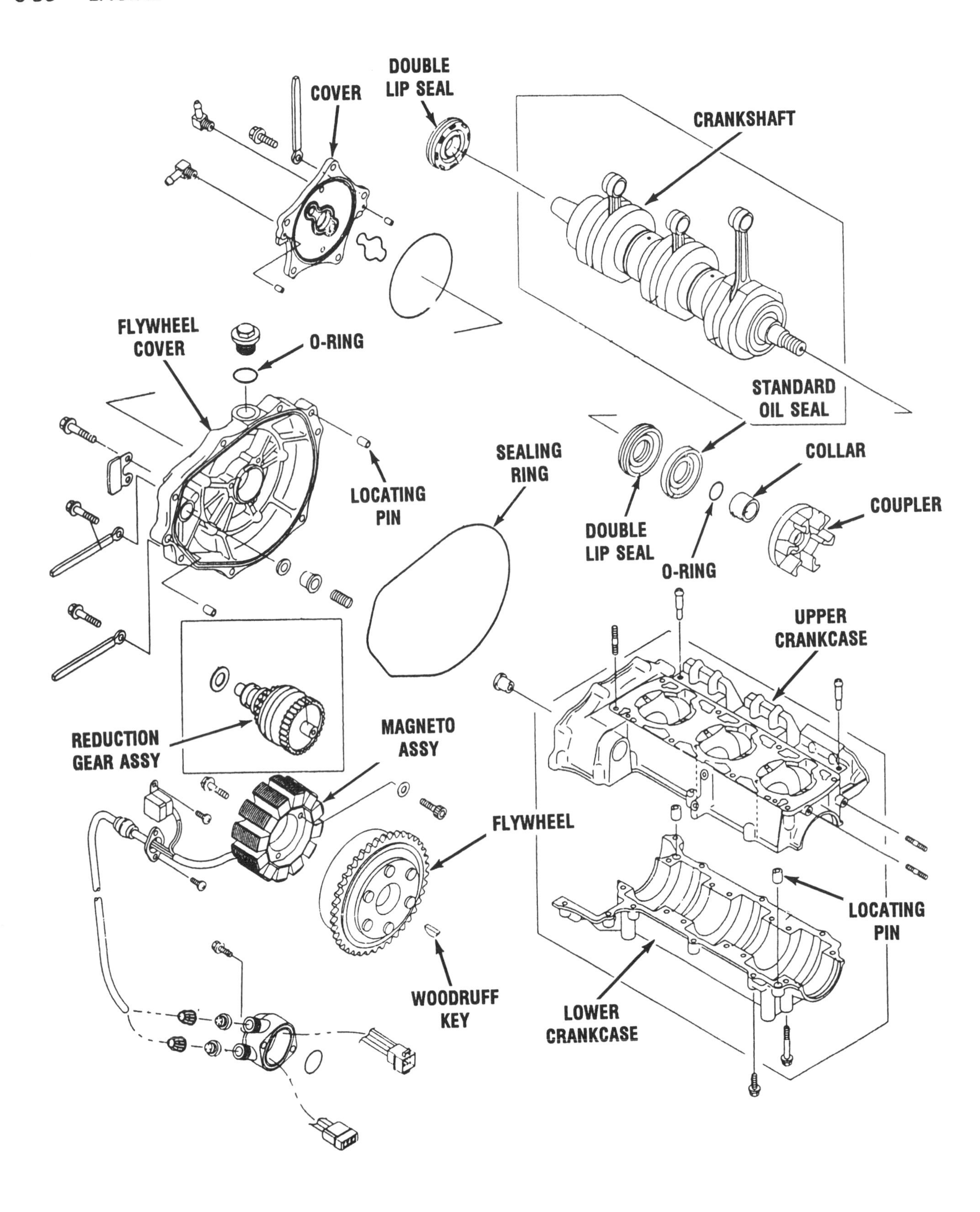

*Exploded drawing of the Model **900** and Model **1100** Series lower crankcase half, with major parts identified.*

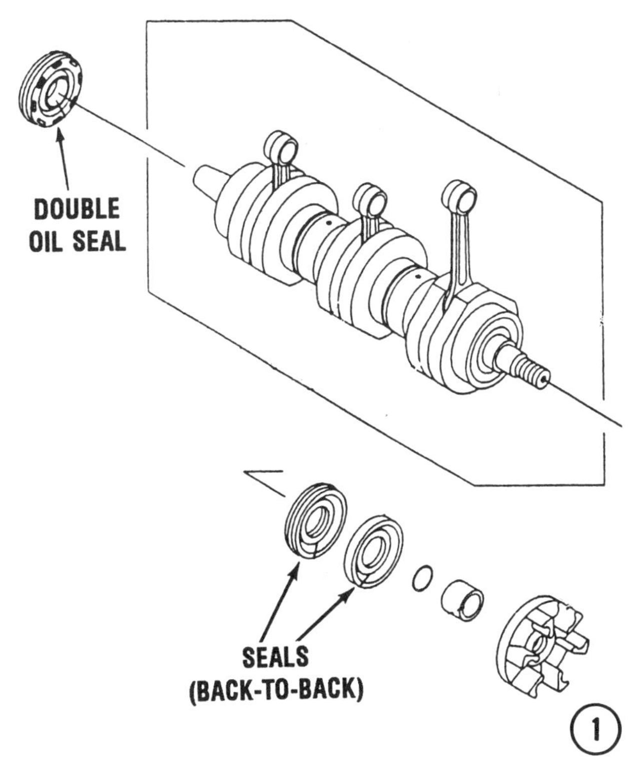

The outer seal prevents water from entering the crankcase and the other seal prevents pressure in the crankcase from escaping.

2- Apply a coating of Liquid Gasket compound to the mating surface of the lower crankcase -- looping around the bolt and locating pin holes.

3- Place the upper crankcase in the inverted position on the work surface on a couple of wood blocks, as shown. The wooden blocks are necessary to provide room for the connecting rods as the crankshaft is lowered into place.

Seat the crankshaft assembly into the lower crankcase half and fit the labyrinth seal circlip into the groove in the crankcase. Check to be sure the locating pins are in place on one of the crankcase halves.

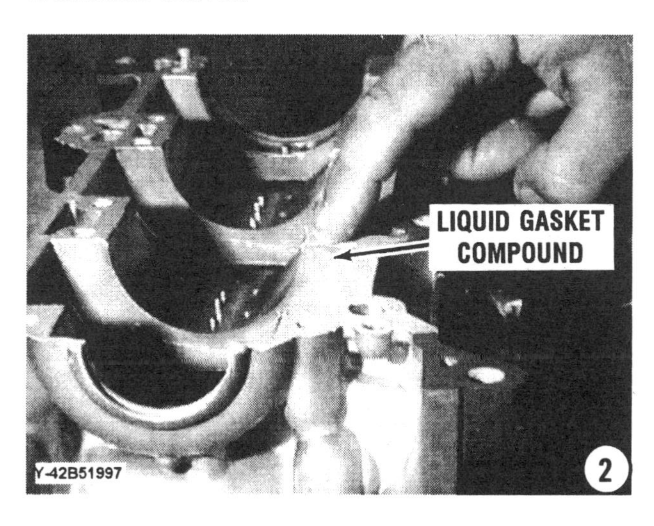

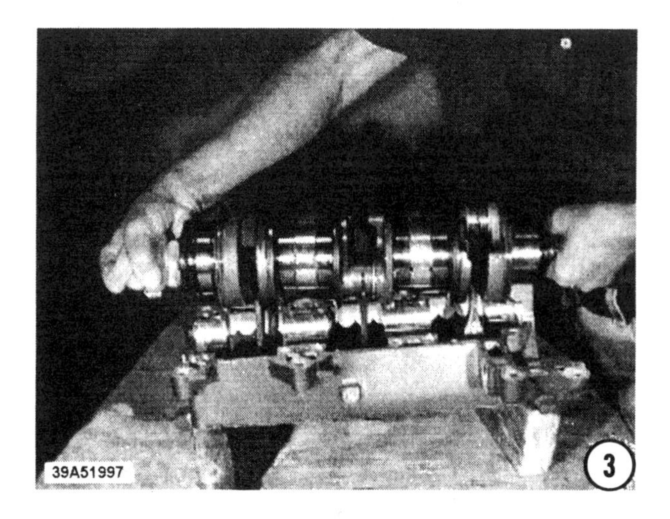

4- Move the lower crankcase half into position over the crankshaft. Bring the two crankcase halves together with the locating pins indexed into the matching holes in the other half. Apply a coating of non-permanent locking compound to the threads of the crankcase bolts.

CRITICAL BOLT WORDS

Refer to the accompanying illustration for this step. Bring the bolts to "just snug". Then, tighten the bolts in the sequence shown to 1/2 torque value. Notice the sequence begins in the center; work across and toward the ends alternately. Once the bolts are tightened to the first increment, repeat the sequence and bring the bolts to full torque.

For the 900 Series -- the 8mm bolts are to be tightened to 22 ft lb (29Nm). The 6mm bolts are to be tightened to only 69 in lb (7.8Nm).

For the 1100 Series -- the 8mm bolts are to be tightened to 22 ft lb (29Nm). The 6mm bolts are to be tightened to only 78 in lb (8.8Nm).

Pay attention -- to ensure the correct bolts are threaded into the proper openings.

Now, rotate the crankshaft by hand to be sure the crankshaft does not bind.

BAD NEWS

If binding is felt, it will be necessary to remove the crankcase and reseat the crankshaft. It will also be necessary to check the positioning of the labyrinth seal circlips and the bearing locating pins. If binding is still a problem after the crankcase has been installed a second time, the cause might very well be a broken piston ring.

Engine Bed Bracket Installation

5- Stand the engine upright with the flywheel end on the work surface, as shown. Position the engine bed bracket over the crankcase. For both the 900 Series and 1100 Series, the small notches of the bracket must face **FORWARD**. Apply a coating of non-permanent locking compound to the threads of the four mounting bolts. Install and tighten the bolts to a torque value of 27 ft lb (36Nm).

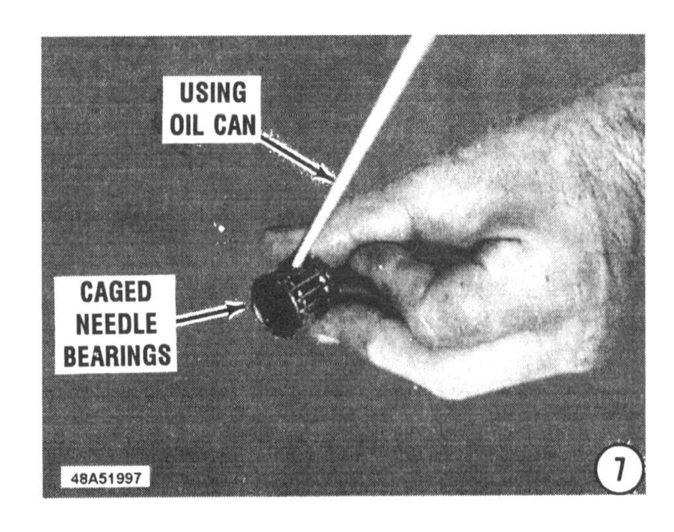

Piston Installation
All Series Engines

6- Shift the engine on the work surface into the normal upright position. Observe the embossed letter on each piston ring. The ring **MUST** be installed with this letter facing **UP.**

Install a new set of piston rings onto the pistons. No special tool is necessary to install the piston rings. **HOWEVER**, take care to spread the ring only enough to clear the top of the piston. The rings are **EXTREMELY** brittle and will snap if spread beyond their limit. Align the ring end gap over the locating pin in each piston groove.

The purpose of the pin is to prevent the ring from rotating in the ring groove, during engine operation. The ring end gap must travel up and down a portion of the cylinder wall where there is no port opening. If the ring was free to rotate, an end might catch on the port opening and break the ring.

7- Dip the caged needle bearings into a container of engine oil, and then insert the cage into the small end of the connecting rod. The

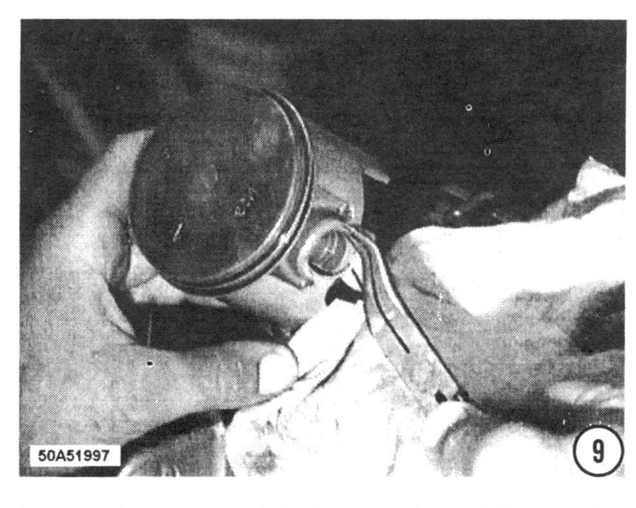

illustration shows oil being applied with a "squirt type" oil can.

CRITICAL WORDS

An arrow is embossed on each piston crown. The direction of this arrow **MUST** be **TOWARD** the exhaust manifold -- port side. Insert the piston pin into the piston and caged needle bearings. Center the pin in the piston.

8- Position the piston over the connecting rod with the arrow embossed on the crown facing **TOWARD** the exhaust manifold -- port side. Insert the piston pin into the piston and caged needle bearings. Center the pin in the piston.

SAFETY WORDS

The piston pin C-lockrings are made of spring steel and may slip out of the groove with considerable force. Therefore, warn other persons in the area and **WEAR** eye protection glasses while installing the piston pin lockrings in the next step.

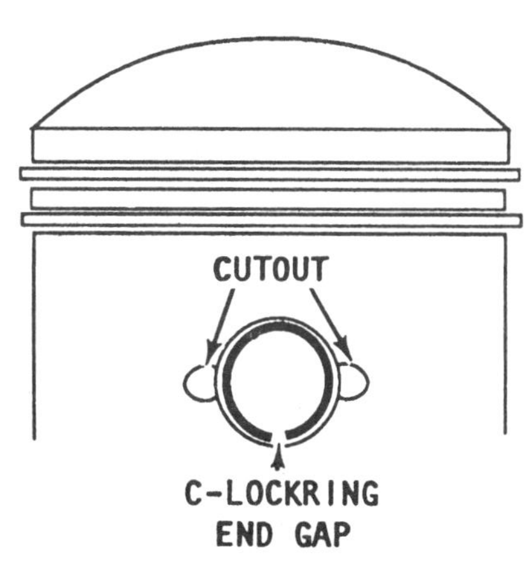

The C-lockring end gap must be positioned away from the two cutouts in the piston, as shown.

9- Install a C-lockring at each end of the piston pin. The lockrings may be worked in by hand. Position the end gap away from the cutout on the piston opening, as shown in the accompanying line drawing.

Piston Installation

Words of Advice

The following piston installation and cylinder block procedures are accomplished with very few problems, if the services of an assistant can be obtained.

10- Place a small amount of oil into each oil hole, as shown in the accompanying illustration. This hole allows oil to seep down into the crankcase. Place a few drops of oil on each rod bearing.

11- Install the base gasket over the crankcase studs, with the "U" -- for "up" -- facing **UP.** The manufacturer does not call for any type material to be applied to either side of this gasket.

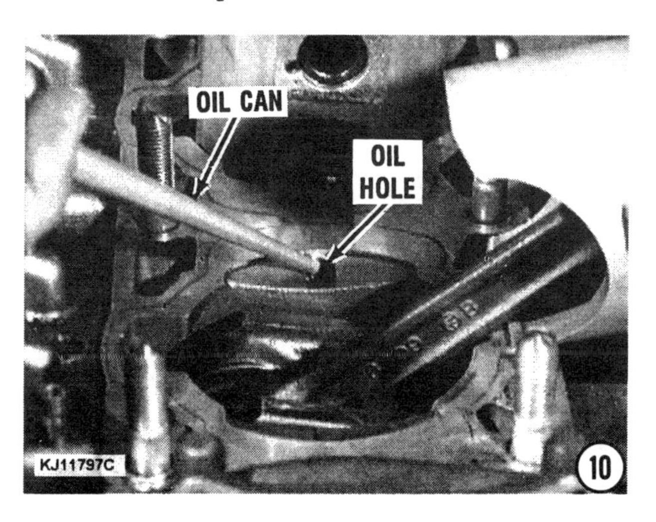

12- Coat the upper portion of each piston and rings with 2-stroke engine oil.

13- Obtain a special ring keeper tool. Secure it to the No. 2 piston. Check to be **SURE** the ring gap straddles the locating pin. **SLOWLY** lower the cylinder block down over the long studs over the piston crown.

14- Guide the piston to start into the bore. Once the piston is positioned within the bore, remove the ring keeper tool. Secure the ring keeper to the No. 3 piston. If a second ring keeper tool is available, secure it to the No. 1 piston. With the help of an assistant, guide the pistons to start into the cylinder block as it is lowered. Once the pistons have started into the bores, remove the ring keeper tools.

If the ring keeper tool is not available, squeeze the rings inward around the locating pin and guide the pistons into the bores.

CRITICAL WORDS

If difficulty is experienced in fitting the pistons into the cylinders, **DO NOT** use force. Such action might result in a broken piston ring. Check to be sure each ring end gap straddles the locating pin.

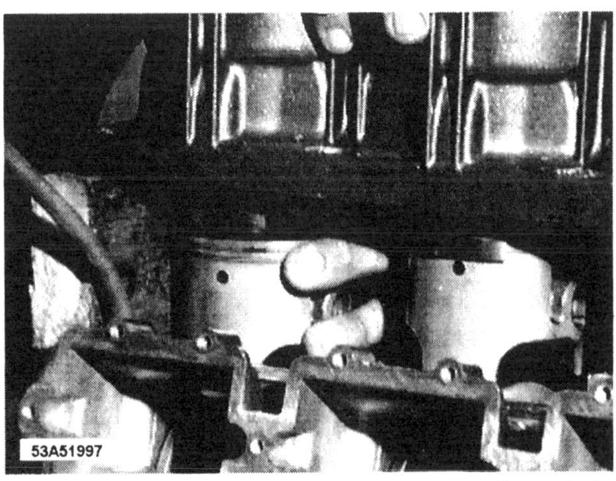

Close view of the piston entering the cylinder with the fingers of an assistant visible.

If more help is needed, obtain a large hose clamp and install the clamp around the piston to compress the rings.

15- Make a final check of the piston installation. The arrow **MUST** face **TOWARD** the exhaust side of the engine.

Install the nuts securing the cylinder block to the upper crankcase. Tighten the nuts to one half the final torque value following the pattern shown in the accompanying illustration. Tighten the nuts to the full torque value of 25 ft lb (34Nm), following the same pattern as the first.

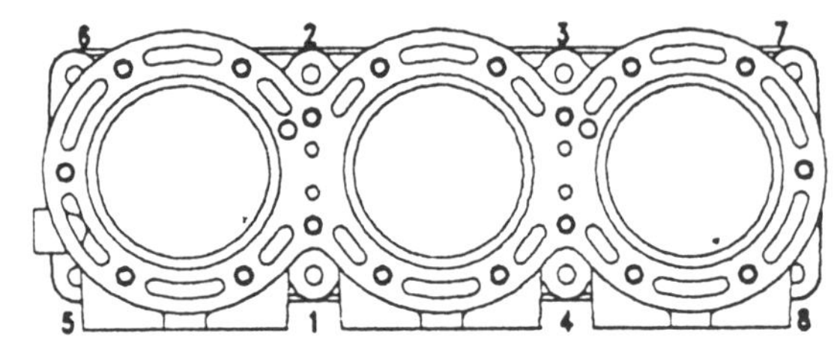

Cylinder block nut tightening sequence.

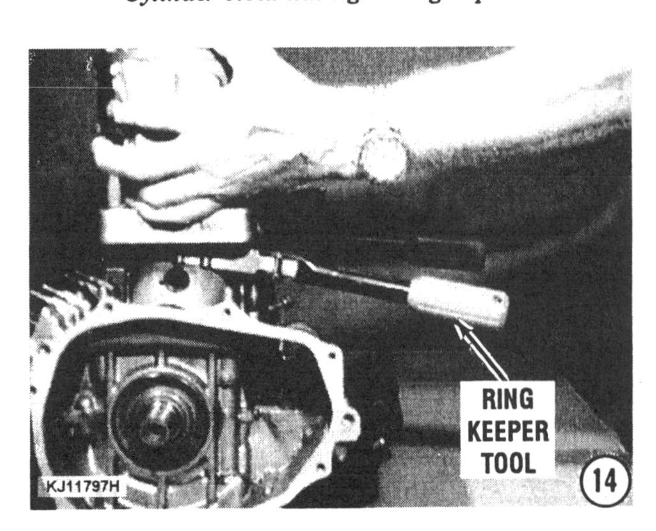

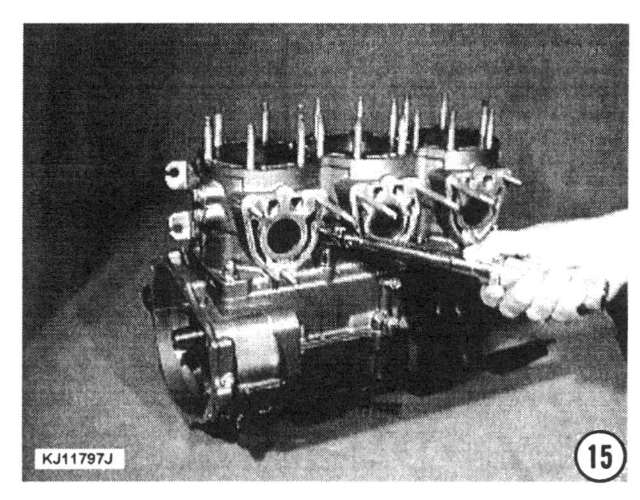

Reed Block/Intake Manifold Installation

16- Take note of the two unique reed block gaskets. The unmarked gasket is positioned on the mounting studs first, followed by the reed block housing. Place the second gasket in position over the mounting studs, with the letters "U" and "P" facing **UP** and readable -- toward the cylinder head, as indicated in the accompanying illustration.

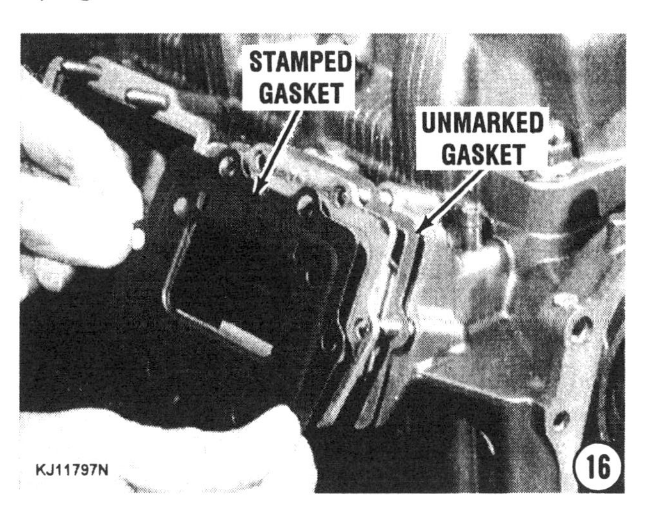

17- Install the intake manifold. Tighten the mounting nuts to a torque value of 87 in lb (9.8Nm).

Exhaust Manifold Installation

18- Position a new exhaust manifold gasket onto the block.

19- Install the exhaust manifold to the engine block. Apply a coating of Loctite Lock N' Seal to the threads of the mounting studs or bolts. Tighten the stud nuts to a torque value of 14.5 ft lb (20Nm).

Cylinder Head Installation

20- Observe the embossed word **"UP"** and arrow on a new cylinder head gasket. The gasket **MUST** be installed with these markings facing **UP** and with the arrow pointing toward the exhaust -- port side -- of the engine. Note the two alignment pins, located between the No. 2 and No. 3 cylinders. Install the gasket as described.

SPECIAL WORDS

The manufacturer recommends **NO** sealing agent be used on either side of the head gasket.

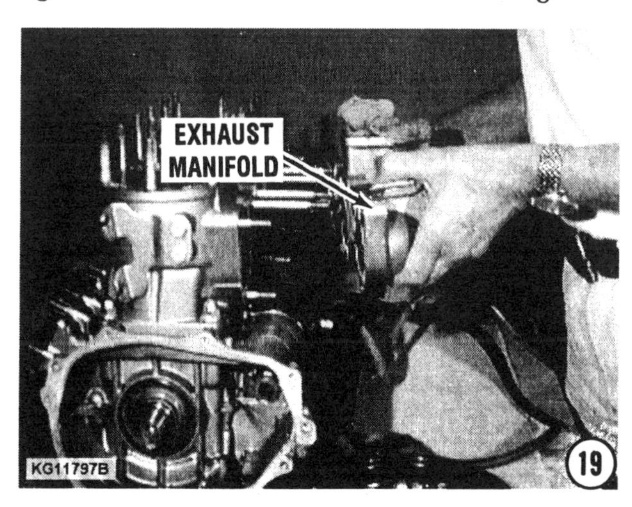

SEALING SURFACE WORDS
CRITICAL READING

Because of the high temperatures and pressures the sealing surfaces of the cylinder head and the block are the most prone to water leaks. No sealing agent is recommended **BECAUSE** it is almost impossible to apply an even coat would be essential to ensure a air/water tight seal.

Some head gaskets are supplied with a "tacky" coating on both surfaces applied at the time of manufacture. This "tacky" substance will provide an even coating all around. Therefore, no further sealing agent is required.

HOWEVER, if a slight water leak should be noticed following completed assembly work and engine start up, **DO NOT** attempt to stop the leak by tightening the head bolts beyond the recommended torque value. Such action will only aggravate the problem and most likely distort the head.

FURTHERMORE, tightening the bolts, which are case hardened aluminum, may force the bolt beyond its elastic limit and cause the bolt to **FRACTURE.** A fractured bolt must usually be drilled out and the hole re-tapped to accommodate an oversize bolt, etc. Avoid such a situation.

Probable causes and remedies of a new head gasket leaking are:

a- Sealing surfaces not thoroughly cleaned of old gasket material. Disassemble and remove **ALL** traces of old gasket.

b- Damage to the machined surface of the head or the block. The remedy for this damage is the same as for the next case "**c**".

c- Permanently distorted head or block. Spray a light **EVEN** coat of any type metallic spray paint on both sides of a new head gasket. Use only metallic paint -- any color will do. Regular spray paint does not have the particle content required to provide the extra sealing properties this procedure requires.

Assemble the block and head with the gasket while the paint is still **TACKY**. Install the head bolts and tighten in the recommended sequence and to the proper torque value and **NO** more!

Allow the paint to set for at least 24 hours before starting the engine. Consider this procedure as a temporary "band aid" type solution until a new head may be purchased or other permanent measures can be performed.

Under normal circumstances, if procedures have been followed to the letter, the head gasket will not leak.

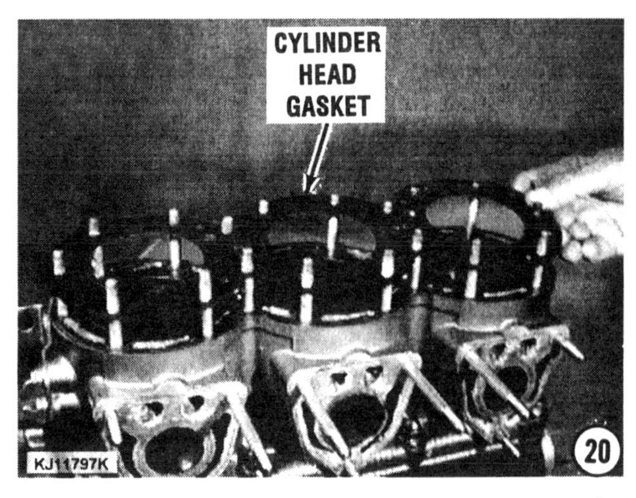

21- Lower the cylinder head down over the mounting studs and onto the engine block.

22- Install the washers and nuts onto the studs. Tighten the nuts to a torque value of 22 ft lb (29Nm).

Flywheel Installation
900 Series and 1100 Series

23- Install the cranking motor reduction gear assembly, with the spring going onto the shaft first, followed by the washer.

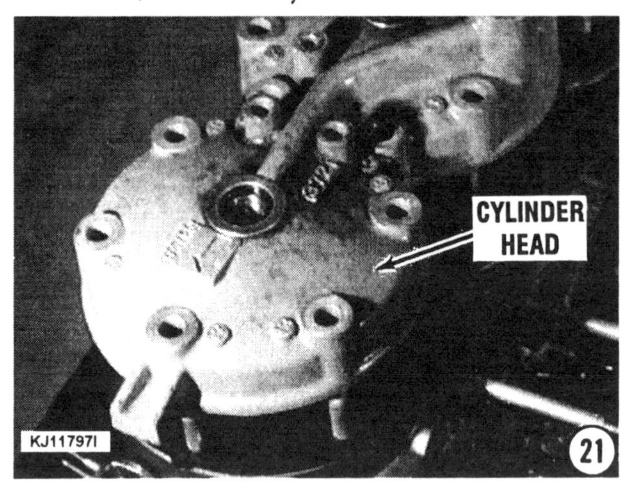

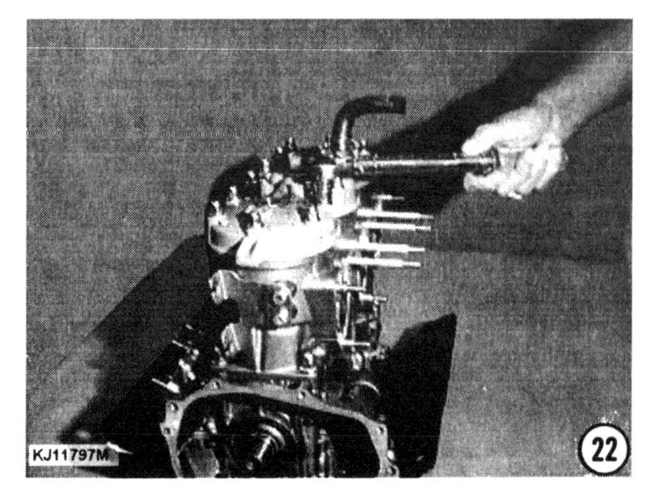

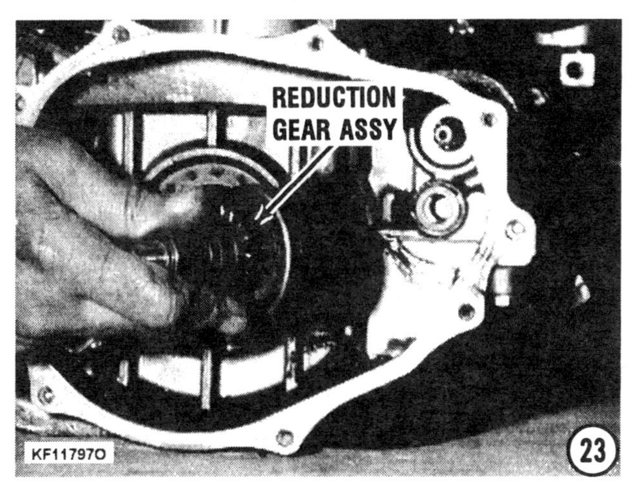

24- Place just a dab of thick lubricant onto the curved surface of the Woodruff key to hold it in place while the flywheel is being installed. Press the Woodruff key into place in the crankshaft recess. Wipe away any excess lubricant to prevent the flywheel from "walking" during engine operation.

Check the flywheel magnets to ensure they are free of any metal particles. Double check the taper in the flywheel hub and the taper on the crankshaft to verify they are clean and free

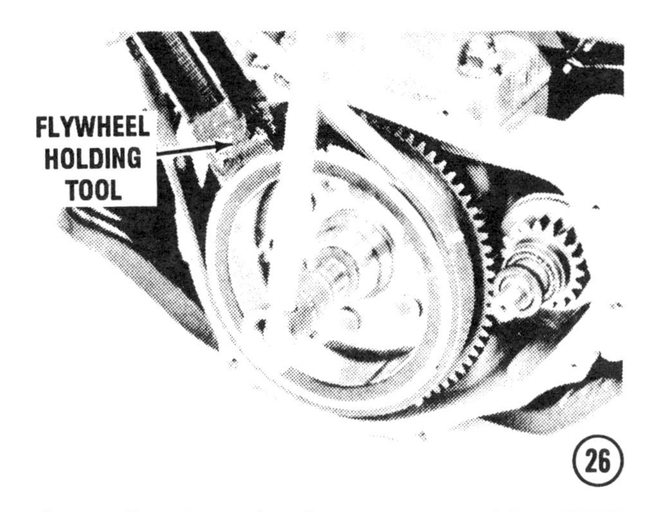

of any oil. The reduction gear assembly **MUST** be installed before the flywheel is in place.

25- Slide the flywheel onto the crankshaft with the keyway in the flywheel aligned with the Woodruff key in place on the crankshaft. Move the flywheel completely onto the crankshaft with the keyway in the flywheel indexed over the Woodruff key.

Thread the special flanged bolt onto the crankshaft -- just "fingertight" -- at this time.

26- Obtain special flywheel holding tool Kawasaki 57001-1313, a spanner tool, or a home made tool to prevent the flywheel from rotating. Tighten the flywheel bolt to a torque value of 95 ft lb (125Nm).

Coupler Installation

FIRST -- CRITICAL COUPLER WORDS

The coupler is made of cast material, therefore, if the special coupler tool is not available, be **EXTREMELY** careful of attempting to use a bar through the coupler slots to tighten the coupler.

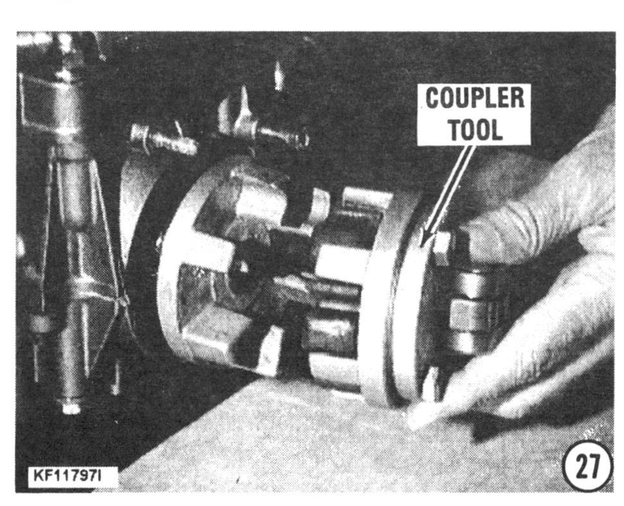

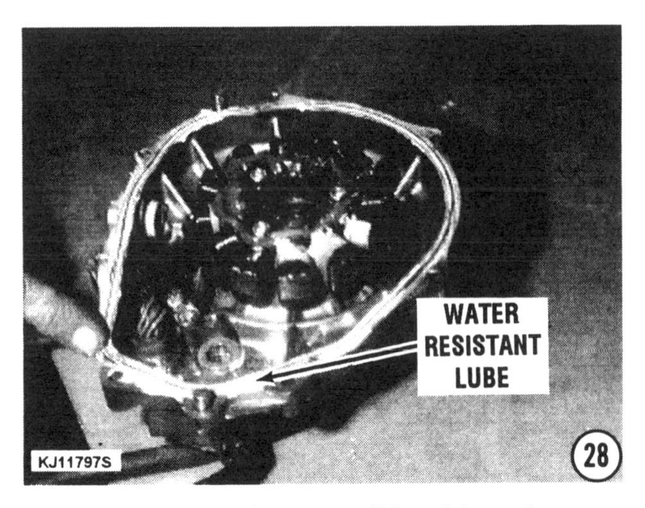

The coupler could very well be chipped or one of the leaves broken, causing an out-of-balance condition, resulting in unwanted vibration **AND** vibration causes expensive parts to wear rapidly.

27- With the flywheel holding tool still installed from the previous step, to prevent the flywheel/crankshaft from rotating, install the special coupler tool, Kawasaki 57001-1230.

Using the proper size wrench with a long handle, tighten the coupler to a torque value of 98 ft lb (130Nm). After the coupler is installed, remove the special tool and install the rubber damper.

Flywheel Cover Installation

28- Apply a coating of multi-purpose water resistant lubricant to the sealing ring. It is strongly recommended a new sealing ring be installed. The lubricant will help hold the ring in position while the cover is installed.

29- Install the flywheel cover over the flywheel.

Apply a coating of Loctite Lock N' Seal to the threads of the cover attaching bolts, and then tighten the bolts in a "cross" pattern to a torque value of 69 in lb (7.8Nm).

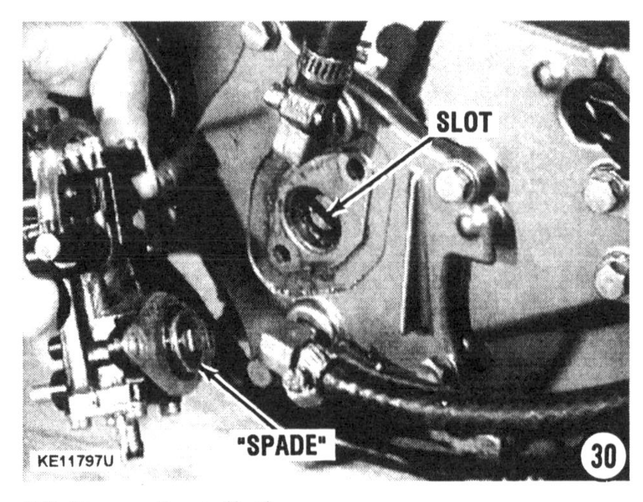

Oil Pump Installation

30- Connect the oil lines to the fittings on the pump, if they were disconnected. The "spade" on the oil pump shaft **MUST** index into the slot in the end of the crankshaft,or the pump will fail to operate. Move the pump into position on the flywheel cover with the "spade" indexed into the slot in the end of the crankshaft.

31- Secure the oil pump with the two bolts tightened to a torque value of 78 in. lbs (8.8Nm).

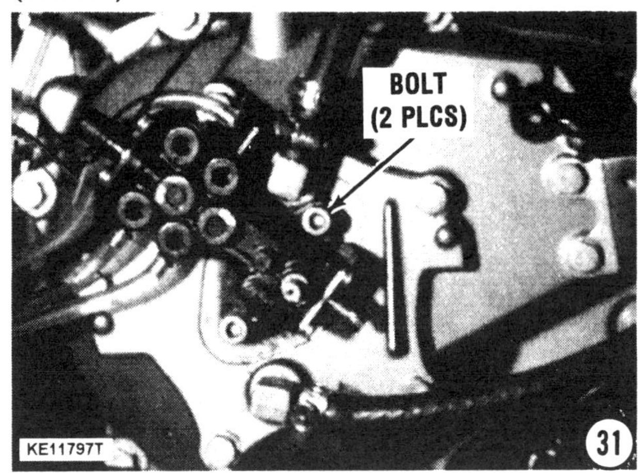

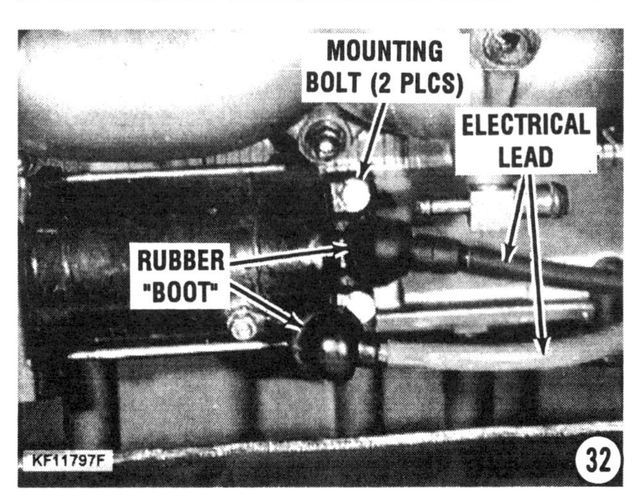

Cranking Motor Installation
900 Series and 1100 Series

32- Clean the cranking motor lugs and the contact surface of the crankcase in the area where the motor is grounded. Apply a coating of engine oil to the O-ring. Position the motor against the crankcase and at the same time, insert the end of the armature shaft into the reduction gear inside the flywheel cover.

If the battery ground cable was originally connected to the crankcase with the lower mounting bolt, place the eye of the cable over the bolt and apply a non-permanent locking compound on both mounting bolts. Install and secure the bolts to a torque value of 69 in lb (7.8 Nm) for the 900 Series; 78 in lb (8.8Nm) for the 1100 Series engine.

Secure the large Red cable to the terminal on the motor frame. Coat the connection with some type of dialectric grease, and then slide the boot over the connection.

33- Install the coupler shroud over the coupler. Actually, this shroud will "hide" the coupler during engine installation, but it must be installed at this time.

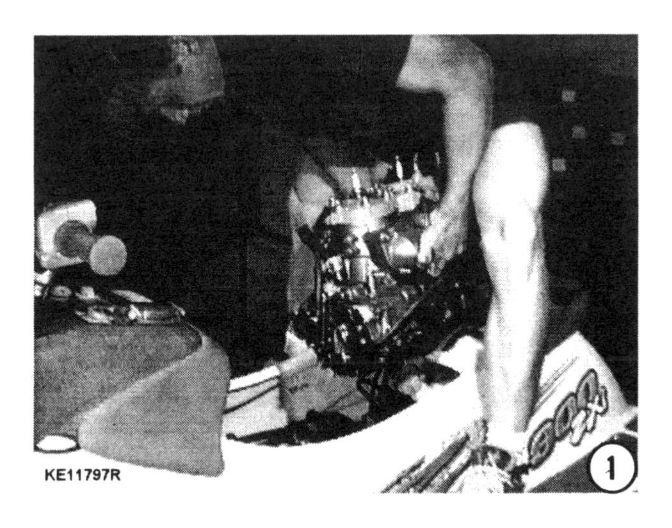

ENGINE INSTALLATION
THREE-CYLINDER SERIES

1- Insert the rubber coupler into the metal coupler on the jet pump driveshaft. Enlist the aid of an assistant and move the engine into the craft slightly forward of the engine mounts. The engine may need to be tilted "nose down" for easier placement in the compartment. Lower the engine slowly onto the rubber dampers and at the same time move the engine aft to engage the engine coupler with the rubber coupler on the jet pump driveshaft.

2- Apply a coating of Loctite #242 to the four engine mounting bolts. Temporarily thread the bolts --just a few turns -- into the fixed engine rubber dampers.

SPECIAL ENGINE
ALIGNMENT WORDS

The 900 Series and 1100 Series Models driveshaft holder incorporates a carrier bearing, which allows for proper alignment of the engine and jet pump without the need of shim material.

Due to the crankshaft coupler cover or shroud which was installed previously, care must be used in confirming the couplers are aligned. This is done by "feel" to ensure the couplers mesh together properly. Care and a little patience will go a long way!

Connect the electrical wires to the electrical box. There are two connectors on the aft side of the electrical box (port side). The top connector is the battery power lead, the bottom connector is the cranking motor power lead.

Reconnect the pin connectors, located on the front side of the electrical box. Install the front electrical case connector with the two

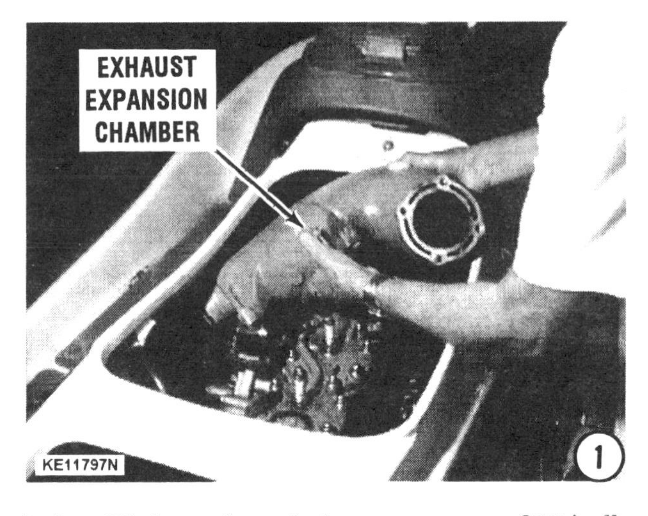

bolts. Tighten these bolts to a torque of 69 in lb (7.8Nm).

Connect the battery leads.

"Bleed" (purge) air from the oil systems according to the illustrated procedures in Chapter 6. Check the Table of Contents for the page number.

ENGINE SUPPORT EQUIPMENT INSTALLATION

Exhaust Expansion Chamber and Elbow Installation

1- Lower the exhaust expansion chamber into the engine compartment forward of the engine and slide it left into position for mounting. Two alignment collars assist in aligning the expansion chamber to the exhaust manifold. Apply a coating of Loctite on the chamber attaching bolts. Install and tighten the bolts to a torque value of 78 in lb (8.8Nm).

2- Connect the cooling hose from the exhaust pipe, to the aft bottom end of the expansion chamber. Attach and tighten the large

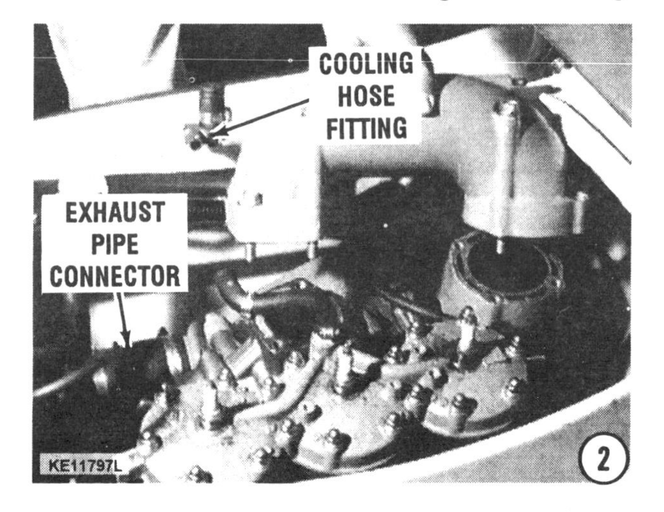

hose connector with a hose clamp at the aft end of the chamber. Connect the cooling hose from the flywheel cover to the top of the expansion chamber.

3- Position the round metal gasket onto the top of the expansion chamber. Position the exhaust manifold top gasket on the manifold. Take note of the embossed **"UP"**. Apply a coating of Loctite on the exhaust elbow bolts and take note of the two lengths. Install the bolts accordingly.

4- Connect the bypass hose to the forward fitting on the exhaust elbow. Use a zip tie to fasten the hose.

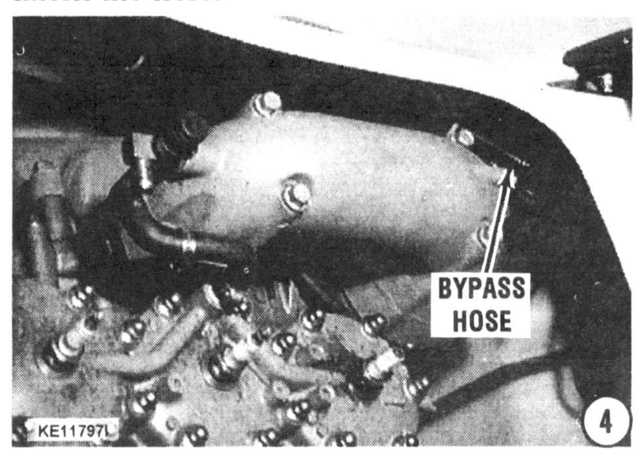

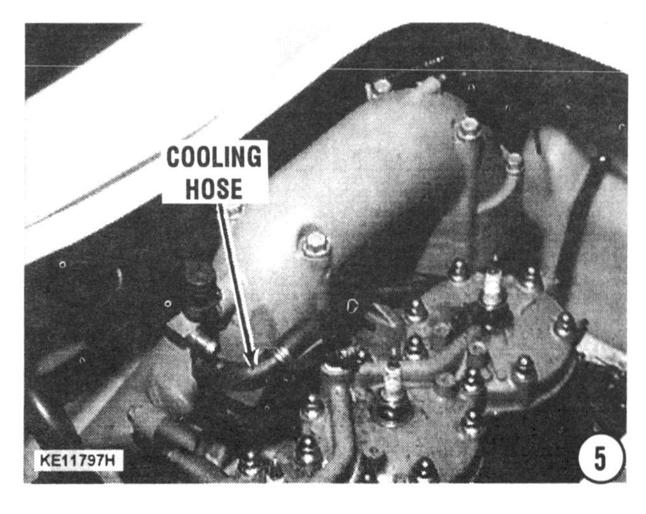

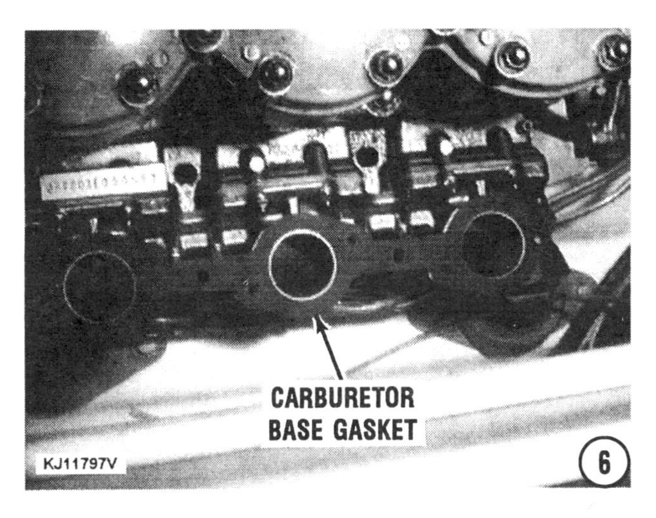

5- Connect the cooling hose from the cylinder head to the aft end of the exhaust elbow and secure it with the hose clamp.

6- Take note of the embossed and/or stamped marking on the carburetor base gasket. Install the gasket accordingly, taking note of the alignment pins. Lower the carburetor rack onto the base -- **SLOWLY** -- to prevent damaging the gasket.

CONNECTION WORDS

Connect the hoses and linkages items per the tags attached during the disassembling. If the tags were not attached, as instructed, in most cases, the shape and length of the hoses and linkage will indicate where the connection is to be made.

7- Connect the primary pulse hose to the intake manifold, by the No. 1 carburetor. Connect the secondary pulse hose to the intake manifold, by the No. 3 carburetor.

Connect the inlet fuel hose to the fitting under the No. 3 pulse fitting. Connect the rest of the supply fuel hoses.

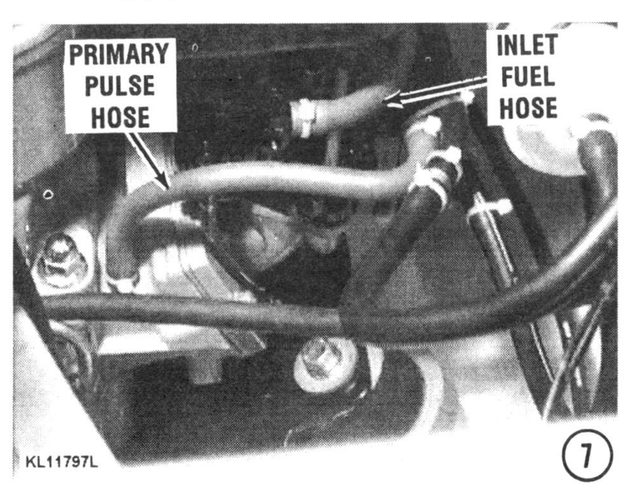

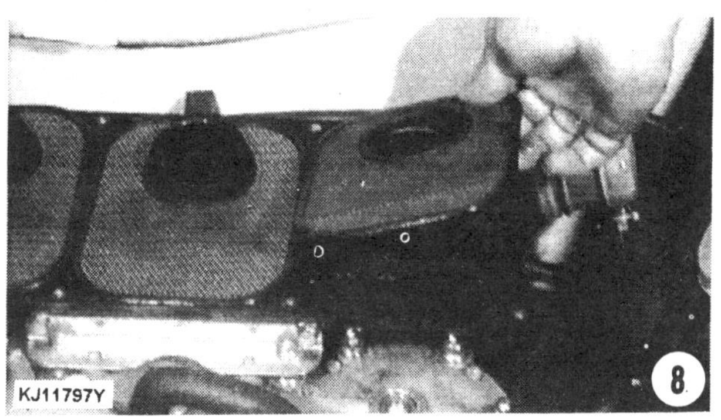

8- Lower the flame arrestor box onto the carburetors. With the arrestor screens removed, install and tighten the long carburetor mounting bolts to a torque value of 69 in lb (7.8Nm). Insert the flame arrestor screens in the arrestor box, and then install and tighten the cover mounting bolts to a torque value of 69 in lb (7.8Nm). Install the stay mounting bolts and tighten these bolts to 69 in lb (7.8Nm).

8-7 CLEANING AND INSPECTING ALL ENGINES

The success of the overhaul work is largely dependent on how well the cleaning and inspecting tasks are completed. If some parts are not thoroughly cleaned, or if an unsatisfactory unit is allowed to be returend to service through negligent inspection, the time and expense involved in the overhaul work will not be justified with peak engine performance and long operating life.

Therefore, the procedures in the following sections should be followed closely and the work perfomed with patience and attention to detail.

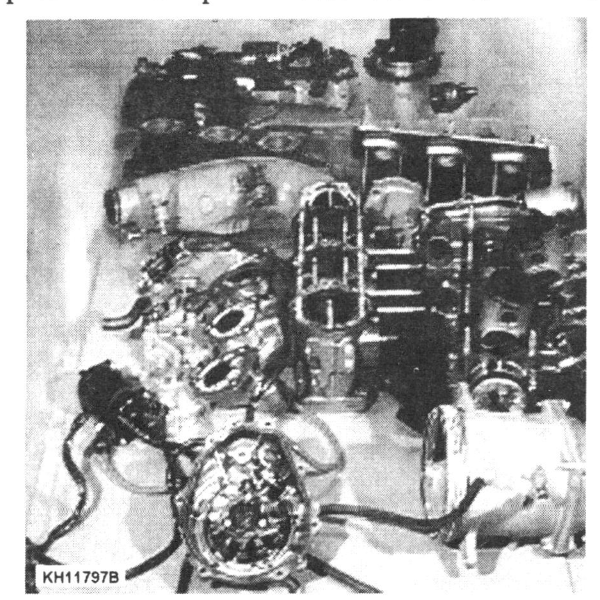

Internal parts laid out on the bench, following a complete disassembly of a Model 750 Series engine.

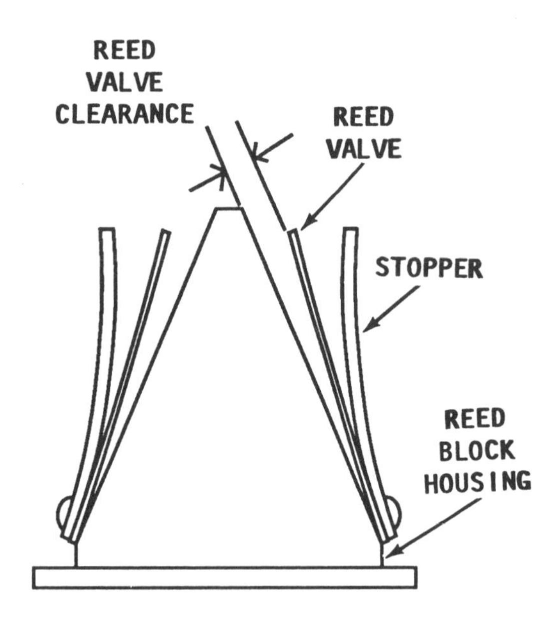

The maximum reed valve clearance is 0.008" (0.2mm), for all engines equipped with reed valves.

Reed Block Service
All Models Except 550

Disassemble the reed block housing by first removing the screws securing the reed stoppers and reed petals to the housing. After the screws are removed, lift the stoppers and petals from the housing.

Clean the gasket surfaces of the housing. Check the surfaces for deep grooves, cracks or any distortion which could cause leakage.

Replace the reed block housing if it is damaged. The reed petals should fit flush against their seats, and not be preloaded against their seats or bend away from their seats.

Petal Clearance

The maximum petal clearance for all 650 Series, 750 Series, 900 Series, and 1100 Series engines is 0.008" (0.2mm).

If the reed petal is distorted, rarely, if ever, can the petal be successfully straightened. **THEREFORE,** it must be replaced.

Do not remove the reed valves unless they, or the stoppers, are to be replaced. The reeds **MUST** be replaced in sets.

Apply Loctite to the threads of the reed retaining screws. Tighten each screw gradually, starting in the center and working outwards across the reed block. Tighten the screws to a torque value of 3.5 ft lbs (5Nm).

TAKE CARE not to mar the special rubber coating on the reed block housing.

Crankshaft Service

Clean the crankshaft with solvent and wipe the journals dry with a lint free cloth. Inspect the main journals and connecting rod journals for cracks, scratches, grooves, or scores. Inspect the crankshaft oil seal surface for nicks, sharp edges or burrs which might damage the oil seal during installation or might cause premature seal wear. **ALWAYS** handle the crankshaft carefully to avoid damaging the highly finished journal surfaces. Blow out all oil passages with compressed air. The oil passageway leads from the rod to the main bearing journal. **TAKE CARE** not to blow dirt into the main bearing journal bore.

Inspect the threads at both ends for signs of abnormal wear. Check the crankshaft for runout by supporting it on two **"V"** blocks at the main bearing surfaces.

Install a dial indicator gauge above the main bearing journals. Rotate the crankshaft and measure the runout (or the out-of-round) and the taper at both ends (and in the center journal on all two-cylinder models).

The following crankshaft runout limits are recommended by the manufacturer.

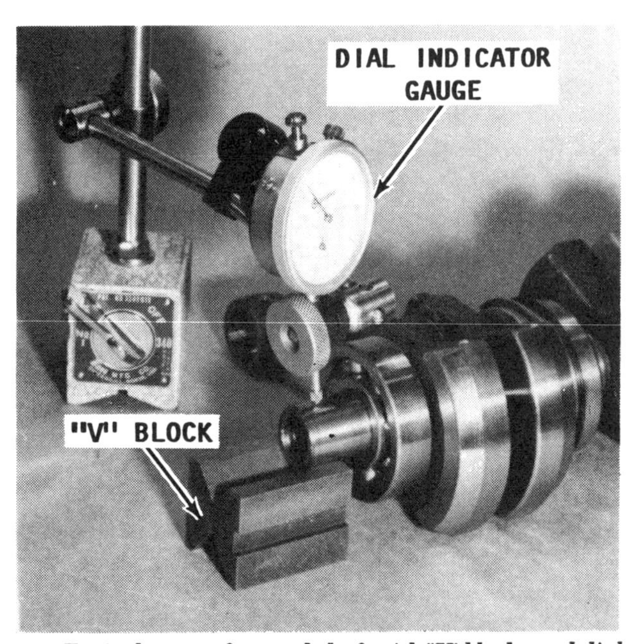

Typical setup of a crankshaft with "V" blocks and dial indicator to measure main bearing journals on an assembled two-stroke crankshaft. The bearings and connecting rod will not affect the reading.

CRANKSHAFT RUNOUT LIMITS

Series	Standard	Service Limit
550	0.0019" (0.05mm)	0.0031" (0.08mm)
650	0.0015" (0.04mm)	0.0039" (0.10mm)
750	0.0015" (0.04mm)	0.0039" (0.10mm)
900 & 1100	0.0015" (0.04mm)	0.0039" (0.10mm)

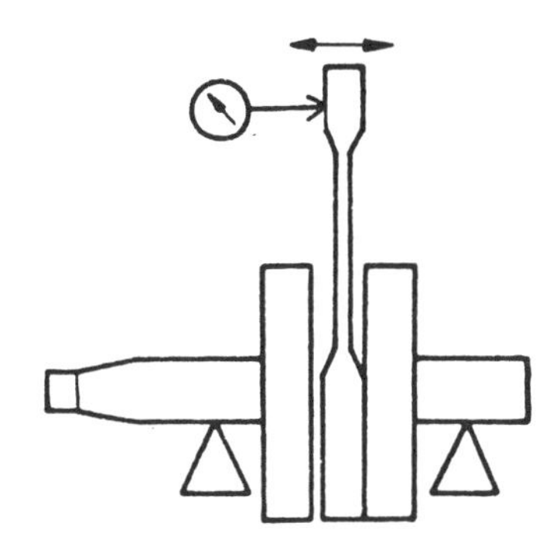

The simplest line drawing possible to depict the setup of the crankshaft and piston rod to check for axial rod "play", as explained in the text.

If **"V"** blocks or a dial indicator are not available, a micrometer may be used to measure the diameter of the journal. Make a second measurement at right angles to the first. Check the difference between the first and second measurement for out-of-round condition. If the journals are tapered, ridged, or out-of-round by more than the specification allows, the journals should be reground, or the crankshaft replaced.

Any out-of-round or taper shortens bearing life. Good shop practice dictates new main bearings be installed with a new or reground crankshaft.

Normally the connecting rods would **ONLY** be presssed from the crankshaft if either the crankshaft and/or the connecting rods were to be replaced. Therefore, the connecting rod axial "play" is checked at the piston end to determine the amount of wear at the crankshaft end of the connecting rod. Checking the "play" is described in the following paragraph.

Setup the crankshaft in the **"V"** blocks. Setup a dial indicator to touch the flat surface of the piston end of the rod. Now, hold the crankshaft steady in the **"V"** blocks and at the same time, rock the piston end of the rod along the same axis as the crankshaft. If the dial indicator needle moves through more than 0.08" (2mm) for all engines covered in this manual, the "play" is considered excessive. The rod must be pressed from the crankshaft, the journal

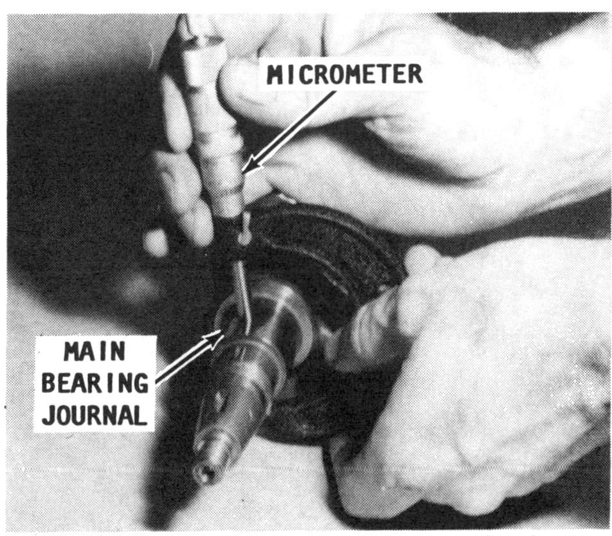

Method of using a micrometer to take the first measurement on a crankshaft main journal for out-of-round and taper.

Taking a second measurement on a crankshaft main bearing journal at 90° to the first measurement.

checked and a determination made as to whether the crankshaft and/or the rod must be replaced. A new rod may be purchased, **BUT** it must be pressed onto the crankshaft throw with a hydraulic press.

To check the connecting rod side clearance at the crankshaft, first insert a feeler gauge between the connecting rod and the counterweight of the crankshaft. Acceptable clearances are as follows:

CONNECTING ROD CLEARANCE

Series	Standard	Service Limit
550	0.0157-0.0196" (0.40-0.50mm)	0.0275" (0.7mm)
650	0.0177-0.0216" (0.45-0.55mm)	0.0314" (0.8mm)
750	0.0177-0.0216" (0.45-0.55mm)	0.0314" (0.8mm)
900 & 1100	0.0177-0.0216" (0.45-0.55mm)	0.0314" (0.8mm)

Inspect the crankshaft oil seal surfaces to be sure they are not grooved, pitted, or scratched. Replace the crankshaft if it is severely damaged or worn. Check all crankshaft bearing surfaces for rust, water marks, chatter marks, uneven wear or overheating. Clean the crankshaft surfaces with 320-grit carborundum cloth. **NEVER** spin-dry a crankshaft ball bearing with compressed air.

Clean the crankshaft and crankshaft ball bearing with solvent. Dry the parts, but not the ball bearing, with compressed air. Check the crankshaft surfaces a second time. Replace the crankshaft if the surfaces cannot be cleaned properly for satisfactory service. If the crankshaft is to be installed for service, lubricate the surfaces with light oil. **DO NOT** lubricate the crankshaft ball bearing at this time.

After the crankshaft has been cleaned, grasp the outer race of the crankshaft ball bearing installed on the lower end of the crankshaft, and attempt to work the race back-and-forth. There should not be excessive "play". A very slight amount of side "play" is acceptable because there is only about 0.001" (.025mm) clearance in the bearing.

Lubricate the ball bearing with light oil. Check the action of the bearing by rotating the outer bearing race. The bearing should have a smooth action and no rust stains. If the ball bearing sounds or feels rough or catches, the bearing should be removed and discarded.

Connecting Rod Service

Inspect the connecting rod bearings for rust or signs of bearing failure. **NEVER** intermix new and used bearings. If even one bearing in a set needs to be replaced, all bearings at that location **MUST** be replaced.

Clean the inside diameter of the piston pin end of the connecting rod with crocus cloth.

Clean the connecting rod **ONLY** enough to remove marks. **DO NOT** continue, once the marks have disappeared.

Assemble the piston end of the connecting rod with loose needle bearings, caged needle bearings, or no needle bearing, depending on the model being serviced. Insert the piston pin and check for vertical "play". The piston pin should have **NO** noticeable vertical "play".

If the pin is loose or there is vertical "play" check for and replace the worn part/s.

Inspect the piston pin and matching rod end for signs of heat discoloration. Overheating is identified as a bluish bearing surface color and is caused by inadequate lubrication or by operating the engine at excessive high rpm.

Piston Service

Inspect each piston for evidence of scoring, cracks, metal damage, cracked piston pin boss, or worn pin boss. Be especially critical during inspection if the craft and

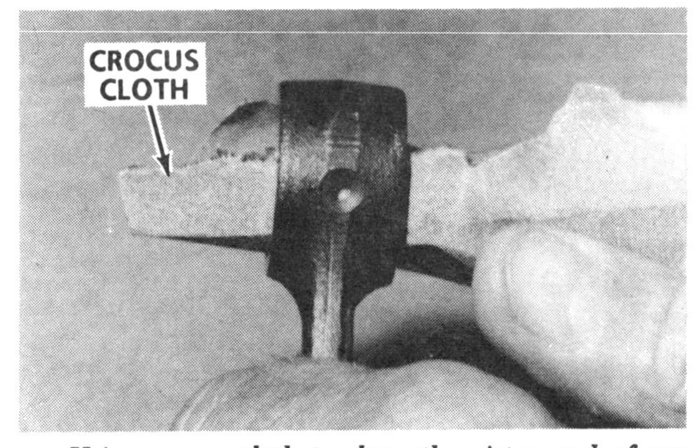

Using crocus cloth to clean the piston end of any connecting rod.

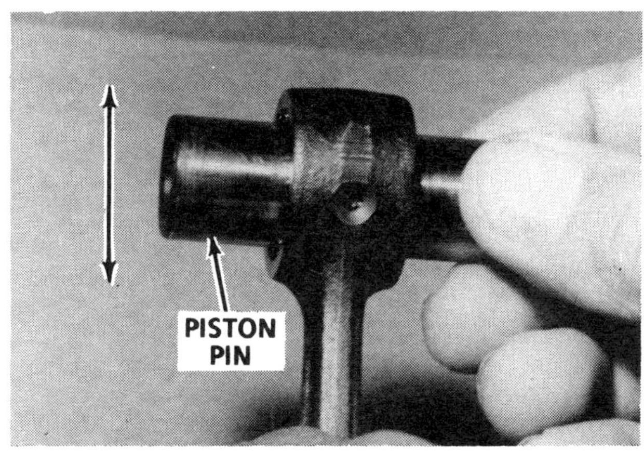

Check the piston end of a connecting rod for vertical free "play" using a piston pin.

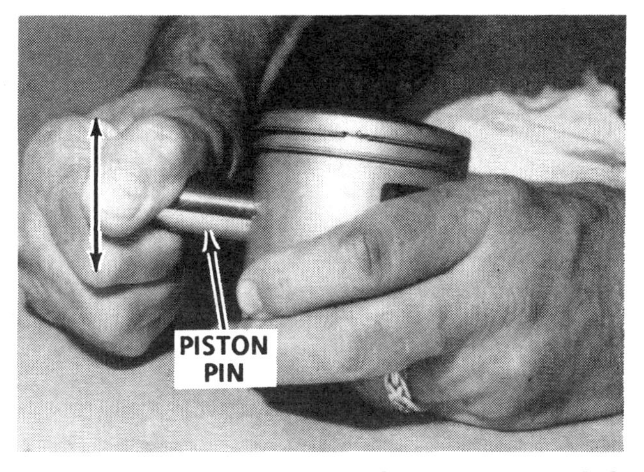

Check free "play" between the piston pin and the piston boss. There should be NO "play".

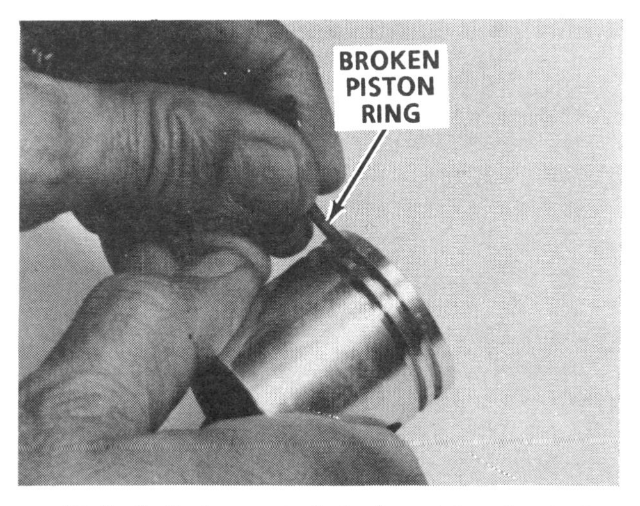

*Method of using part of a broken piston ring to clean the ring groove. Exercise **CARE** not to disturb the locating pin in each groove.*

It is believed, this crown seized with the cylinder wall when the unit was operated at high rpm and the timing was not adjusted properly. At the same instant, the rod apparently pulled the lower part of the piston downward, severing it from the crown.

engine has been submerged. If the piston pin is bent, the piston and pin **MUST** be replaced as a set for two reasons. First, a bent pin will damage the boss when it is removed. Secondly, a piston pin is not sold as a separate item.

Check the piston ring grooves for wear, burns distortion, or loose locating pins. During an overhaul, the rings should be replaced to ensure lasting repair and proper powerhead performance after the work is completed. Clean the piston dome, ring grooves and the piston skirt. Clean carbon deposits from the ring grooves using the recessed end of a broken piston ring.

NEVER use a rectangular ring to clean the groove for a tapered ring, or use a tapered ring to clean the groove for a rectangular ring.

Locating pins prevent the rings from rotating and catching on a port in the cylinder wall.

The pitted damage to this piston crown was probably caused by a broken piston ring working its way into the combustion chamber. The little "hills" then became "hot" spots on the crown, contributing to "dieseling" after the engine was shut down.

NEVER use an automotive-type ring groove cleaner, because such a tool may loosen the piston ring locating pins.

Clean carbon deposits from the top of the piston using a soft wire brush, carbon removal solution or by sand blasting. If a wire brush is used, **TAKE CARE** not to burr or round machined edges. Clean the piston skirt with crocus cloth.

The rings on this piston became stuck due to lack of adequate lubrication, incorrect timing, or overheating.

The rings on this piston were broken, possibly during installation, and then caused extensive damage to the piston. Ring parts found their way into the combustion chamber and caused damage to the piston crown.

Install the piston pin through the first boss only. Check for vertical free "play". There should be **NO** vertical free "play". The presence of "play" is an indication the piston boss is worn. The piston is manufactured from a softer material than the piston pin. Therefore, the piston boss will wear more quickly than the pin.

Excessive piston skirt wear **CANNOT** be visually detected. Therefore, good shop practice dictates, the piston skirt diameter be measured with a micrometer.

Piston diameters are are measured at a definite distance up from the bottom of the skirt at right angles to the piston pin axis. Size limits are as follows:

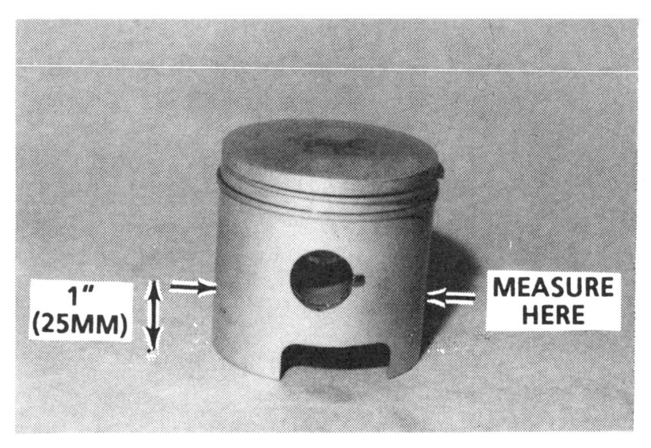

Measure the piston diameter at the specified distance above the bottom of the piston skirt.

STOCK PISTON DIAMETER

Series	Standard	Service Limit
550 3/16" Up	2.9496-2.9503" (74.92-74.94mm)	2.9441" (74.78mm)
650 3/4" Up	2.9905-2.9911" (75.96-75.97mm)	2.9846" (75.81mm)
750 3/16" Up	3.1457-3.1463" (74.92-74.94mm)	3.1398" (74.78mm)
900 3/16" Up	2.8699-2.8705" (72.90-72.91mm)	2.8642" (72.75mm)
1100 3/16" Up	3.1443-3.1449" (79.87-79.88mm)	3.1386" (79.72mm)

Ring End Gap

Before the piston rings are installed onto the piston, the ring end gap clearance for each ring must be determined. The purpose of the piston rings is to prevent the blowby of gases in the combustion chamber. This cannot be achieved unless the correct oil film thickness is left on the cylinder wall.

This thin coating of oil acts as a seal between the cylinder wall and the face of the piston ring. An excessive end gap will allow blowby and the cylinder will lose compression. An inadequate end gap will scrape too much oil from the cylinder wall and limit lubrication. Lack of adequate lubrication will cause excessive heat and wear.

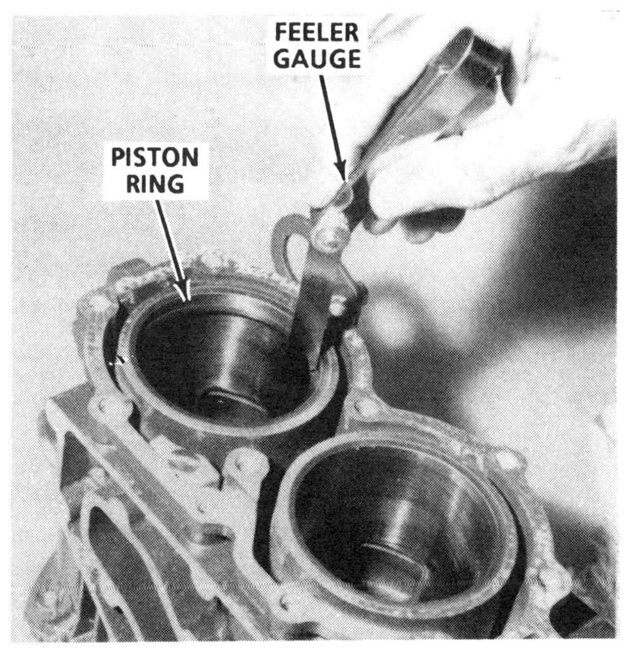

Example of measuring the piston ring end gap with a feeler gauge.

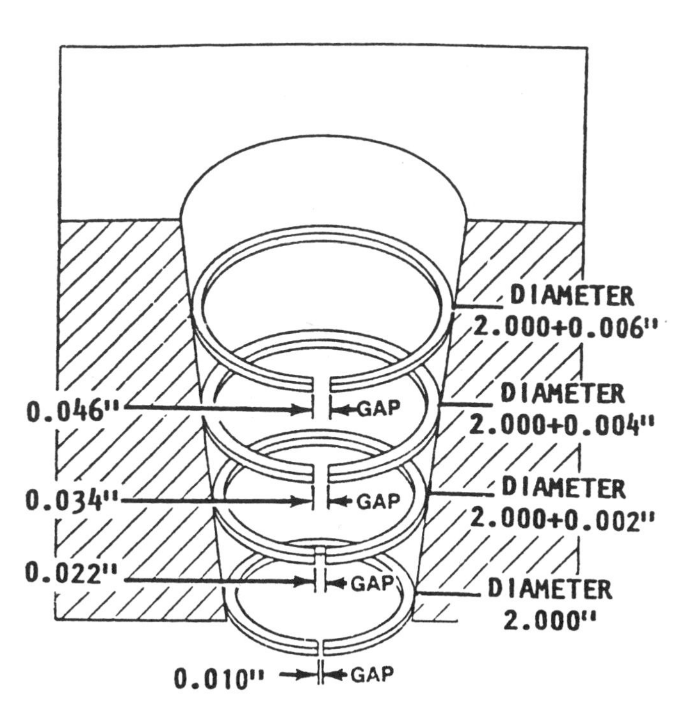

Simple line drawing to illustrate cylinder taper. The taper drastically affects ring and end gap, as indicated.

IDEALLY the ring end gap measurement should be taken **AFTER** the cylinder bore has been measured for wear and taper **AND** after any corrective work, such as boring or honing, has been completed.

IF the ring end gap is measured with a taper to the cylinder wall, the diameter at the lower limit of ring travel will be smaller than the diameter at the top of the cylinder.

IF the ring is fitted to the upper part of a cylinder with a taper, the ring end gap will not be great enough at the lower limit of ring travel. Such a condition could result in a broken ring and/or damage to the cylinder wall and/or damage to the piston and/or damage to the cylinder head.

IF the cylinder is to be only honed, not bored, **OR** if only cleaned, not honed, the ring end gap should be measured at the lower limit of ring travel.

The ring should be inserted into the cylinder just above the ports to perform this measurement, using the piston crown. Be sure the ring is parallel to the surface of the block, then measure the ring end gap with a feeler gauge.

If the end gap is greater than the amount listed, replace the entire ring set.

If the end gap is less than 0.008-0.016" (0.20-0.40mm) for both top and bottom rings, carefully file the ends of the ring --

just a little at a time -- until the correct end gap is obtained.

Inspect the piston ring locating pins to be sure they are tight. There is one locating pin in each ring groove. If the locating pins are loose, the piston **MUST** be replaced.

Oversized Pistons and Rings

Scored cylinder blocks can be saved for further service by reboring and installing oversize pistons and piston rings. **HOWEVER,** if the scoring is over 0.0075" (0.13mm) deep, the block cannot be effectively rebored for continued use.

Check with the local dealer for oversize piston availability.

If oversize pistons are not available, the local marine shop may have the facilities to "knurl" the piston, making it larger.

Cylinder Block Service

Inspect the cylinder block and cylinder bores for cracks or other damage. Remove carbon with a fine wire brush on a shaft attached to an electric drill or use a carbon remover solution.

STOP: If the cylinder block is to be submerged in a carbon removal solution, the crankcase bleed system **MUST** be removed from the block to prevent damage to hoses and check valves.

Use an inside micrometer or telescopic gauge and micrometer to check the cylinders for wear. Check the bore for out-of-round and/or oversize bore. If the bore is tapered, out-of-round or worn more than the wear limit specified by the manufacturer, the cylinders should be rebored -- provided oversize pistons and rings are available.

Check with the dealer **PRIOR** to reboring. If oversize pistons and matching rings are not available, the block **MUST** be replaced.

GOOD WORDS

Oversize piston weight is approximately the same as a standard size piston. Therefore, it is **NOT** necessary to rebore all cylinders in a block just because one cylinder requires reboring.

Cylinder sleeves are an integral part of the die cast cylinder block and **CANNOT** be replaced. In other words, the cylinder cannot be "resleeved".

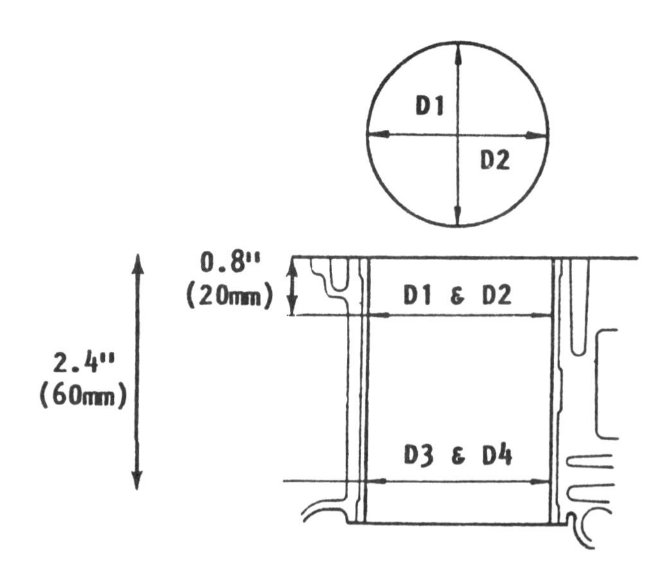

Top view diagram and cross-section of a typical cylinder to indicate where measurements are to be taken for wear limit, taper, and out-of-round limits.

Four inside cylinder bore measurements must be taken for each cylinder to determine an out-of-round condition, the maximum taper, and the maximum bore diameter.

In the accompanying illustration, measurements D1 and D2 are diameters measured at the top of the cylinder at right angles to each other. Measurements D3 and D4 are diameters measured from the top of the cylinder at right angles to each other.

Out-of-Round

Measure the cylinder diameter at D1 and D2. The manufacturer requires the difference between the two measurements should be less than 0.002" (0.050mm) for **ALL** models.

Maximum Taper

Measure the cylinder diameter at D1, D2, D3, and D4. Take the largest of the D1 or D2 measurements and subtract the smallest measurement at D3 or D4. The answer to the subtraction -- the cylinder taper -- should be less than 0.003" (0.08mm) for **ALL** models.

Bore Wear Limit

The maximum cylinder diameter D1, D2, D3, and D4 must not exceed the bore wear limits indicated in the following table **BEFORE** the bore is rebored for the **FIRST** time. These limits are only imposed on original parts because the sealing ability of the rings would be lost, resulting in power loss, increased engine noise, unnecessary vibration, piston slap, and excessive oil consumption.

The limits indicated are usually 0.003 to 0.005" (0.83 to 0.127mm) above the standard bore. Therefore, if the bore is resized, it may sustain another 0.003 to 0.005" (0.83 to 0.127mm) wear before a second reboring is required -- provided the oversize pistons and matching rings are available.

CYLINDER BORE WEAR LIMIT

Series	Standard	Service Limit
550	2.9557-2.9564" (75.07-75.09mm)	2.9759" (75.59mm)
650	2.9940-2.9946" (74.92-74.94mm)	2.9960" (76.10mm)
750	3.1496-3.1502" (80.00-80.02mm)	3.1535" (80.10mm)
900	2.8740-2.8746" (73.00-73.02mm)	2.8787" (73.12mm)
1100	3.1496-3.1502" (80.00-80.02mm)	3.1535" (80.10mm)

Piston Clearance

Piston clearance is the difference between a maximum piston diameter and a minimum cylinder bore diameter. If this clearance is excessive, the engine will develop the same symptoms as for excessive cylinder bore wear -- loss of ring sealing ability, loss of power, increased engine noise, unnecessary vibration, and excessive oil consumption.

Maximum piston diameter was described earlier in this section. Minimum cylinder bore diameter is usually determined by measurement D3 or D4 also described earlier in this section.

If the piston clearance exceeds the limits outlined in the following table, either the piston or the cylinder block **MUST** be replaced.

Calculate the piston clearance by subtracting the maximum piston skirt diameter from the maximum cylinder bore measurement and compare the results for the model being serviced.

PISTON CLEARANCE LIMITS

Series	Clearance
550	0.0053-0.0068" (0.13-0.17mm)
650	0.0033-0.0037" (0.084-0.094mm)
750	0.0037-0.0043 (0.095-0.110mm)
900	0.0039-0.0043 (0.100-0.110mm)
1100	0.0051-0.0055" (0.130-0.140mm)

HONING CYLINDER WALLS

Hone the cylinder walls lightly to seat the new piston rings, as outlined in this section. If the cylinders have been scored, but are not out-of-round or the bore is

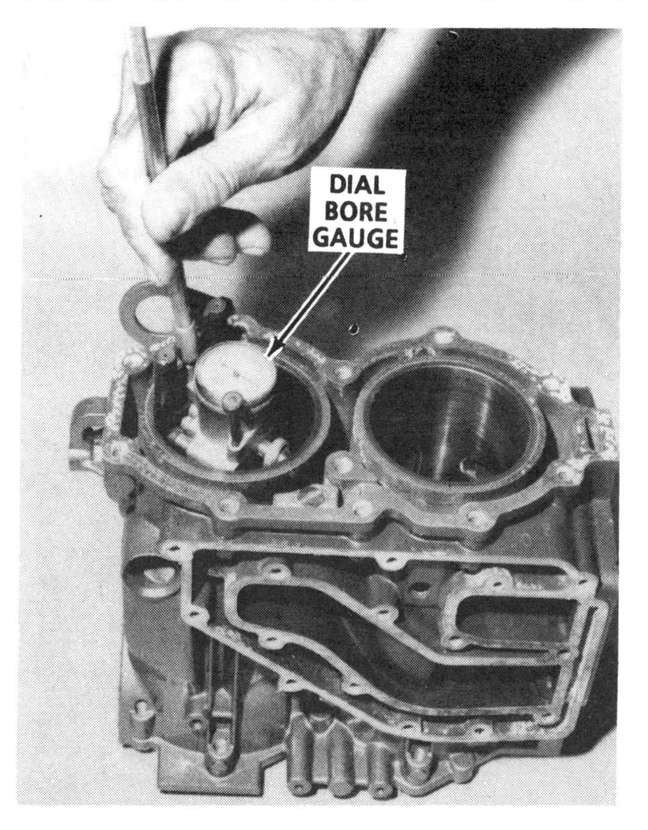

Effective method of checking the cylinder taper of any cylinder using an inside micrometer. One measurement should be taken near the top and another near of the bottom of the cylinder. The difference between the two is the amount of taper.

rough, clean the surface of the cylinder with a cylinder hone as described in the following procedures.

SPECIAL WORDS

If overheating has occurred, check and resurface the spark plug end of the cylinder block, if necessary. This can be accomplished with 240-grit sandpaper and a small flat block of wood.

To ensure satisfactory engine performance and long life following the overhaul work, the honing work should be performed with patience, skill, and in the following sequence:

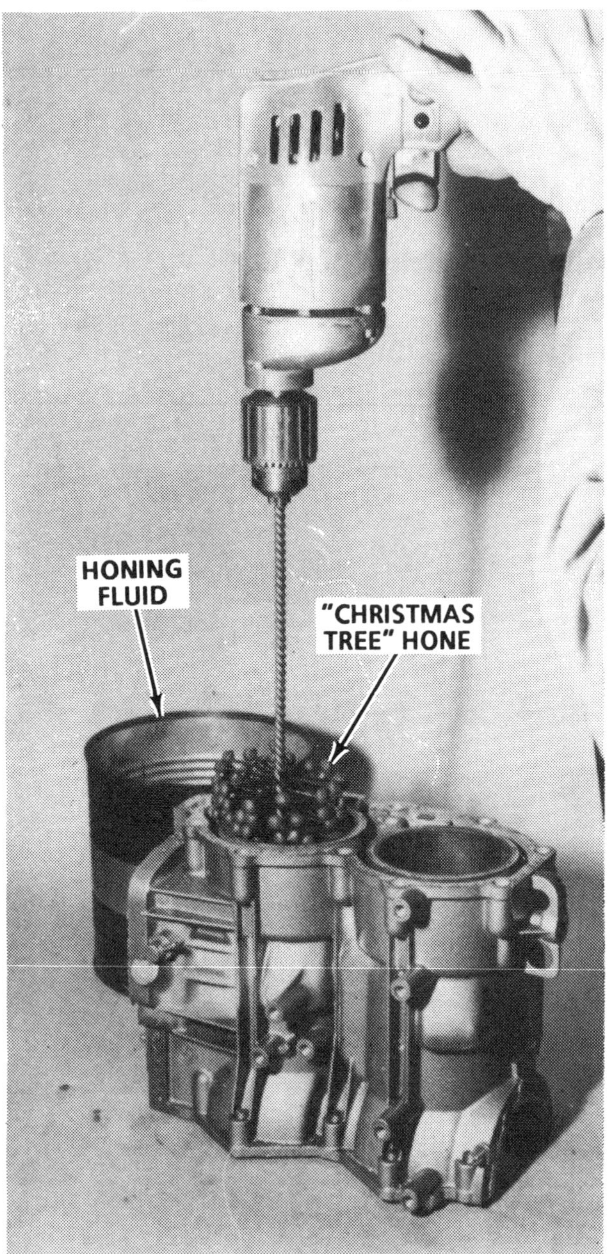

*Excellent example of refinishing a cylinder wall using an electric drill and "Christmas Tree" hone. **ALWAYS** keep the tool moving in long even strokes the entire cylinder depth. Use a continuous **LIBERAL** amount of honing fluid.*

a- Follow the hone manufacturer's recommendations for use of the hone and for cleaning and lubricating during the honing operation. A "Christmas tree" hone may also be used.

b- Pump a continuous flow of honing oil into the work area. If pumping is not practical, use an oil can. Apply the oil generously and frequently on both the stones and work surface.

c- Begin the stroking at the smallest diameter. Maintain a firm stone pressure against the cylinder wall to assure fast stock removal and accurate results.

d- Expand the stones as necessary to compensate for stock removal and stone wear. The best cross-hatch pattern is obtained using a stroke rate of 30 complete cycles per minute. Again, use the honing oil generously.

e- Hone the cylinder walls **ONLY** enough to de-glaze the walls.

f- After the honing operation has been completed, clean the cylinder bores with hot water and detergent. Scrub the walls with a stiff bristle brush and rinse thoroughly with hot water. The cylinders **MUST** be thoroughly cleaned to prevent any abrasive material from remaining in the cylinder bore. Such material will cause rapid wear of new piston rings, the cylinder bore, and the bearings.

The wall of this cylinder were damaged beyond repair when a piston ring broke and worked its way into the combustion chamber.

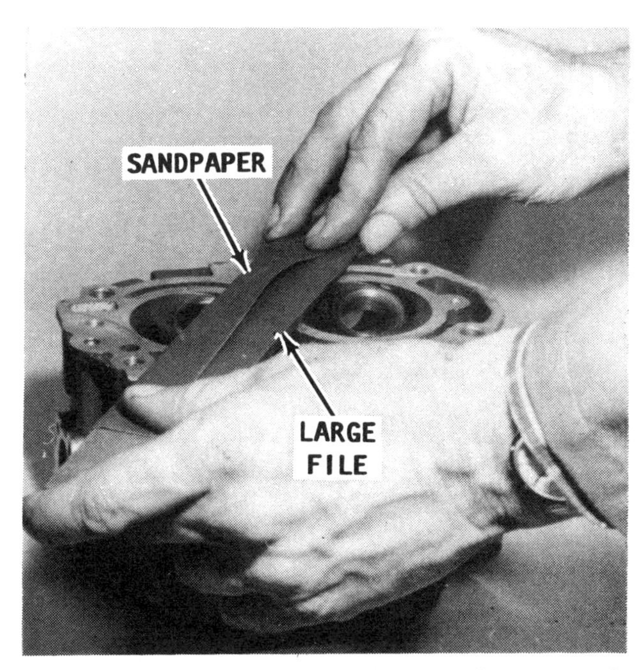

*A **LARGE** file and piece of wet sandpaper may be used to resurface a head or cylinder block when a suitable flat surface is not available.*

g- After cleaning, swab the bores several times with engine oil and a clean cloth, and then wipe them dry with a clean cloth. **NEVER** use kerosene or gasoline to clean the cylinders.

h- Clean the remainder of the cylinder block to remove any excess material spread during the honing operation.

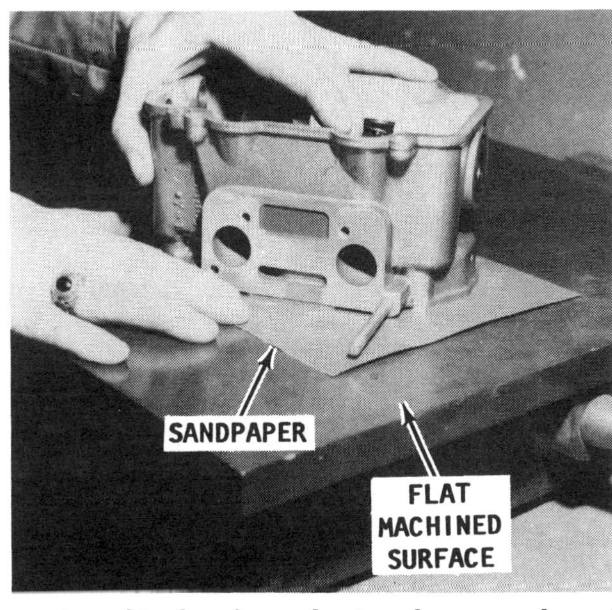

*A machined surface and wet sandpaper may be used to resurface a cylinder head, as explained in the text. A piece of glass or mirror may also be used, but do **NOT** attempt this task on a workbench.*

Block and Cylinder Head Warpage

First, check to be sure all old gasket material has been removed from the contact surfaces of the block and the cylinder head. Clean both surfaces down to shiny metal, to ensure a true measurement.

Next, place a straight edge across the gasket surface. Check under the straight edge with a 0.004" (0.1mm) feeler gauge. Move the straight edge to at least eight different locations. If the feeler gauge can pass under the straight edge -- anywhere contact with the other is made -- the surface will have to be resurfaced.

The block or the cylinder head may be resurfaced by placing the warped surface on 400-600 grit **WET** sandpaper, with the sandpaper resting on a **FLAT MACHINED** surface. If a machined surface is not available a large piece of glass or mirror may be used. **DO NOT** attempt to use a workbench or similar surface for this task. A workbench is never perfectly flat and the block or cylinder head will pickup the imperfections of the surface and the warpage will be made worse.

Sand -- work -- the warped surface on the wet sandpaper using large figure **"8"** motions. Rotate the block, or head, through 180° (turn it end-for-end) and spend an equal amount of time in each position to

This cylinder head was removed from the same engine as the pitted piston shown on Page 8-72. In order to return this head to a serviceable condition, the combustion chamber surface must be refinished to a smooth surface using a stone. If left in its present condition, the sharp corners of the gouges will become "hot" spots resulting in pre-ignition and possible "dieseling" after the engine is shut down.

avoid removing too much material from one side.

If a suitable flat surface is not available, the next best method is to wrap 400-600 grit wet sandpaper around a **LARGE** file. Draw the file as evenly as possible in one sweep across the surface. Do not file in one place. Draw the file in many, many directions to get as even a finish as possible.

As the work moves along, check the progress with the straight edge and feeler gauge. Once the 0.004" (0.1mm) feeler gauge will no longer slide under the straight edge, consider the work completed.

If the warpage cannot be reduced using one of the described methods, the block or cylinder head should be replaced.

ONE LAST CHANCE

If the warpage cannot be reduced and it is not possible to obtain new items -- and the warped part must be assembled for further use -- there is a strong possibility of a water leak at the head gasket. In an effort to prevent a water leak, follow the instructions outlined in Assembly on Page 8-34.

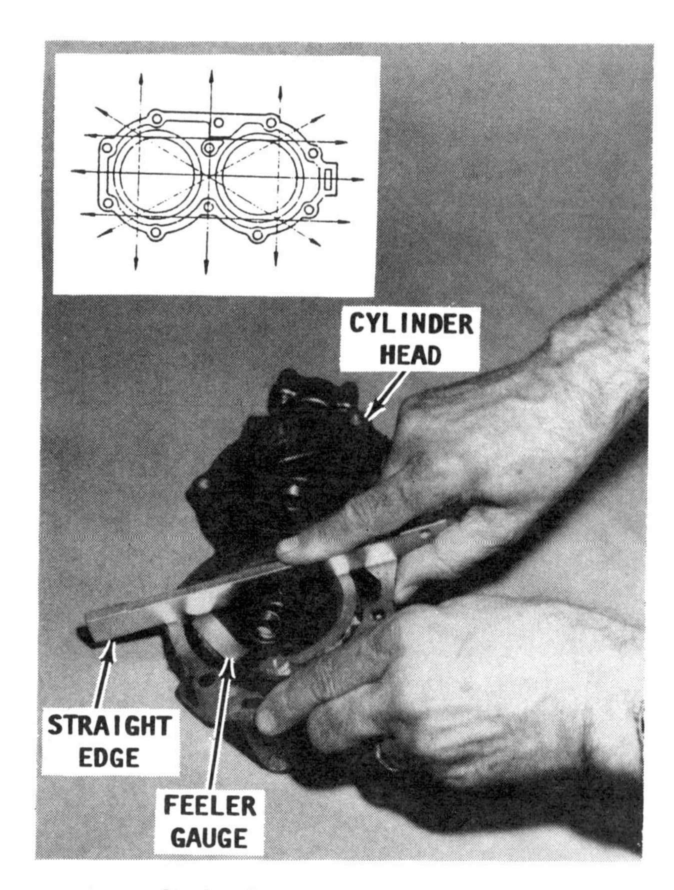

Any cylinder head surface may be checked for warpage using a straight edge and feeler gauge. Move the straight edge to at least eight different positions, as indicated in the superimposed line drawing.

8-8 SEALANTS, ADHESIVES, LUBRICANTS, AND FUEL STABILIZERS

It is common practice for the larger manufacturers of personal watercraft to

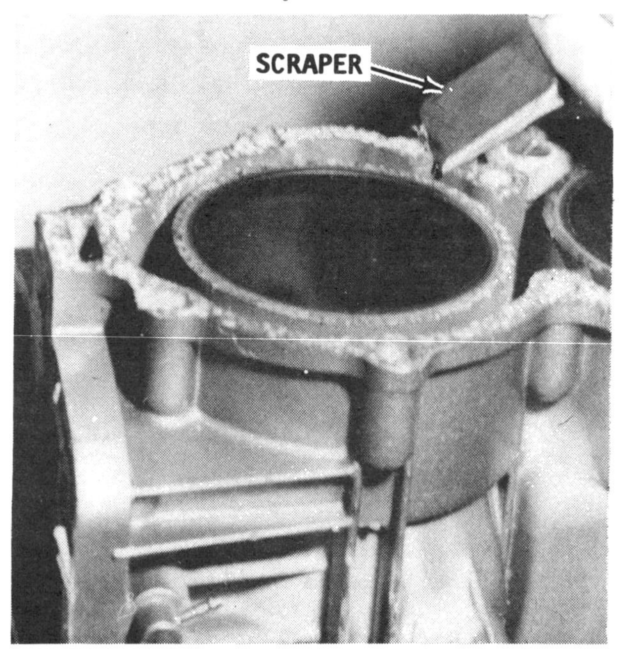

*When servicing any block, all traces of gasket material **MUST** be removed to ensure a good seal with a new gasket. If checking for warpage, even the smallest trace of foreign material would give a false reading.*

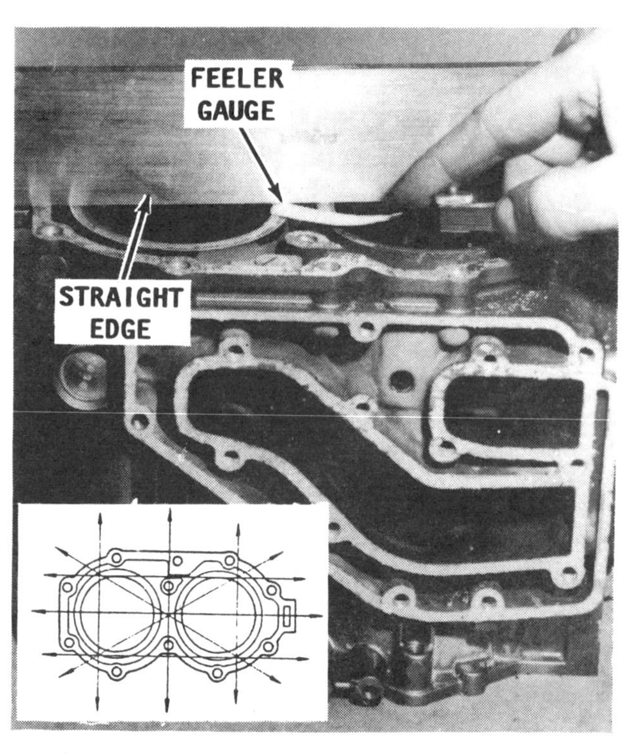

A straight edge and feeler gauge must be used to check engine-to-head mating surface on any block for warpage.

market their own line of products for use on their craft. Kawasaki chemical engineers have developed such a line available for use with Kawasaki outboard units, personal watercraft, snowmobiles, and motorcycles. Throughout this manual, the authors recommend the application of the manufacturer's products as a first choice. All products listed are alternatives of equal value, and may be used with confidence if the manufacturer's line is not available.

Sealants and Adhesives

Four sealants are recommended and they are **NOT** interchangeable. Each is designed to perform under a different set of conditions. Follow the directions on the package for cleaning and preparing surfaces. Sealants and adhesives **MUST** be applied **ONLY** to clean and dry parts. Apply sparingly -- excessive amounts may block oil passageways and cause serious damage.

Loctite Lock N' Seal is a non-hardening, non-permanent locking compound. This material is recommended for application to the threads of load bearing fasteners. Loctite helps prevent loosening of the bolt due to vibration, thread wear, and corrosion.

Loctite Stud N' Bearing Mount compound is also a non-hardening, non-permanent locking compound. This material is recommended for application to the threads of load bearing fasteners submerged under water.

Loctite Superflex is a water resistant silicone sealer which provides a very effective flexible seal and is able to withstand high temperatures. This sealer is recommended for use where gaskets are required next to a metal surface at high temperature, for example, on the exhaust manifold cover gaskets. Loctite Superflex is blue in color.

Kawasaki Bond is a non-hardening material and is recommended for metal-to-metal joints such as the crankcase halves. This substance is highly resistant to oil and gasoline.

Lubricants

Different lubricants are recommended in the lubrication procedures presented in this manual. These lubricants are **NOT** interchangeable, each is designed to perform under varying conditions.

Shell Alvania EP1 is a general marine lubricant, chemically formulated to resist salt water. This lubricant is recommended for application to bearings, bushings, and oil seals.

Kawasaki Lubricant is a two-stroke engine oil. It has a petroleum base, and is considered a clean burning lubricant. Yamalube reduces carbon deposits and ensures maximum protection against engine wear. This lubricant also contains an ashless detergent to keep piston rings "free". Oil additives are usually not recommended by the manufacturer, and in some cases the use of such a substance may invalidate the warranty.

Fuel Stabilizer

"Sta-Bil" Fuel Conditioner and Stabilizer is recommended during engine operation and during the storage period. This fluid absorbs water in the fuel system and protects against corrosion.

If used during operation, this fuel additive will prevent the formation of gum and varnish deposits and greatly extend the period between required carburetor overhauls.

When added to the fuel during storage, the additive will prevent the fuel from "souring" for up to twelve full months.

NOTES

9 ELECTRICAL

9-1 INTRODUCTION

The battery, cranking system, starter interlock, stop or "kill" switch, electric bilge pump and fan, temperature warning system, and trim motor -- if equipped -- are all considered subsystems of the electrical system. Each of these areas will be covered in this chapter beginning with the battery. The charging system is considered a subsystem of ignition and is covered in detail in Chapter 7.

9-2 BATTERIES

The battery is one of the most important parts of the electrical system. Because of its job and the consequences, (failure to perform when needed), the best advice is to purchase a well-known brand, with an extended warranty period, from a reputable dealer.

Personal Watercraft Batteries

The battery developed for use in a personal watercraft is the same kind used for many years on motorcycles. Personal watercraft batteries are required to perform under the most rigorous conditions.

Personal watercraft batteries have a much heavier exterior case than the usual automobile battery to withstand the violent pounding and shocks imposed on it as the craft moves through rough water. The battery case must also be vibration and seepage resistant. Therefore, a personal watercraft battery should always be the best the owner can afford.

The caps of marine batteries are "spill proof" to prevent acid from spilling into the bilges when the craft is subjected to violent maneuvers. These special caps are designed to vent gases regardless of the battery position, even when upside down.

Because of these features, the personal watercraft battery will recover from a low charge condition and give satisfactory service over a much longer period of time.

Battery Construction

A battery consists of a number of positive and negative plates immersed in a solution of diluted sulfuric acid. The plates contain dissimilar active materials and are kept apart by separators. The plates are grouped into what are termed elements. Plate straps on top of each element connect all of the positive plates and all of the negative plates into groups.

The battery is divided into cells which hold a number of the elements apart from

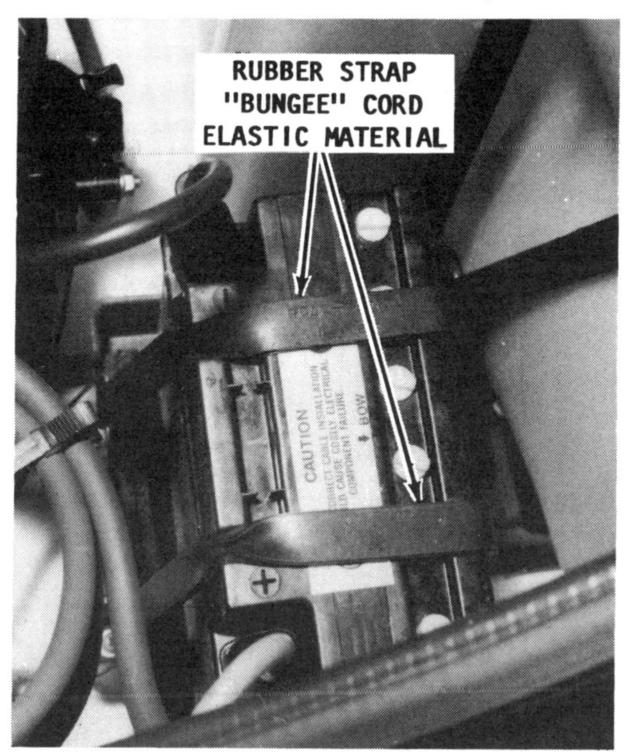

On a Personal Watercraft, two heavy-duty elastic rubber straps, or "bungee" type cords do an excellent job of holding the battery in place, during violent maneuvers.

the others. The entire arrangement is contained within a hard plastic case. The top is a one-piece cover and contains the filler caps for each cell. The terminal posts protrude through the top where the battery connections for the craft are made. Each of the cells is connected to its neighbor in a positive-to-negative manner with a heavy strap called the cell connector.

Battery Ratings

Four different methods are used to measure and indicate battery electrical capacity:

1- Ampere-hour rating
2- Cold cranking performance
3- Reserve capacity
4- Watt-hour rating

Ampere-Hour

The ampere-hour rating of a battery refers to the battery's ability to provide a set amount of amperes for a given amount of time under test conditions at a constant temperature of 80° (27°C). Amperes x hours equals ampere-hour rating. Therefore, if the battery is capable of supplying 3 amperes of current for 5 consecutive hours, the battery is rated as a 15 ampere-hour battery.

The ampere-hour rating is useful for some service operations, such as slow charging or battery testing.

The recommended ampere-hour rating for a personal watercraft battery is 15-19 ampere-hours.

Cold Cranking Performance

Cold cranking performance is measured by first cooling a fully charged battery to 0°F (-18°C), and then testing it for 5 seconds to determine the maximum current flow. In this manner the cold cranking amperes rating is the number of amperes available to be drawn from the battery before the voltage drops below 9.8 volts. A typical cold cranking ampere rating for a personal watercraft battery is 240 amps.

Reserve Capacity

The reserve capacity of a battery is considered the length of time -- in minutes --at 80°F (27°C) a 25 ampere current can be maintained before the voltage drops below 10.5 volts. This test is intended to provide an approximation of how long the engine, including electrical accessories such as bilge pump, could operate satisfactorily if the stator assembly or charging coil did not produce sufficient current. A typical rating is 25 minutes.

Watt-Hour

Watt-hour is a very useful rating of battery power. It is determined by multiplying the number of ampere hours times the voltage. Therefore, a 12-volt battery rated at 16 ampere-hours would be rated at 192 watt-hours (16 x 12 = 192).

If possible, the new battery should have a power rating equal to or higher than the unit it is replacing.

Battery Installation

Every battery installed in a personal watercraft must be secured in a well-protected ventilated area. If the battery area lacks adequate ventilation, hydrogen gas which is given off during charging could become very explosive. This is especially true if the gas is concentrated and confined.

Battery Service

The battery requires periodic servicing and a definite maintenance program will ensure extended life. If the battery should

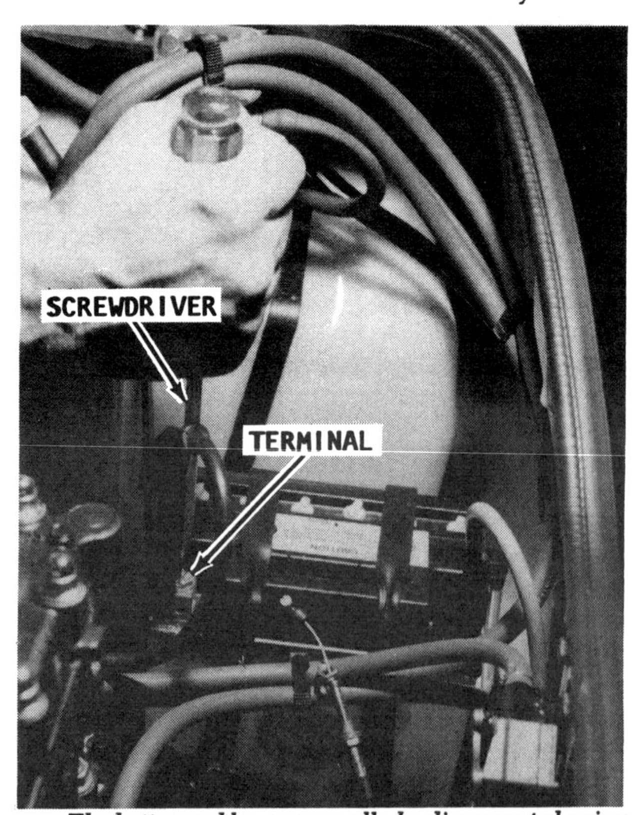

The battery cables can usually be disconnected using a slotted screwdriver or socket wrench.

test satisfactorily, but still fails to perform properly, one of five problems could be the cause.

1- Slow speed engine operation for long periods of time resulting in an undercharged condition.

2- A defect in the charging system. A faulty stator assembly or charging coil, defective rectifier, or high resistance somewhere in the system could cause the battery to become undercharged.

3- Failure to maintain the battery in good order. This might include a low level of electrolyte in the cells; loose or dirty cable connections at the battery terminals; or possibly an excessively dirty battery top.

Remove the battery from the engine compartment for service. Inspect and service the battery, cables and connections. Check for signs of corrosion. Inspect the battery case for cracks or bulges, dirt, acid, and electrolyte leakage.

Make sure the breather pipe is attached to the battery and not pinched by any part of the engine compartment.

Cleaning

Dirt and corrosion should be cleaned from the battery just as soon as it is discovered. Any accumulation of acid film or dirt will permit current to flow between the terminals. Such a current flow will drain the battery over a period of time. In time an acid film on top of the battery will rot the two rubber hold down straps.

Clean the exterior of the battery with a solution of diluted ammonia or a soda solution to neutralize any acid which may be present. Flush the cleaning solution off with clean water. **TAKE CARE** to prevent any of the neutralizing solution from entering the cells, by keeping the caps tight.

A poor contact at the terminals will add resistance to the charging circuit. This resistance will cause the voltage regulator to register a fully charged battery, and thus cut down on the stator assembly or charging coil output adding to the low battery charge problem.

Scrape the battery posts clean with a suitable tool or with a stiff wire brush. Clean the inside of the cable clamps to be sure they do not cause any resistance in the circuit.

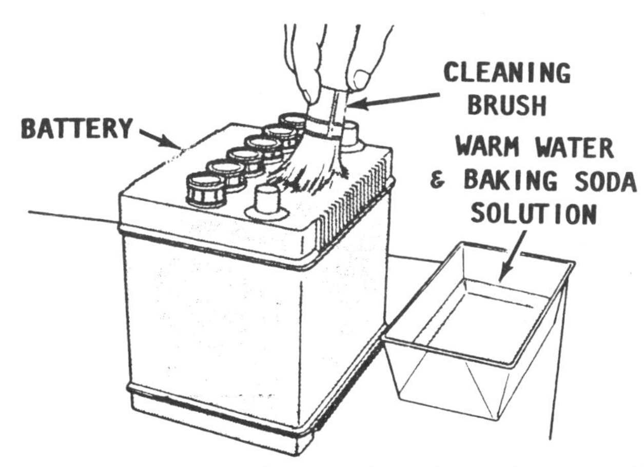

The battery top should be kept clean. A water and baking soda solution applied with a brush will neutralize battery electrolyte and clean the top nicely.

Battery Sulfation

Battery sulfation can be identified by the formation of white crystals forming on top of exposed battery plates. This usually occurs when a battery has been stored in a discharged condition over a period of time and the electrolyte has evaporated leaving the tops of the plates exposed to the air. This condition may also occur if the battery is not serviced at regular intervals during the season and the electrolyte not replenished when low.

Once white crystals form, the plates are permanently damaged and the battery will never hold a charge. It must be replaced.

Filling the Battery

Remove each filler cap using a pair of pliers. Fill each cell to the proper level with distilled water. The level of electro-

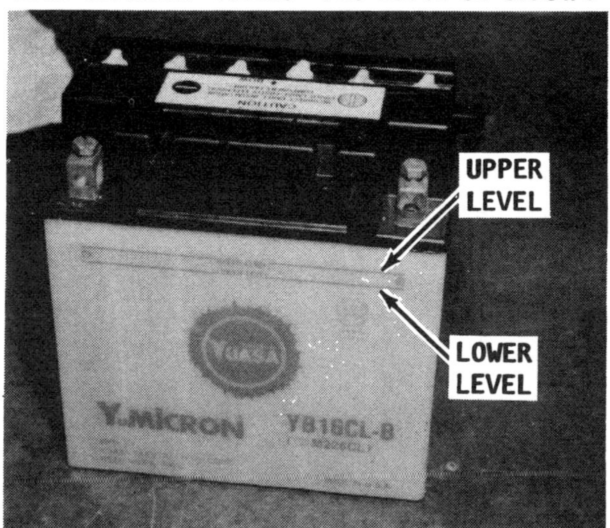

*If the battery is not a "Maintenance Free" type, it **MUST** be filled correctly. The proper procedure for this task is detailed in the text.*

lyte should be between the upper and lower level marks during operation. When filling the battery, **WITHOUT CHARGING,** fill to the upper line and allow the battery to stand for twenty minutes. Check the level again and replenish, as necessary. If filling the battery **PRIOR** to charging, fill to the lower lever only, because the electrolyte level will rise during charging.

During hot weather and periods of heavy use, the electrolyte level should be checked more often than during normal operation. Add distilled water to bring the level of electrolyte in each cell to the proper level. **TAKE CARE** not to overfill, because adding an excessive amount of water will cause loss of electrolyte and any loss will result in poor performance, short battery life, and will contribute quickly to corrosion. **NEVER** add electrolyte from another battery. **NEVER** add sulphuric acid because such acid would make the electrolyte too stong and permanently damage the battery. **NEVER** add ordinary tap water. The minerals in tap water will shorten battery life.

Battery Testing

A hydrometer is a device to measure the percentage of sulphuric acid in the battery electrolyte in terms of specific gravity. When the condition of the battery drops from fully charged to discharged, the acid leaves the solution and enters the plates, causing the specific gravity of the electrolyte to drop.

Use a temperature corrected hydrometer to test the specific gravity of the electrolyte. At 20°C (68°F) the hydrometer reading should be 1.20-1.26 on the scale.

It may not be common knowledge, but hydrometer floats are calibrated for use at 80°F (27°C). If the hydrometer is used at any other temperature, hotter or colder, a correction factor must be applied. Remember, a liquid will expand if it is heated and will contract if cooled. Such expansion and contraction will cause a definite change in the specific gravity of the liquid, in this case the electrolyte.

A quality hydrometer will have a thermometer/temperature correction table in the lower portion, as shown in the accompanying illustration. By knowing the air temperature around the battery and from the table, a correction factor may be applied to the specific gravity reading of the

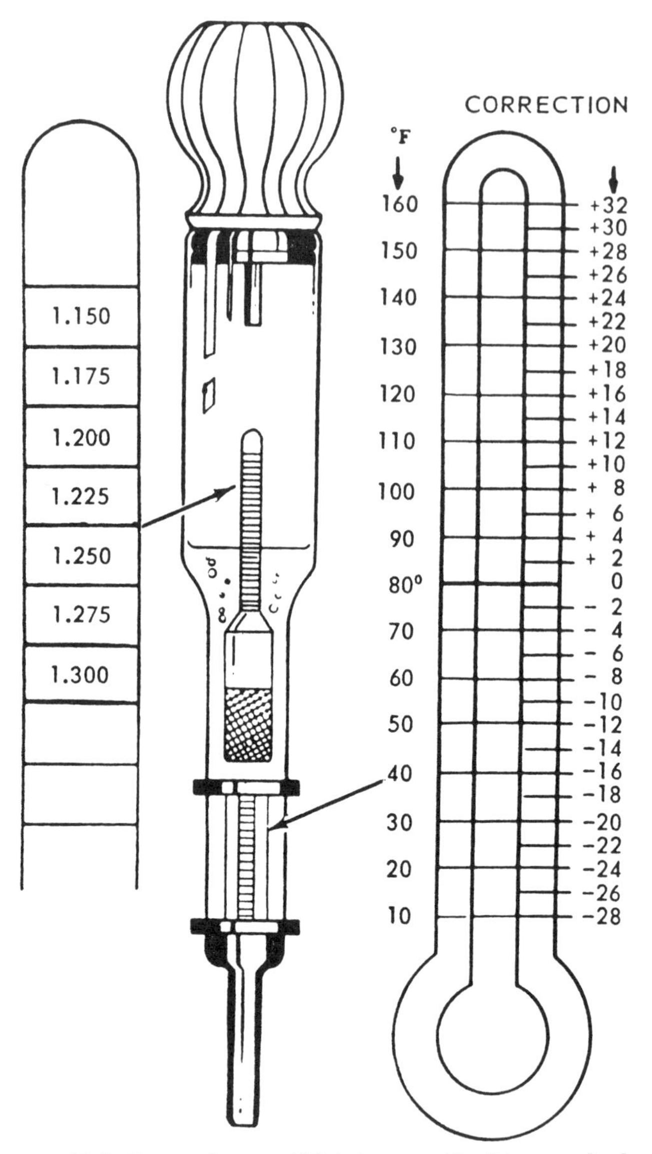

If the battery is not a "Maintenance Free" type, a check of the electrolyte in the battery should be on the maintenance schedule of any craft. A hydrometer reading of 1.300, or in the Green band, indicates the battery is in satisfactory condition. If the reading is 1.150 or in the Red bank, the battery must be charged. Observe the six safety points listed in the text when using a hydrometer.

hydrometer float. In this manner, an accurate determination may be made as to the condition of the battery.

The following six points should be observed when using a hydrometer.

1- NEVER attempt to take a reading immediately after adding water to the battery. Allow at least 1/4 hour of charging at the recommended rate to thoroughly mix the electrolyte with the new distilled water. This time will also allow for the necessary gases to be created.

2- ALWAYS be sure the hydrometer is clean inside and out as a precaution against contaminating the electrolyte.

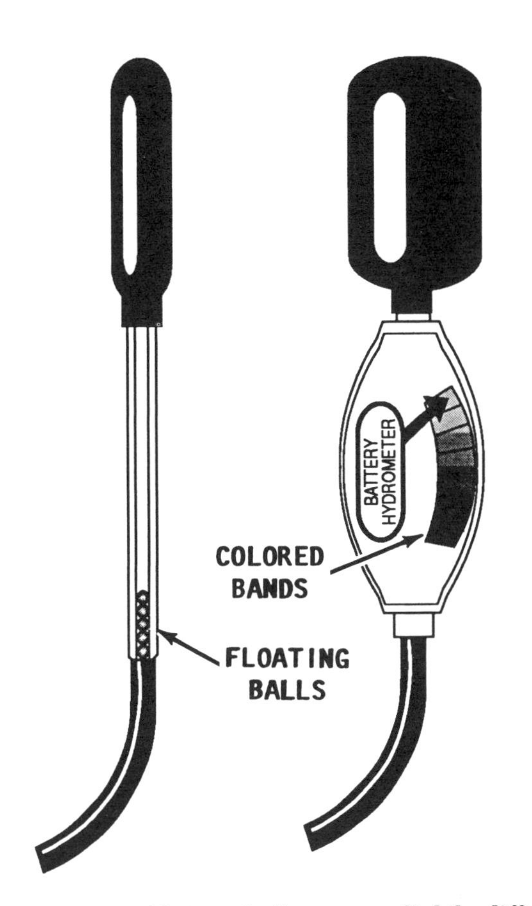

Two types of battery hydrometers, slightly different than the "old style". Their action is explained in the text.

3- If a thermometer is an integral part of the hydrometer, draw liquid into it several times to ensure the correct temperature before taking a reading.

4- **BE SURE** to hold the hydrometer vertically and suck up liquid only until the float is free and floating.

5- **ALWAYS** hold the hydrometer at eye level and take the reading at the surface of the liquid with the float free and floating.

Disregard the light curvature appearing where the liquid rises against the float stem. This phenomenon is due to surface tension.

6- **DO NOT** drop any of the battery fluid on the boat or on your clothing, because it is extremely caustic. Use water and baking soda to neutralize any battery liquid that does accidentally drop.

After withdrawing electrolyte from the battery cell until the float is barely free, note the level of the liquid inside the hydrometer.

Reading the Hydrometer

Different manufacturers of battery hydrometers use different methods of displaying specific gravity readings. Three methods are used:

Red, white, and green bands
Small floating balls
Graduated scale

Colored Band Hydrometer

When using a battery hydrometer with colored bands of red, white, and green, the electrolyte level is interpreted as follows:

Draw electrolyte from the battery cell until the float is barely free, note the level of the liquid inside the hydrometer.

If the level is within the green band range for all cells, the condition of the battery is satisfactory. If the level is within the white band for all cells, the battery is in fair condition.

If the level is within the green or white band for all cells except one, which registers in the red, the cell is shorted internally. No amount of charging will bring the battery back to satisfactory condition.

If the level in all cells is about the same, even if it falls in the red band, the battery may be recharged and returned to service. If the level fails to rise above the red band after charging, the only solution is to replace the battery.

Small Float Ball Hydrometer

If using a battery hydrometer with small floating balls, the electrolyte level is interpreted as follows:

If four or five balls are floating on top of the electrolyte drawn into the hydrometer, the battery is 100% charged. If three balls are floating, the battery is 75% charged. If two balls are floating, the battery is 50% charged. If only one ball is floating the battery is 25% charged. If no balls float, the battery is 0% charged.

If the same number of balls float for each cell, the condition of the battery is satisfactory.

If all the cells but one float balls, the cell is shorted internally. No amount of charging will bring the battery back to satisfactory condition.

If the number of floated balls for all cells is about the same, even if it is only one or two, the battery may be recharged and returned to service. If the number of balls fails to increase after charging, the only solution is to replace the battery.

Graduated Scale Hydrometer

If using a battery hydrometer with a graduated scale, note the level of the liquid inside the hydrometer. If the level is between the 1.280 and 1.265 mark, the battery is 100% charged. If the level is closer to the 1.210 mark, the battery is 75% charged. If the level is closer to the 1.160 mark, the battery is 50% charged. If it is closer to the 1.120 mark the battery is only 25% charged. If the level is below 1.100 the battery is at 0% charge.

If the level is the same and above 1.160 for all cells, the condition of the battery is satisfactory. If the level is the same and above 1.120 for all cells, the battery is in fair condition.

If the level is the same for all cells except one, which registers low on the scale, the cell is shorted internally. No amount of charging will bring the battery back to satisfactory condition.

If the level in all cells is about the same, even if it registers low on the scale, the battery may be recharged and returned to service. If the level fails to rise above 1.210 after charging, the only solution is to replace the battery.

Charging the Battery

If it is necessary to charge the battery, leave the filler caps lightly resting on the cell openings to allow the gases to escape.

The charging current should not exceed 1.9 amps for 10 hours.

NEVER use a fast charger, such as those found at automotive service stations, to charge a personal watercraft battery. These chargers operate at too high an output, ideal for automotive batteries, but very damaging to personal watercraft batteries. If such a battery is fast charged, excessive heat will be generated which will warp the plates and cause internal shorting. The plates will also shed active material which will accumulate at the bottom of the battery and cause internal shorting.

If, under the recommended charging procedures, the temperature of the electrolyte rises above 115°F (45°C) during charging, reduce the charging rate and increase the charging time proportionately. If possible, stop, allow the battery to cool and then resume charging for the recommended length of time.

After charging, push each filler cap into place and wipe the top of the battery clean before installation.

Installing the Battery

Check to be sure the battery is fastened securely in position. The hold-down rubber straps should be tight enough to prevent any movement of the battery in the holder.

Clean the battery posts and battery cable ends with a wire brush to ensure good, clean connections. If the battery posts or cable terminals are corroded, the cables should be cleaned separately with a baking soda solution and a wire brush. Apply a thin coating of Multi-purpose Lubricant to the posts and cable clamps before making the connections. The lubricant will help to prevent corrosion.

Identify the "POS" or "+" and "NEG" or "-" embossed symbols. Correctly connect the battery cables to the battery terminals. If the cables are connected incorrectly, the ignition system **WILL** be destroyed the first time the engine is started.

Check to be sure the breather pipe is attached to the battery and not pinched by any part of the engine compartment.

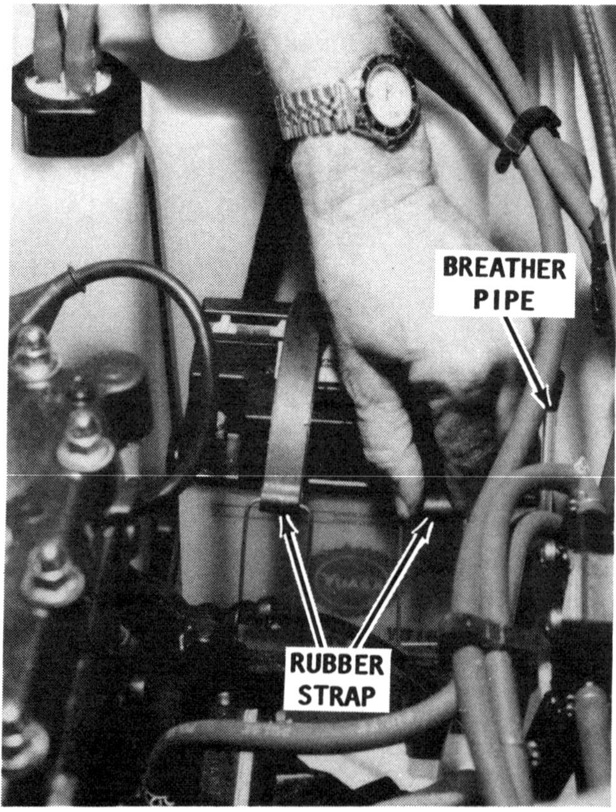

The breather pipe must not be pinched by any item in the engine compartment, after the holding straps are secured.

Jumper Cables

If a booster battery is used for starting an engine, the jumper cables must be connected correctly and in the proper sequence to prevent damage to either battery, or to the diodes in the rectifier/regulator.

ALWAYS connect a cable from the positive terminal of the dead battery to the positive terminal of the good battery **FIRST. NEXT,** connect one end of the other cable to the negative terminal of the good battery and the other end to a good ground on the engine. **DO NOT** connect the negative jumper from the good battery to the negative terminal of the low battery. Such action will almost always cause a spark which could ignite gases escaping through the vent holes in the battery filler caps. Igniting the gases may result in an explosion destroying the battery and causing severe personal **INJURY.**

By making the negative (ground) connection on the engine, if an arc is created, it will not be near the battery.

DISCONNECT the battery ground cable before replacing a stator assembly or charging coil, or before connecting any type of meter to the ignition system.

If it is necessary to use a trickle charger on a dead battery, **ALWAYS** disconnect one of the cables from the battery **FIRST,** to prevent burning out the diodes in the regulator/rectifier.

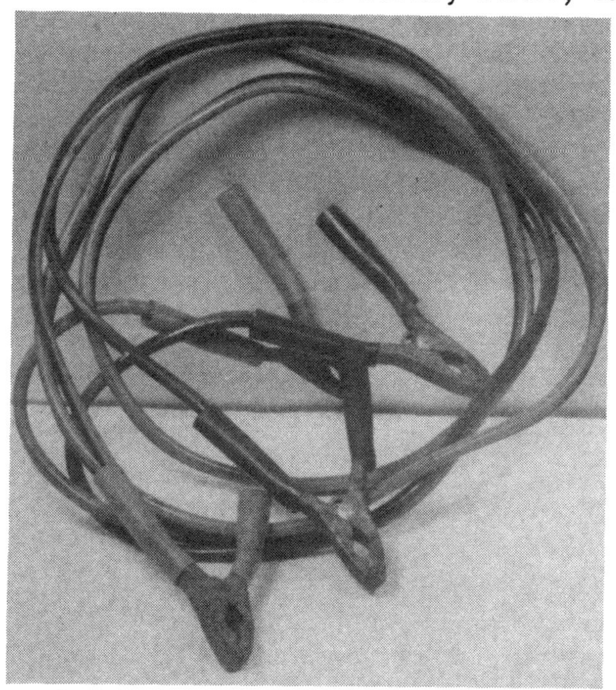

A common set of heavy-duty jumper cables. The booster battery must be connected correctly and in the proper sequence, as outlined in the text. Proper procedure will prevent damage to the battery or the diodes in the rectifier/regulator.

NEVER use a trickle charger as a booster to start the engine because the diodes will be **DAMAGED.**

Storage

If the craft is to be laid up for the winter or for more than a few weeks, special attention must be given to the battery to prevent complete discharge or possible damage to the terminals and wiring. Before putting the craft in storage, disconnect and remove the battery. Clean the battery thoroughly of any dirt or corrosion, and then charge it to full specific gravity reading. After the battery is fully charged, store the battery in a clean cool dry place where it will not be damaged or knocked over, preferably on a couple blocks of wood. Storing the battery up off the deck, will permit air to circulate freely around and under the battery and will help to prevent condensation.

NEVER store the battery with anything on top of it or cover the battery in such a manner as to prevent air from circulating around the filler caps. All batteries, both new and old, will discharge during periods of storage, more so if they are hot than if they remain cool. Therefore, the electrolyte level and the specific gravity should be checked at regular intervals. A drop in the specific gravity reading is cause to charge the battery back to a full reading.

One of the largest personal watercraft battery manufacturers recommends that a battery in storage should be checked every two weeks and charged if necessary. The electrolyte level can also be replenished at this time to prevent sulfation occurring.

In cold climates, care should be exercised in selecting the battery storage area. A fully charged battery will freeze at about 60° below zero. A discharged battery, almost dead, will have ice forming at about 19° above zero.

9-3 TACHOMETER

An accurate tachometer can be connected to any engine. Such an instrument provides an indication of engine speed in revolutions per minute (rpm). This is accomplished by measuring the number of electrical pulses per minute generated in the primary circuit of the ignition system.

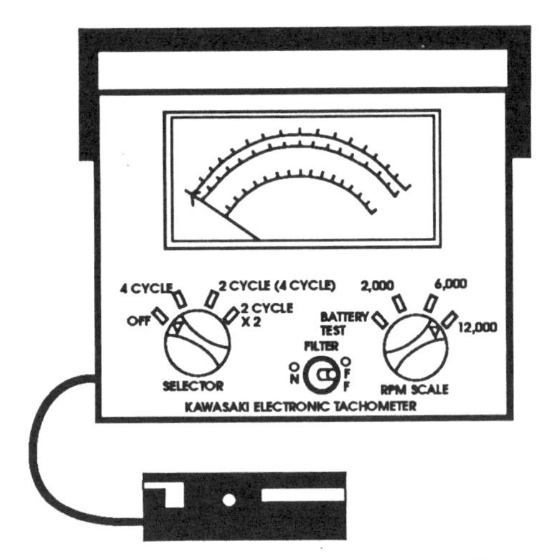

Maximum engine performance can only be obtained through proper tuning using a tachometer.

The meter readings range from 0 to 6,000 rpm, in increments of 100. Tachometers have solid-state electronic circuits which eliminates the need for relays or batteries and contributes to their accuracy. The electronic parts of the tachometer susceptible to moisture are coated to prolong their life.

The manufacturer recommends the use of a Kawasaki Electronic Tachometer. Instructions on how to connect the leads are provided with the meter. However, as a general rule, on a two stroke engine with CDI type ignition, the tachometer leads are usually connected across the charging coil leads -- refer to Chapter 7 for charging coil lead color identification. The color code differs for each model.

Components of the ignition circuit and the charging circuit on a Model 550 and 650 are housed on the back side of the flywheel, covered and protected by the flywheel.

9-4 ELECTRICAL SYSTEM GENERAL INFORMATION

The electrical system consists of three separate circuits:

Charging circuit
Cranking motor circuit
Ignition circuit

Charging Circuit

Components of the charging circuit include:

- Permanent magnets attached to the inside perimeter of the flywheel.
- A charging coil installed on the stator plate.
- A rectifier/regulator located elsewhere on the engine.
- An external battery
- Necessary wiring to connect it altogether.

The negative side of the rectifier/regulator is grounded. The positive side of the rectifier/regulator is connected to the battery. The negative side of the battery is connected to a good ground on the engine.

The alternating current generated in the stator windings passes to the regulator/rectifier. The rectifier/regulator changes the alternating current (AC) to direct current (DC) to charge the 12-volt battery.

The stator is located behind and protected by, the flywheel. Therefore, the stator, including the charging coil, seldom causes problems in the charging circuit. Most problems in the charging circuit can usually be traced to the rectifier/regulator or to the battery. If either the stator or the rectifier/regulator fails the troubleshooting tests, the defective unit cannot be repaired, it **MUST** be replaced.

See Chapter 7 to service and test components in the charging system.

Cranking Motor Circuit

The cranking motor circuit consists of a cranking motor and a starter-engaging mechanism. A starter relay is used as a heavy-duty switch to carry the heavy current from the battery to the cranking motor. The starter relay is actuated by depressing the **START** button.

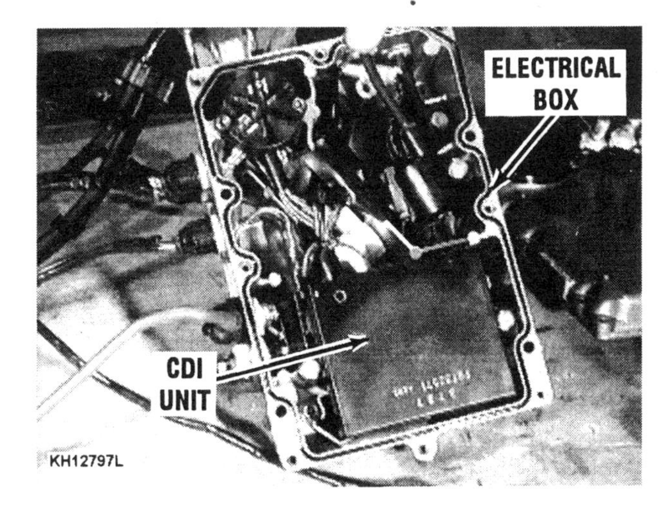

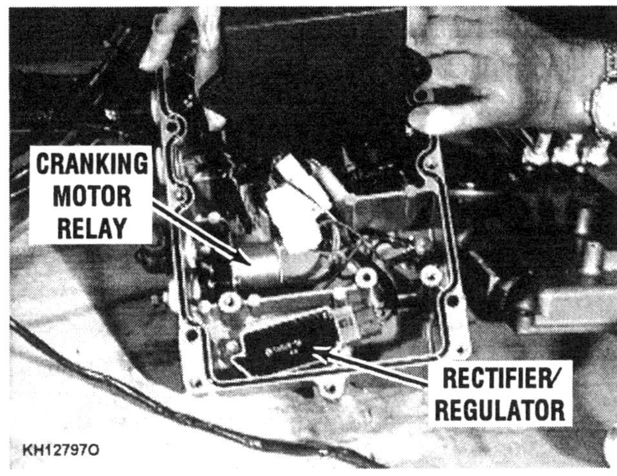

The cover of the electrical box on a Model 750, 900 and 1100 Series engines removed (top). The CDI unit released and lifted to expose the cranking motor relay and the rectifier/regulator (bottom).

Ignition Circuit

The ignition circuit is covered extensively in Chapter 7, Ignition.

9-5 CRANKING MOTOR CIRCUIT SERVICE

DESCRIPTION

As the name implies, the sole purpose of the cranking motor circuit is to control operation of the cranking motor to crank the engine until the air/fuel mixture ignites and the flywheel develops enough torque to continue under its own power. The circuit includes a solenoid or magnetic switch to connect or disconnect the motor from the battery. The operator controls the switch with a push button.

The cranking motor is a series wound electric motor which draws heavy current from the battery. It is designed to be used only for short periods of time to crank the engine for starting. To prevent overheating the motor, cranking should not be continued for more than 30-seconds without allowing the motor time to cool -- say at least three minutes. Actually, this time can be spent in making preliminary checks to determine why the engine fails to start.

Theory of Operation

On Kawasaki Personal Watercraft, three different methods are used to transmit power from the cranking motor to the engine crankshaft through a combination of meshed gears or clutch mechanisms.

Model 550 Series

The 550 Model Series are equipped with a Bendix drive on the end of the armature shaft. When the motor is operated, the pinion gear moves forward and meshes with teeth on the flywheel ring gear. A rubber cushion is built into the Bendix drive to absorb the shock when the pinion gear meshes with the flywheel ring gear.

After the engine starts, the pinion gear is driven faster than the armature shaft, and as a result, it spins out of mesh with the flywheel. The parts of the drive **MUST** be properly assembled for efficient operation.

If the drive is removed for cleaning, **TAKE CARE** to assemble the parts in the

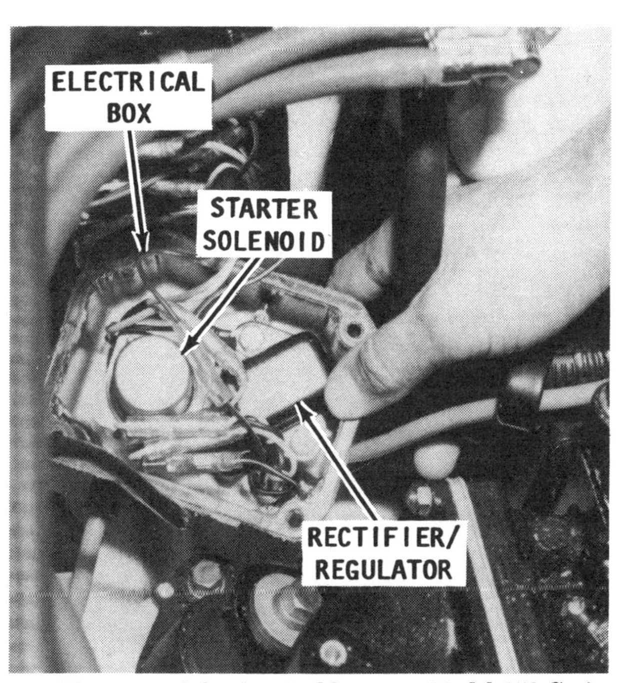

The cover of the electrical box on a Model 650 Series removed exposing the internal parts housed inside.

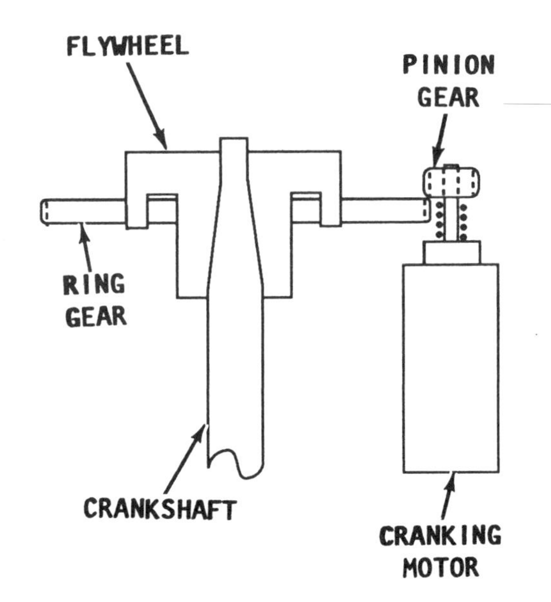

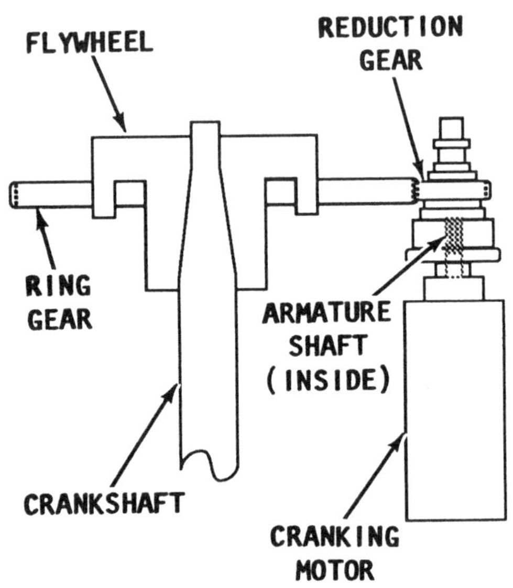

Line drawings to depict the two power train methods for transferring energy from the cranking motor to the engine flywheel. The Model 550 Series is shown (top), and all other models is depicted (bottom).

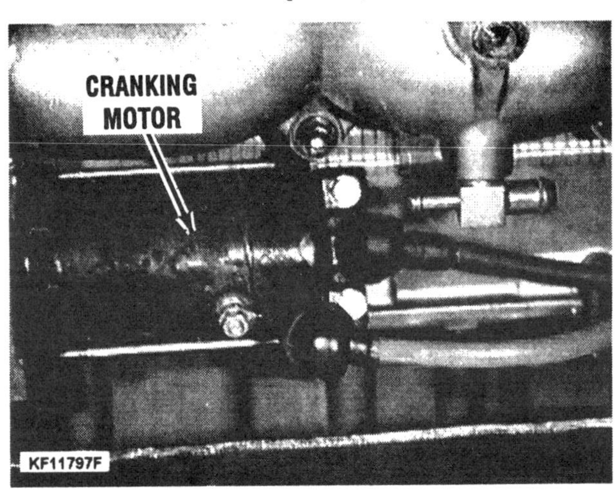

Location of the cranking motor for all engines covered in this manual. The installation differs slightly, as explained in the text.

correct order. If the screw shaft assembly is reversed, it will strike the splines and the rubber cushion will not absorb the shock.

All Models Except Series 550

All Models except the Model 550 Series have a reduction gear housed in the flywheel cover containing the Bendix drive. The armature shaft has a splined end which meshes inside the reduction gear. When the cranking motor is operated, the pinion gear -- a part of the reduction gear assembly -- moves forward and meshes with the teeth on the flywheel ring gear. The simple line drawing on this page illustrates this arrangement.

When the start button is depressed, the idle gear is driven by the cranking motor. A one-way clutch on the idle gear shaft slides forward, engaging the pinion gear -- at the forward end of the shaft -- with the flywheel ring gear.

After the engine starts and the start button is released, engine rpm rises above cranking rpm. The rollers inside the clutch move in a different direction at increased speed and no longer contact the outer race of the one-way clutch. Therefore, the one-way clutch ceases to rotate and slides back on the shaft disengaging the pinion gear from the flywheel ring gear.

Cranking Motor Noises

The sound of the motor during cranking is a good indication of how the cranking motor is operating -- properly or not. Naturally, temperature conditions will affect the speed at which the cranking motor is able to crank the engine. The speed of cranking a cold engine will be much slower than when cranking a warm engine. An experienced operator will learn to recognize the favorable sounds of the engine cranking under various conditions.

Close view of the reduction gear assembly used on the Model 650 Series engines.

Faulty Symptoms

If the cranking motor spins, but fails to crank the engine, the cause is usually a corroded or gummy Bendix drive. The drive should be removed, cleaned, and given an inspection.

If the cranking motor cranks the engine too slowly, the following are possible causes and the corrective actions that may be taken:

a- Battery charge is low. Charge the battery to full capacity.

b- High resistance connections at the battery, solenoid, or motor. Clean and tighten all connections.

c- Undersize battery cables. Replace cables with sufficient size.

Maintenance

The cranking motor does not require periodic maintenance or lubrication, except for lubrication of the pinion gear. If the motor fails to perform properly, the checks outlined in the previous paragraph should be performed.

Naturally, the motor will have to be removed if the corrective actions outlined under **Faulty Symptoms** above, does not restore the motor to satisfactory operation.

Otherwise, the cranking motor is not normally removed from the engine until it fails.

CRANKING MOTOR TROUBLESHOOTING

Before wasting too much time troubleshooting the cranking motor circuit, the following checks should be made. Many times, the problem will be corrected.

a- Battery fully charged.
b- Main fuse is "good" (not blown).
c- All electrical connections clean and tight.
d- Wiring in good condition -- insulation not worn or frayed.

The following troubleshooting procedures are presented in a logical sequence.

Do not operate the cranking motor for more than 15 seconds. Prolonged cranking motor operation will cause overheating and damage the motor.

After each test, allow the cranking motor to cool for a minute or so.

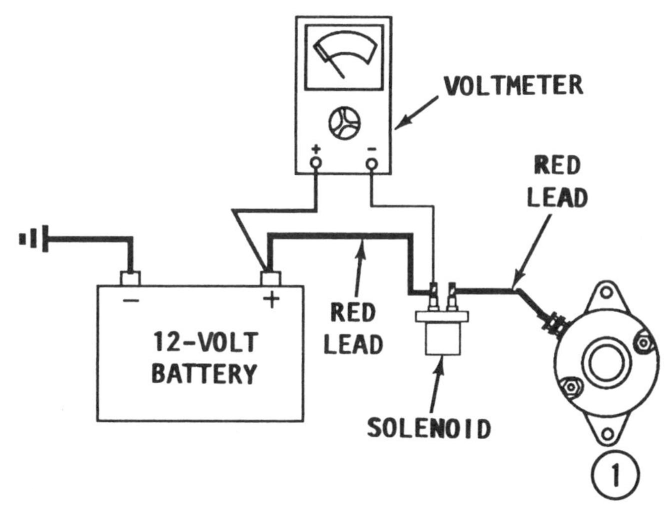

Never depress the **START** button to activate the cranking motor while the engine is operating. Such action will damage the pinion and/or flywheel gears.

Cranking Circuit Tests

1- Remove both spark plug leads from the plugs and ground the ends to the engine to prevent accidental engine start. Keep the leads grounded for the following tests.

Obtain a voltmeter and select the Vx1 DC scale. Observe the starter solenoid. Both large terminals have Red leads connected. One terminal has the battery cable connected and the other terminal has the lead from the cranking motor connected. Make contact with the negative Black meter lead to the terminal on the starter solenoid with the battery cable connected.

Make contact with the positive Red meter lead to the positive battery terminal. The voltmeter should register less then 0.25V DC. If the reading exceeds 0.25V DC, clean the connections at both ends of the

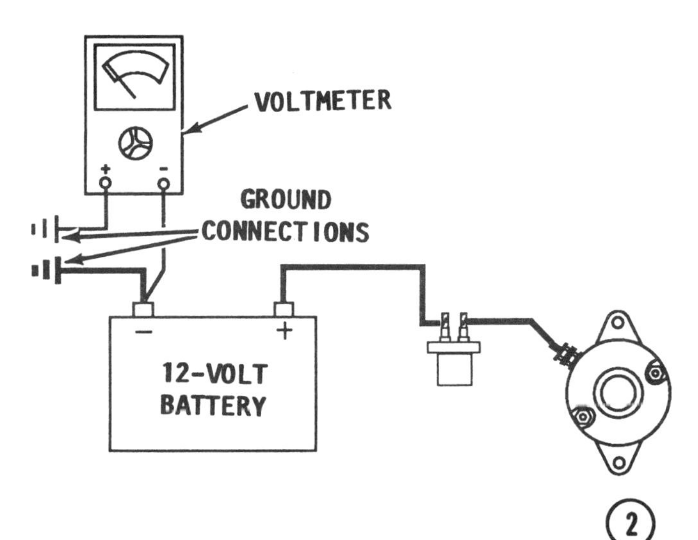

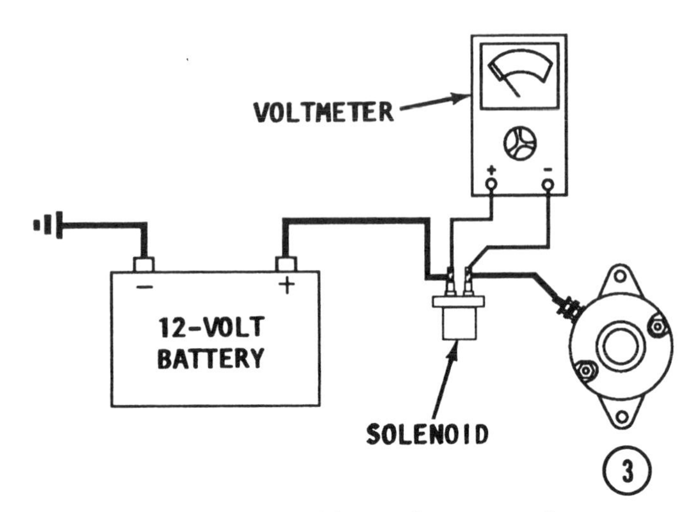

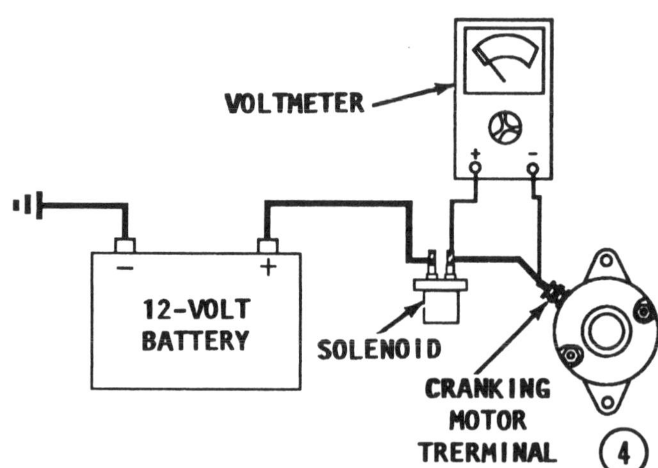

positive battery cable and repeat the test. If the reading does not change, replace the cable.

2- Keep the meter on the same scale. Make contact with the negative Black voltmeter lead to the negative terminal on the battery. Make contact with the positive Red voltmeter lead to a good engine ground. The voltmeter should register less then 0.25V DC. If the reading exceeds 0.25V DC, clean the connections at both ends of the negative battery cable and repeat the test. If the reading does not change, replace the cable.

CRITICAL WORDS

During the following tests -- Step 3 and Step 4, both meter leads must be placed on the solenoid terminals **WHILE** the engine is being cranked. **ALSO,** both meter leads must be removed from the solenoid terminals **WHILE** the engine is being cranked and **BEFORE** the engine has stopped cranking. If these precautions are not followed, the voltmeter may be damaged.

3- Keep the meter on the same scale. Obtain the services of an assistant.

Signal the assistant to start cranking the engine. While the engine is being cranked, make contact with the negative Black voltmeter lead to the solenoid terminal which has the lead from the cranking motor connected, and then quickly make contact with the positive Red voltmeter lead to the solenoid terminal which has the battery cable connected.

If the meter reading exceeds 0.25V DC, the starter solenoid must be replaced. Remove the meter leads and signal the assistant to cease cranking.

4- Keep the meter on the same scale and the assistant at the **START** button. Observe the same precautions as before in Step 3. While the engine is being cranked, make contact with the negative Black voltmeter lead to terminal on the cranking motor, and then quickly make contact with the positive Red voltmeter lead to the solenoid terminal which has the Red cranking motor connected.

If the meter reading exceeds 0.25V DC, the connections at both ends of the solenoid-to-cranking motor lead should be cleaned and the test repeated. If the reading does not change, the lead must be replaced.

CRANKING MOTOR RELAY REMOVAL FOR TESTING

Description

The cranking motor relay is actually a switch between the battery and the engine. The switch cannot be serviced. Therefore, if testing indicates the switch to be faulty, it **MUST** be replaced. The relay is housed inside the electrical case. The case must be removed from the engine to gain access to the relay. The relay must be removed from the case for testing purposes.

Electric Case Removal
Model 550 Series

1- Disconnect the battery ground cable from the battery. Remove the spark plug lead from the spark plug. Remove the two bolts securing the connector cover to the electric case. Pull off the cover. Unplug the 4-prong connector and the four free leads at their quick disconnect fittings.

Pull back the two boots over the large cables at the cranking motor relay. Discon-

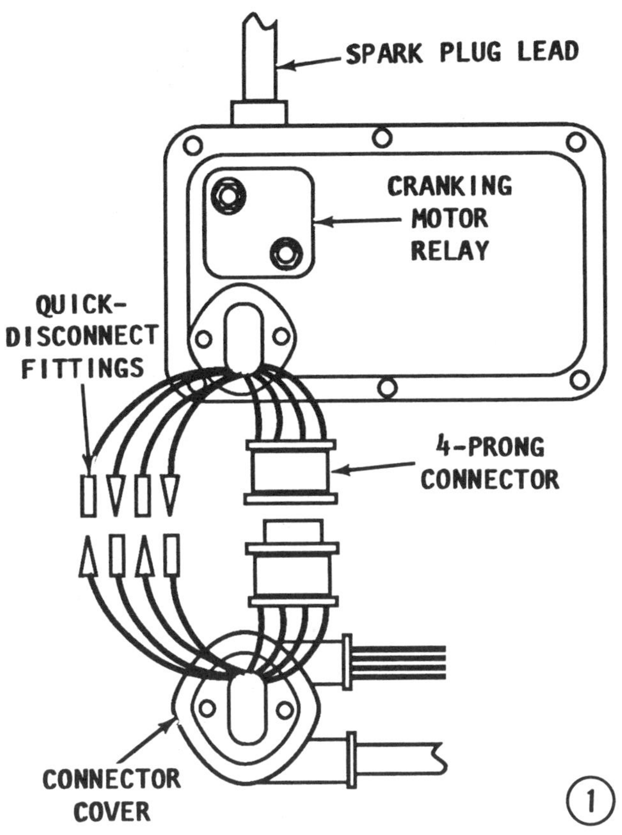

nect both large cables from the relay terminals. Remove the six bolts securing the electrical box to the engine compartment bulkhead. Remove the two bolts securing the two halves of the case together, and then open the case.

Electric Case Removal
All Other Models

2- Disconnect the battery ground cable from the battery. Remove the spark plug leads from the spark plugs. Pull back the large boot over

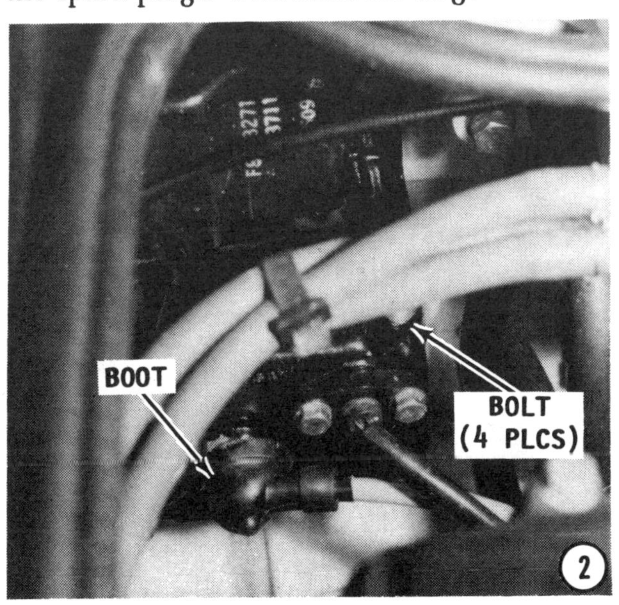

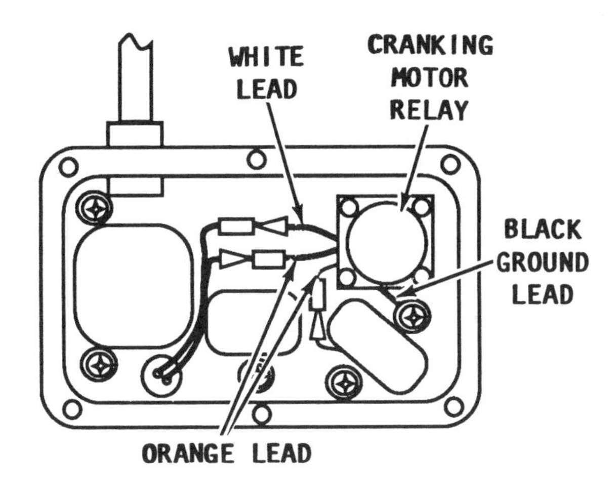

Line drawing to depict the cranking motor relay on the Model 550 Series engine.

both large cables at the cranking motor relay. Disconnect both cables from the relay terminals. Remove the bolts securing the cover to the electrical box engine compartment bulkhead or hull. Note the Model 650 Series has two different bolt sizes.

Cranking Motor Relay Removal
All Series

3- Inside the case -- disconnect the ground switch from the relay at the regulator/rectifier. Disconnect the two Orange leads and the White lead at their quick-disconnect fittings.

Outside the case -- remove the nuts from the two large relay terminals. Pull the relay out of the case, taking care to observe the arrangement of washers and insulating grommets.

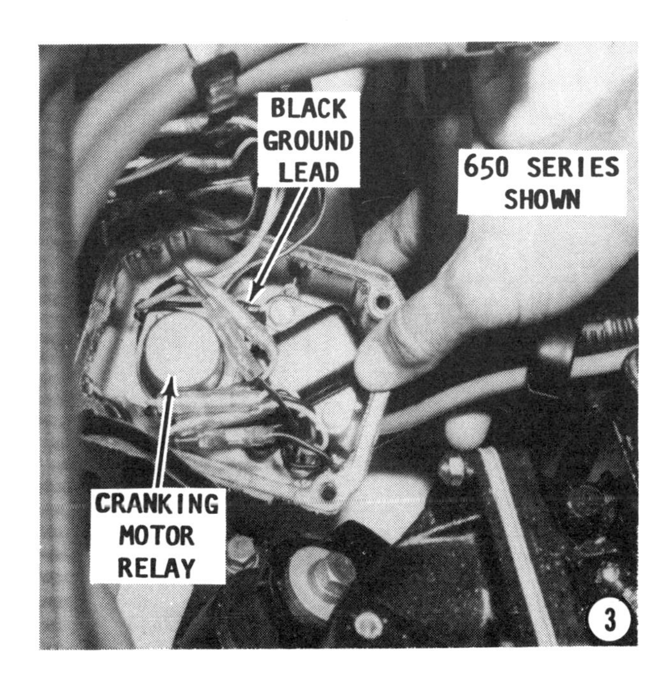

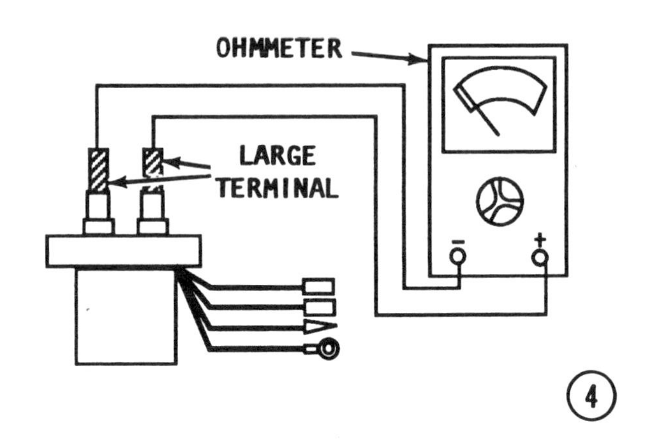

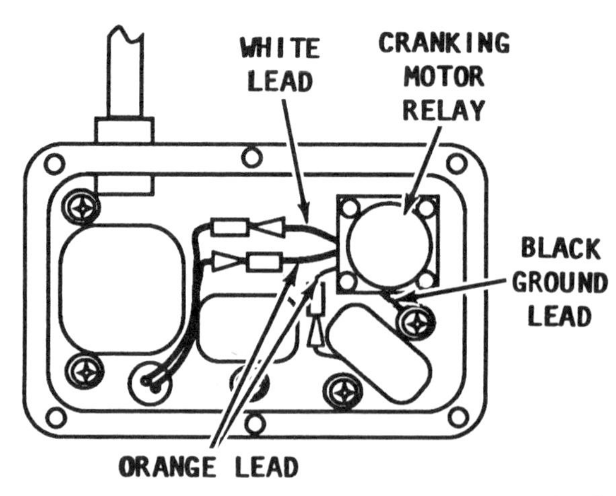

Location of the cranking motor relay inside the electrical box on a Model 750, 900 and 1100 Series engine.

Cranking Relay Testing

4- Select the Rx1 ohm scale on an ohmmeter. Make contact with the two meter leads across the two large relay terminals. The meter should read an infinite resistance. If the meter shows a resistance of less than infinity, the relay is defective and must be replaced.

5- Keep the ohmmeter on the same scale. Obtain a 12V battery. Connect the small Black ground lead from the relay to the negative battery terminal. Connect the small White lead from the relay to the positive battery terminal. Make contact with the two meter leads across the two large relay terminals. If the relay emits a "click" and the ohmmeter registers zero ohms, the relay is in good condition. If no "click" is heard, or if the meter registers a high or infinite resistance, the relay is defective and must be replaced.

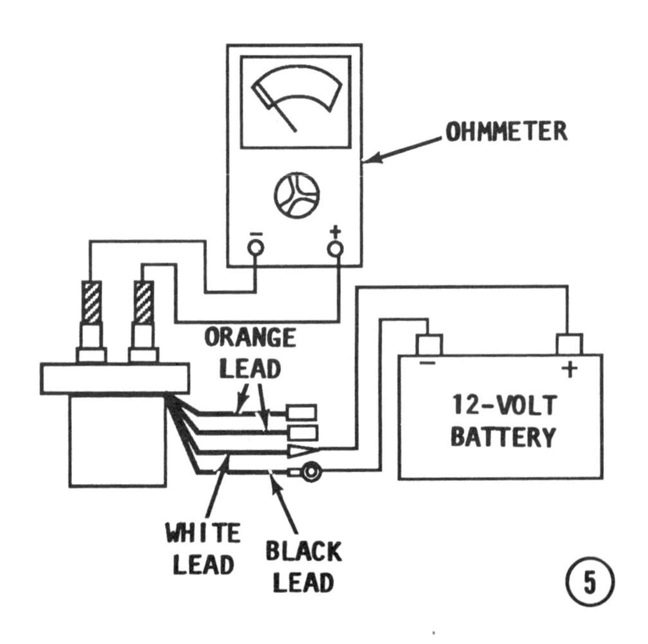

Cranking Relay Installation

6- Apply a coat of multi-purpose water resistant lubricant to all the insulating washers and grommets. Slide a large flat metal washer over each of the two large terminals, followed by a flat white plastic insulator. Insert the relay into the electric case.

Install the following items over each of the relay terminals protruding from the case in the following order: First, the large Black plastic insulator, then the large White plastic insulator, followed by the small Black plastic insulator, and then the small metal washer, and finally, the locknut. Tighten both locknuts securely. Connect the Orange and White leads, matching color-to-color and secure the relay Black round lead under the rectifier/regulator mounting screw.

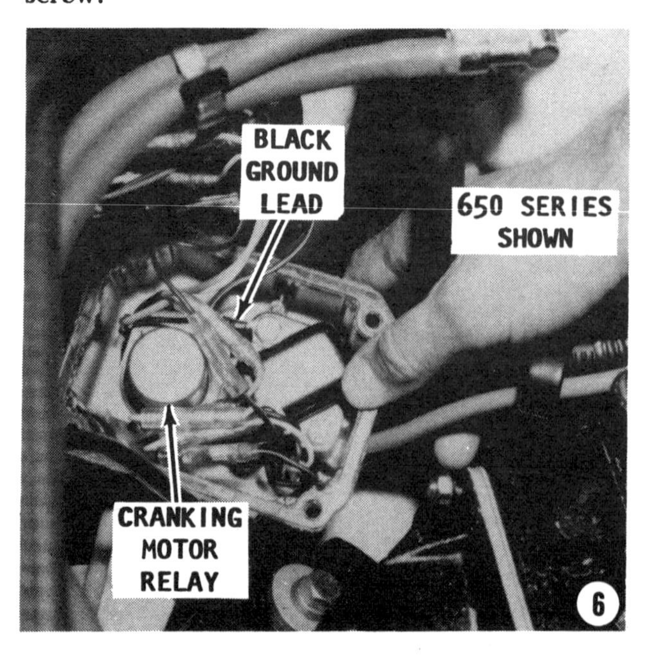

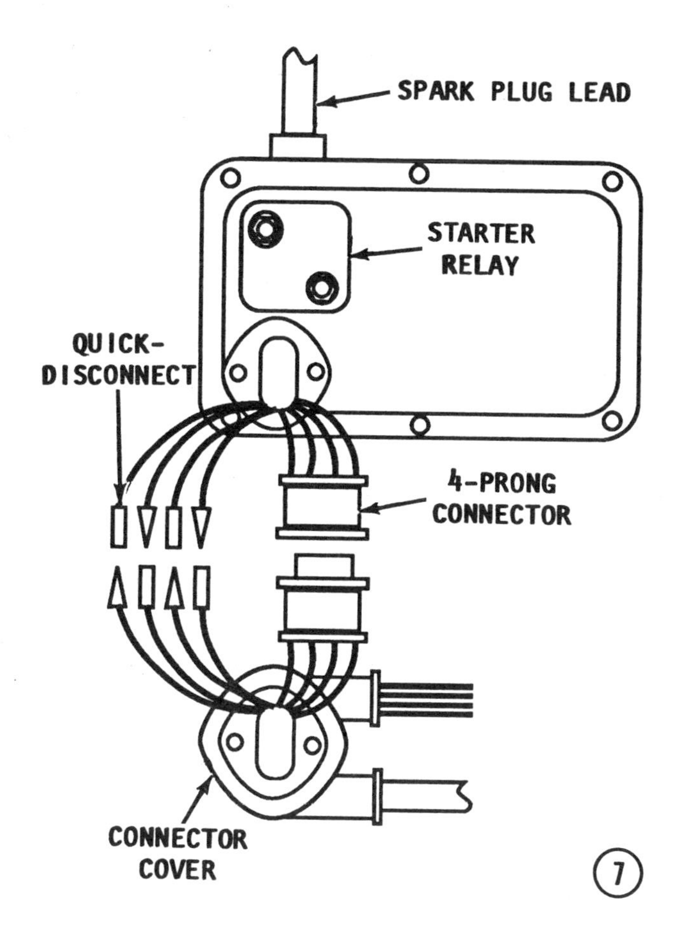

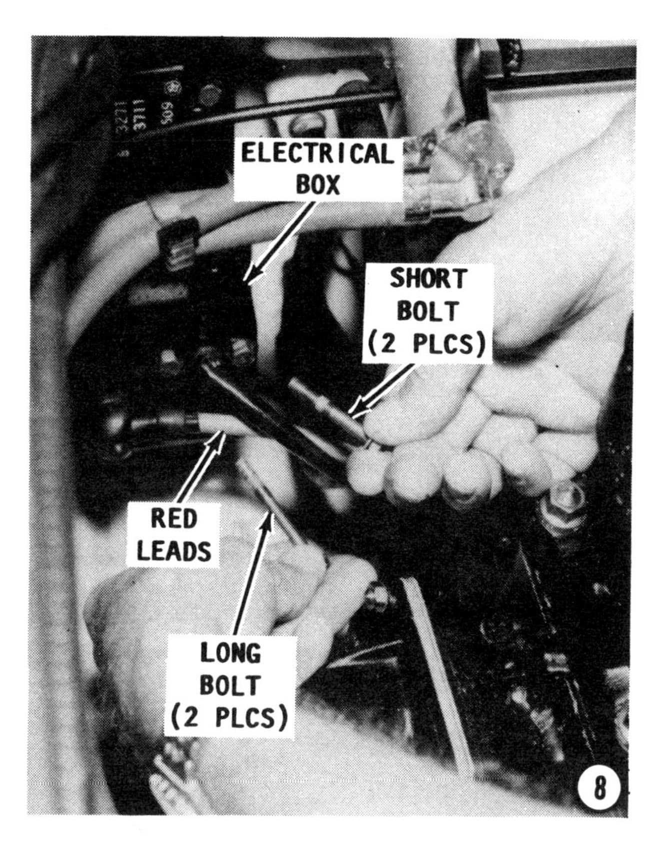

Electrical Box (Case) Installation
Model 550 Series Only

7- Secure the halves of the case together with the two bolts and washers. Position the case against the engine compartment bulkhead and secure it in place with the six bolts. Connect the two large cables to the relay terminals -- one from the battery and one to the cranking motor. Cover the terminals with the rubber boots -- one on each cable.

Connect the two halves of the 4-prong connector. Connect the four free leads, matching color-to-color, at their quick-disconnect fittings. Install the connector cover over the electrical box and secure it in place with the two bolts.

Install the spark plug leads and connect the battery ground cable.

Electrical Box (Case) Installation
All Others

8- Install the cover over the electrical case mounted on the engine compartment bulkhead -- Model 650 Series -- or on the hull -- all others. Secure the cover to the box with the four bolts and washers. Notice there are two different length bolts, as shown -- Model 650 Series only. Connect the two large cables to the relay terminals -- one from the battery and one to the cranking motor. Cover the terminals with the single large boot.

Install the spark plug leads and connect the battery ground cable.

9-6 CRANKING MOTOR SERVICE

Description

Marine cranking motors are very similar in construction and operation to the units used in the automotive industry.

NEVER operate a cranking motor for more than 30 seconds without allowing it to cool for at least three minutes. Continuous operation without the cooling period can cause serious damage to the cranking motor.

Two types of power train arrangements are used on the cranking motors covered in this manual. Stating this fact in other terms -- Kawasaki uses two methods of transferring energy from the cranking motor to the engine flywheel.

Method one -- Model 550 Series Only -- uses a pinion gear.

Method two -- Model 650 thru 1100 Series-- uses a reduction gear.

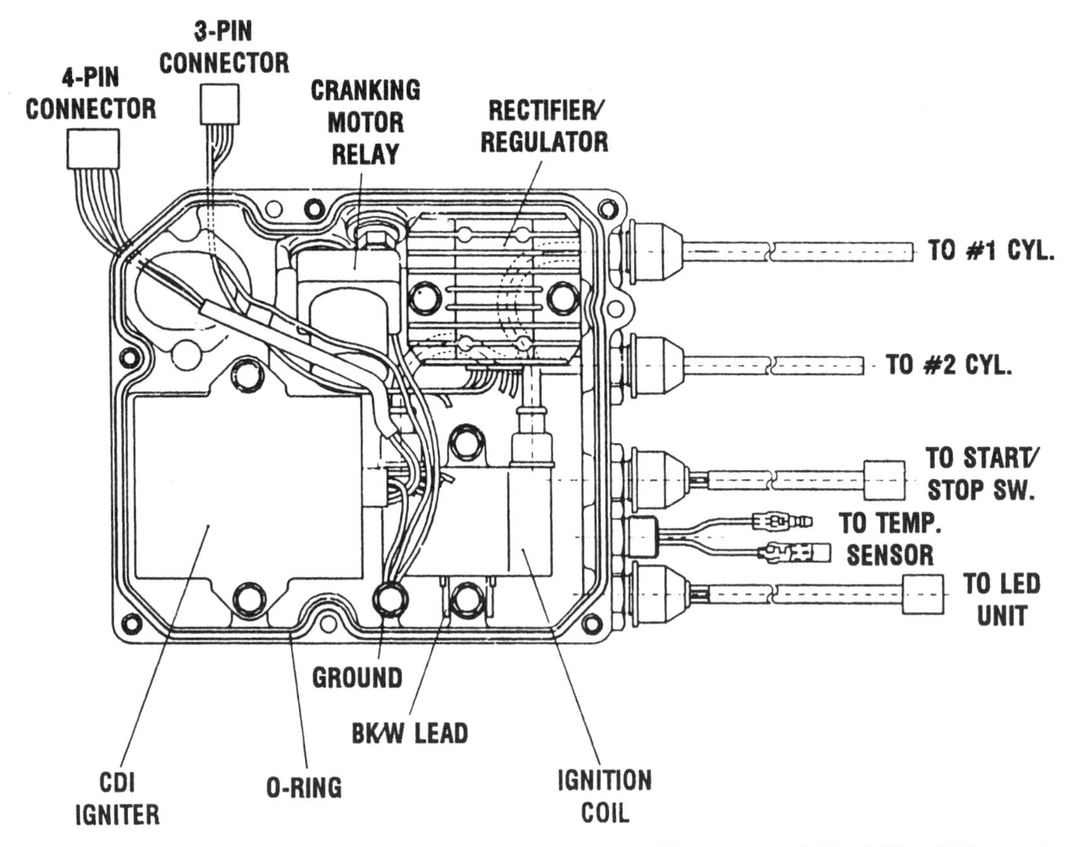

Functional diagram of the interior of the electrical box on a ***STANDARD*** *Model 750 Series engine. Principle components housed inside have been identified.*

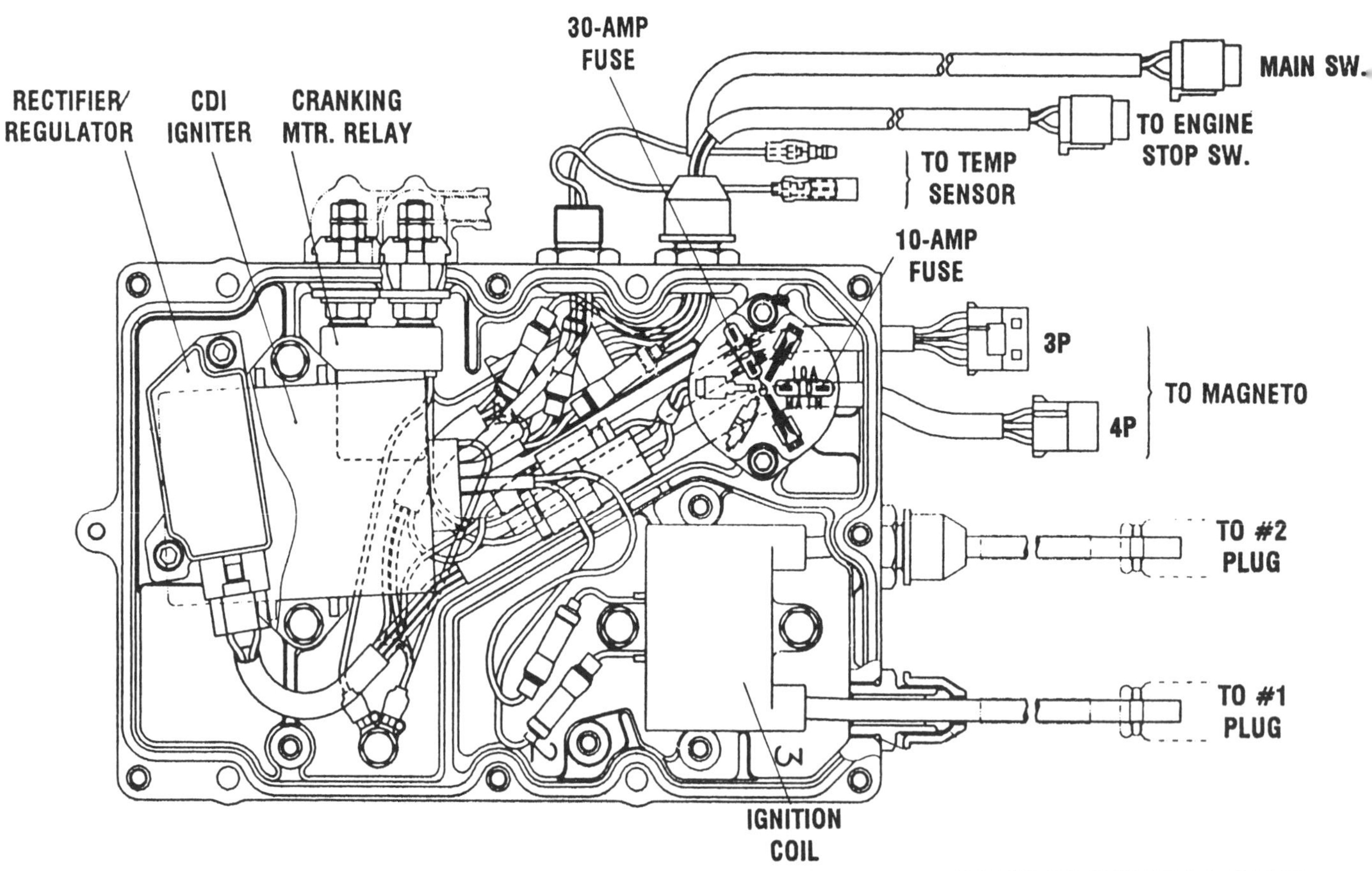

Functional diagram of the interior of the electrical box on a ***HIGH PERFORMANCE*** *Model 750 Series engine. Principle components housed inside are identified.*

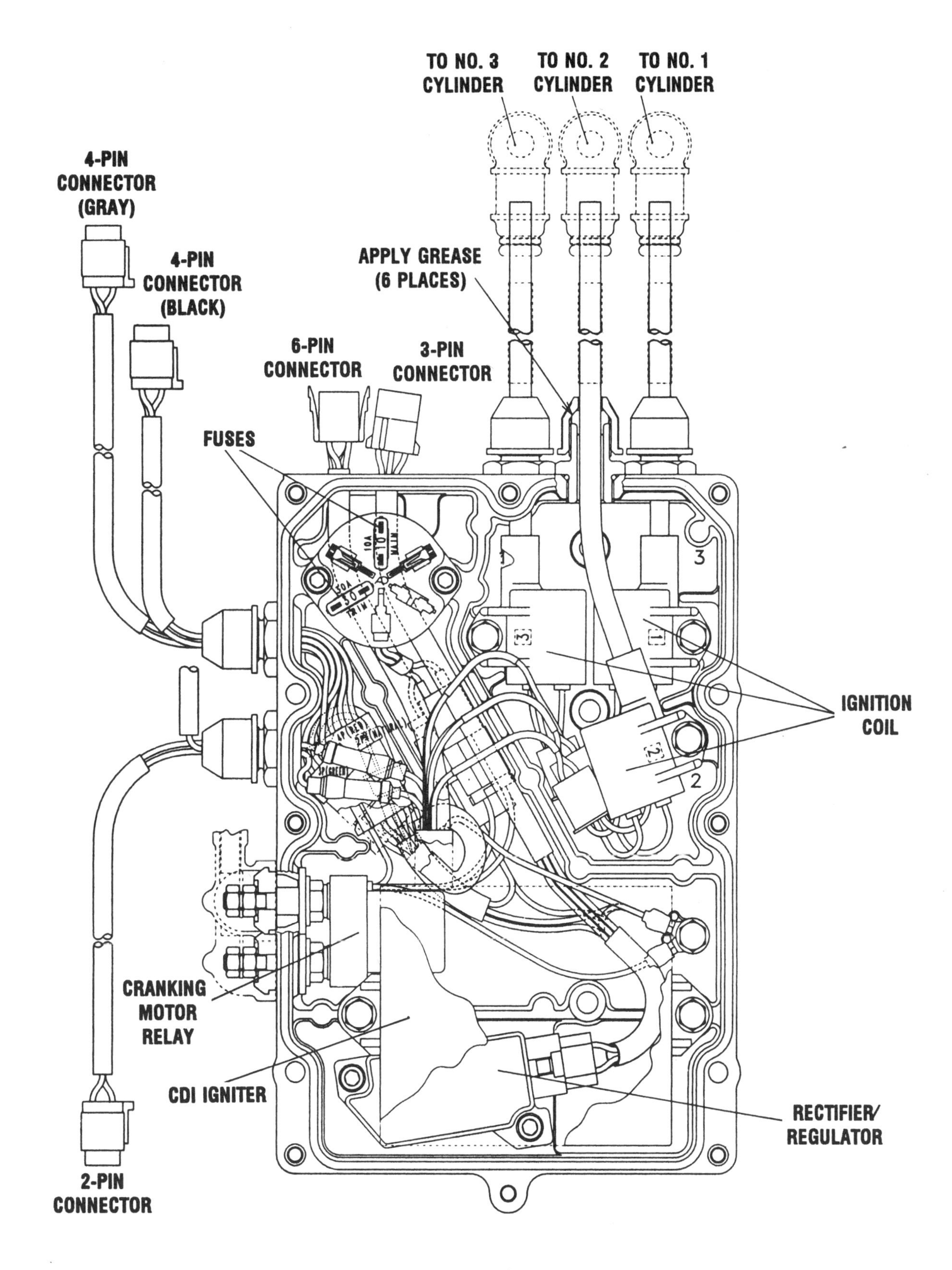

*Functional diagram of the interior of the electrical box on a **STANDARD** Model 900 Series engine. On a **HIGH PERFORMANCE** Model 900 and Model 1100 Series engine, the leads to the spark plugs are numbered -- No. 1, No. 2 and No.3 from left to right -- in reverse, as shown in this illustration for a standard Model 900.*

Cranking Motor Removal
All Models Except 550 Series

SPECIAL WORDS

The cranking motor may be removed without removing the reduction gear assembly located inside the flywheel cover. However, if the reduction gear needs service, see Chapter 7 for removal of the flywheel cover, flywheel, and reduction gear.

The exhaust pipe and exhaust manifold may have to be removed on most Model 650 Series. These items, including the expansion chamber, must be removed on most Model 750 Series, and all Model 900 Series and Model 1100 Series.

1- Disconnect the negative cable from the battery.

2- Push back the protective boot from the terminal on the cranking motor and remove the large Red cable.

3- Remove the two mounting bolts securing the cranking motor to the crankcase. On some 650 Series, the lower mounting bolt also secures the battery ground cable to the crankcase.

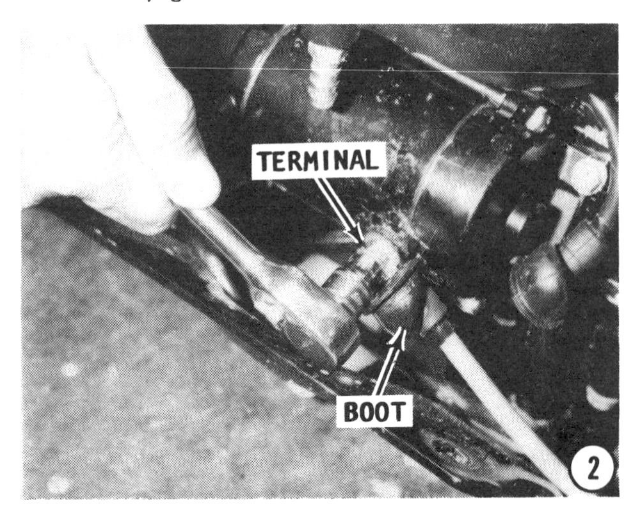

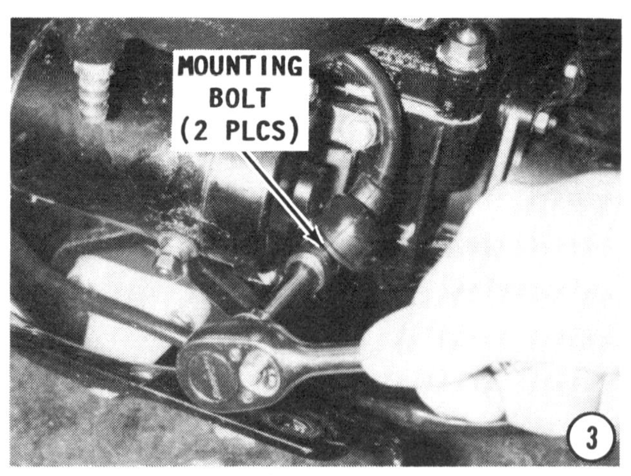

4- Ease the armature shaft back out of the reduction gear inside the flywheel cover.

SPECIAL WORDS

Without some type of alignment marks, the task of inserting the thru bolts and starting the threads into the end cap is almost impossible.

Cranking Motor Disassembling
All Models Except 550 Series

1- Remove and discard the O-ring around the armature shaft. Make a mark across the mating surface of the front cover and motor frame and another mark across the mating surface of the motor frame and the rear cover, as an aid during assembling.

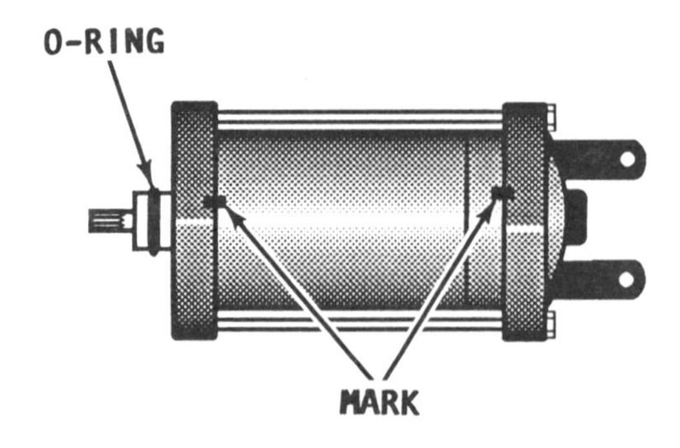

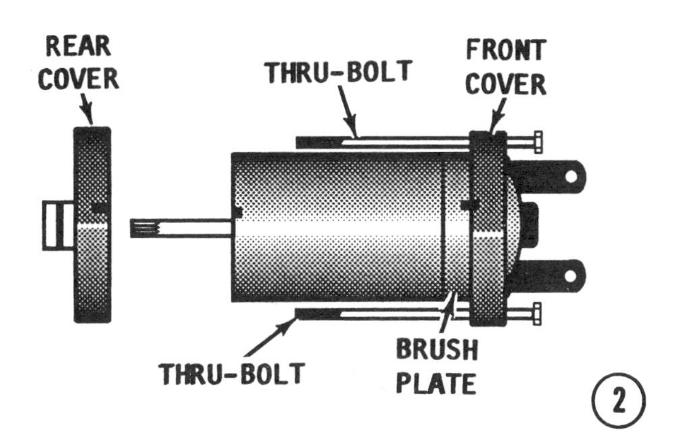

2- Remove the two long thru bolts and pull off the rear and front covers. Remove and discard the two O-rings.

3- Pull the armature shaft free of the motor frame. Make a note of the type and number of washers at each end of the shaft. Place them in order on the workbench in order of removal as an assist in later assembly. Incorrect placement of these washers will cause the armature to be misaligned with the brushes and the field magnets during assembling.

4- Identify the two positive brushes. These brushes have insulated leads between the carbon brush and the positive terminal in the motor frame. Lift off the retaining

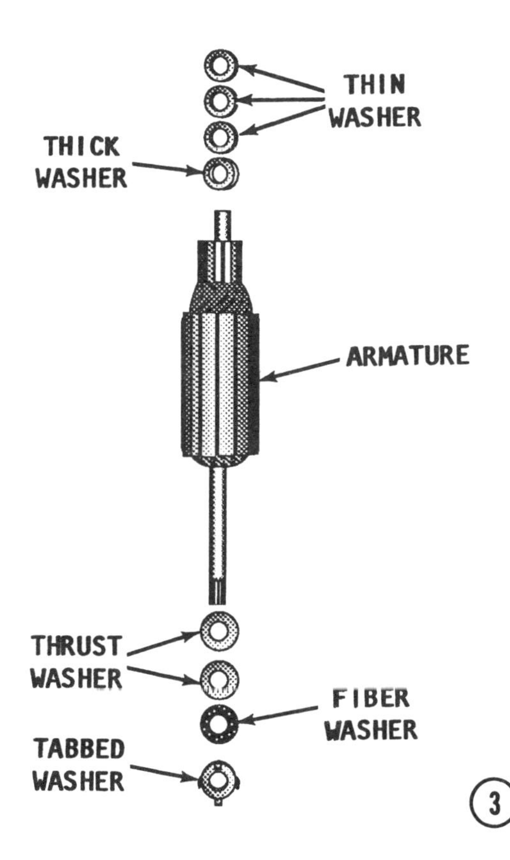

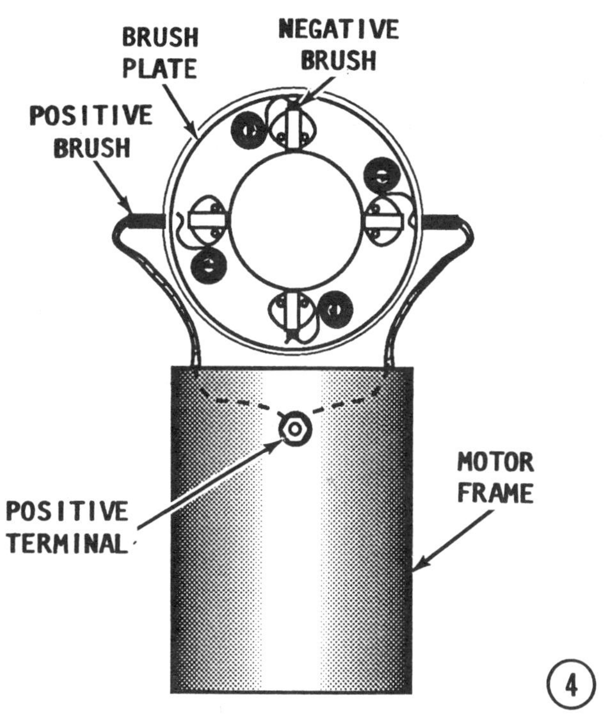

springs and pull out the carbon brushes from their holders to release them from the brush plate. Slide the brush assembly from the armature shaft.

Proceed to Cleaning and Inspecting on Page 9-21.

Cranking Motor Removal
Model 550 Series Only

1- Disconnect the battery ground cable at the battery.

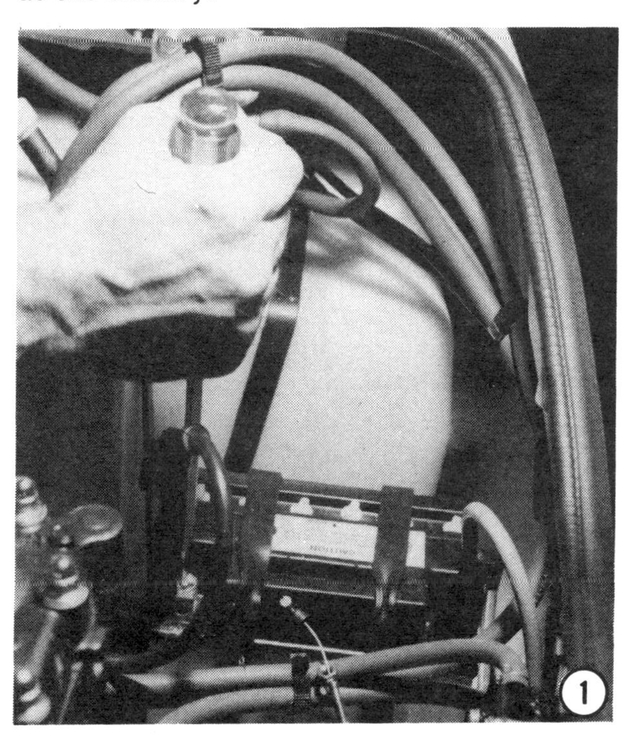

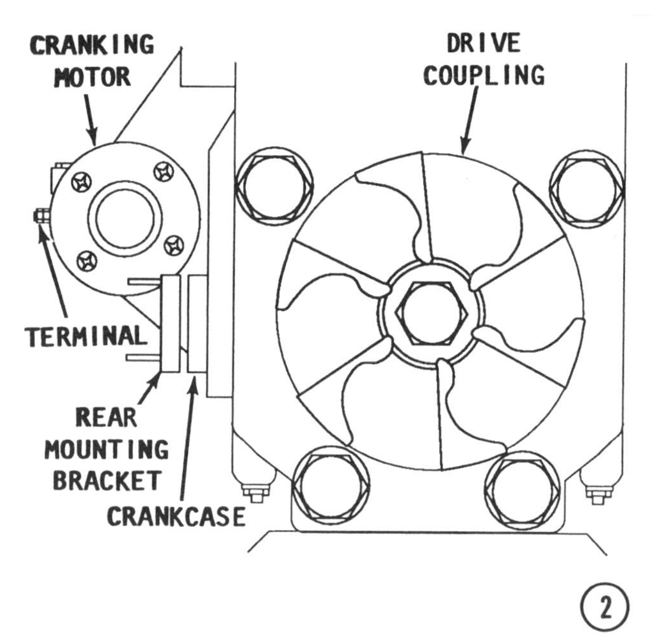

2- Remove the two rear cranking motor mounting bolts. Remove and **SAVE** any shim material found between the rear mounting bracket and the crankcase.

Remove the two front cranking motor mounting bolts.

Slide the cranking motor aft -- toward the rear of the engine. Disconnect the large Red cable from the terminal on the motor.

Armature Shaft End Play

Before disassembling the cranking motor, check the armature end play. Clamp the motor horizontally in a vise equipped with soft jaws. Make a set up with a dial indicator against the armature end. Zero the dial indicator. Now, push the armature shaft in and out. At the same time, observe the deflection of the dial indicator needle.

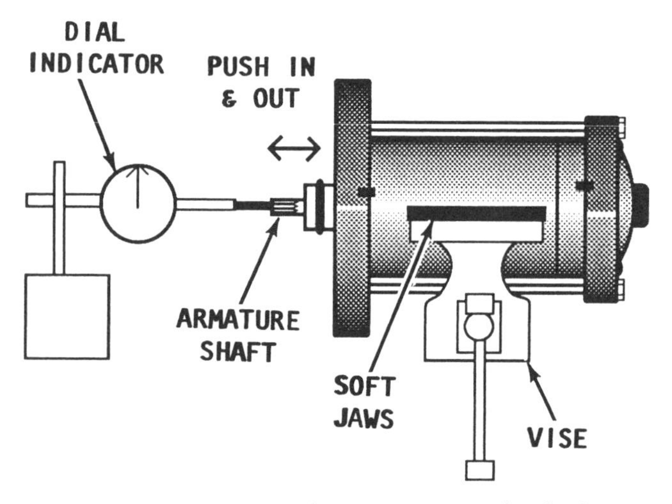

Setup to measure the armature end "play", as explained in the text.

The needle should move through a total range of 0.05-0.30mm. If the needle moves through more than 1.00mm, make a note to add thrust washers from the rear of the armature shaft, as required. If there is no needle movement, remove one of the thrust washers from the rear of the armature shaft.

Cranking Motor Disassembling
Model 550 Series Only

1- With the cranking motor on a work surface, pry the snap ring out of its groove and free of the armature shaft.

After the snap ring is free, slide the collar, spring, and pinion gear free of the armature. Remove the large O-ring around the upper end cap. Discard this O-ring **ONLY** if a replacement is available.

2- Make a mark on both end caps and matching marks on the field frame assembly as an aid to alignment and installation of the thru bolts, during assembling.

SPECIAL WORDS

Without some type of alignment marks, the task of inserting the thru bolts and starting the threads into the end cap is almost impossible.

Remove the two thru bolts and the two small brush plate retaining screws from the rear end cap. Observe the four small O-rings around these bolts and screws. Take care not to lose them. If these O-rings are

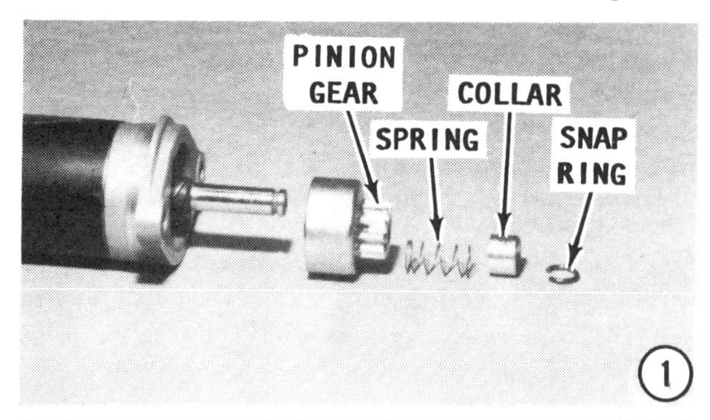

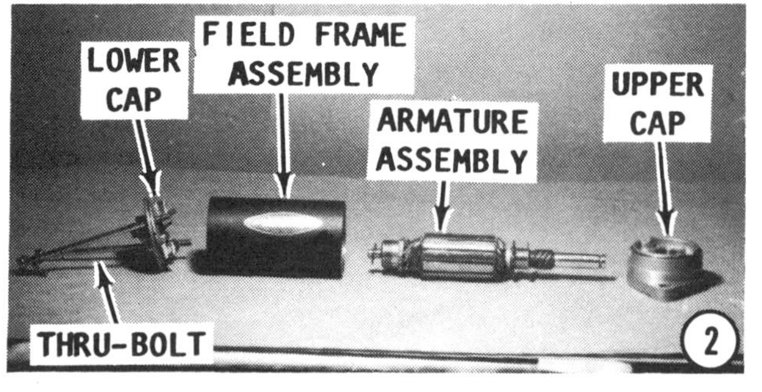

The cranking motor on a Model 650 Series engine can be removed without disturbing the reduction gear. However, if the reduction gear must be replaced, the flywheel must first be removed, see Chapter 8.

no longer fit for service they must be replaced. Remove both end caps. If the end caps are stuck, tap around the end cap circumference with a soft head mallet. Use a small screwdriver and pry off the curved springs from the positive brush holders.

Pull out the two carbon brushes and slide the brush plate free of the armature. Note the number of washers on each end of the armature. Incorrect placement of these washers will cause the armature to be misaligned with the brushes and the field magnets during assembling.

Remove the O-ring around each end cap. Discard this O-ring **ONLY** if a replacement is available.

The internal splines can only be inspected after the reduction gear has been removed and disassembled.

*The flywheel **MUST** be removed before the reduction gear can be withdrawn from the cranking motor.*

CLEANING AND INSPECTING

Pinion Gear Assembly Model 550 Series Only

Inspect the pinion gear teeth for chips, cracks, or a broken tooth. Check the splines inside the pinion gear for burrs and to be sure the gear moves freely on the armature shaft.

BAD NEWS

This type cranking motor pinion gear assembly cannot be repaired if the unit is defective. Replacement of the pinion gear assembly -- as a unit -- is the only answer.

Check to be sure the return spring is flexible and has not become distorted. Check both ends of the spring for signs of damage.

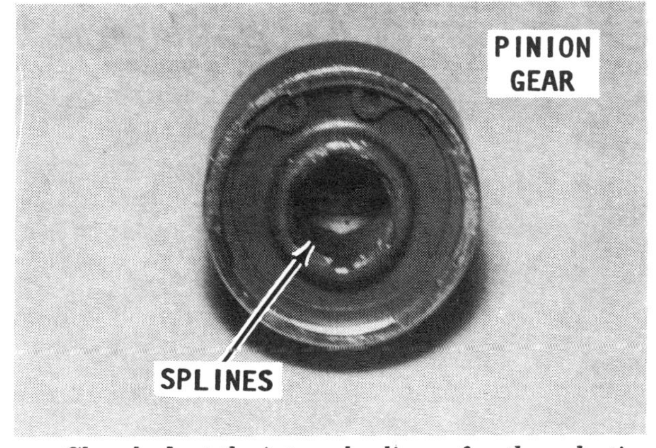

Close look at the internal splines after the reduction gear has been disassembled.

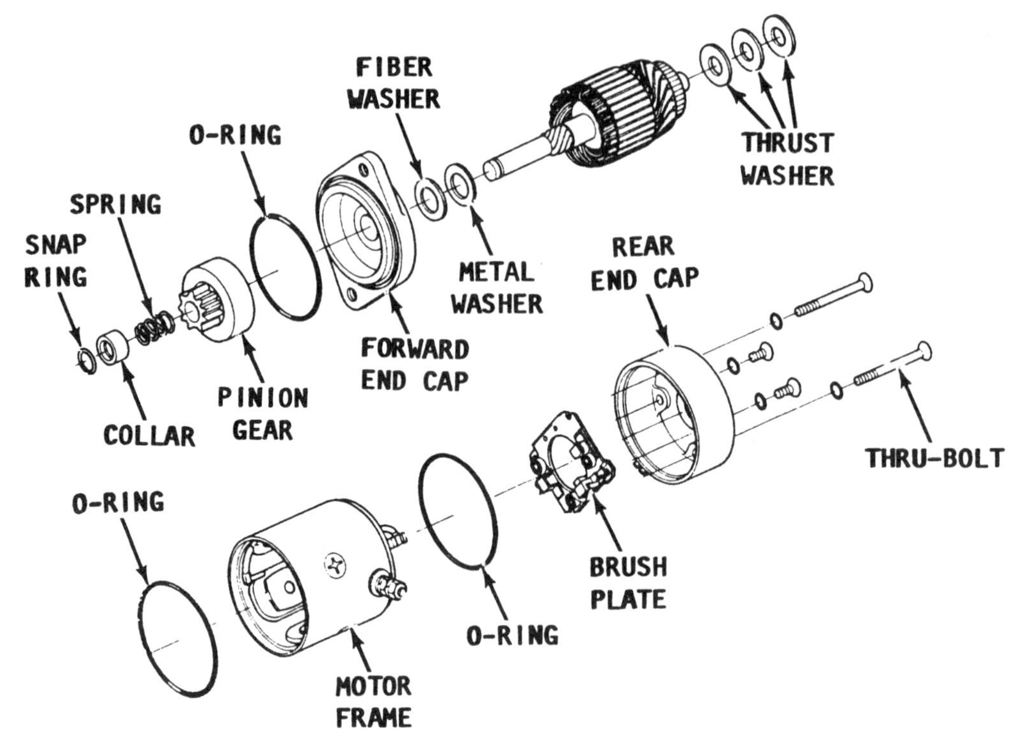

Exploded drawing of the cranking motor installed on the Model 550 Series engines. Major parts have been identified.

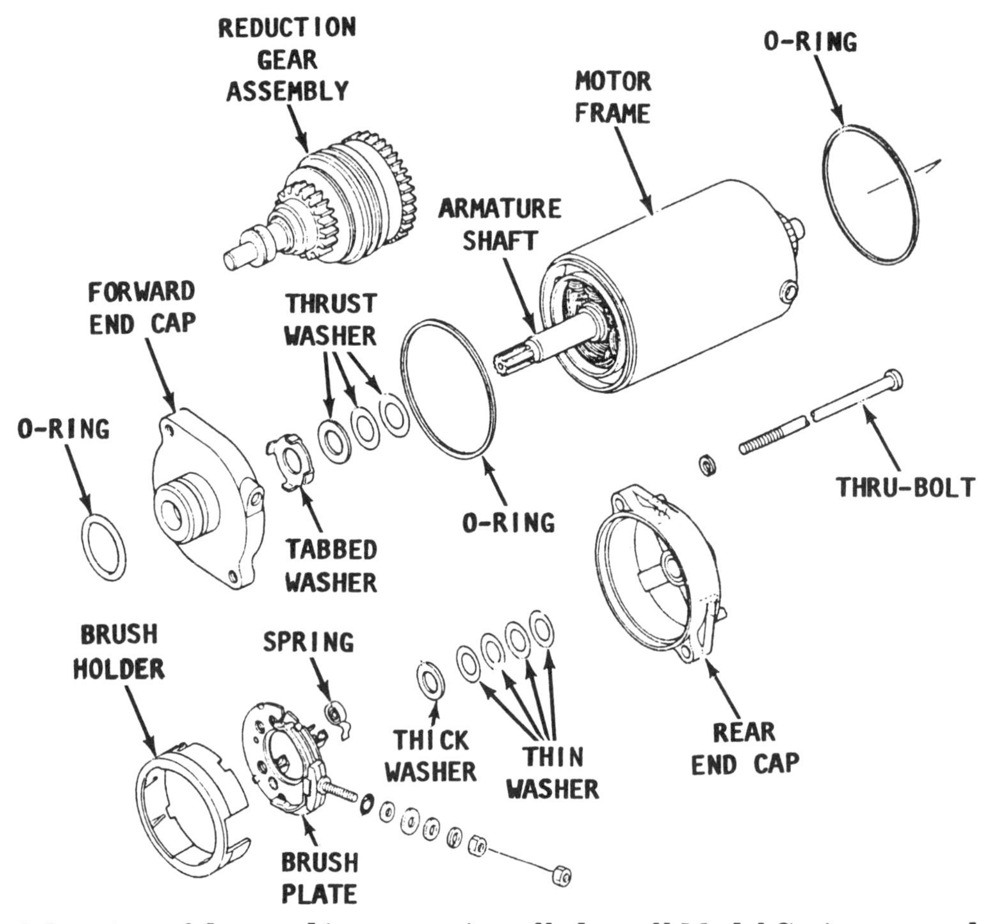

Exploded drawing of the cranking motor installed on all Model Series except the 550 series, which is shown above. Major parts have been identified.

All Series

Clean the armature shaft and check to be sure the shaft is free of any burrs. If burrs are discovered, they may be removed with crocus cloth.

Clean the commutator on the armature shaft and check to be sure the shaft is free of any burrs. If burrs are discovered, they can be removed with crocus cloth.

Clean the field coils, armature, commutator, armature shaft, brush assembly, and end caps with a brush or compressed air. Wash all other parts in solvent and blow them dry with compressed air.

Inspect the insulation and the unsoldered connections of the armature windings for breaks or burns.

Perform electrical tests on any suspected defective part, according to the procedures outlined in Section 9-6, beginning on on the following page.

Check the commutator for runout. Inspect the armature shaft and both bearings for scoring.

Turn the commutator in a lathe, if it is out-of-round by more than 0.005" (0.13mm).

Check the springs in the brush holder to be sure none are broken. Check the insulated brush holders for shorts to ground.

The armature, fields, and brush holders must be checked before assembling the cranking motor. See Section 9-7, beginning on Page 9-24 for detailed procedures to test cranking motor parts.

To check the action of the reduction gear on a 650 Series: Rotate the smaller pinion gear counterclockwise. It should rotate freely. Now, rotate the pinion gear clockwise. The pinion gear should rotate on the curved splines to the top of the shaft and then stop. Release the gear. The gear should quickly return back down to its original position. If the pinion gear does not function as expected, the reduction gear assembly must be replaced.

The negative brushes are mounted on the brush assembly plate. The positive brushes are attached to the positive terminal. When all the brushes are installed onto the plate, each brush is held in place with wound spring steel coils mounted horizontally to provide the tensioning force against the commutator. These coils are permanently riveted to the brush plate and are not removable. If they are broken or missing, a new plate must be purchased.

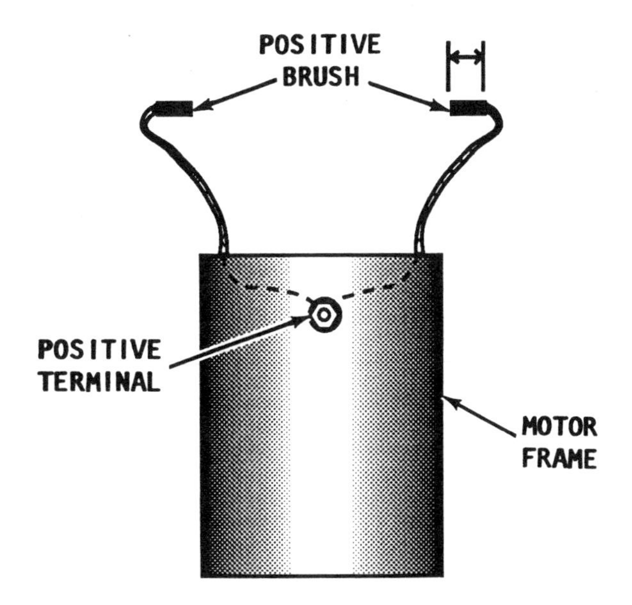

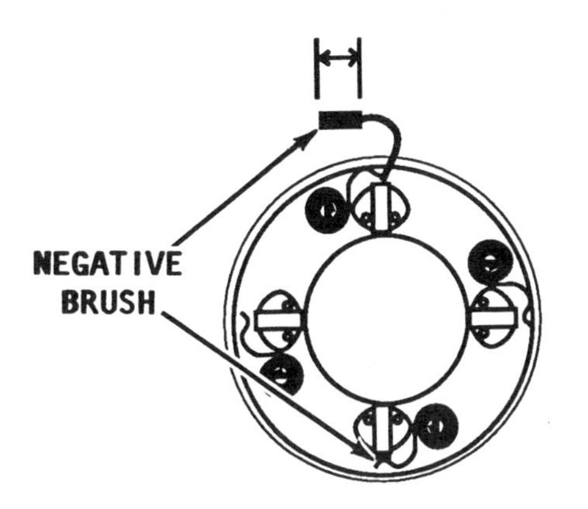

Measuring the length of the positive brushes (top) and the negative brushes (bottom). The text gives dimensions of new brushes for all Series and the minimum limits.

Measure the length of each brush.

If servicing a Model 550 Series -- a new brush measures 0.67 in. (17.0mm). A brush less than 0.5 in. (13.0mm) in length, must be replaced.

For all other Models covered in this manual, a new brush measures 0.49 in. (12.5mm). A brush less than 0.25 in. (6.5mm) in length must be replaced.

To replace the positive brushes, the positive terminal must be removed from the motor frame. Remove the following parts and place them in order on the workbench to ensure they will be installed in exactly the same order: a nut, a spring washer, a large insulator, and finally an O-ring -- all installed on the terminal on the outside of the motor frame. Inside the motor frame, the brush leads are soldered to the terminal.

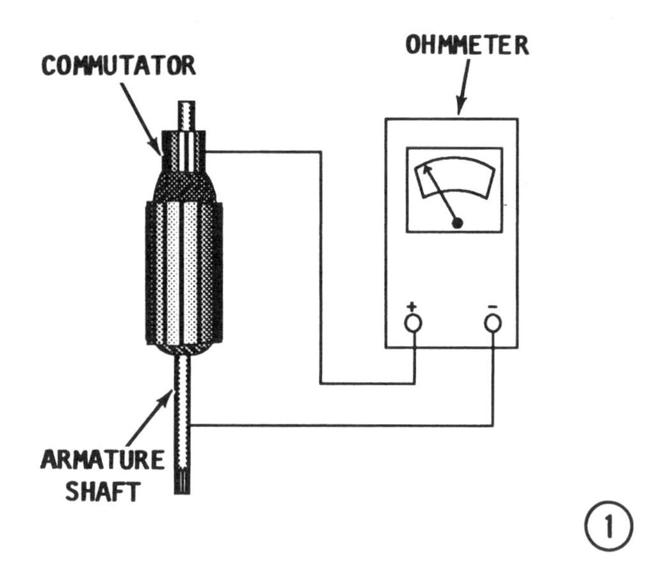

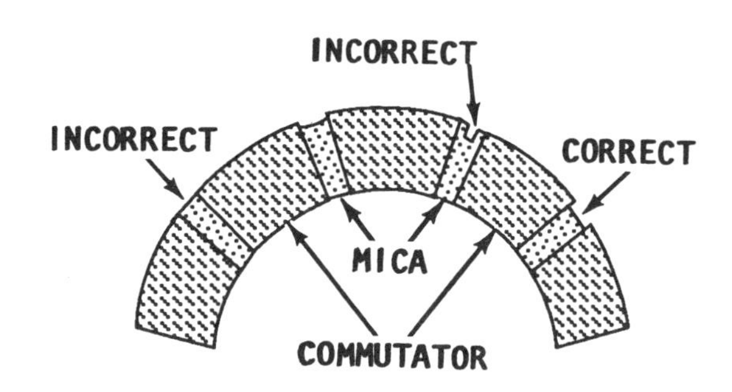

Simple line drawing to illustrate correct and incorrect mica cutting on a commutator.

TESTING CRANKING MOTOR PARTS

SPECIAL WORDS

Most marine shops and all electric motor rebuild shops will test an armature for a very modest charge. If the armature has a short, it **MUST** be replaced.

Check the armature for a short circuit by placing it on a growler and holding a hack saw blade over the armature core while the armature is rotated. If the saw blade vibrates, the armature is shorted. Clean between the armature bars, and then check again on the growler. If the saw blade still vibrates, the armature must be replaced. Occasionally carbon dust from the brushes will short the armature. Therefore, blow the slots in the armature clean with compressed air.

1- Make contact with one probe of a test light or ohmmeter on the armature core or shaft. Make contact with the other probe on the commutator. If the light comes on, or the meter indicates continuity, the armature is grounded and must be replaced.

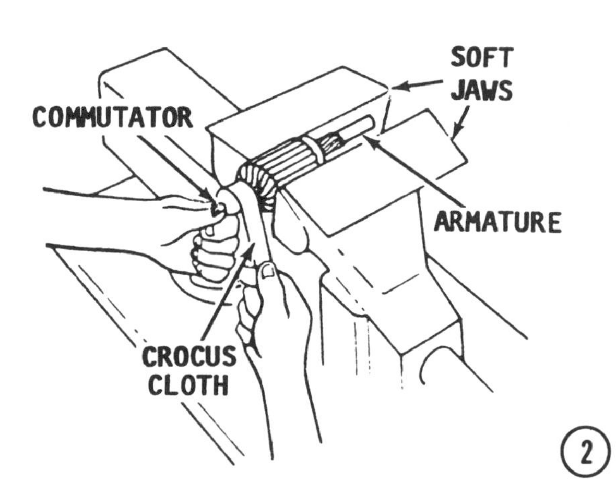

Turning the Commutator

2- True the commutator, if necessary, in a lathe. **NEVER** undercut the mica because the brushes are harder than the insulation. Undercut the insulation between the commutator bars 1/32" (0.80mm) to the full width of the insulation and flat at the bottom. A triangular groove is not satisfactory. After the undercutting work is completed, clean out the slots carefully to remove dirt and copper dust. Sand the commutator lightly with No. 500 or No. 600 sandpaper to remove any burrs left from the undercutting.

3- Test light probes, placed on any two commutator bars, should light and indicate continuity.

4- Check the armature a second time on the growler for possible short circuits.

Positive Brush

5- Obtain an ohmmeter. Make contact with one lead of the ohmmeter to the positive brush. Make contact with the other

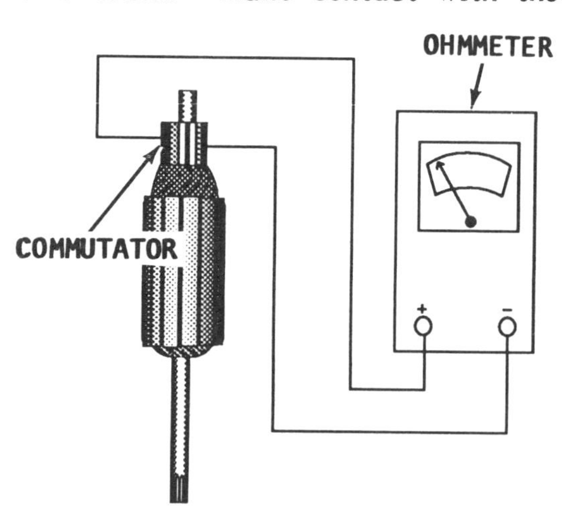

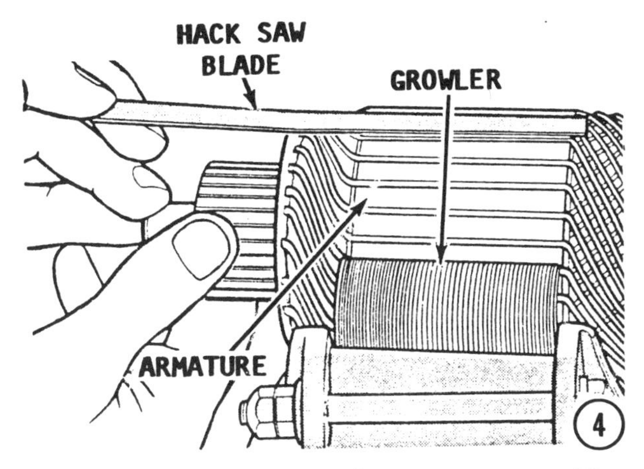

test lead to the end cap/motor frame. The ohmmeter should indicate **NO** continuity. If the meter indicates continuity, the positive lead is shorted to the motor frame and must be replaced.

6- Make contact with one ohmmeter lead to a positive brush. Make contact with the other ohmmeter lead to the positive terminal. The meter should register continuity. Repeat this test for the other positive brush. If the meter registers no continuity in either test, the positive brushes have an open in the circuit and must be replaced.

Negative Brush

7- Make contact with one ohmmeter lead to the negative brush. Make contact with the other test lead to the brush plate. The ohmmeter should indicate continuity. If the meter indicates no continuity, there is an open in the negative lead or the lead is not grounded properly to the brush plate.

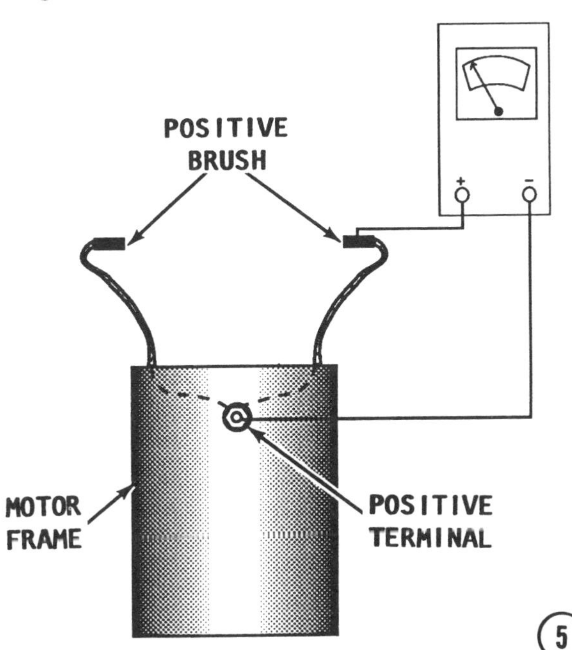

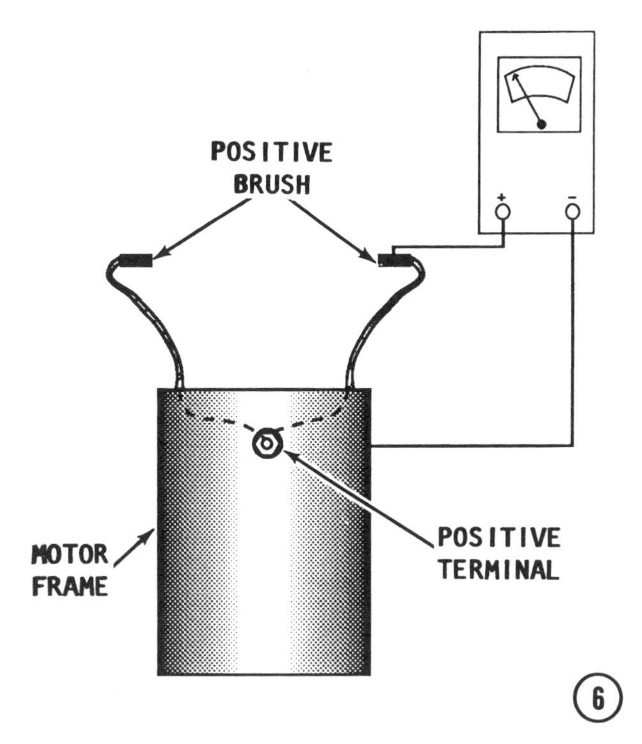

8- Make contact with the ohmmeter leads across the positive and negative brush holders on the brush plate. The meter should register no continuity. Repeat the test for the other pair of positive and negative brush holders. If the meter registers continuity in either test, the positive brush holders are shorted to the plate and the plate must be replaced.

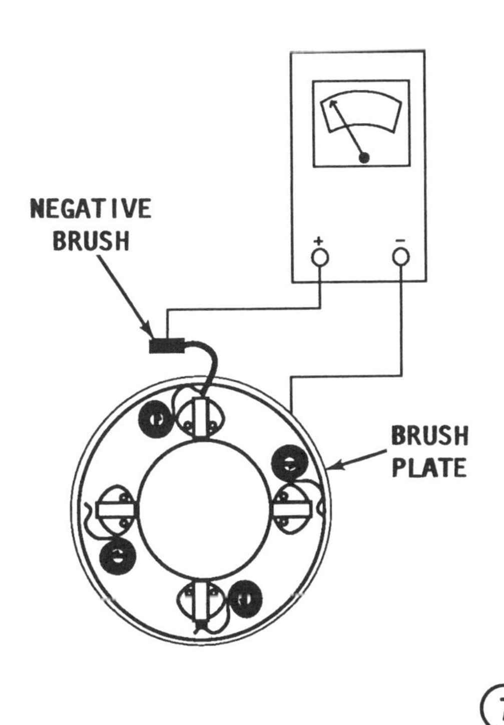

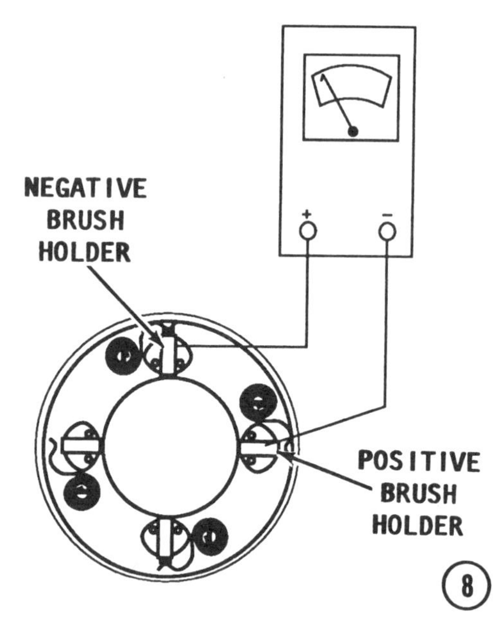

CRANKING MOTOR ASSEMBLING

Assembling All Except Model 550 Series

1- If the positive brushes were removed or replaced, apply Loctite Superflex to the bolt hole in the motor frame. Insert the terminal, with positive brushes attached, through the bolt hole from the inside. Install a washer and an insulator over the terminal and then secure it in place with a locknut.

2- Install the brush plate over the motor frame, with the two positive brush leads indexed into the two slots in the plate. Install the positive brushes into their respective holders. Insert the armature shaft into the other end of the motor frame. If necessary, push back the brushes to allow the plate to slide over the commutator.

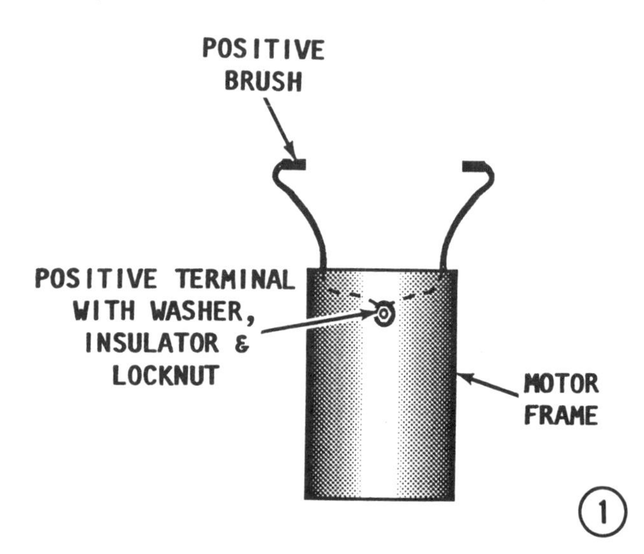

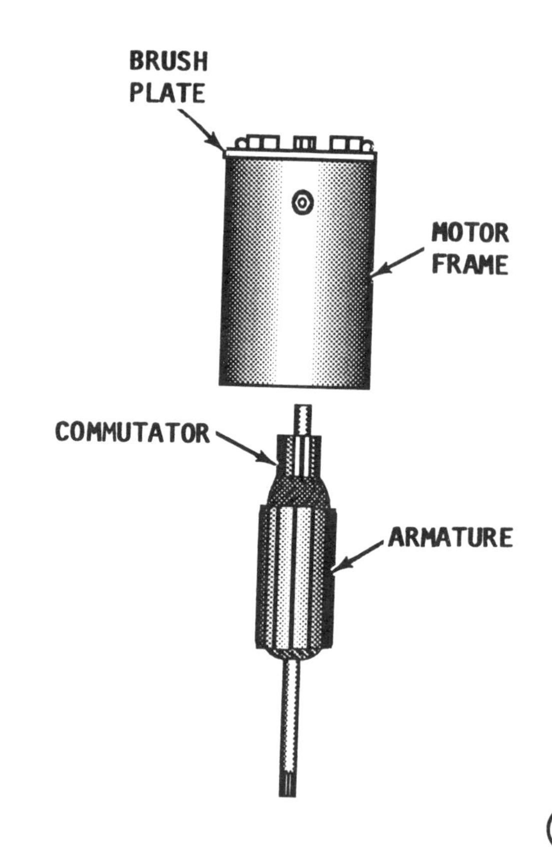

3- Check each of the four brush springs for tension. Rotate the armature shaft. The brushes must stay in contact with the commutator surface without binding.

4- Install the thick washer, followed by the thin washers over the commutator end of the armature. Install a **NEW** O-ring around the motor frame, and then install the front cover, aligning the indexing mark made during disassembling. Hold all the pieces of the motor together until the long thru-bolts are installed in a later step.

5- Install the two thrust washers, the fiber washer and the tabbed washer -- with tabs facing up, over the other end of the armature. Install a **NEW** O-ring around the motor frame, followed by the rear cover, aligning the marks made during disassembling. The tabs on the tabbed washer must

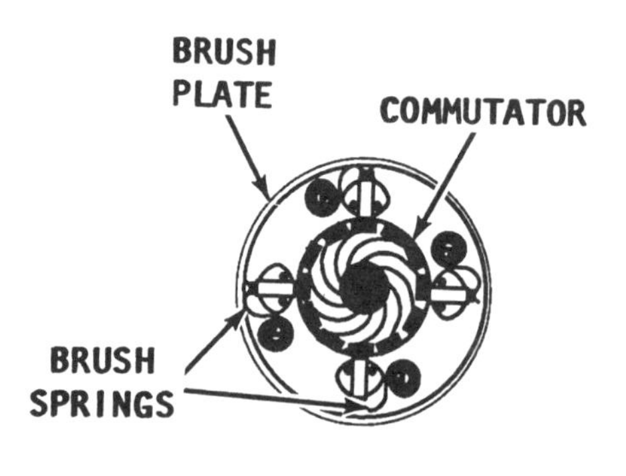

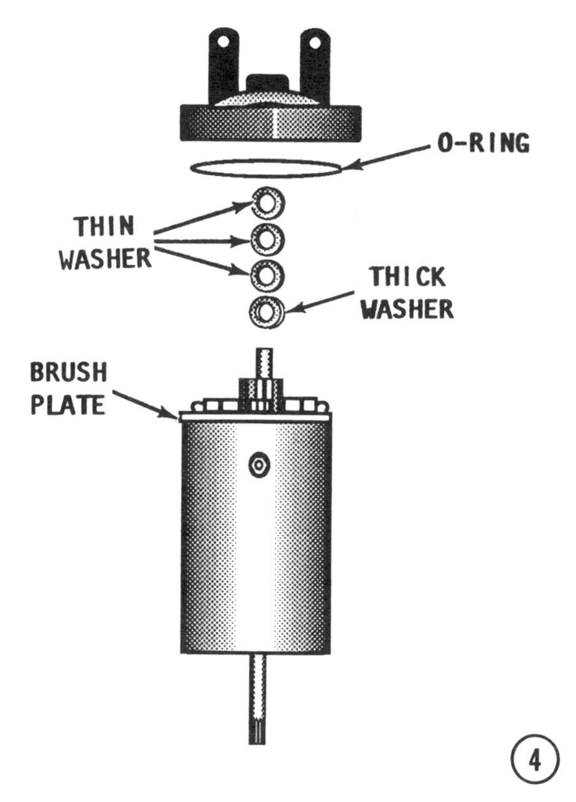

index into the four slots cut into the hub of the rear cover.

6- Install the two thru-bolts and tighten them securely. Install a new O-ring around the outside of the rear cover.

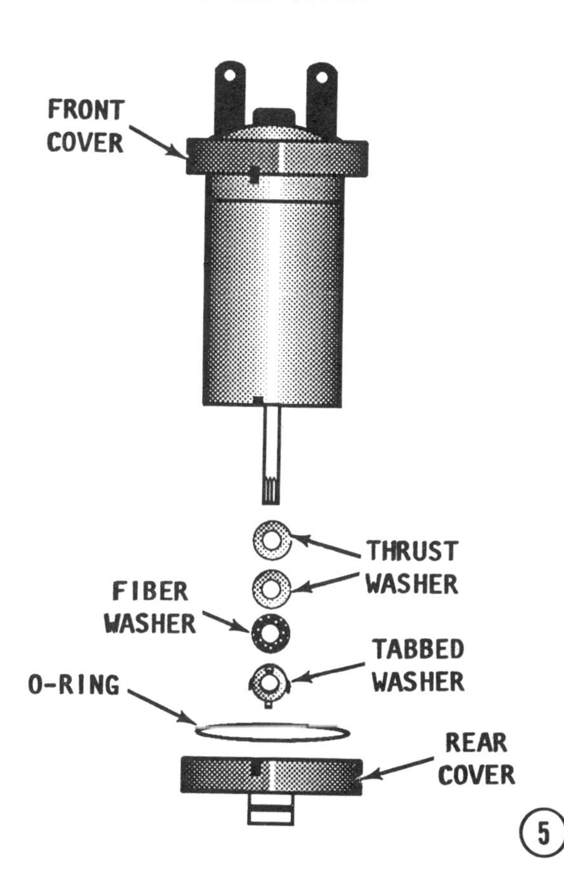

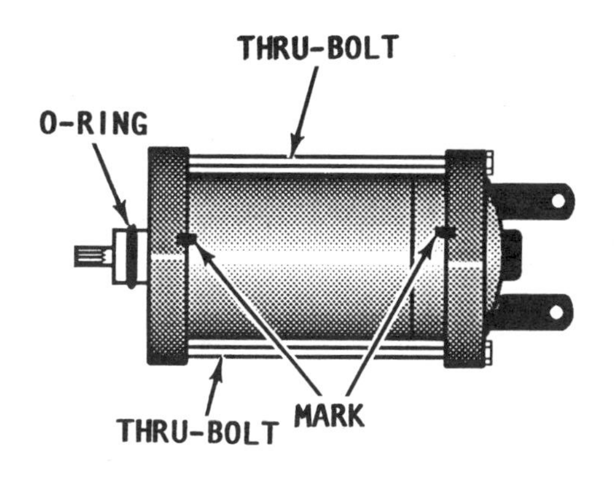

CRANKING MOTOR INSTALLATION ALL MODELS EXCEPT 550 SERIES

1- Clean the cranking motor lugs and the contact surface of the crankcase in the area where the motor is grounded. Coat the large O-ring on the forward end cap with engine oil. Position the motor against the crankcase and at the same time insert the end of the armature shaft into the reduction gear inside the flywheel cover.

2- If the battery ground cable was originally connected to the crankcase with the lower mounting bolt, then place the eye of the cable over the bolt and apply a non-permanent locking substance on both mounting bolts. Install and tighten the bolts securely.

3- Secure the cranking motor cable to the terminal on the motor frame. Pull the boot over the connection.

4- Connect the negative cable to the battery and tighten the fastener securely.

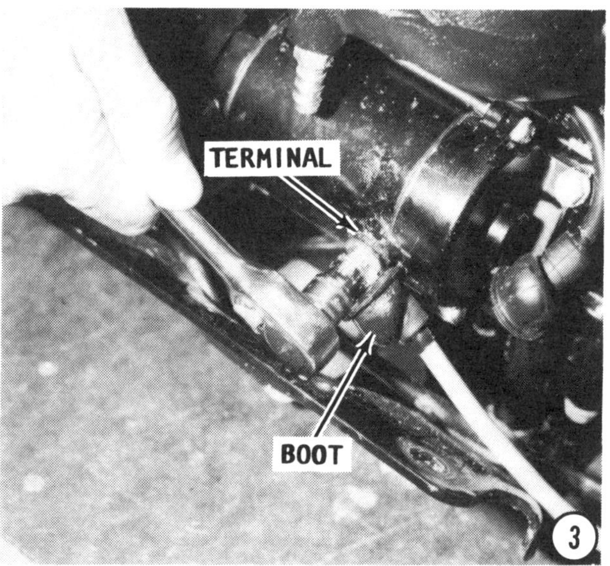

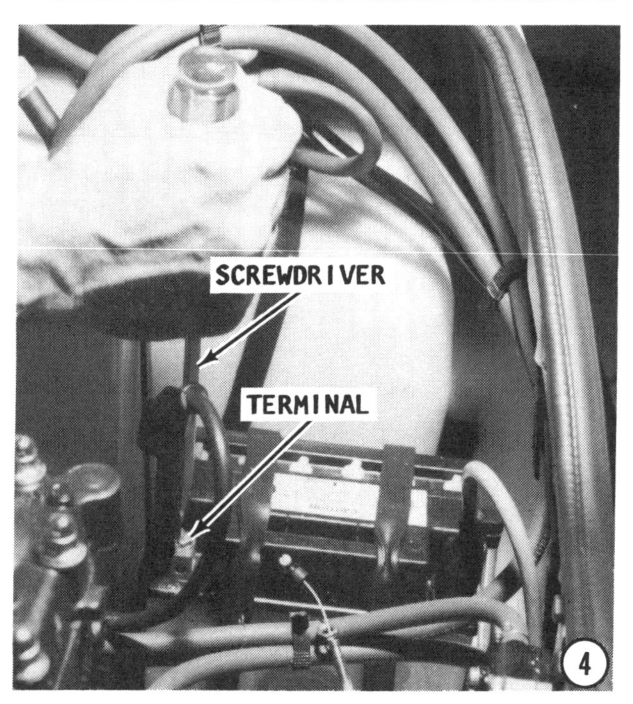

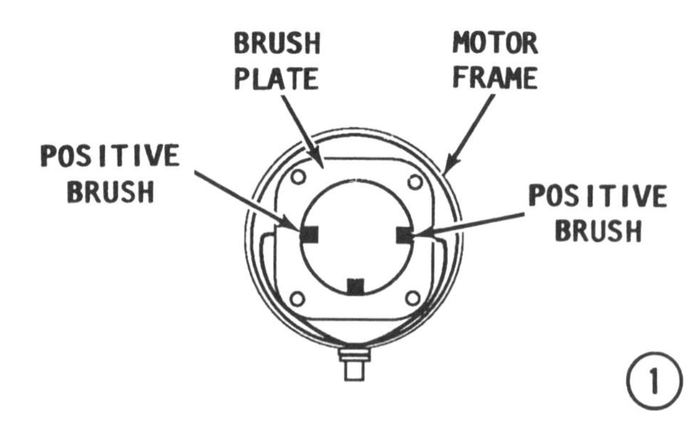

CRANKING MOTOR ASSEMBLING MODEL 550 SERIES ONLY

1- Position the two positive brushes outside the motor frame. Place the brush plate inside the frame. Push the tensioning spring aside, and at the same time, insert each brush into its holder.

2- Slide the metal washer, followed by the fiber washer onto the spiral end of the armature shaft.

Stretch a new **O**-ring around the circumference of the front end cap. Apply a coating of water resistant lubricant to the armature shaft and install the end cap over the spiral end of the armature shaft.

3- Push the commutator end of the armature shaft through the motor frame. Hold back the three brushes to allow the commutator to pass through. **TAKE CARE** not to distort the brush springs. Once assembled, rotate the shaft inside the bushing and check to be sure there is

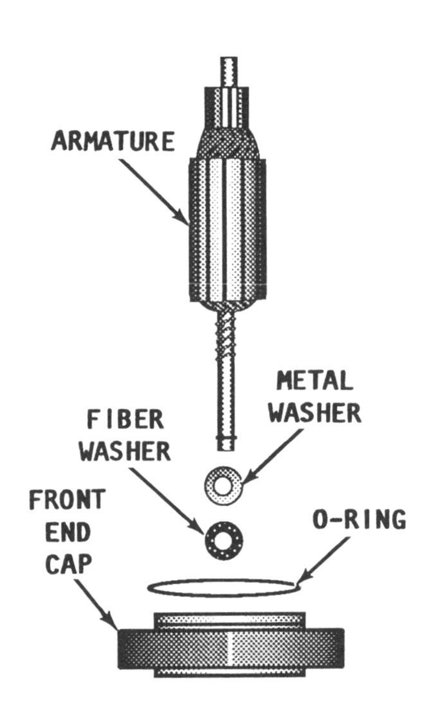

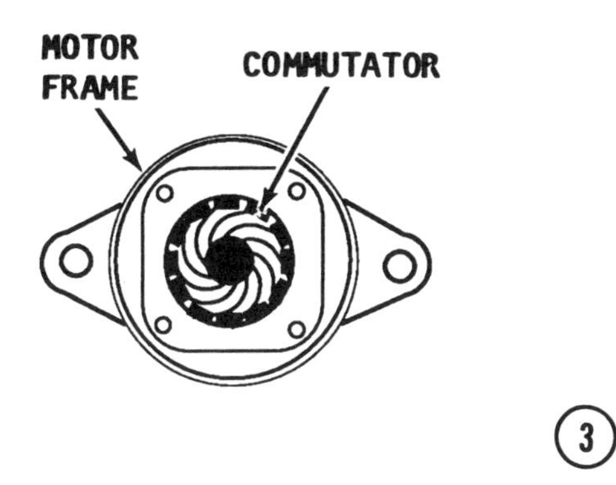

no binding and the brushes sweep across the commutator smoothly.

4- Install the original number of thrust washers over the rear of the armature shaft. If the armature end play was measured and found to be out of specification, then adjust the number/thickness of thrust washers accordingly.

Apply a thin coating of Loctite Superflex, or equivalent waterproof anti-seize lubricant, onto the bushing in the end cap.

Stretch a new **O**-ring around the circumference of the rear end cap and install it over the brush plate. **TAKE CARE** not to lose the brushes from the commutator. Seat the end cap into the motor frame with the marks made prior to disassembly aligned and the bolt and screw holes in the rear end cap aligned with those in the brush plate.

Make sure each of the two thru-bolts and two screws have **O**-rings in good condition. Install the two brush plate securing screws and tighten them securely.

Check the alignment of the rear end cap with the marks made during disassembly.

Now, hold it all together and slide the two thru-bolts through the rear end cap, through the motor frame, and then thread them into the forward end cap.

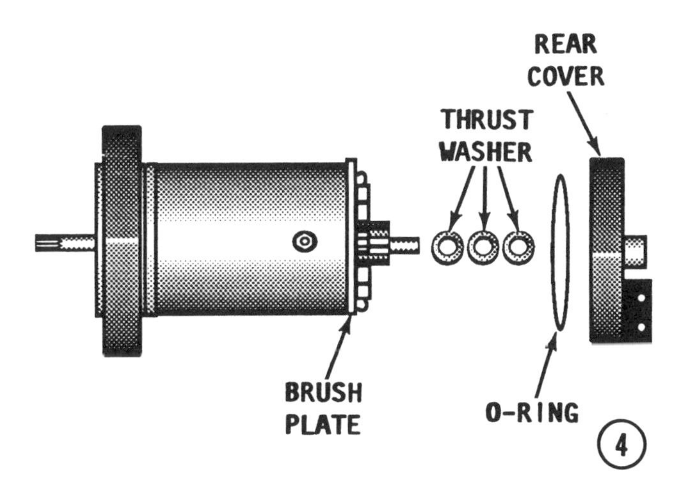

SPECIAL WORDS

Installing the two thru-bolts is not as simple as it sounds. The reason: as the bolts pass the magnets inside the motor frame, the magnets will attract the bolts, "pulling" them out of line. With a bit of patience, the bolts can be made to align with the holes in the upper end cap and the threads started.

After the bolts are successfully threaded into the end cap, tighten them securely.

Position the large **O**-ring into the groove around the upper end cap. Install the following components onto the armature shaft in the sequence given: first, the pinion gear, followed by the spring, the collar, and finally the snap ring. It will be necessary to hold the collar down tight against the spring while the snap ring is installed. The ring must be crimped around the groove in the shaft with a pair of pliers. When properly installed the collar will completely hide the snap ring.

Cranking Motor Installation
Model 550 Series Only

1- Apply a coating of Loctite Superflex to the **O**-ring around the forward end cap. Apply engine oil to the pinion gear. Connect the Red lead to the cranking motor terminal. Position the motor from the rear of the engine up against the rear flywheel housing. Apply Locktite Lock N' Seal to the threads of all four mounting bolts. Install the front two mounting bolts and tighten them to a torque value of 12 ft lb (16Nm).

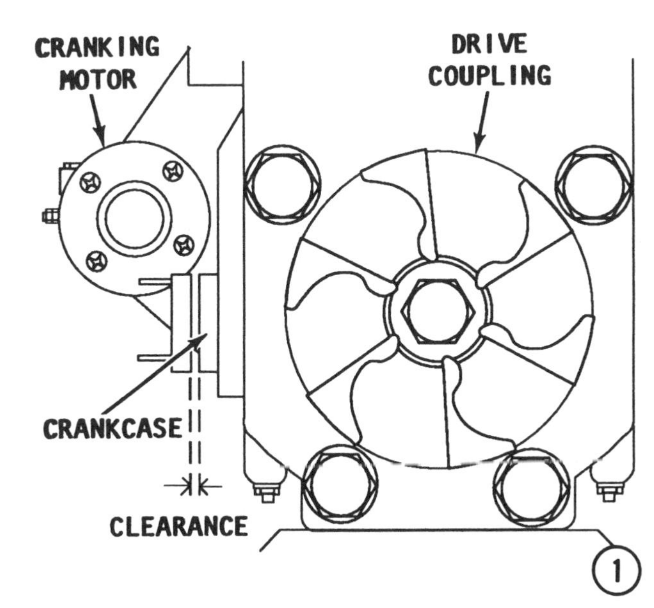

Before installing the two rear mounting bolts, measure the distance between the rear mounting bracket and the crankcase. Fill this space with the original pieces of shim material or purchased pieces. Shim material is available in thicknesses of 0.4, 0.6 and 0.8mm. Install the two rear mounting bolts and tighten them to a torque value of 52 in lb (6Nm).

2- Connect the battery ground cable to the battery.

ALL SERIES

Check Completed Work

Secure the craft in a test tank or body of water. If a body of water or tank is not possible, connect a flush attachment and hose to the jet pump.

CAUTION

Water must circulate through the jet pump -- to and from the engine, anytime the engine is operating. Circulating water will prevent overheating -- which could cause damage to moving engine parts and possible engine seizure.

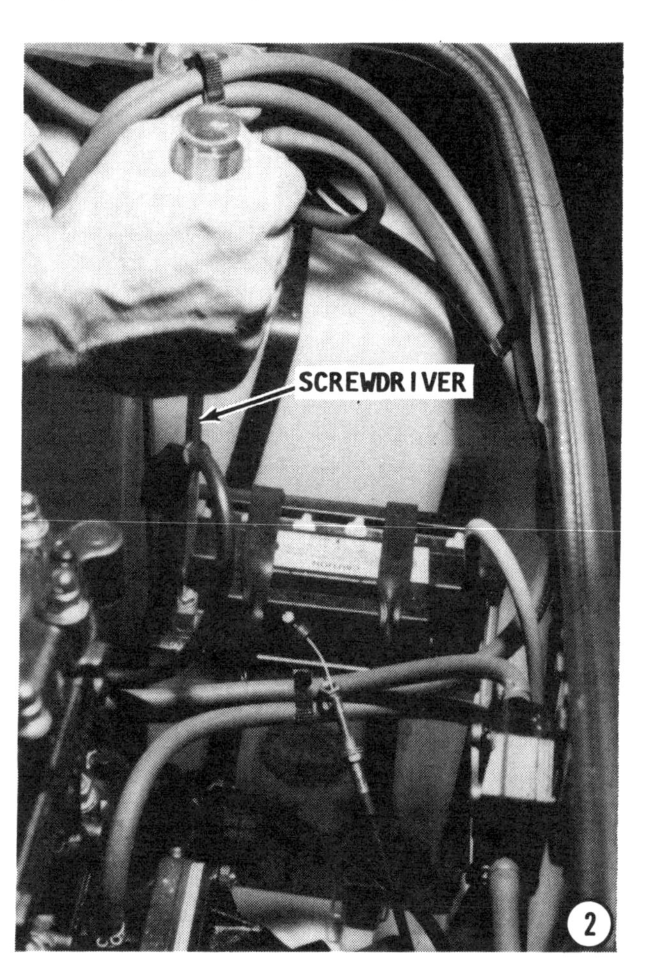

NEVER, AGAIN, NEVER operate the engine at high speed with a flush device attached. The engine, operating at high speed with such a device attached, would **RUNAWAY** from lack of a load on the impeller, causing extensive damage.

9-7 TESTING OTHER ELECTRICAL COMPONENTS

This short section provides testing procedures for other electrical parts installed on the engine. If a unit fails the testing, the faulty part **MUST** be replaced. In most cases, removal and installation is through attaching hardware.

Colored Lead Identification

Test leads are identified by their color to aid in locating the correct lead. However, the manufacturer has been known to change lead colors. Therefore, identify the component and consult the wiring diagram for the correct component connections if colors vary.

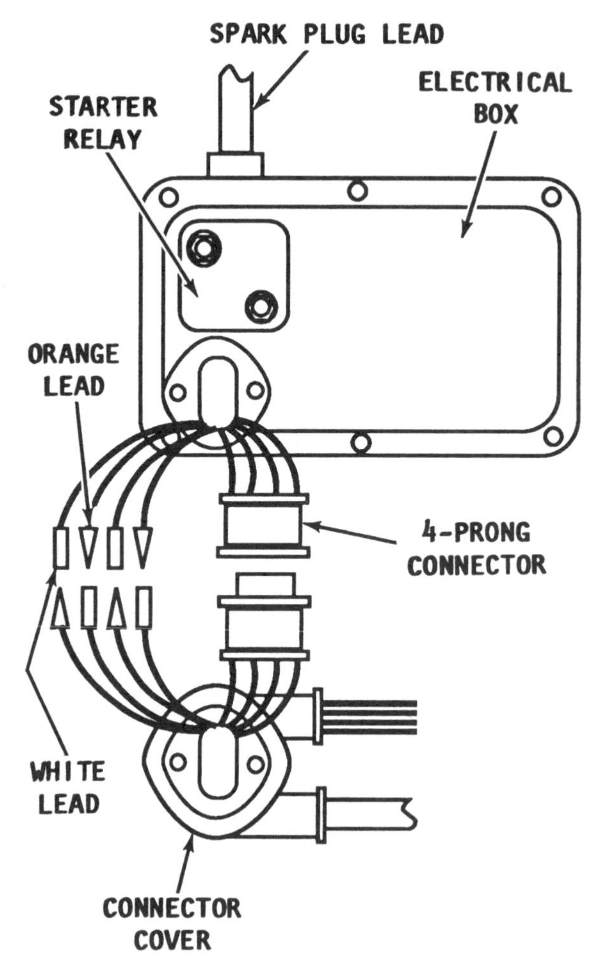

Simple line drawing of a typical electrical box to illustrate the connections for resistance tests outlined in this section.

Start Button Test

All Models

Remove the connector cover from the electrical box, see Page 9-12. Disconnect the Orange and White start switch leads at their quick-disconnect fittings.

Obtain an ohmmeter. Select the Rx1 ohm scale. Connect the ohmmeter leads across the two disconnected leads from the connector cover or from the control stick. Position the starter interlock switch (if equipped), to the far right -- the small lever under the start button, on the right.

With the start button depressed, the meter should register almost zero ohms. With the start button released, the meter should register no continuity. Both tests **MUST** be successful. If the tests are not successful, the start button **MUST** be replaced. The stop/start button assembly is a one piece sealed unit and cannot be serviced.

Starter Interlock Switch Test
(Also Known As the Safety Switch)

Using the same two disconnected leads as in the previous test, position the starter interlock switch to the far left. With the start button depressed or released, the meter should register no continuity. If the tests are not successful, the stop/start button **MUST** be replaced. The starter interlock switch is an integral part of the stop/start button assembly. The stop/start button assembly is a one piece sealed unit and cannot be serviced. Reconnect the White and Orange leads, color to color.

Stop Switch Test

Remove the connector cover from the electric case, see Page 9-12. Disconnect the Blue and Black stop switch leads at their quick disconnect fittings.

Obtain an ohmmeter. Select the Rx1 ohm scale. Connect the ohmmeter leads across the two disconnected leads from the connector cover. Position the starter interlock switch to the far right. With the stop button depressed, the meter should register almost zero ohms. With the stop button released, the ohmmeter should register no continuity.

Both tests **MUST** be successful. If the tests are not successful, the stop button **MUST** be replaced. The stop/start button assembly is a one piece sealed unit and cannot be serviced.

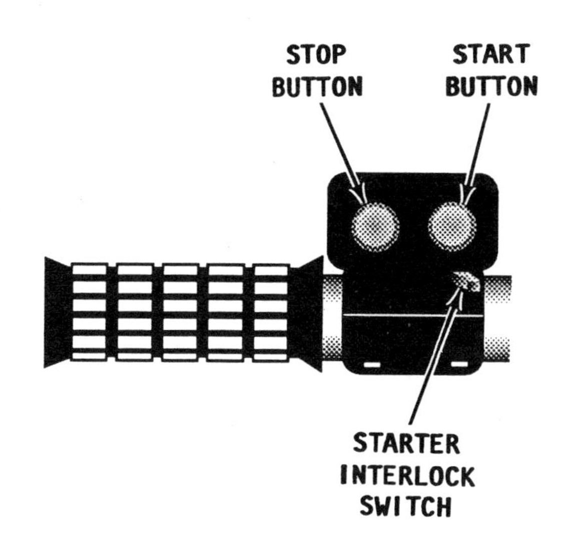

Tests for the stop, start, and the starter interlock switch are given on this page.

Stop Switch Relay Test
If Equipped

The stop switch relay is an electronic solid state device to prevent the engine from restarting after the stop switch has been released. If an engine starts up again immediately after the stop switch has been released, the stop switch relay is defective and must be replaced. A defective relay will allow the engine to restart due to the momentum of the flywheel.

On most models, the stop switch relay is located inside the electrical box and has Black, Brown, Blue, and Black/White leads. Open the electrical box, see Page 9-12. Disconnect the four leads to the stop switch relay at the quick-

POSITIVE METER LEAD CONNECTION

NEGATIVE METER LEAD CONNECTION	BLACK/ WHITE	BLUE	BLACK	BROWN
BLACK/ WHITE		SOME READING	∞ NO READING	∞ NO READING
BLUE	SOME READING		∞ NO READING	∞ NO READING
BLACK	∞ NO READING	∞ NO READING		SOME READING
BROWN	∞ NO READING	∞ NO READING	SOME READING	

Simple chart to visually indicate the stop switch relay tests, listed in the text.

disconnect fittings. Obtain an ohmmeter. Select the Rx100 ohm scale. The manufacturer does not give specific resistance values between different colored leads. There is either "no" reading, or "some" reading on this resistance scale. Observe the correct meter lead connections as shown in the accompanying table. If any of the test values are not as expected, replace the relay.

Electric Bilge Pump Test
Model 650 Series Only
(If So Equipped)

To test the electric bilge pump: Obtain a fully charged 12 volt battery and two leads. Disconnect the two pump leads -- the Black ground lead and the Brown lead from the bilge pump switch. Connect the battery across the two disconnected leads -- positive battery terminal to the Brown disconnected lead and negative battery terminal to the Black ground lead. Check to be sure adequate water is in the bilge area for the pump to function. The pump should pump out the water from the bilge, if not, replace the pump.

Blower Fan Test
Model 650 Series Only
(If So Equipped)

If the fan does not operate, the fault could lie in one or more of three areas:

Fan 5 amp fuse
Fan switch
Fan motor

To inspect the fuse, remove the fuse from the fuse holder. If the fuse is suspect, replace it with a new one.

To test the fan switch, obtain an ohmmeter and select the Rx1 ohm scale. Connect the meter leads across the Red and Yellow leads to the fan switch. With the switch in the **OFF** position, the meter should register no continuity. With the switch in the **ON** position, the meter should register almost zero ohms. Replace the switch if the results are not as expected.

To test the fan motor: Obtain a fully charged 12 volt battery and two leads. Disconnect the two fan motor leads -- the Black ground lead and the Blue lead from the fan switch. Connect the battery across the two disconnected leads -- positive battery terminal to the Blue disconnected lead and negative battery terminal to the Black ground lead. The fan should operate. If not, replace the fan.

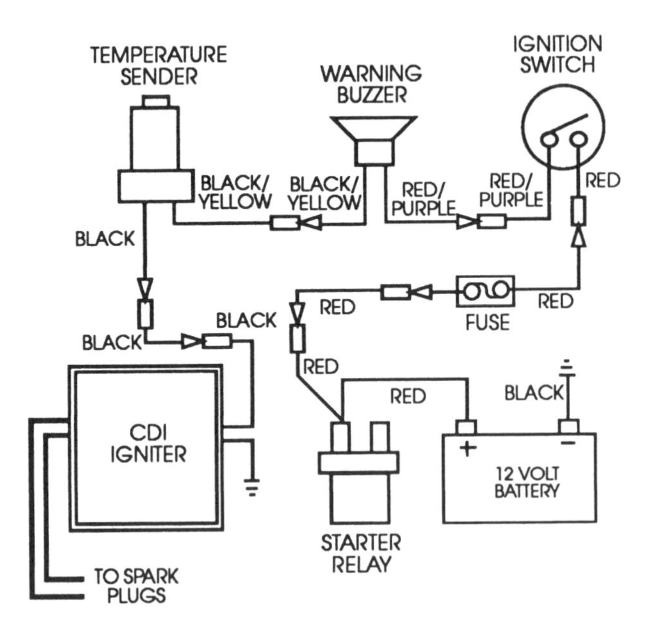

Functional diagram of the temperature warning system circuit installed on some Model 650 Series craft.

Temperature Warning System
Some Model 650 Series Only

The temperature sender is located inline with the by-pass hose to measure the temperature of the cooling water. If the cooling water temperature rises above 203°F (95°C), the sender transmits a signal to the warning buzzer to alert the operator of the craft. If the warning buzzer sounds, shut down the engine immediately to prevent damage from overheating.

To test operation of the temperature sender, disconnect the Black/Yellow and Black leads at their quick-disconnect fittings. Remove the sender from the by-pass hose. Obtain an ohm-

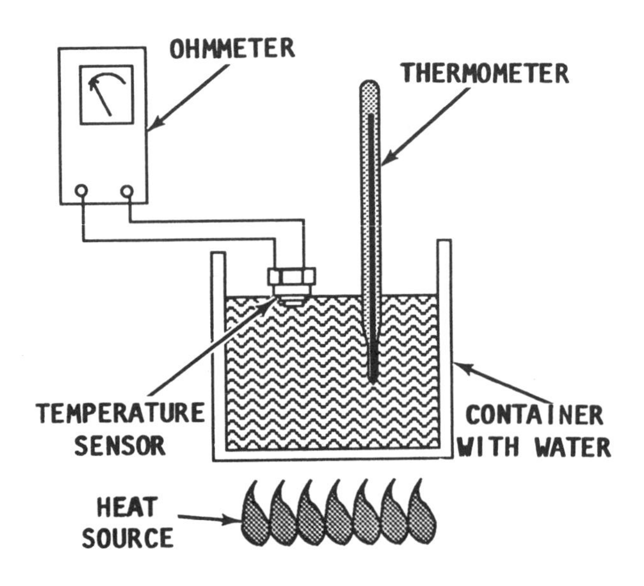

Line drawing to depict the setup for testing the operation of the temperature sensor, as outlined in the text.

meter and select the Rx1000 (Rx1K) ohm scale. Connect the meter leads across the sender leads. Heat a pan of water on the stove and have a thermometer handy to measure the water temperature. Submerge only the sensor end in the water, as shown, without the sensor touching the sides or the bottom of the pan. Observe the meter reading as the water temperature rises. Stir the water with the thermometer to evenly distribute the heat.

At temperatures above approximately 190°F (88°C), the meter should register more than 1,000,000 ohms (1,K ohms). At temperatures less than 190°F (88°C), the meter should register almost zero ohms.

Overheat Buzzer Test
Model 650 Series Only

To test the overheat buzzer: Obtain a fully charged 12 volt battery and two leads. Disconnect the Red/Purple and Black/Yellow leads from the buzzer. Connect the battery across the two disconnected leads -- either lead to either lead. The buzzer should sound, if not, replace the buzzer.

9-8 ELECTRIC TRIM SYSTEM

At press time, the publisher has been advised that replacement parts for the trim motor system may be difficult to obtain. However, this short illustrated section covers removal and opening of the trim system motor assembly.

Remove the flame arrestor "box". For complete instructions for removing the flame arrestor assembly, see Chapter 8, Section 8-1.

Turn the craft on its port side. Remove the pump cover.

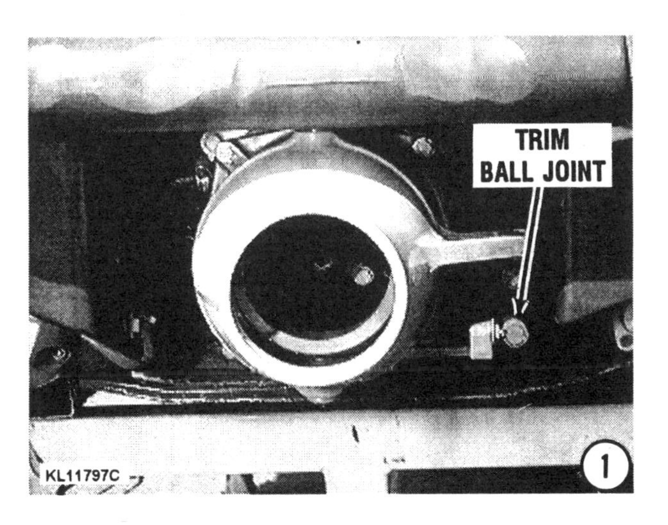

1- Disconnect the trim cable ball joint. Unscrew the cable nut -- located just forward of the jet pump -- while holding the fittings with a wrench. Record the proper sequence of which the cable attaching hardware -- O-ring, washers, etc. -- is removed from the cable and fitting. Improper assembling of this hardware may result in water entering the fittings and causing damage to the trim motor.

2- Remove the bolts securing the trim motor to the hull. Lift the trim motor housing and attached control cable from the craft and set it on a suitable work surface.

3- Carefully disconnect the hardware securing the control cable to the trim motor housing. Take note of the positioning of the keeper on the end of the housing fitting. This must be positioned properly during assembly to ensure water does not enter the trim motor housing.

4- Remove the bolts securing the upper half of the motor housing to the lower half. Pull the upper housing away from the lower half and set it aside. Inspect the housing for signs of moisture.

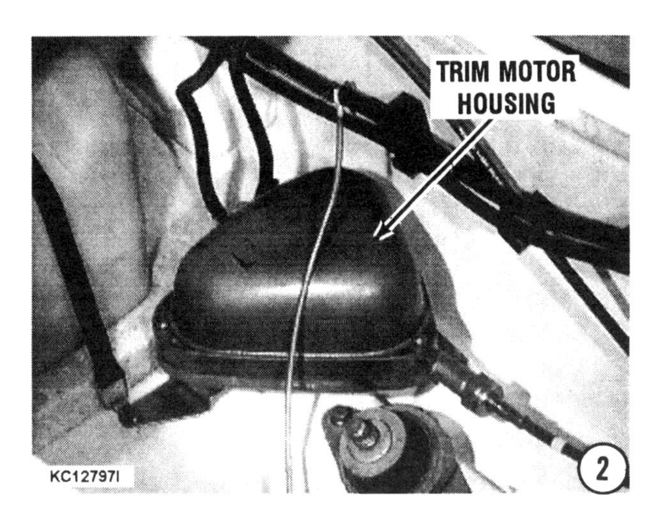

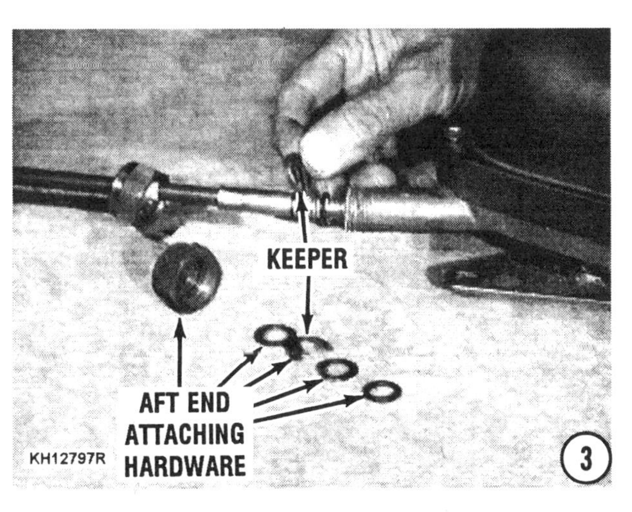

Trim Motor Assembling

Position the upper trim motor housing onto the lower half. Install and tighten the attaching bolts securely.

Install the keeper onto the control cable and motor housing fitting. Be sure it is positioned properly.

Position the trim motor assembly and attached cable in place within the engine compartment. Install and tighten the fastening bolts securely.

Obtain the attaching hardware for the aft cable connection. Install the hardware -- in the proper order -- and secure the cable nut to the aft fitting. Install the pump cover and tighten the bolts to a torque value of 69 in lb (7.8Nm). Connect the trim cable to the ball joint.

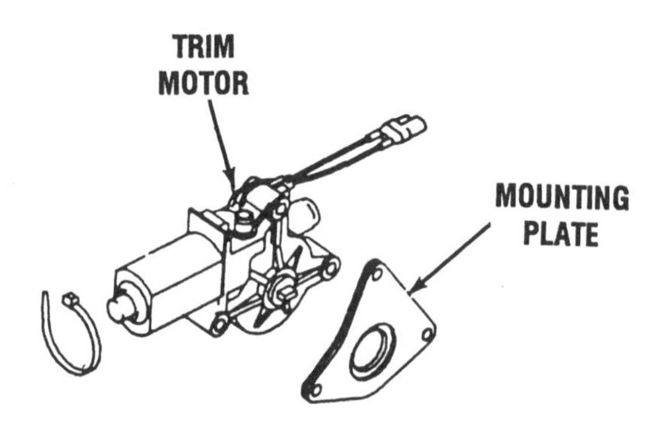

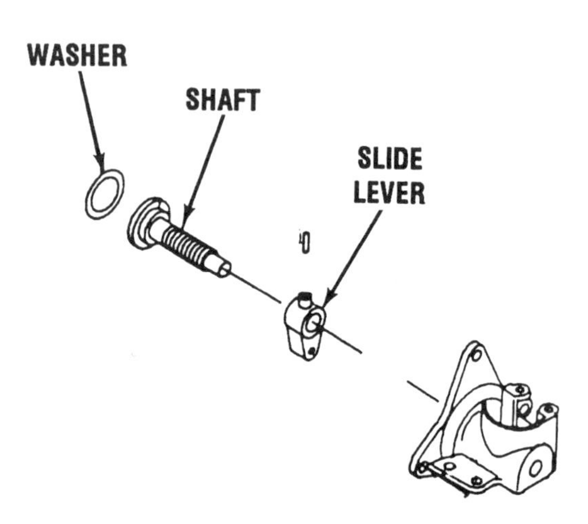

Quick simple line drawing to indicate arrangement of trim motor parts inside the trim motor housing. Bear-in-mind, replacement parts may be very difficult to obtain.

10
JET PUMP

10-1 INTRODUCTION

Model Identification and Chapter Coverage

Kawasaki personal watercraft use both axial and mixed flow jet pumps. The 550 Series models use a mixed flow pump design. All other models covered in this manual use the axial flow pump.

The axial flow pump used on all models except the 550 Series, has a two-piece impeller shaft -- actually a long drive shaft and a short impeller shaft -- each shaft typically having two ball bearings at each end. Be sure to note the number of bearings and proper positioning of oil seals during disassembly.

The engineers at Kawasaki provide impeller-to-pump case clearances for **ALL** craft equipped with axial or mixed flow jet pumps. Therefore, a special section -- 10-2 -- deals with impeller-to-pump case clearance.

Shimming procedures for the mixed pump installed on the 550 Series is covered in Section 10-5. No shimming is required on the axial flow pumps -- due to the design feature of the "carrier" bearing or bearing housing. However, it is important that impeller-to-pump case clearances **MUST** be observed.

One set of service procedures covering the axial flow pump and the mixed flow pump -- one set for each type -- is presented in Section 10-4. Where differences occur between the designs or the type flow pump, these differences, will be clearly identified.

The bearing housing supporting the driveshaft at the bulkhead is covered in Section 10-6.

Engine crankshaft-to-impeller shaft alignment is covered in Chapter 8 -- Engine -- Page 39.

SPECIAL WORDS ON JET PUMPS

Note that both the mixed flow and axial flow pumps have the same basic components.

Pump parts are **NOT** interchangeable, because many of the castings are of unique design even though they share the same function and are similar in appearance. The main difference between the two pumps, regarding maintenance, concerns positioning the impeller on the impeller shaft.

The mixed flow pump is equipped with shim material to position the impeller at a specific distance from the pump case.

The axial flow pump does not have shim material to position the impeller.

Neither the axial or the mixed flow pump has a removable liner, just a solid aluminum casting. Therefore, if the impeller-to-pump case clearance becomes exces-

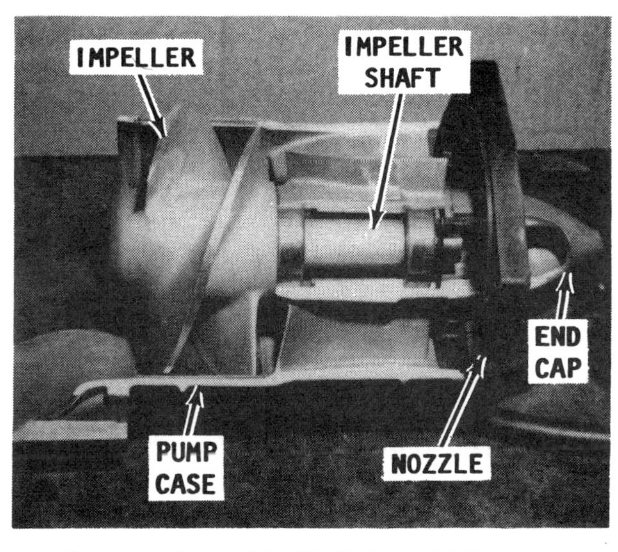

Cutaway of a Model 650 Series axial flow jet pump, with a few major parts identified.

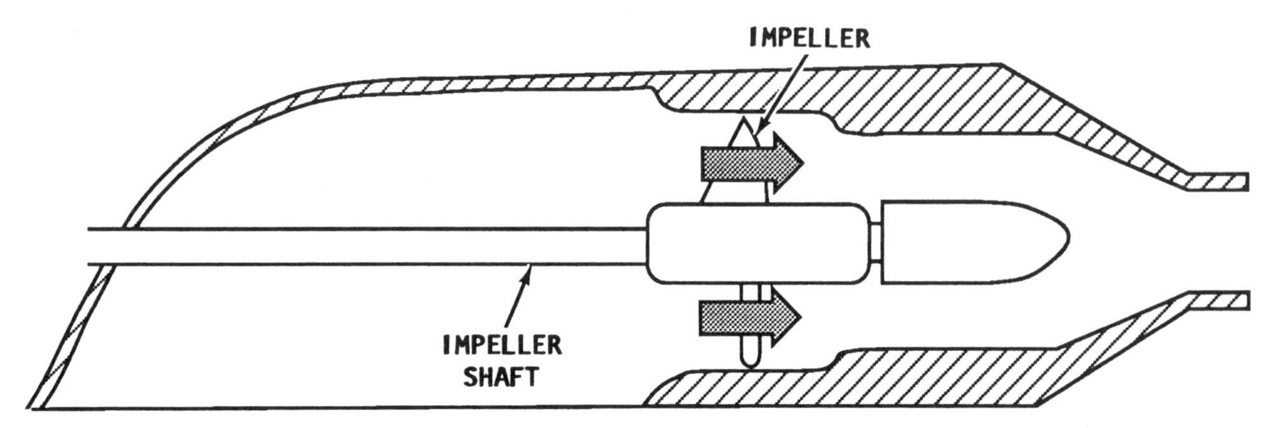

Simple cross-section line drawing to depict water flow through an axial flow jet pump. The water passes through parallel to the axis of the impeller rotation.

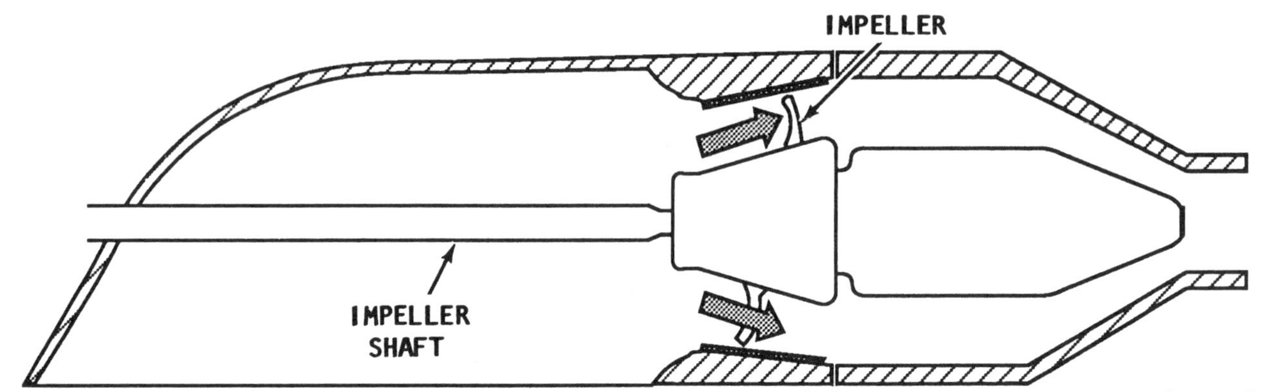

Line drawing to illustrate water flow through a mixed flow jet pump. Notice how the water is directed upward at an angle to the axis of impeller rotation.

sive, and cannot be corrected by the addition of shim material -- on the mixed flow pump -- or the pump case is deeply grooved, the case **MUST** be replaced.

JET PUMP DESCRIPTION

Axial Flow

Water in an axial pump moves on a single axis, as depicted in the adjacent illustration, thus the word "axial". In simple terms, water is ingested and discharged parallel to the axis of impeller rotation. The inside of the pump case is parallel to the water flow.

Mixed Flow

Many experts consider the mixed flow system to be the most efficient for personal watercraft application. The water flow is discharged at an inclined angle to the axis of impeller rotation, as indicated by the arrows in the accompanying illustration. The impeller blades are inclined in such a manner to cause the water to be thrown off at a 45 degree angle to the impeller shaft within the pump case.

From the pump case, the water then flows through the guide vane casting and forced out through the outlet nozzle and steering nozzle. The vanes in the guide vane casting help straighten the water flow through the nozzle. The inside of the pump case is inclined at an angle to the water flow. Therefore, this design pump incorporates shim material to space the impeller blades at a specified distance from the pump case. Mixed flow pump efficiency may decrease as much as 30% if the impeller-to-pump case clearance increases by only 0.04" (1mm).

Axial or Mixed?

Axial and mixed flow pumps are intended for different applications. One is not necessarily better than the other. The axial pump can be operated at higher engine rpm and is more suitable for racing. A mixed flow pump adds thrust at low engine rpm due to the pump design. Therefore, the mixed flow pump performs more efficiently at low engine rpm. This type pump is more suitable for a craft equipped to carry extra weight or intended to pull a water skier.

A jet pump seldom requires service except for the following:

Impeller-to-pump case clearance becomes excessive.
Impeller is damaged.
The oil seal around the impeller shaft fails -- allowing water to enter hull.
Silicone seal around the perimeter of the pump case allows the pump to ingest air.

10-2 IMPELLERS

Impellers provide thrust using a combination of water flow and water pressure in a closed environment.

The blades of an impeller overlap, thus water is trapped and forced through the impeller into the pump case and outlet nozzle. The water moving past the impeller resists cavitation because it is under pressure. Blades in the pump case are angled exactly opposite to those of the impeller. These blades redirect the water to establish a concentrated flow through the nozzle.

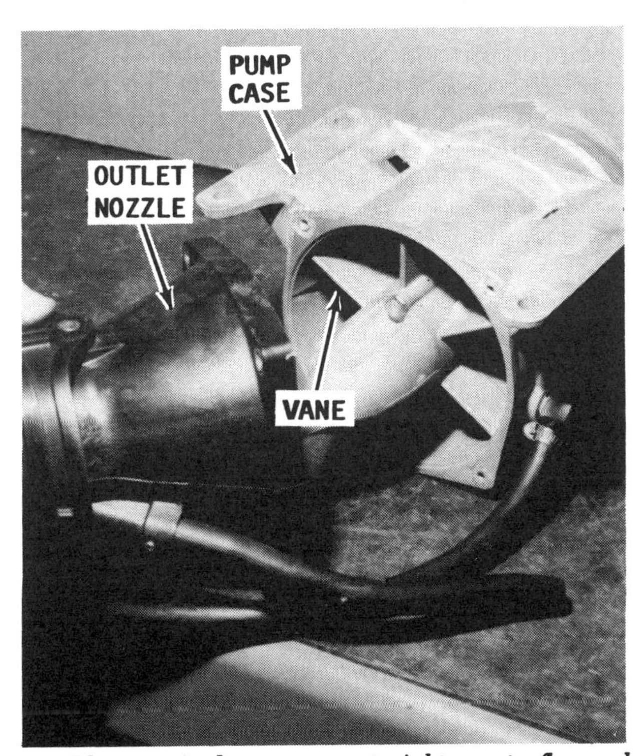

The vanes in the pump case straighten water flow and the conical shape of the outlet nozzle increases flow velocity.

The manufacturer recommends a radius of about 1/64" (0.3-0.5mm) for the leading edge of the impeller blade. A sharper radius will result in cavitation and a greater radius will reduce pump efficiency.

Cavitation Burns

Cavitation burns are the worst enemy of an impeller. These burns literally "eat away" material from the impeller, leaving holes and weakening the impeller structure. In extreme cases entire blades have been known to "depart" from the impeller hub because the cavitation burns at the base of the blade were so severe.

Cavitation burns are the result of imperfections (damage) on the impeller blades or air mixed with the water flow.

Cavitation burns may be caused by:

Wave jumping - air is sucked into the pump as the craft leaves the surface of the water.

Worn seals between the impeller and intake housings -- allowing air to enter.

Leading edge of the impeller is damaged.

To explain exactly what causes cavitation burns, first, let us examine the last cause. If the impeller leading edge is damaged and "mushrooms" over the blade, a low pressure area will form under the mushroomed lip.

Under atmospheric pressure -- at sea level -- water boils at 212°F. Anytime there is an imperfection on the surface of an impeller blade and water passes over that imperfection, the water pressure is lowered. When the pressure is lowered, water will boil at a much lower temperature. Therefore, air bubbles will form in the boiling water under the mushroomed lip. The bubbles will then creep down the surface of the blade and accumulate at the impeller hub. A high pressure area is formed at the base of the blade. Here, the air bubbles will collapse and reform back into water with a release of energy. This energy is absorbed by the impeller and the end result is impeller material is "eaten away".

The impeller should be dressed to a slightly rounded edge. It is impossible to

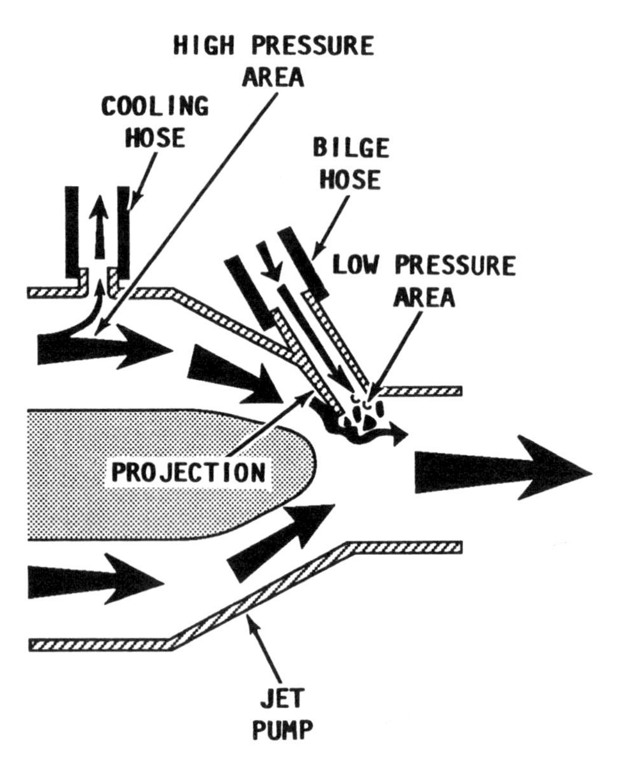

The direction of water flow through the cooling hose and the bilge hose is opposite. The cooling hose routes water from the pump to the engine. Water is "vacuumed" from the bilge through the bilge hose to the pump where it is forced out through the nozzle.

achieve a perfectly straight sharp edge, as one side will always be convex and the other side concave. The concave side will form an area of low pressure and encourage cavitation as described above.

The other two causes of cavitation -- defective seals and wave jumping -- occurs when air is sucked in with the water flow. The same principles apply, as just described, in the previous paragraphs. Air bubbles will creep down the surface of the blade and accumulate at the impeller hub. A high pressure area is formed at the base of the blade. Here, the air bubbles will collapse and reform back into water with a release of energy. As explained earlier, this energy is absorbed by the impeller and the end result -- impeller material is "eaten away".

Cooling Water and Bilge Hoses

On most pumps, two hoses are attached to the jet pump housing. One hose channels some of the water flowing through the pump to the exhaust manifold -- the hottest part of the engine -- to cool the block during operation. The other hose siphons water from the bottom of the hull, and is referred to as the bilge system. The cooling hose has water flowing **FROM** the pump. The bilge hose has water flowing **TO** the pump. Both hoses attach to the pump housing in a similar manner.

What determines the direction of water flow?

The answer to this question is in a simple explanation of high and low pressure areas along the inside surface of the housing.

Close inspection of the area inside the housing at the cooling water hose fitting reveals a smooth rounded shoulder with no obstruction or obstacle to impede the flow of water. A high pressure area develops here and draws the water down the hose attached to the fitting.

Further inspection of the area inside the housing at the bilge hose fitting reveals a small protrusion around the opening. This protrusion causes disturbance to the water flow and creates an area of low pressure. This low pressure area will have the effect of emptying air and water from the hose aft with the impeller water flow -- similar to the action of a vacuum cleaner, as depicted in the accompanying illustration. If the other end of the hose is submerged in water inside the hull, the water will be "vacuumed" out. In this manner, the bilge system drains the bilge.

Many an owner -- with good intentions -- thinking his action would smooth water flow and increase pump performance -- has filed the protrusion described in the previous paragraph smooth with the surrounding area of the housing. During operation of the water craft, the bilge system worked in **REVERSE.** Water was pumped from the impeller hous-

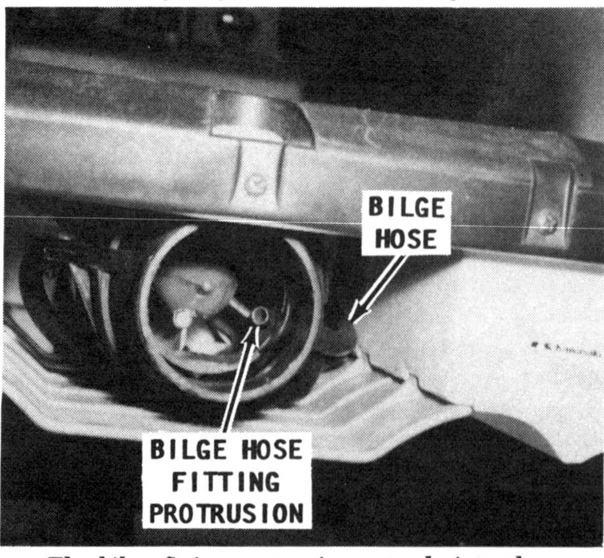

The bilge fitting protrusion extends into the water flow area, causing a low pressure area and a "vacuuming" effect.

ing into the bilge area, quickly filling the bilge and possibly sinking the craft before the rider had time to realize what was happening.

10-3 IMPELLER-TO-PUMP CASE CLEARANCE — ALL AXIAL AND MIXED FLOW JET PUMPS

SPECIAL WORDS

The following procedure may be performed with the watercraft elevated on saw horses enabling a person to work underneath, or with the watercraft on its side allowing access to the rock grate and impeller.

If the impeller-to-pump case clearance is to be determined with the watercraft raised on saw horses, the battery may be left secured in the craft.

If the watercraft is to be positioned on its side, first disconnect the electrical leads at both battery terminals, and then remove the battery from the watercraft.

Remove the three bolts and washers securing the rock grate to the jet pump. Insert a feeler gauge between each impeller blade and the pump case and determine the clearance. Make a note of all three clearances, one for each blade. Calculate the average of the three clearances.

On a new craft the clearance will be what the manufacturer refers to as the "standard" clearance -- the first column. The second column is the "service limit". This value is the maximum allowable clearance.

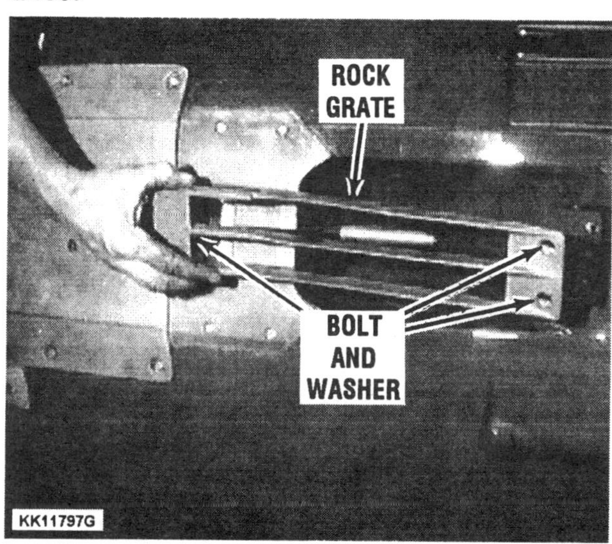

Removing the rock grate to gain access to the jet pump in order to measure clearance between the impeller and the pump case, as explained in the text.

IMPELLER-TO-PUMP CASE CLEARANCE

Series	Standard Clearance	Service Limit
550	0.0079-0.0118" (0.20-0.30mm)	0.0350" (0.60mm)
650	0.0079-0.0118" (0.20-0.30mm)	0.0236" (0.60mm)
750	0.0079-0.0118" (0.20-0.30mm)	0.0236" (0.60mm)
750H.P., 900, 1100	0.0059-0.0118" (0.15-0.30mm)	0.0236" (0.60mm)

Note: H.P. - High Performance 750 Models

Axial Flow Pump
All Models Except 550 Series

If the clearance is less then the maximum value, no action is required.

If the clearance is more than the service limit, inspect the condition of the pump case. If the pump case has scratches deeper than 0.04" (1mm), replace the pump case. If the pump case is satisfactory, the problem must be with the impeller. Visually inspect the impeller for nicks, scratches, pitting or a "mushroomed" edge. If the cause of the excessive clearance cannot be determined visually, the pump must be disassembled and the pump case measured and compared with specifications.

The jet pump must be removed from the craft to perform this work.

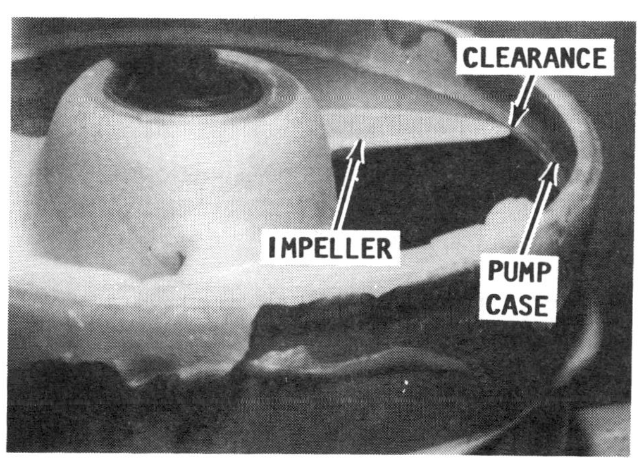

On both axial and mixed flow Kawasaki jet pumps, a feeler gauge is inserted between the impeller blade and the pump case to measure clearance.

Mixed Flow Pump
550 Series

If the clearance is less then the maximum value, no adjustment is required.

If the clearance is greater than the maximum value, and the pump case is in good condition, the impeller clearance must be adjusted. Shim material must be added between the pump case and the impeller. The addition of 0.04" (1mm) of shim material will decrease the impeller clearance by 0.005" (0.144mm).

Mixed flow pump efficiency decreases by as much as 30% when the blade to pump case clearance increases by only 0.04" (1mm).

If the clearance is greater than the maximum value, but the pump case has scratches deeper than 0.04" (1mm), replace the pump case.

The jet pump must be removed from the craft to add shim material behind the impeller. Procedures for pump removal are presented in the next section -- 10-4. The impeller-to-pump case clearance is **NOT** used to determine the total amount or the additional amount of shim material used behind the impeller. The thickness of shim material used depends entirely on the height of the end cap shoulder and the depth to which the bearings seat inside the guide vane casting.

A sample calculation and a bearing shim material chart are provided in the assembling procedures to correctly space the impeller on the pump shaft for the specified impeller-to-pump case clearance.

10-4 JET PUMP SERVICE

SPECIAL WORDS ON ACCOMPANYING ILLUSTRATIONS

The following series of illustrations were taken during service of an axial flow pump. The mixed flow pump differs slightly in appearance, and has some additional components. Where differences occur, these differences have been clearly identified.

SPECIAL WORDS ON SPECIAL TOOLS

The following list of special tools are required in order to perform some of the necessary work on the jet pump. Some tasks become extremely difficult, if not impossible, without the special tools or some type of equivalent. Manufacturers' part numbers are given. In specific steps of the procedures -- whenever possible -- a special effort has been made to give some type of equivalent or homemade tool which may be substituted. However, the cost of a special tool may be offset by money saved if an expensive part is not damaged.

SPECIAL TOOLS FOR JET PUMP SERVICE

550 Series
Mixed Flow Pump

Oil Seal and Bearing Remover	P/N 57001-1058
Bearing Driver Set	P/N 57001-1229

650 Series

Impeller Wrench	P/N 57001-1228
Oil Seal and Bearing Remover	P/N 57001-1058
Bearing Driver Set	P/N 57001-1129

750 Series
Standard Models

Impeller Wrench	P/N 57001-1228
Oil Seal and Bearing Remover	P/N 57001-1058
Bearing Driver Set	P/N 57001-1129

900 Series, 1100 Series and 750 Series
High Perf. Models

Impeller Wrench	P/N 57001-1228
Impeller Holder	P/N 57001-1393
Oil and Bearing Remover	P/N 57001-1058
Bearing and Driver Set	P/N 57001-1129

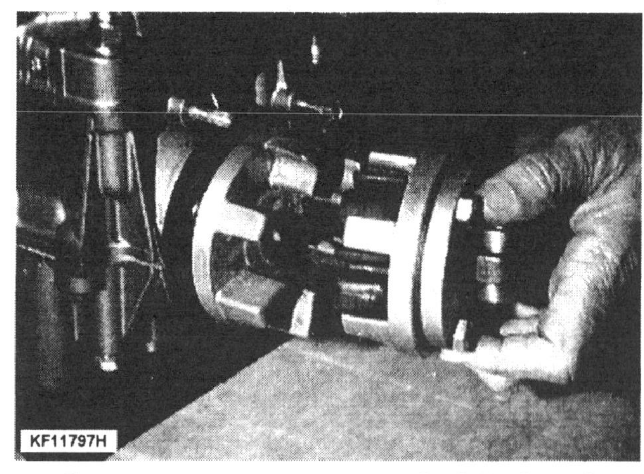

Servicing a jet pump is extremely difficult -- almost impossible without the use of some special tools. Such tools are called out by number in the text.

REMOVAL

Disconnect both battery cables from the battery and remove the battery from the craft.

CAUTION

The battery **MUST** be removed to avoid spilling electrolyte into the bilge through the battery vent holes when the craft is tilted to one side.

Rotate the fuel control valve, if equipped, to the **"OFF"** position. Drain or remove the fuel tank. If working on a model other than the 550 Series engine, drain the oil from the oil injection tank.

Place an old blanket or equivalent on the workshop floor to cushion the edges of the craft and prevent the craft from sliding while the work progresses. Rotate the craft on the blanket to expose the rock grate, and then block the hull to hold the desired position.

GOOD WORDS

Before starting service work the impeller-to-pump case clearance should be checked, as described in the previous section -- 10-2.

1- If the clearance is checked, the rock grate will have been removed. If the clearance was not checked -- remove the three bolts and washers securing the grate to the hull.

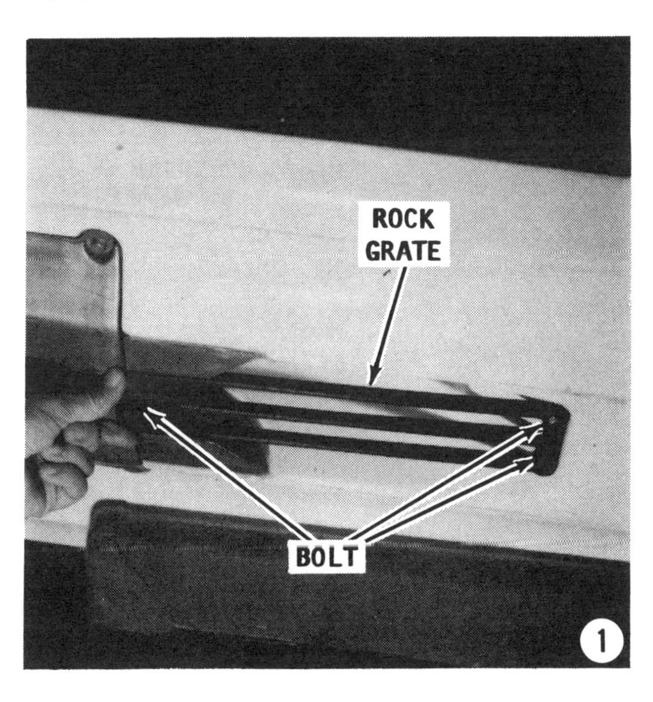

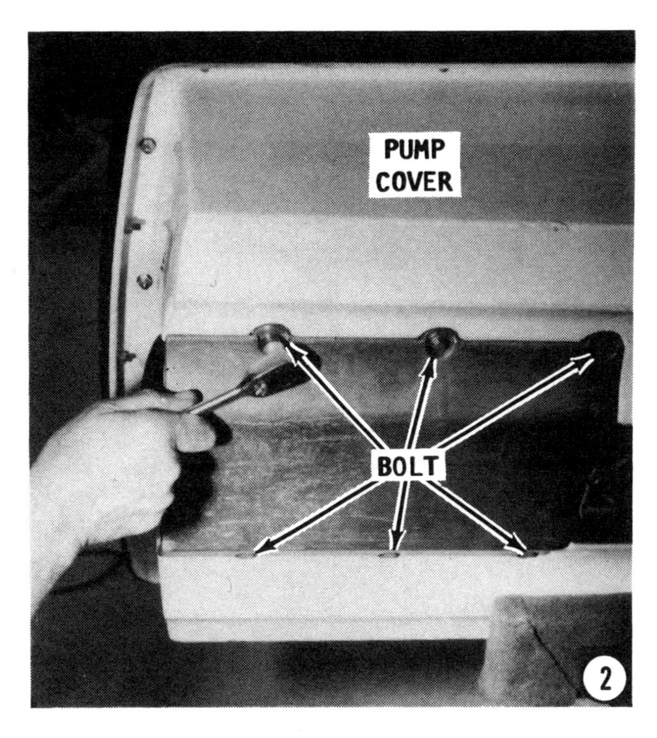

On some recent "high performance" models, a separate "ride plate" is located forward of the jet pump cover. It is not necessary to remove this plate to gain access to the jet pump.

2- Remove the bolts and washers securing the cover over the jet pump.

550 Series	Four bolts
650 Series	Four bolts
750 Standard Series	Six bolts
750H.P. & 900 Series	Eight bolts
1100 Series	Eight bolts

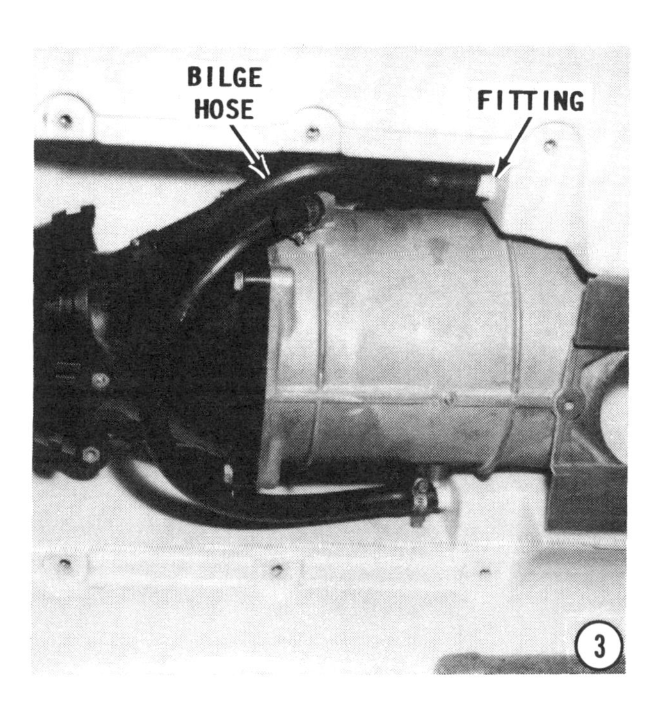

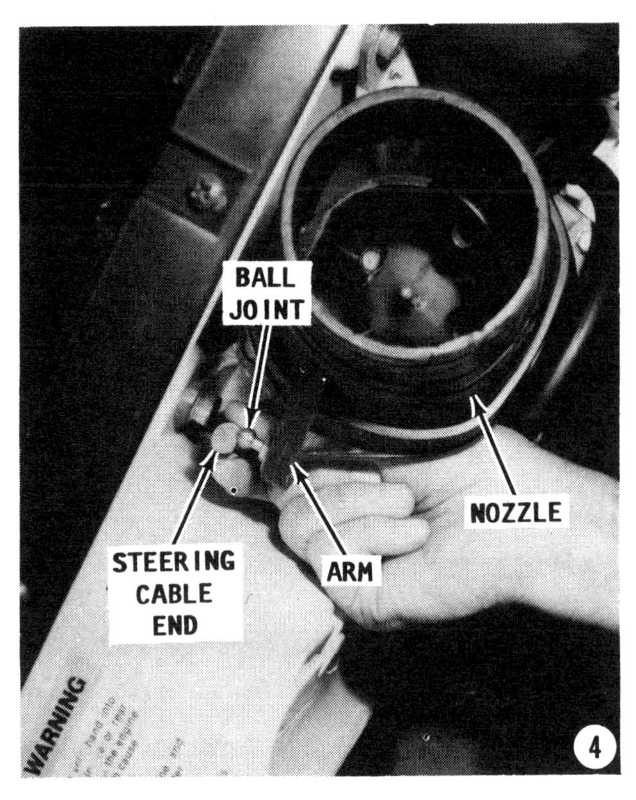

3- Loosen the hose clamp around the bilge hose fitting on the craft hull. Work the hose free of the fitting.

4- Pull back the steering cable end and disengage the cable end from the ball joint on the steering nozzle arm.

5- Loosen the hose clamp around the cooling hose fitting on the pump case. Work the hose free of the fitting.

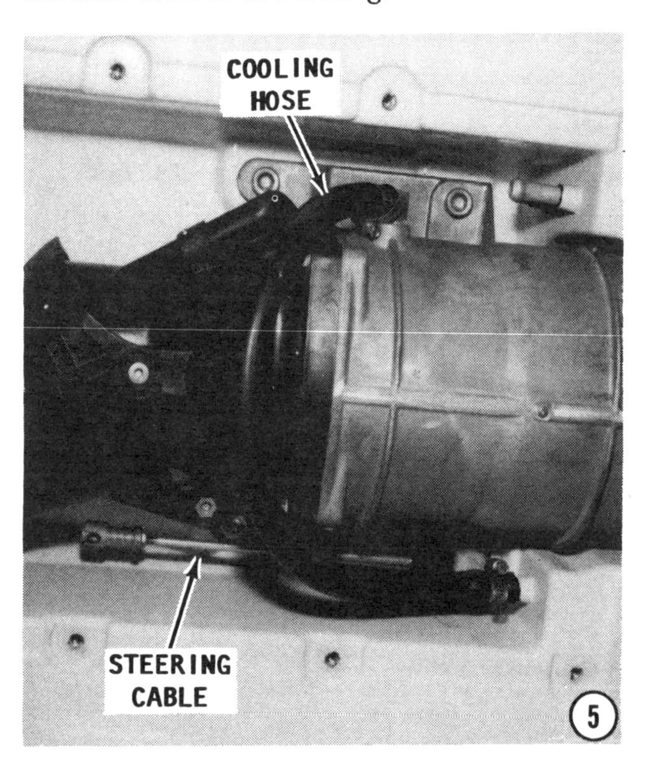

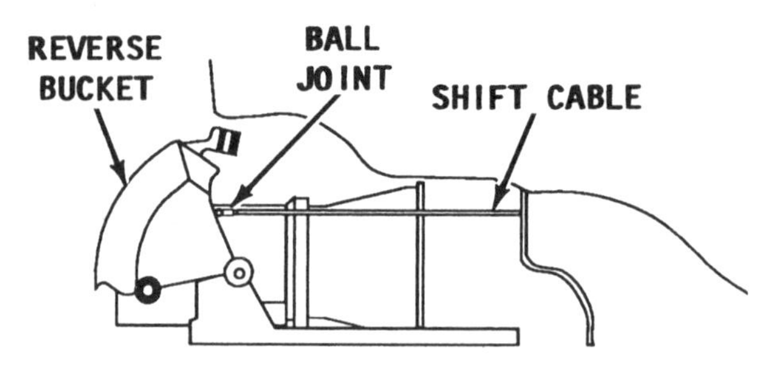

Simple line drawing to show the relationship of the reverse "bucket", ball joint, and the shift cable on a craft equipped with reverse capability.

SPECIAL WORDS
REVERSE BUCKET SYSTEM

If equipped with a reverse gate mechanism, pull back the reverse cable end and disengage the cable from the ball joint on the reverse bucket.

All Models

The jet pump is now "disconnected" and ready to be pulled from the craft.

6- Obtain a razor knife and cut into the silicone seal around the perimeter of the pump case casting. This is not an easy task, especially if working with the sealant installed at the factory. The entire perimeter must be cut several times -- many times -- before the pump will come free of the hull.

7- Using the correct size allen wrench, remove the four allen head bolts securing the jet pump to the craft hull.

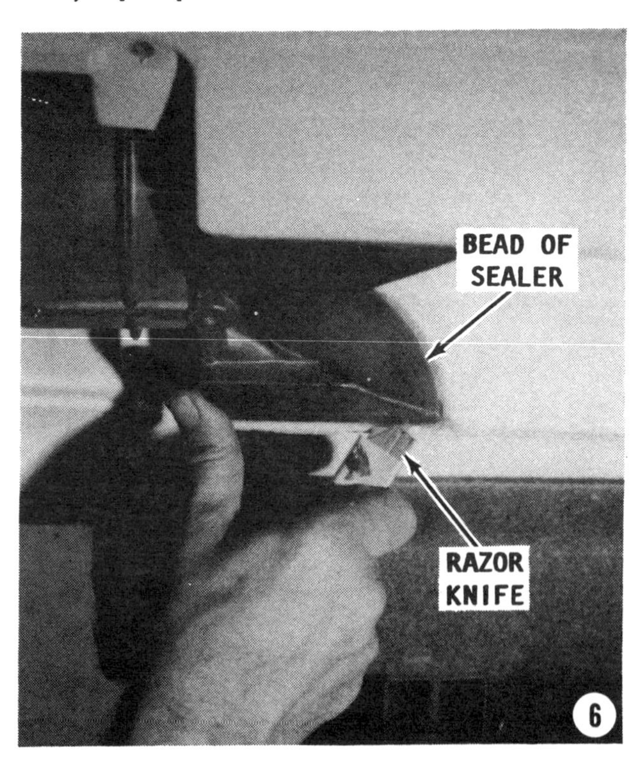

SPECIAL WORDS ON SHIM MATERIAL AND CRANKSHAFT-TO-IMPELLER SHAFT ALIGNMENT

Observe the location and thickness of any shim material placed between the pump case ears and the craft hull. Be sure to identify each of the four stacks to **ENSURE** each stack will be installed back in its original location.

Unless the coupler showed signs of wear, the impeller shaft was properly aligned with the crankshaft. Therefore, it is crucial to install the four stacks of shim material to keep the same alignment.

If the coupler shows signs of wear, as evidenced by the edges of the "petals" being worn away by the fingers of the metal coupler, the impeller shaft was misaligned with the crankshaft. Care must be exercised to achieve proper alignment when a new coupler is installed.

The axial flow pump installed on all models except the 550 Series, has a long driveshaft with external splines at one end and a metal coupler at the other end. The coupler "links" the engine crankshaft with the impeller shaft.

The mixed flow pump installed on the 550 Series models also has a long driveshaft. However, on this series, the shaft is referred to as the impeller shaft, because it is the only shaft in the pump.

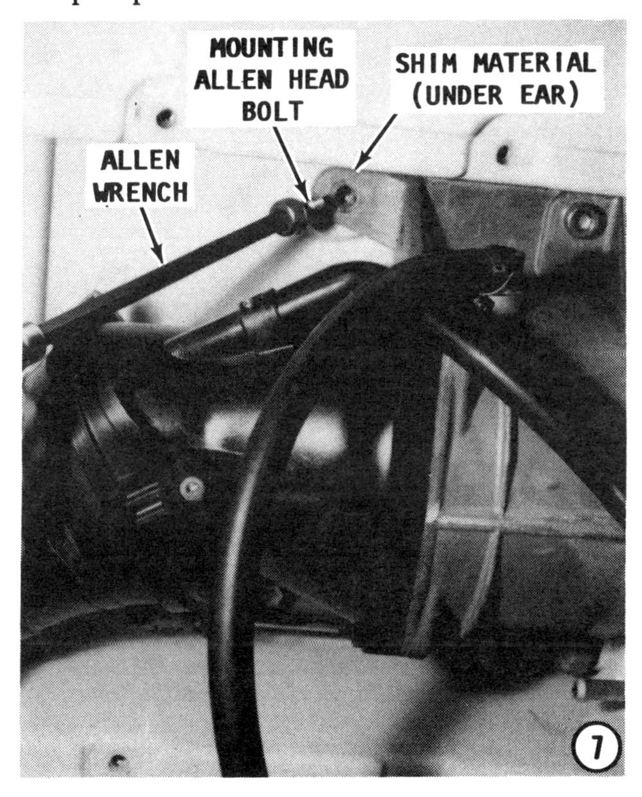

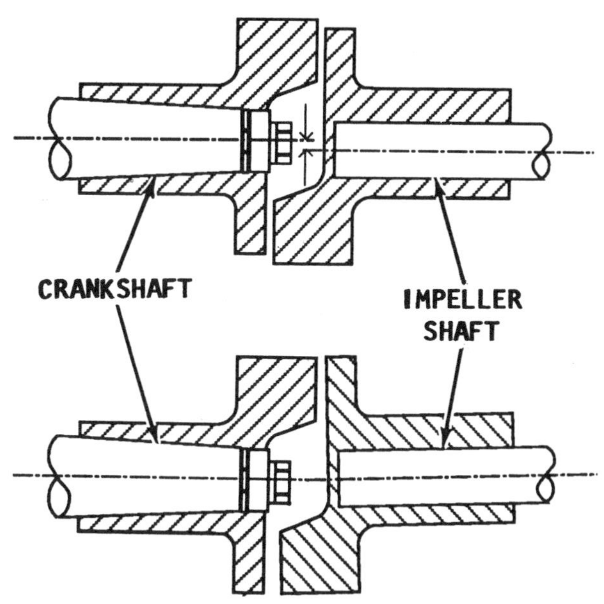

Two types of crankshaft/impeller shaft misalignment which will contribute to excessive rubber coupler wear.

series, the shaft is refered to as the impeller shaft, because it is the only shaft in the pump.

The external splines of the long shaft mesh with the internal splines inside the forward half of the impeller. The metal coupler indexes into a rubber coupler. The crankshaft is also equipped with a metal coupler which indexes into the forward face of the rubber coupler. Therefore, any flexing -- due to misalignment of the drive line -- takes place at the rubber coupler --and is evidenced by wear described earlier in this step.

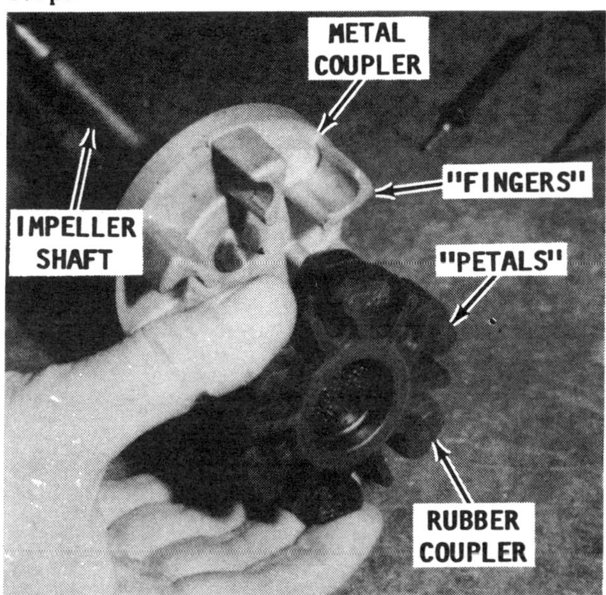

A used rubber coupler still in good condition because the engine crankshaft was properly aligned with the jet pump impeller shaft.

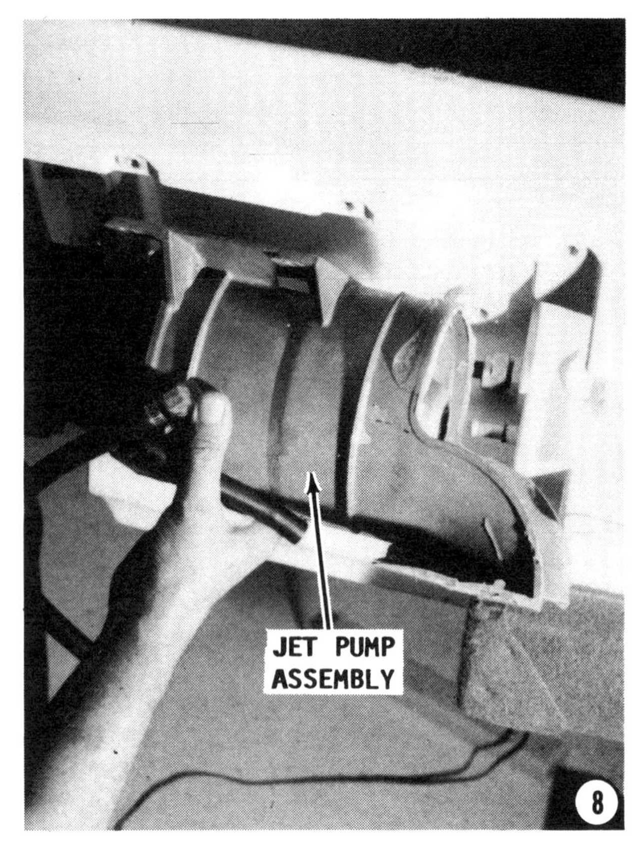

8- Slide the jet pump aft off the impeller/driveshaft and then ease the pump out and clear of the craft.

9- If the engine has been removed: Pull the driveshaft (all models except 550 Series), or the impeller shaft (550 series only), straight out of the bearing housing. Remove the rubber coupler from the metal coupler on the shaft end.

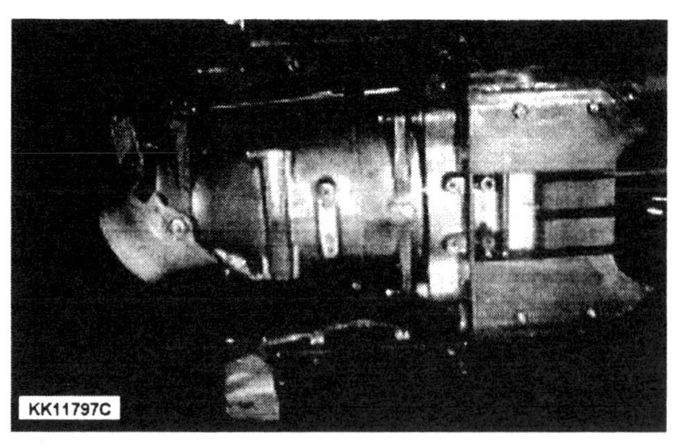

Jet pump on a Model 900 Series watercraft ready for removal and service work.

The work continues with the jet pump on a suitable work surface.

GOOD WORDS ON THE BEARING HOUSING

If the only work to be performed is to replace or service the bearing housing, the housing is now freely accessible with the pump moved out of the way. If no work is to be performed on the jet pump, proceed to Section 10-4 to service the bearing housing.

DISASSEMBLING

1- Remove the four bolts securing the outlet nozzle and the steering nozzle to the pump case.

SPECIAL WORDS 550 SERIES MIXED FLOW PUMP

These four bolts are very long. They not only secure the outlet and steering nozzles

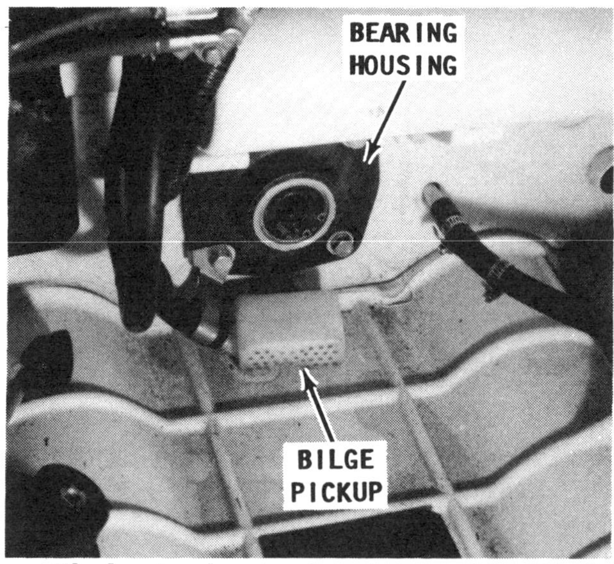

The bearing housing is only accessible after the engine and jet pump have been removed. Detailed illustrated procedures to service the bearing housing are presented in Section 10-6, beginning on Page 10-31.

to the pump case, but also pass though the guide vane casting, between the outlet nozzle and the pump case.

Look for and **SAVE** the single locating pin at the contact surfaces of the pump case and guide vane casting. Another pin is located between the guide vane casting and the outlet nozzle. These pins may be left on either surface when the pump is split.

2- Separate the outlet nozzle -- with steering nozzle attached -- from the pump case, or guide vane casting (550 Series only). The two surfaces are probably stuck together, because they should have been assembled with a sealing agent. Tap lightly all around the perimeter of the outlet nozzle with a soft head mallet. **DO NOT PRY** between the two surfaces. If a deep enough gouge is made, air will be injested into the pump and result in a substantial loss of performance.

Clean all traces of sealant from both contact surfaces.

All Models
Except 550 Series

3- Remove the two bolts and washers securing the end cap to the pump case and remove the cap. If a replacement is available, remove and discard the O-ring around the cap. Remove the old lubricant and wipe the cap clean.

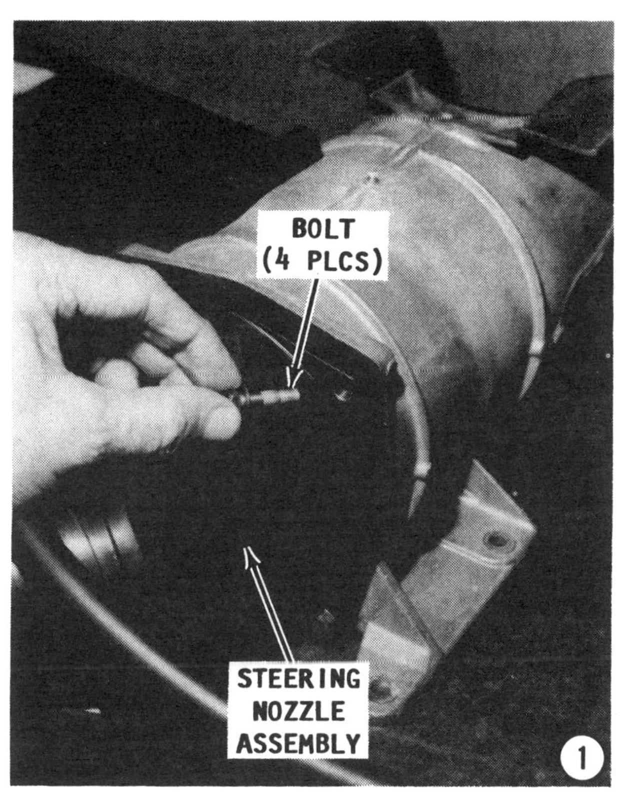

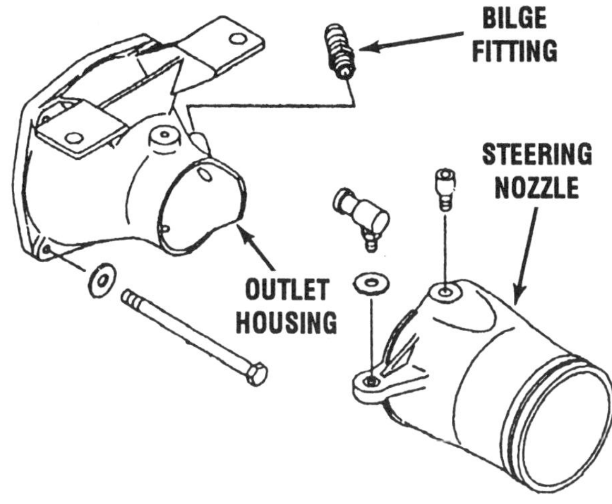

Exploded line drawing of the outlet housing and steering nozzle with associated parts identified.

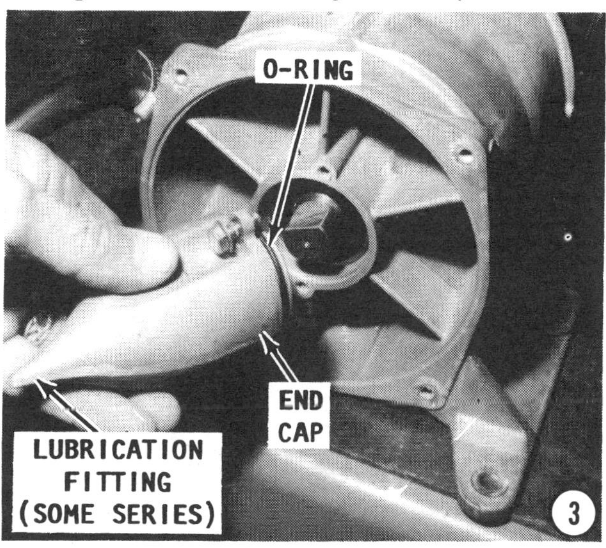

550 Series
Mixed Flow Pump Only

Using the correct size allen wrench, remove the lubrication fitting cover on the end cap. If a replacement is available, remove and discard the O-ring around the cap.

Remove the three bolts and washers securing the end cap to the guide vane casting and remove the cap. Remove and discard the O-ring around the cap, if a replacement is available. Remove the White plastic plug over the end cap. Remove and **SAVE** the pieces of shim material between the plastic plug and the bearing on the impeller shaft. The pieces of shim material found in this location are a "constant" thickness and do not affect the impeller-to-pump case clearance. Be sure to save them and keep them **SEPARATE** from the shim material located behind the impeller.

Impeller Removal
All Models
Except 550 Series

4- Clean the slotted end of the impeller shaft of any lubricant remaining after the end cap was removed. Clamp this end of the impeller shaft into a vise equipped with soft jaws. Insert the special impeller removal tool into the internal splines inside the impeller. With the correct size wrench on the tool, loosen the impeller from the threads on the impeller shaft.

5- Remove the impeller from the impeller shaft. If the oil seal at the forward face of the impeller needs to be replaced, pry it out with a screwdriver.

Impeller Shaft Removal

6- Pull out the impeller shaft from the rear of the pump case. If replacements are available, remove and discard the two O-rings around the shaft.

Impeller Removal
550 Series
Mixed Flow Pump Only

Clean the nut cast into the impeller of any lubricant remaining after the end cap was removed. Position the correct size open end wrench over the impeller nut. Insert the special impeller removal tool into the internal splines inside the impeller. Loosen the impeller from the threads on the impeller shaft.

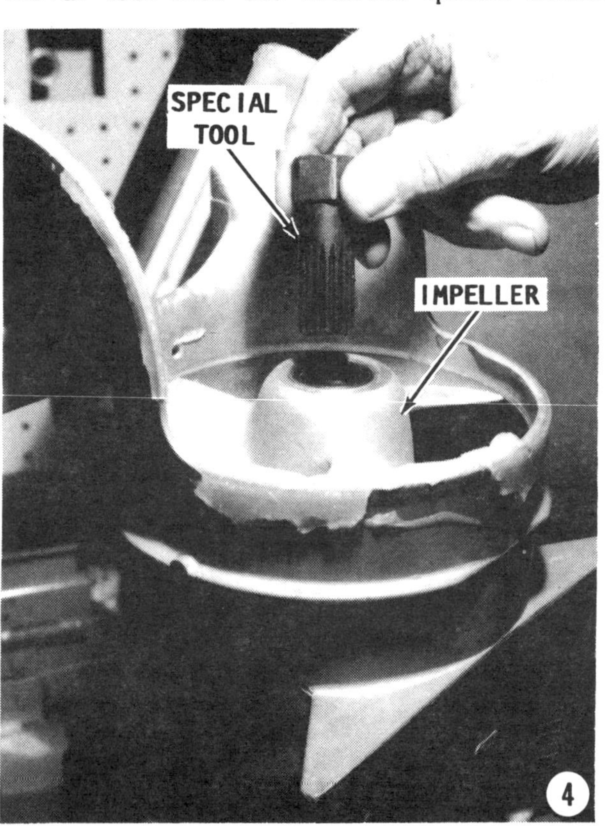

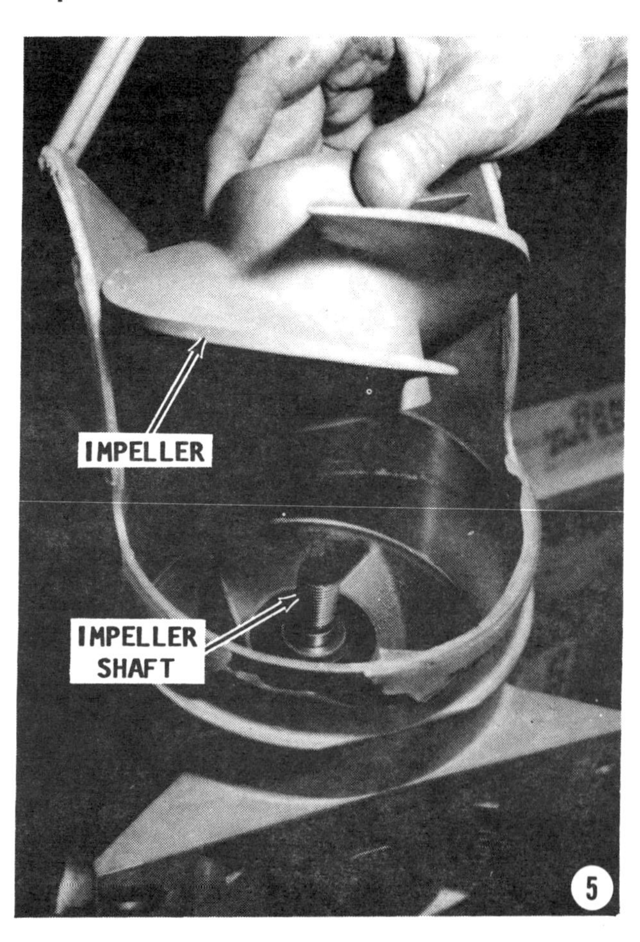

SPECIAL WORDS
ALL Models

All of the impellers covered in this manual have right-hand threads -- both mixed flow jet pumps and axial flow jet pumps. The manufacturer incorporated this into all models after 1991.

SPECIAL WORDS
550 SERIES ONLY

These mixed flow pumps have pieces of shim material between the impeller and the oil seal in the guide vane casting. **SAVE** the shim material for later installation. The thickness of shim material is critical to properly position the impeller with a specific clearance from the pump case. If the clearance was correct no adjustment is required and the original thickness of shim material may be installed. If the clearance was excessive, the thickness of shim material will need to be increased.

Remove the impeller from the impeller shaft. Pull out the impeller shaft from the guide vane casting. If a replacement is available, remove and discard the small **O**-ring at the threaded end of the shaft.

All Models
Except 550 Series

7- Lift out the short bushing from the forward end of the pump case.

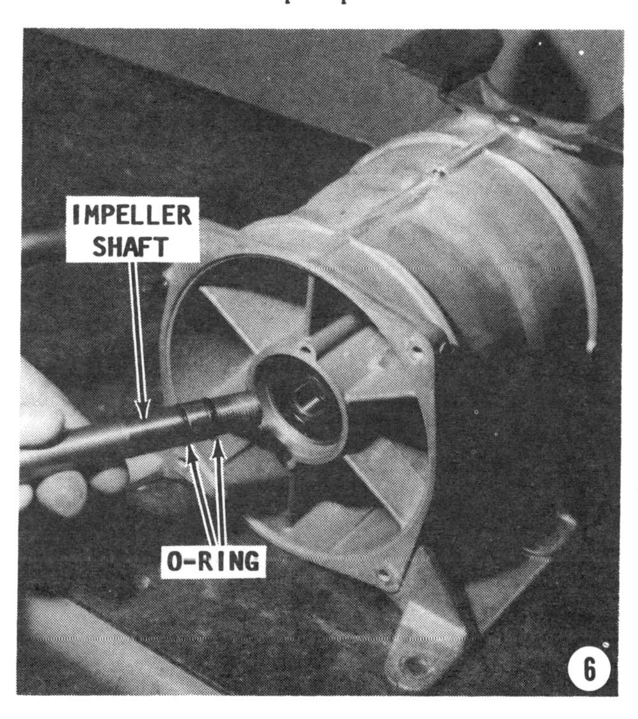

SAFETY WORDS

The snap ring mentioned in the following step is under tremendous pressure when installed. Therefore, wear a pair of safety glasses and exercise extra **CARE** when removing the snap ring to prevent injury to self or others in the area.

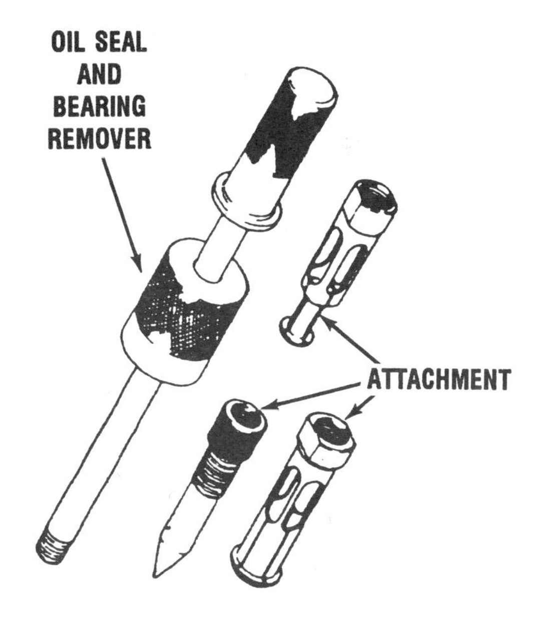

8- Using a screwdriver, pry out the two oil seals at the forward end of the pump case. If the seals are frozen to the hub, obtain the special tool indicated in Step 9. Otherwise, proceed with Step 10.

Pump Case Oil Seal and Bearing Removal All Models Except 550 Series

9- Obtain special oil seal and bearing remover tool, Kawasaki Part Number 57001-1058. Remove the seals at the forward end of the pump using the proper attachment supplied with the tool.

10- Use a slim long punch and tap out the aft bearing first. Next, remove the shaft-to-case spacer -- collar -- then, tap out the forward bearing.

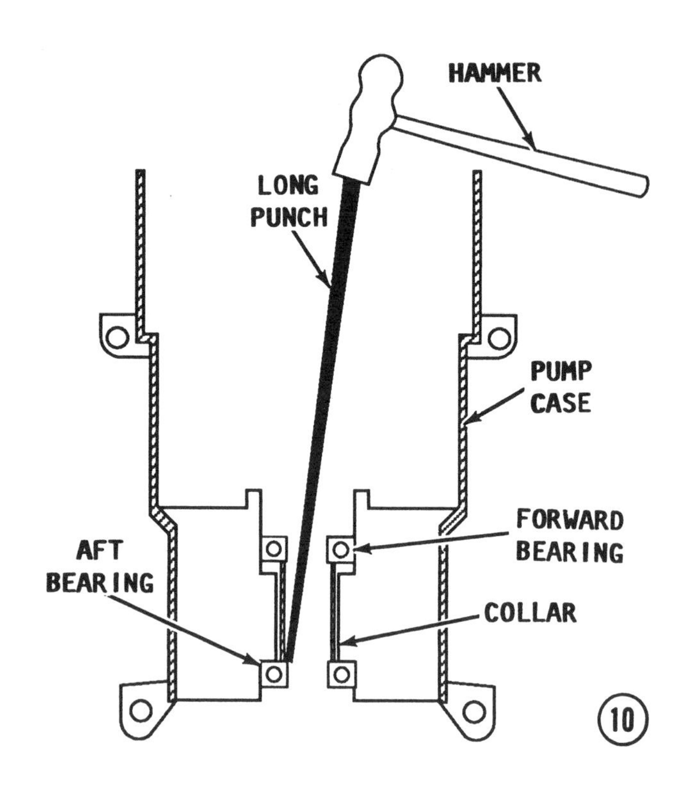

Impeller Shaft Removal 550 Series Mixed Flow Pump Only

11- If the two oil seals in the forward face of the guide vane casting are still satisfactory, obtain an oil seal protector to remove the impeller shaft. However, if the seals have failed, a protector is not necessary. Nevertheless, a seal protector **MUST** be used when installating the impeller shaft.

If a small split occurs in the lip of one of these oil seals, water will find its way -- under pressure -- from the impeller past the two seals. This water will destroy the two bearings supporting the impeller shaft and the pump will seize.

Slide the seal protector tool over the pump shaft threads. If this work is performed without the use of the protector, the threads on the pump shaft may damage the oil seal lips inside the guide vane casting. Use a soft head mallet and tap the shaft free of the casting.

12- If the seals are to be removed, use a screwdriver and pry out the two oil seals from the forward face of the guide vane casting.

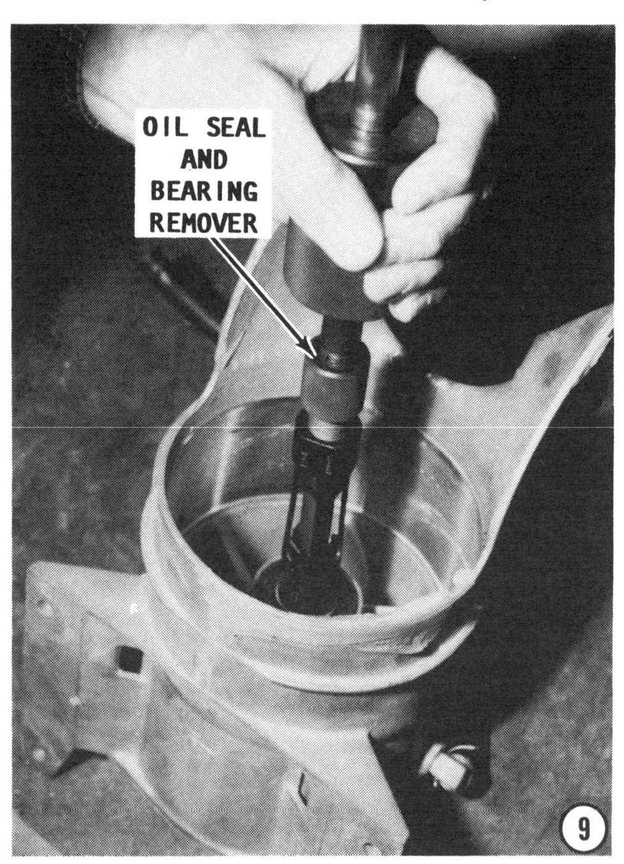

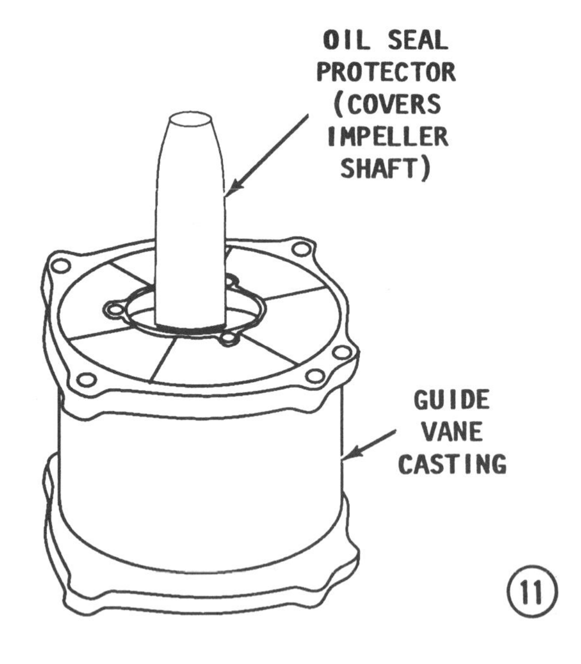

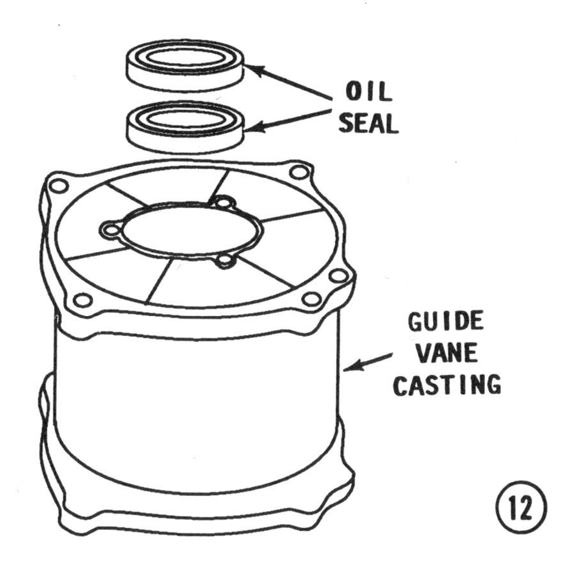

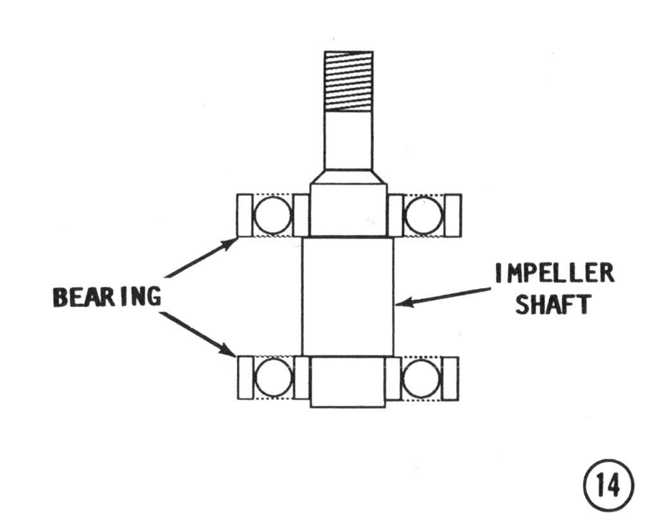

Impeller Shaft Disassembling
550 Series
Mixed Flow Pump Only

13- Insert a slim long punch into the open (threaded), end of the impeller shaft. Tap on the punch to remove the shaft plug from the other end of the shaft.

14- Make a setup in an arbor press, and carefully press the impeller shaft free of the two roller bearings -- one at each end of the shaft. Take care not to allow the shaft to free fall to the floor.

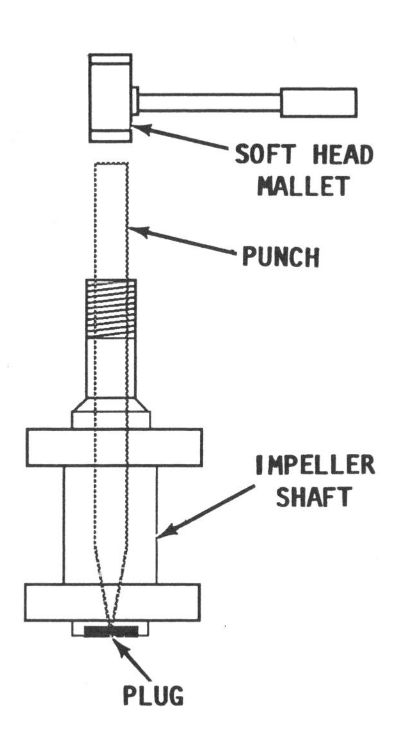

CLEANING AND INSPECTING

Inspect the leading edge of each impeller blade. Ideally, the edge should be rounded to a 1/32" radius, **NOT** a knife-edge! The cross-section shape of an impeller blade is similar to that of an airplane wing.

If the craft is consistently operated in shallow water with a sandy or "pebbly" bottom, the leading edge of the impeller blades will become rounded or will become chipped with missing pieces.

Very little material may be removed from the impeller blades during reconditioning work. Therefore, if the edge is rounded off or corroded, a file can be carefully used to "dress" the edge.

If an impeller blade is chipped more than 1/4", the impeller should be replaced. Any chip smaller than 1/4" can be evenly filed away along the length of the blade. Whatever work is done to one blade must be done to all blades in order to maintain impeller balance. If the impeller becomes excessively out of balance, a definite vibration condition will be quite noticeable as the craft moves through the water. Such vibration will also cause premature bearing and shaft wear.

Measure the outside diameter of the impeller. If the impeller is worn close to **OR** beyond the service limit, it must be replaced.

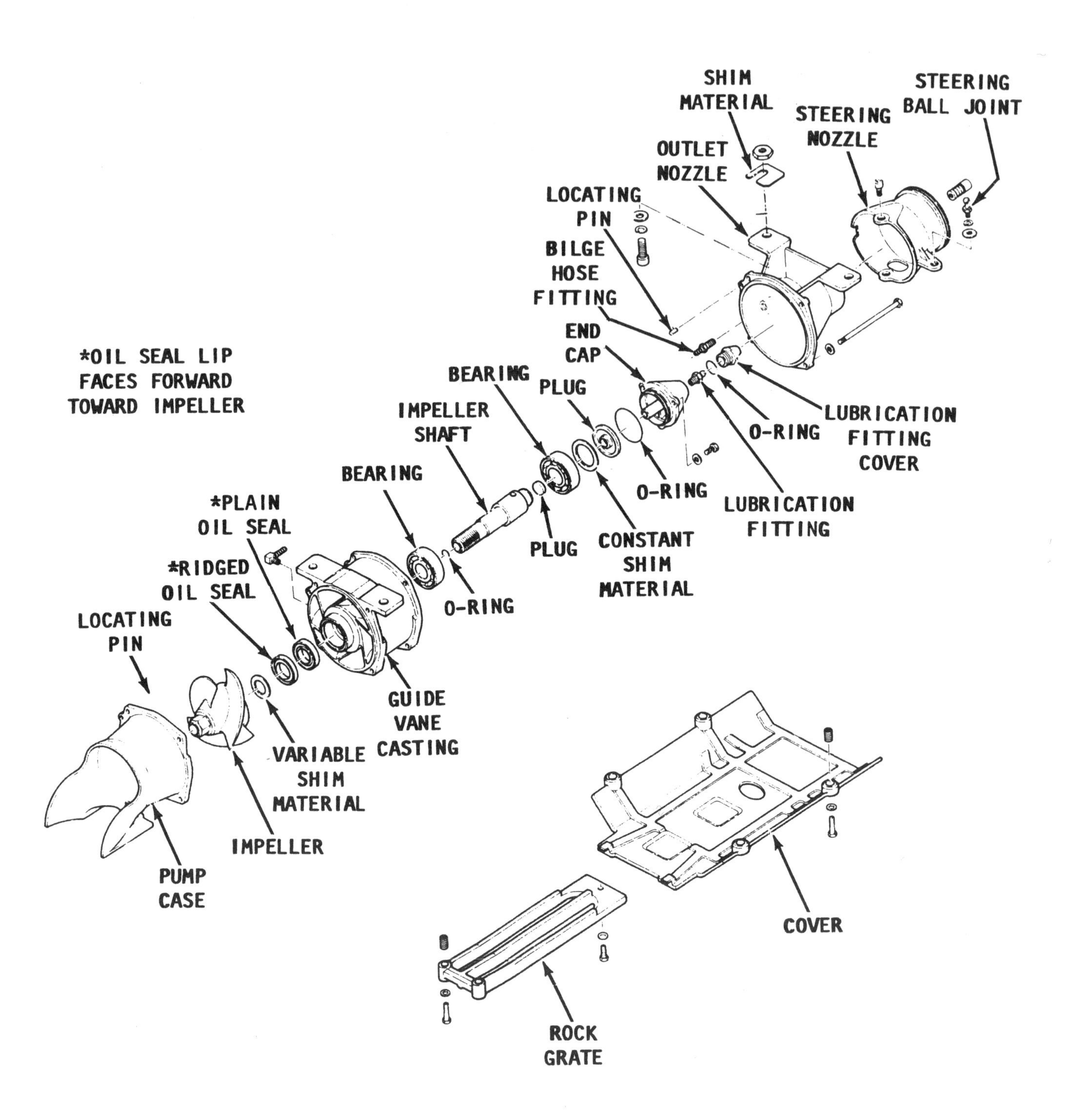

Exploded drawing of the mixed flow jet pump installed on the Model 550 Series watercraft. Major parts have been identified.

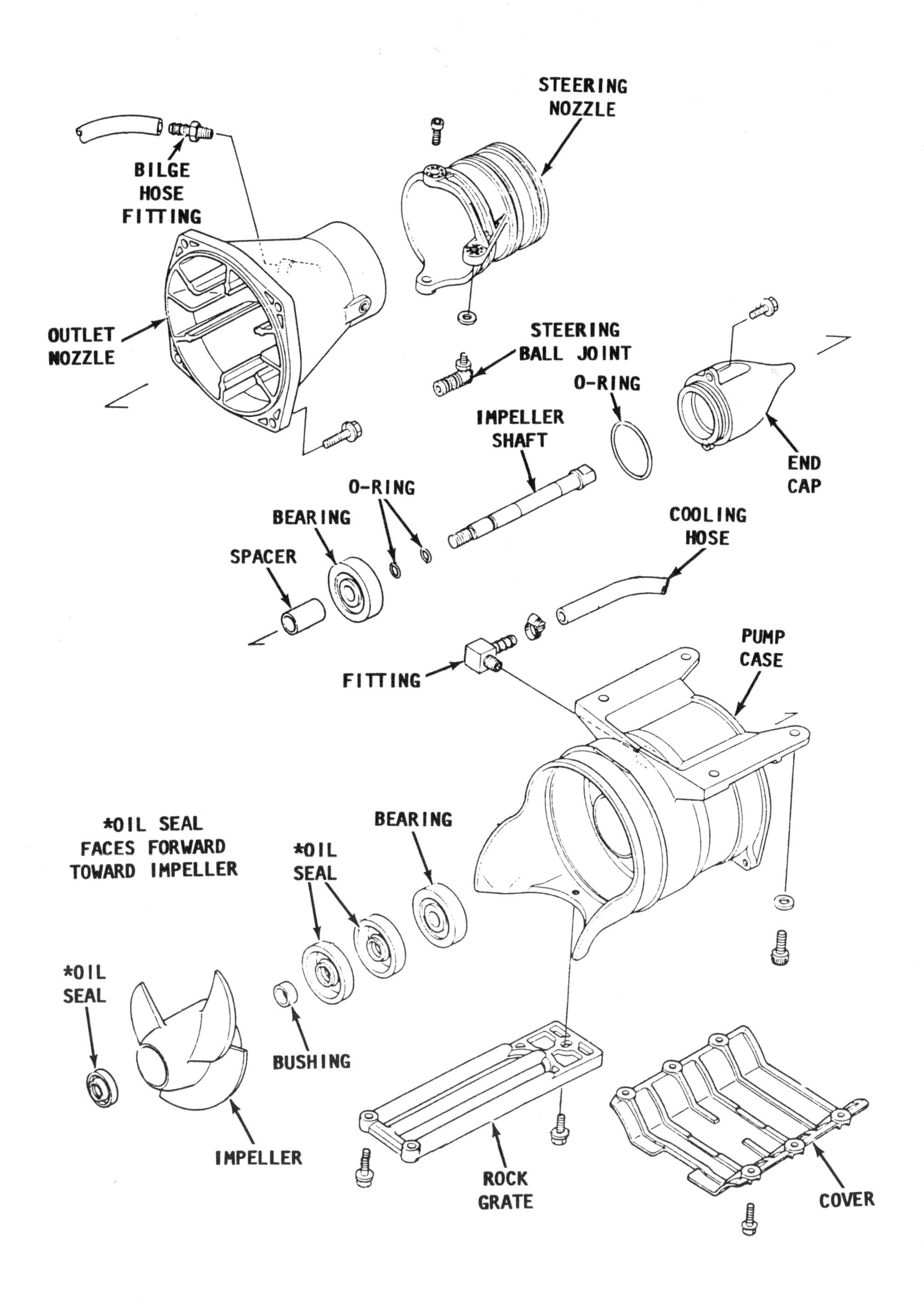

Exploded drawing of the axial flow jet pump installed on the Model **650** *series watercraft. Major parts have been identified.*

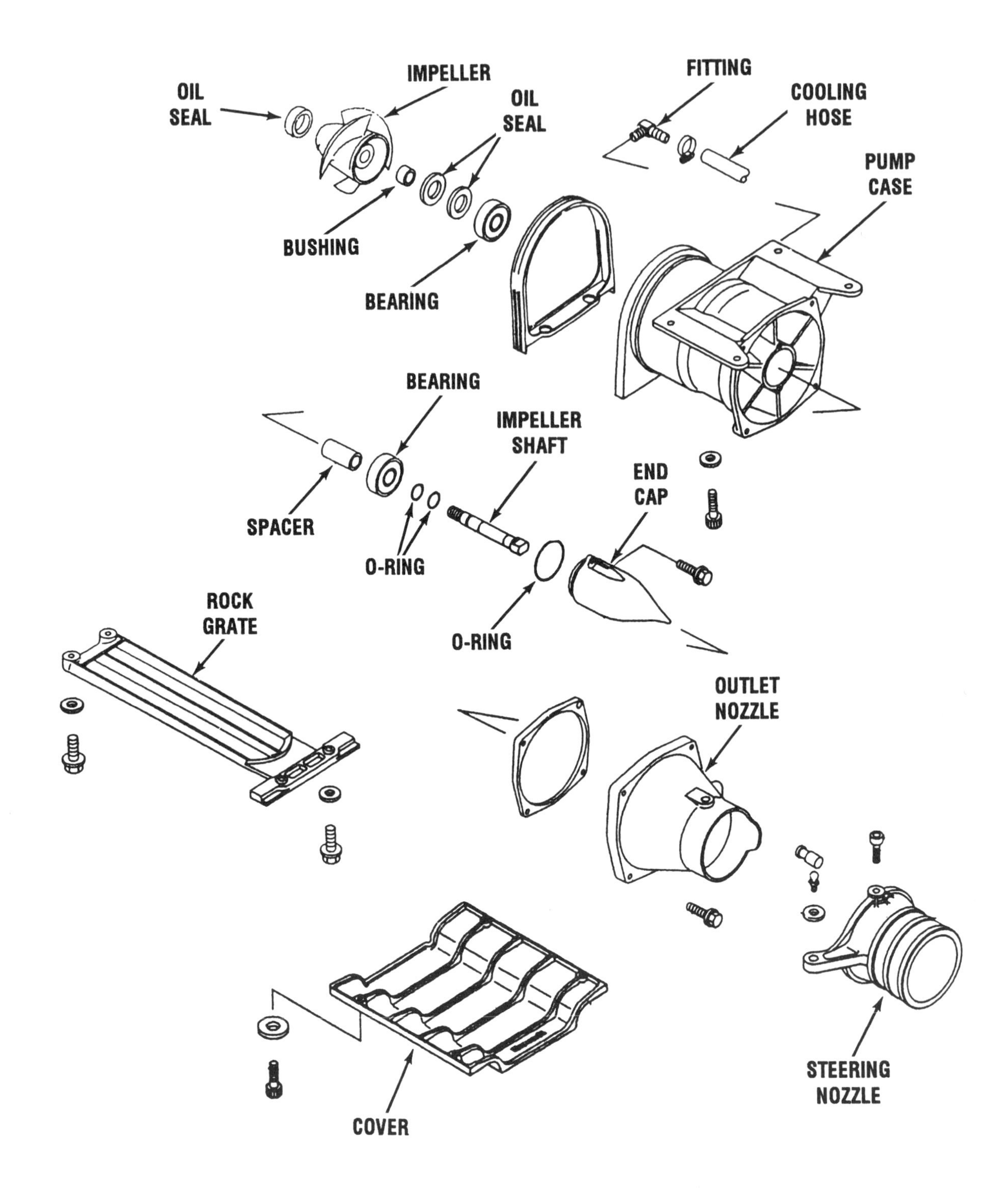

Exploded drawing of the axial flow jet pump installed on the **Standard** *Model* **750** *-- non high-performance -- Series. Major parts have been identified.*

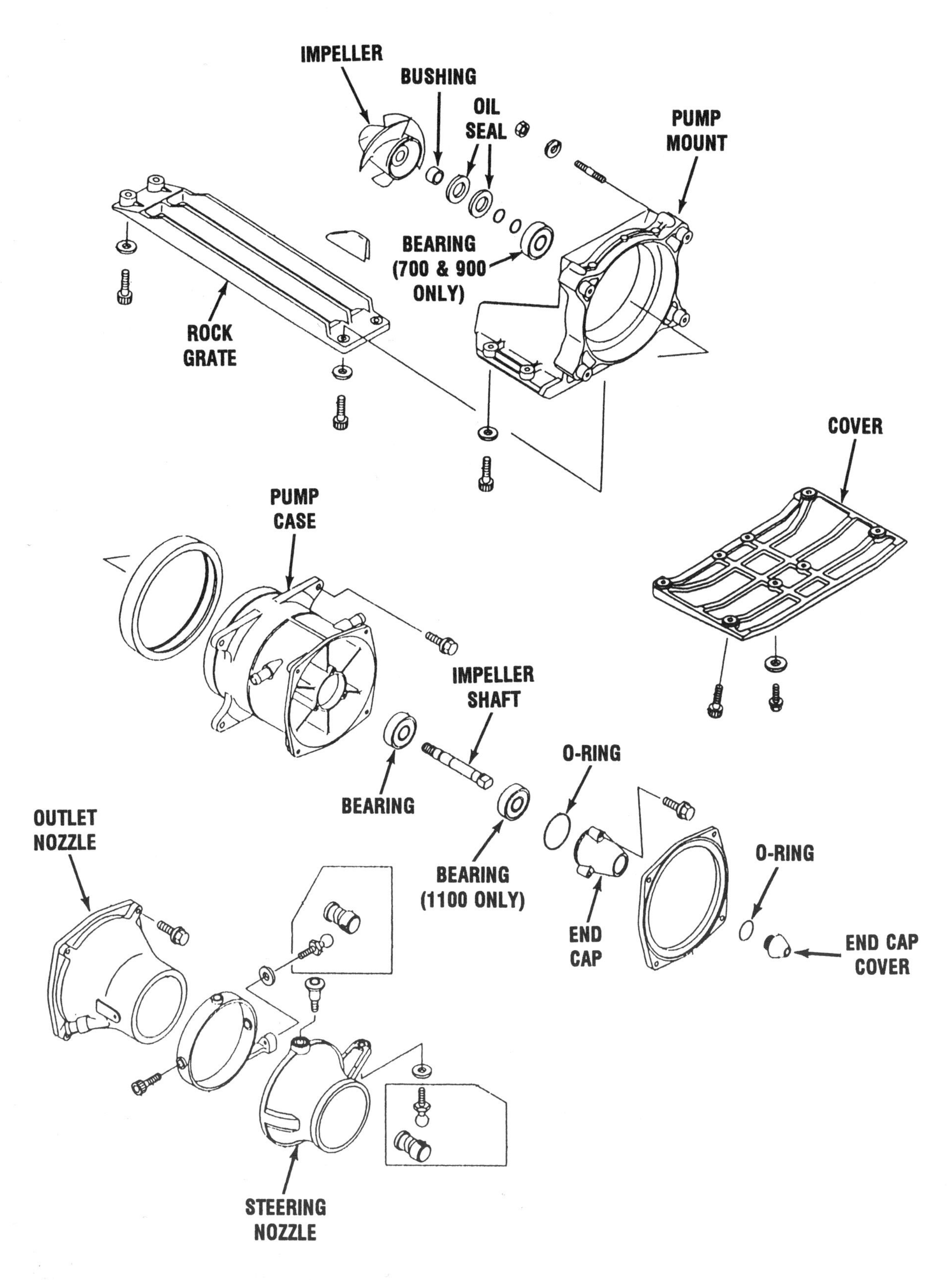

*Exploded drawing of the axial flow jet pump installed on the **High-performance** Model **750** Series, all Model **900** Series and Model **1100** Series. Major parts have been identified.*

IMPELLER OUTSIDE DIAMETER

Series	Standard Diameter	Service Limit
550	Not Applicable	Not Applicable
650	5.4921-5.4960" (139.5-139.6mm)	5.4527" (138.5mm)
750	5.4921-5.5000" (139.5-139.7mm)	5.4527" (138.5mm)
900	5.4921-5.5000" (139.5-139.7mm)	5.4527" (138.5mm)
1100	5.4921-5.5000" (147.5-147.7mm)	5.4527" (146.5mm)

Inspect the interior of the pump case. If the case surface has deep scratches, the pump case must be replaced. Measure the inside diameter of the pump case. If the diameter is greater than the service limit -- the case should be replaced.

PUMP CASE INSIDE DIAMETER

Series	Standard Diameter	Service Limit
550	Not Applicable	Not Applicable
650	5.5120-5.5159" (139.5-139.6mm)	5.5553" (138.5mm)
750	5.5118-5.5157" (140.0-140.1mm)	5.5551" (141.1mm)
900	5.5118-5.5157" (140.0-140.1mm)	5.5551" (141.1mm)
1100	5.8268-5.8307" (148.0-148.1mm)	5.8701" (149.1mm)

Set the long drive shaft on two V-blocks -- one at each end. Position a dial indicator gauge over the center portion of the shaft and measure the runout. The maximum allowable runout is 0.020" (0.5mm).

Check to be sure the rock grate blades are not bent or missing. Such a condition might allow a large rock to pass through and cause damage to the impeller.

Inspect the vanes in the pump case. File the vanes to a sharp edge on the forward end. Be sure to keep all the edges the same distance from the front of the bowl.

ASSEMBLING

The following procedures cover assembly of the complete jet pump -- all parts. The following procedures outline detailed instructions to assemble virtually all parts of jet pumps installed on the personal watercraft covered in this manual.

As the assembly work progresses, if a particular item was not removed or disassembled -- simply skip the steps involved for the part and proceed directly to the work required.

FIRST THESE WORDS

If working on a 550 Series mixed flow pump, proceed with Step 1.

If working on any other model pump (axial flow), proceed directly to Step 4.

Impeller Shaft Assembly
550 Series Mixed Flow

1- Obtain a suitable mandrel which contacts the impeller shaft bearing on its inner race and has an inside diameter larger then the impeller shaft. If the mandrel rests on the bearing cage or the outer race, the bearing will become distorted under the force of the arbor press.

Press the bearing onto the shaft until the inner race rests against the shoulder of the

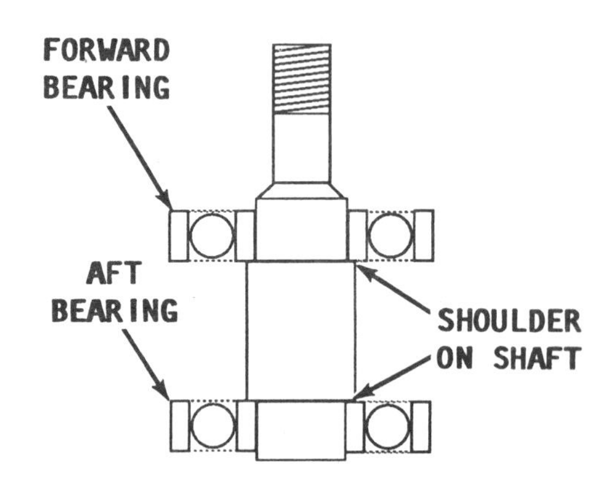

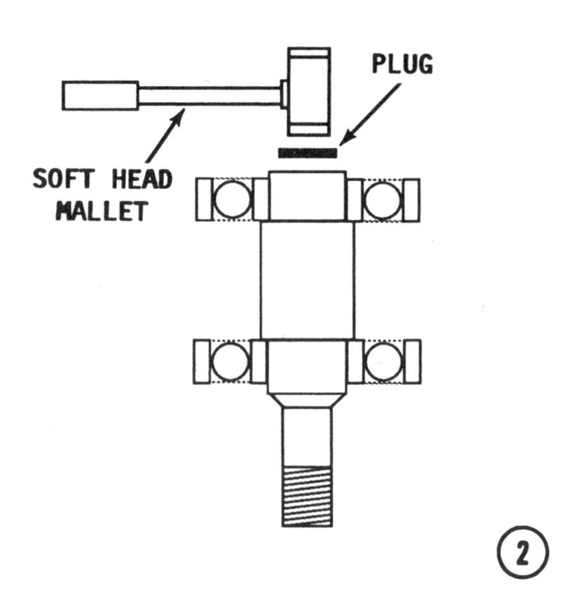

shaft. Install the bearing at the other end of the shaft in the same manner.

Pack both bearings with Shell Alvania EP1 or equivalent water resistant multi-purpose lubricant.

2- Install a **NEW** plug into the threaded end of the impeller shaft, with the flat side facing down.

Guide Vane Casting Seal Installation

550 Series
Mixed Flow Pump

3- Pack the lips of both oil seals with the same lubricant as the bearings. Install the plain oil seal into the forward end of the casting with the seal lip facing **FORWARD** -- up -- toward the impeller -- when installed. The seal is correctly installed when the

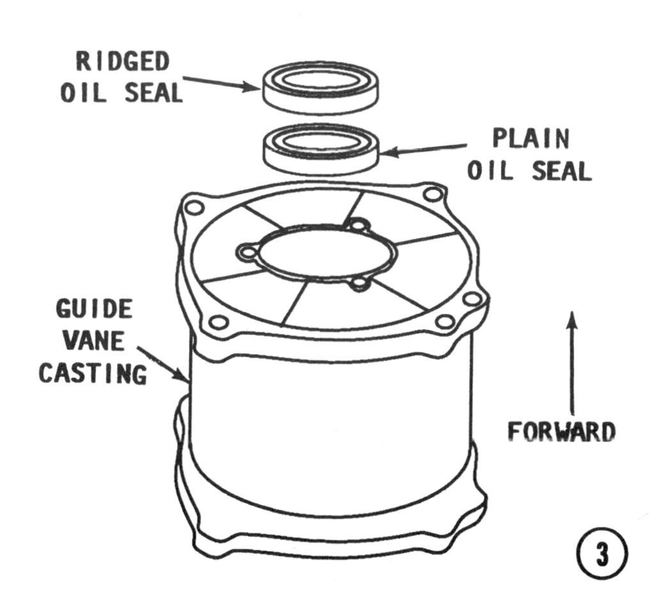

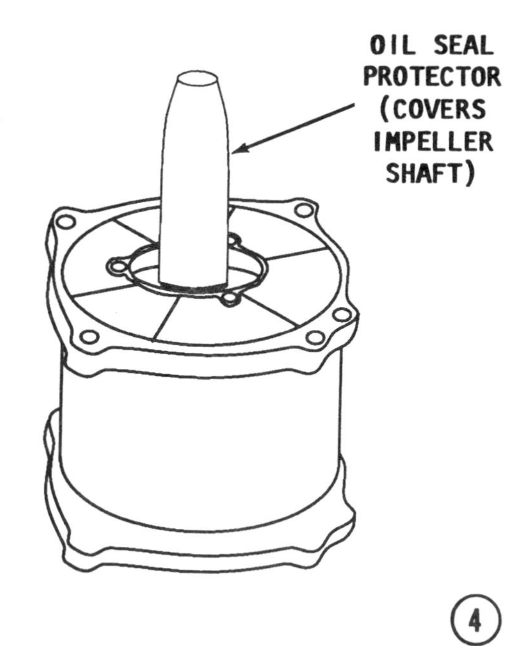

back side of the seal makes contact with an inner shoulder in the hub of the casting.

Install the ridged oil seal in the same manner -- until the back side of the seal rests against the first oil seal. The lip of the seal must face **FORWARD** -- toward the impeller -- after installation.

Impeller Shaft Installation
550 Series
Mixed Flow Pump

4- An oil seal protector **MUST** be used when installing the impeller shaft to avoid damage to the seal lips. If a small split occurs in the lip of one of these oil seals, water will find its way -- under pressure from the impeller -- past the two seals. This water will destroy the two bearings supporting the impeller shaft and the pump will seize.

Slide the protector over the end of the pump shaft and push the shaft into the guide vane casting from the aft end. Use a soft head mallet and drive the shaft into the casting until the lower bearing seats against a shoulder in the casting hub.

10-5 SHIMMING PROCEDURES
550 SERIES
MIXED FLOW PUMPS

Using vernier calipers with a millimeter scale, measure the height of the shoulder on the end cap, as shown in the accompanying illustration -- dimension "A". Now, measure the depth of the shoulder in the aft face of

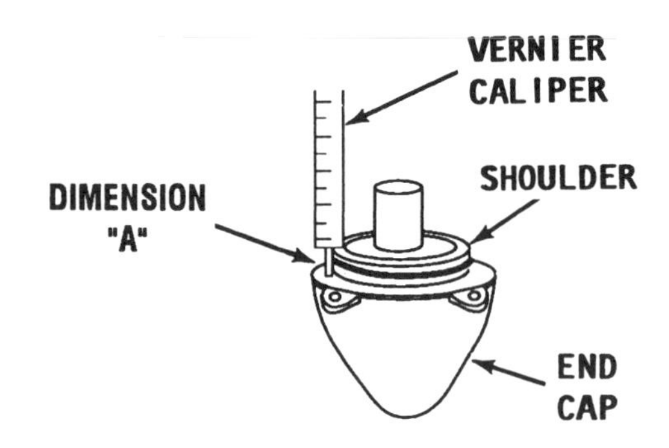

Simple line drawing to illustrate how dimension "A" is determined as the height of the end cap shoulder.

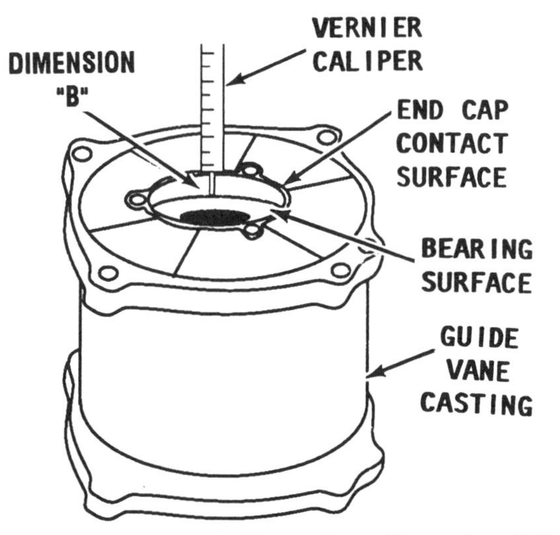

Line drawing to indicate how dimension "B" is obtained as the depth of the bearing. The dimension "A" above and Dimension "B" are explained in the text.

the guide vane casting hub -- from the outer bearing race to the mating surface of the end cap, as shown in the accompanying illustration -- dimension **"B"**.

Subtract the height of the end cap from the depth of the bearing to determine the clearance between the cap and the bearing, when the cap is installed.

B - A = Clearance

Example:
If the depth of the bearing -- **"B"** -- is 13mm, and the height of the cap is 12.5mm, the clearance will be as follows:

13.0 - 12.5 = 0.5mm

Compare the calculated clearance to the chart to determine the amount of shim material to be placed behind the impeller for the correct impeller-to-pump case clearance.

Shim material pieces are only available in two thicknesses -- 0.15mm and 0.30mm. A combination of pieces provide the required thickness to obtain the correct calculated clearance in mm. The following table lists calculated clearances and the total amount of shim material required, including the number of pieces for each thickness.

SHIM MATERIAL AMOUNT FOR CALCULATED CLEARANCE

Calculated Clearance (mm Only)	Total Shim Thickness (mm Only)	No. Pieces Ea. Size (mm Only)
0.10-0.25	None	None
0.26-0.40	0.15	One 0.15
0.41-0.55	0.30	One 0.30
0.56-0.70	0.45	One 0.15 One 0.30
0.71-0.85	0.60	Two 0.30
0.86-0.94	0.75	One 0.15 Two 0.30

In the given example, a clearance of 0.5mm falls in the 0.41-0.55mm range. Therefore, 0.30mm of shim material must be used.

Set aside the correct amount of shim material to be installed later -- Step 9, this section -- behind the impeller.

Pump Case Oil Seal and Bearing Installation All Models Except 550

5- Pack both bearings with Shell Alvania EP1 or equivalent water resistant multipurpose lubricant. Install the aft pump bearing into the pump case hub using the bearing driver set. The bearing is correctly installed when the outer bearing race seats against a shoulder in the hub. Insert the spacer from the forward end of the pump case. Install the forward bearing using the same special tools. The bearing is correctly positioned inside the hub when the inner bearing race contacts the spacer.

Pack the lips of the two oil seals with the same lubricant and install them into the forward face of the pump case. **BOTH** oil seals are installed with the lip facing **UP** -- forward -- when installed. The first seal is correctly installed when the back side seats against the bearing. The back side of the second seal must seat against the lip of the first.

6- Insert the short bushing into the forward end of the pump case hub.

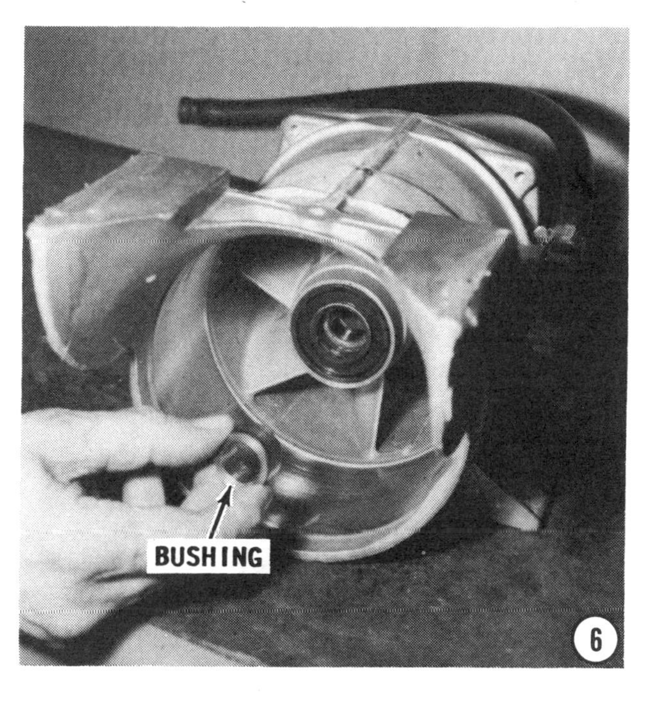

SAFETY WORDS

The snap ring mentioned in the following step is under tremendous pressure while it is being installed. Therefore, wear a pair of safety glasses and exercise extra **CARE** while installing the snap ring to prevent injury to self or others in the area, if the ring should accidently "pop" out of the pliers.

Next, use a pair of internal snap ring pliers and install the snap ring into the groove above the thin oil seal.

Impeller Installation
550 Series
Mixed Flow Pump

7- Install the large White plastic plug into the end cap with the recessed side facing **OUTWARD**. Apply a coating of lubricant to a new **O**-ring, and then install the ring around the groove in the cap.

Apply a coating of Loctite Lock 'N Seal onto the threads of the attaching bolts.

Place the "constant" amount of shim material -- removed in Step 3 of disassemb-

Snap rings are under tremendous tension during removal and installation, presenting a potential hazard. As an eye protection measure, safety glasses or a shield should ALWAYS be worn during work with such snap rings. Warn others in the area such work is in progress.

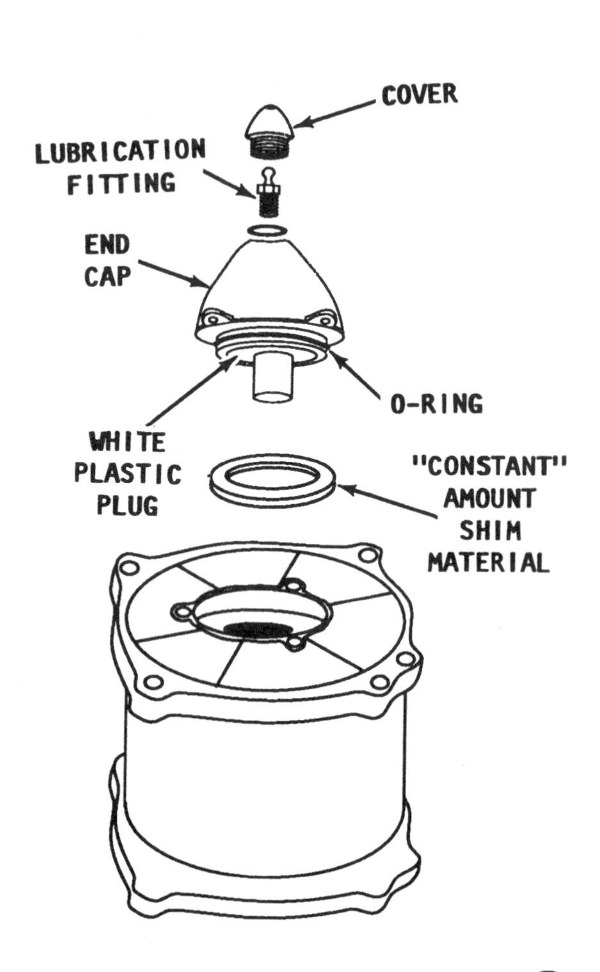

ly -- over the rear bearing in the hub. Install the cap over the guide vane casting hub. Tighten the bolts to a torque value of 52 in lb (6 Nm).

Next, install the lubrication fitting into the end of the end cap and tighten it securely. Install the lubrication fitting cover and new O-ring over the fitting. Tighten the cover with an allen wrench.

Using a grease gun with a cartridge filled with Shell Alvania EP1 or multi-purpose water resistant equivalent lubricant, pump the end cap full of lubricant until it oozes from the seal lip on the other side of the hub. Apply some of the excess lubricant onto the threads of the impeller shaft. Install the small O-ring onto the end of the impeller shaft.

8- Install the shim material required to obtain the correct impeller blade-to-pump case clearance onto the impeller shaft. This thickness of shim material was determined earlier in section 10-5, on Page 10-22. Install the impeller.

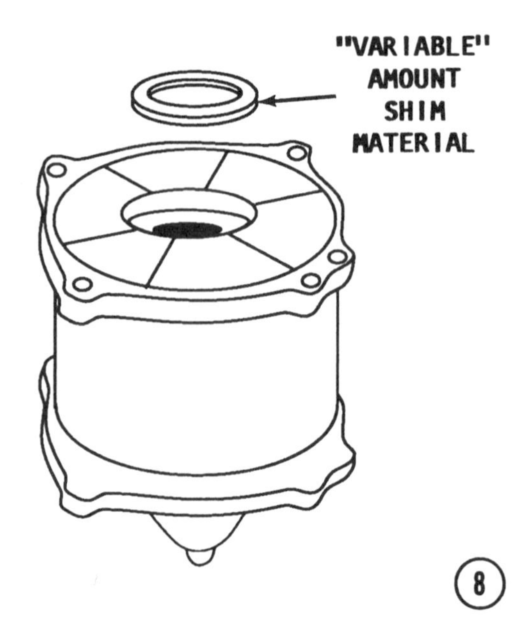

SPECIAL WORDS
ALL Models

All of the impellers covered in this manual have right-hand threads -- both mixed flow jet pumps and axial flow jet pumps. The manufacturer incorporated this into all models after 1991.

Hold the impeller shaft with the special tool and tighten the impeller to a torque value of 14 ft lb (20 Nm).

A "crowsfoot" wrench the same size as the "cast" nut may be used to tighten the nut to the specified torque value. The crowsfoot wrench has a square hole to accommodate a torque wrench.

Impeller Shaft Installation
All Models
Except 550

9- Install two new **O**-rings around the impeller shaft. Apply lubricant to both rings and the entire shaft, with the exception of the slotted nut.

Insert the shaft into the aft end of the pump case hub.

10- Thread the impeller onto the shaft.

11- Use the special tool inserted into the impeller splines to hold the impeller while the slotted nut is tightened. Tighten the slotted nut on the impeller shaft to a torque value of 72 ft lb (100Nm).

If the small oil seal was removed, pack the lip of a new seal with lubricant and tap the seal into place with the seal lip facing **UP** -- forward when installed.

12- Pack the end cap with Shell Alvania EP1 lubricant, or equivalent. Apply a coating of the lubricant to the new **O**-ring around the cap. Apply a coating of Loctite Lock 'N Seal to the attaching bolts. Install the end cap over the hub. Secure it with the two bolts and washers. Tighten the bolts to a torque value of 53 in lb (6Nm).

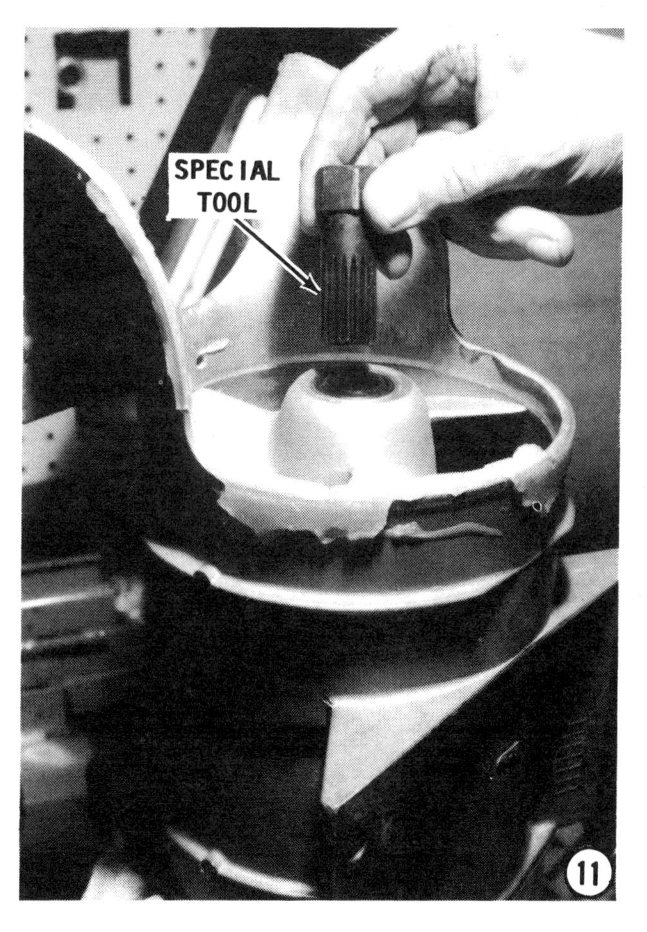

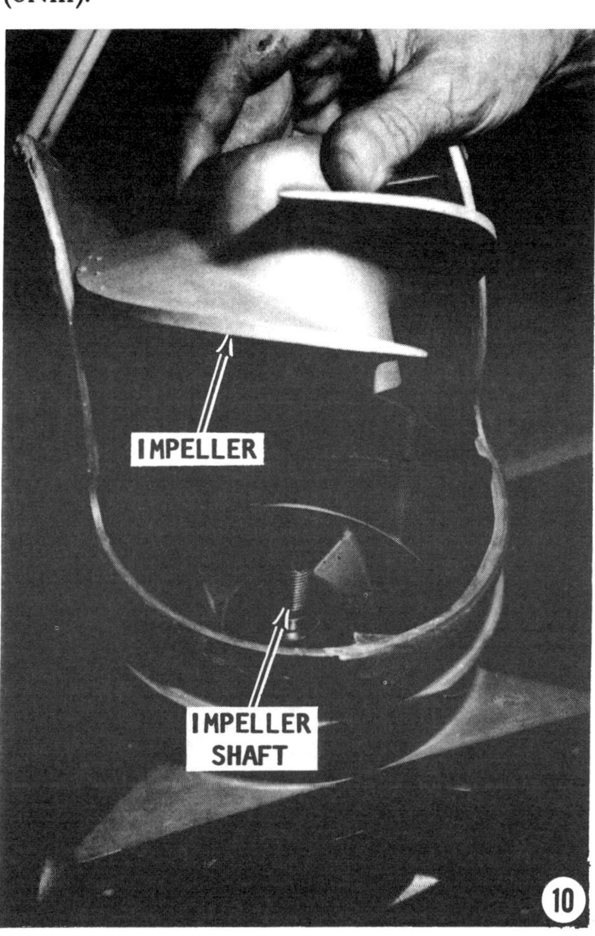

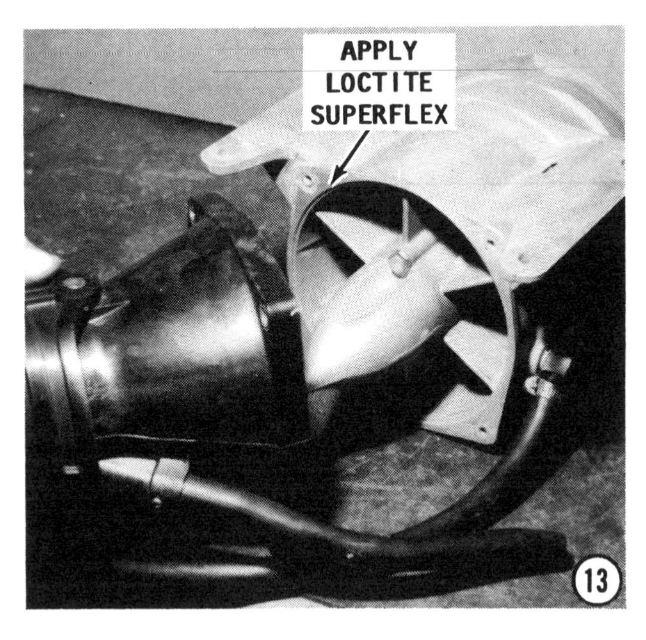

All Units
Except 550 Series

13- Apply a light even coating of Loctite Superflex, or a good brand of silicone sealer, around the perimeter of the pump case and outlet nozzle mating surfaces.

14- Apply a coating of Loctite Stud N' Bearing Mount compound to the threads of the attaching bolts. Install the outlet nozzle, with the steering nozzle attached, onto the pump case with the ball joint on the steering nozzle pointing **UP**. Install and tighten the securing bolts in a cross-tightening sequence to a torque value of 12 ft lb (16 Nm).

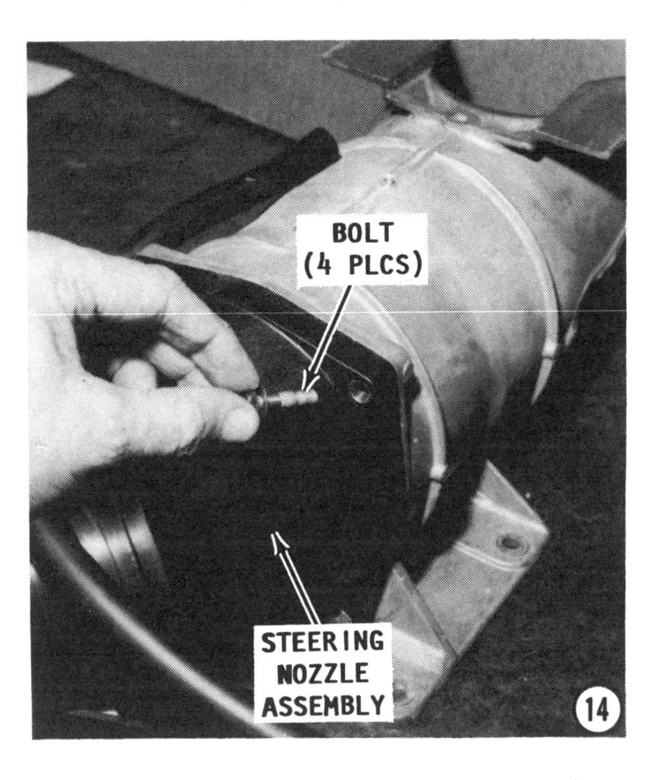

Check the action of the steering nozzle. The nozzle **MUST** swing smoothly from side to side.

550 Series
Mixed Flow Pump

Apply a light even coating of Loctite Superflex, or a good brand of silicone sealer, around the perimeter of the four mating surfaces -- the pump case, both sides of the guide vane casting, and the outlet nozzle. Check to be sure the two locating pins are in place at the two mating surfaces.

Apply a coating of Loctite Stud N' Bearing Mount compound to the threads of the four long attaching bolts. Hold the three pieces together -- with the locating pins indexed into the appropriate holes. All mating surfaces will come together in only one position -- the correct position -- with the locating pins properly indexed.

Install and tighten the securing bolts in a cross-tightening sequence to a torque value of 48 in lb (5.5 Nm).

Check the action of the steering nozzle. The nozzle **MUST** swing smoothly from side to side.

SPECIAL WORDS

Make one last check of the impeller blade-to-pump case clearance before the jet pump is installed into the craft.

Refer to Page 10-5 for specifications.

MORE SPECIAL WORDS

If the bearing housing, supporting the driveshaft at the bulkhead, needs service, this is the time to perform the necessary work. See Section 10-6.

Detailed service procedures for the bearing housing are presented in Section 10-6, beginning on Page 10-30.

INSTALLATION

1- If the engine has been removed and the driveshaft was pulled from the bearing housing, proceed as follows: Insert the driveshaft into the bearing housing. Insert a new rubber coupler into the crankshaft metal coupler and index the two metal couplers with the rubber coupler in the middle. Detailed comprehensive engine crankshaft-to-driveshaft alignment procedures are presented in the last section of the engine chapter -- Chapter 8.

Impeller Installation
Axial Pump
All Models Except 550 Series

Thread the impeller onto the driveshaft. Insert the pin in the adaptor into the hole in the driveshaft. Clamp the driveshaft holder tool around the adaptor and driveshaft. Index the pins of the impeller wrench into the holes in the impeller and tighten the impeller **VERY SECURELY.** The manufacturer does not provide a torque specification for this series axial pump.

2- Apply a coating of Loctite Stud N' Bearing Mount Compound to the threads of the attaching bolts. Bring the jet pump into position within the hull cavity and install the pump over the driveshaft.

3- Insert the stacks of shim material back into their original locations between the mounting ears of the pump case and the craft hull. Install the four allen head bolts and tighten them alternately and evenly to a torque value of 16 ft lb (22 Nm).

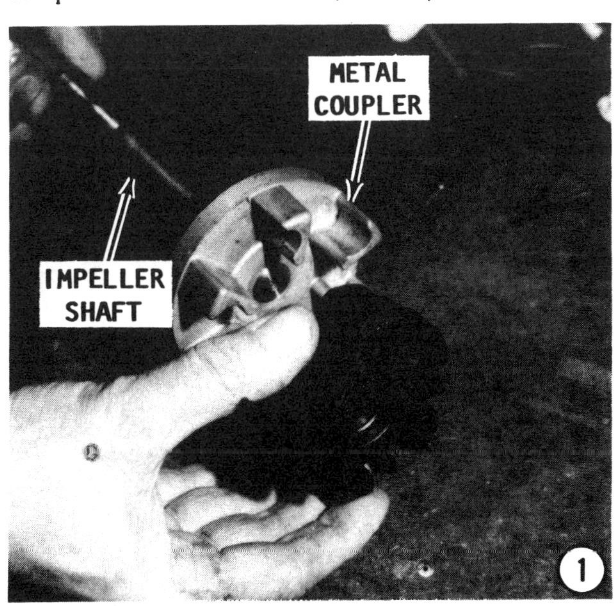

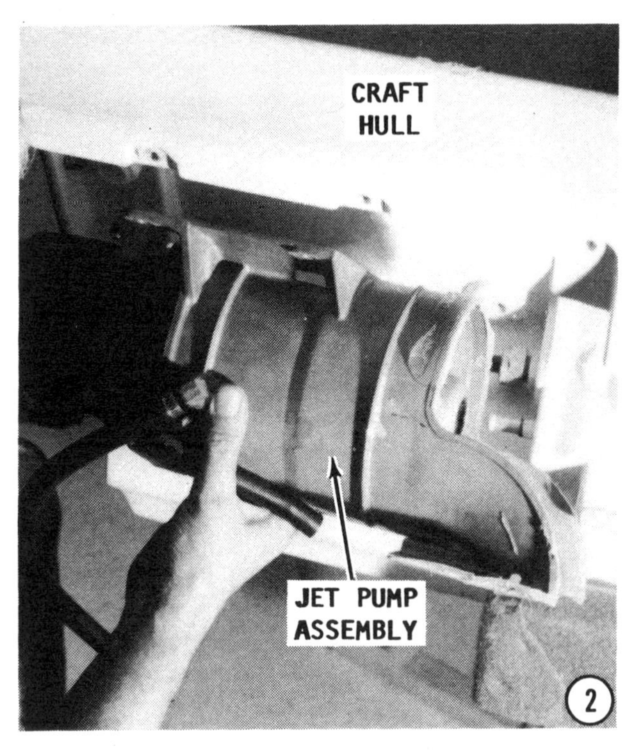

Some Series
Axial Pump

Install the lubrication fitting at the end of the end cap -- if so equipped. This fitting was not installed until this time, to prevent hydraulic lock while the driveshaft was being installed.

4- Apply a solid bead of RTV Silicone around the perimeter of the pump case intake, as shown in the accompanying illustration. If there is an air gap between the pump case and hull, air will be sucked into

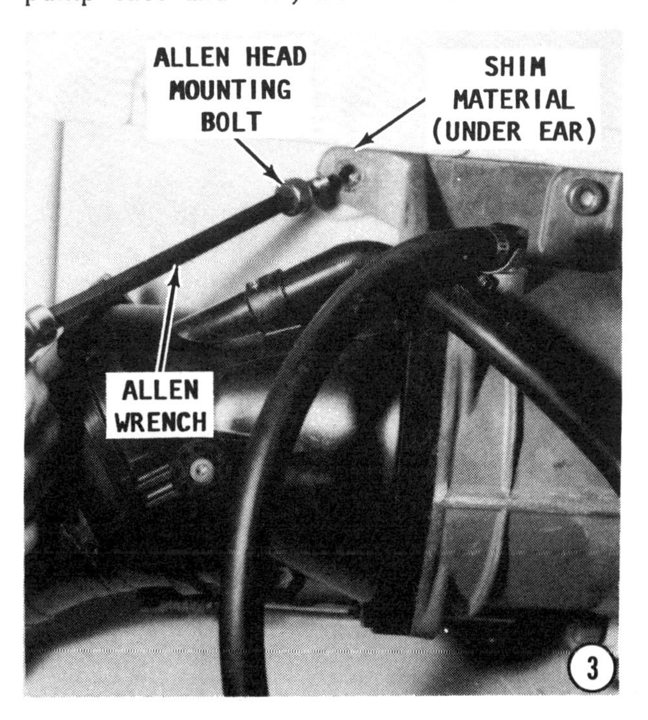

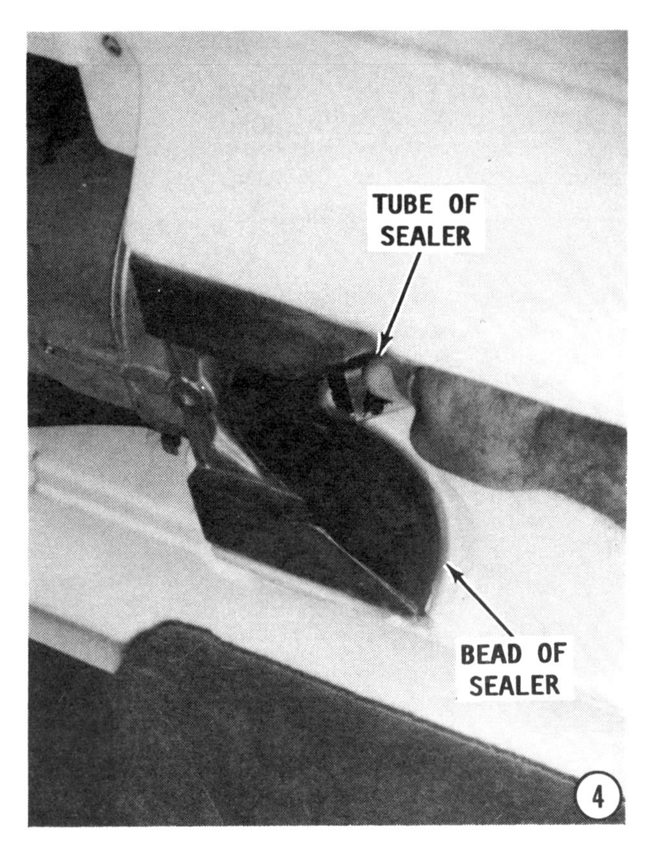

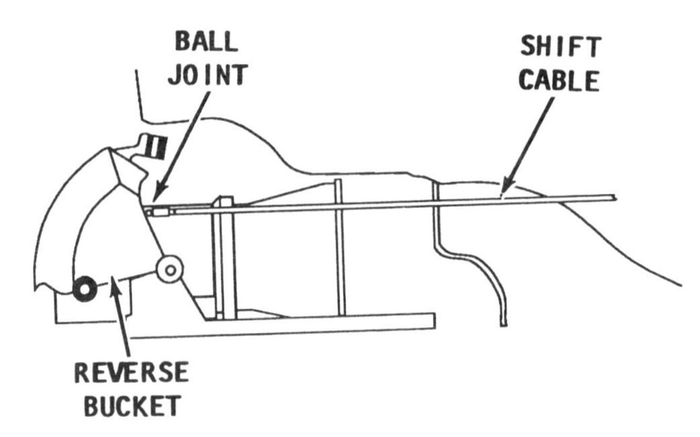

Simple line drawing to reveal the reverse "bucket" installation arrangement, on watercraft equipped with reverse capability.

the pump and performance will be greatly reduced.

REVERSE BUCKET

Pull back the reverse cable end and snap the cable onto the ball joint on the reverse bucket -- if so equipped.

5- Connect the cooling hose to the fitting on the pump case. Secure the hose with a band type hose clamp.

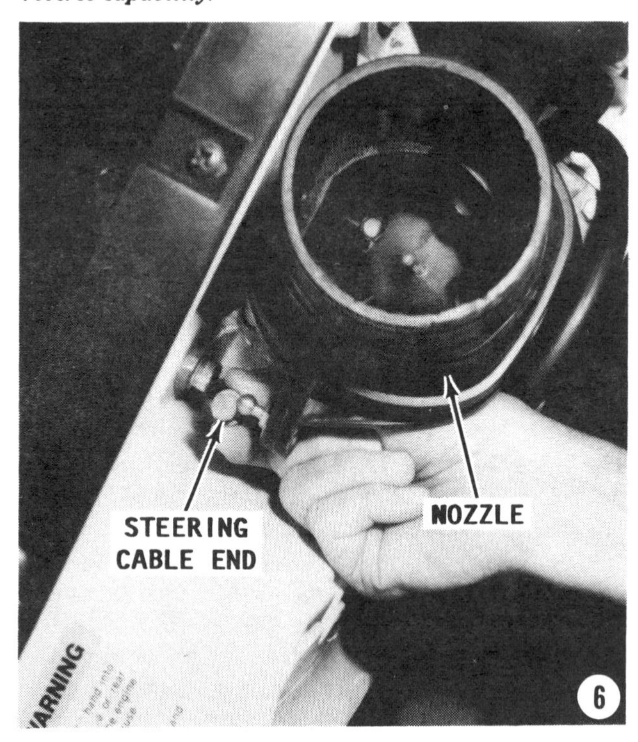

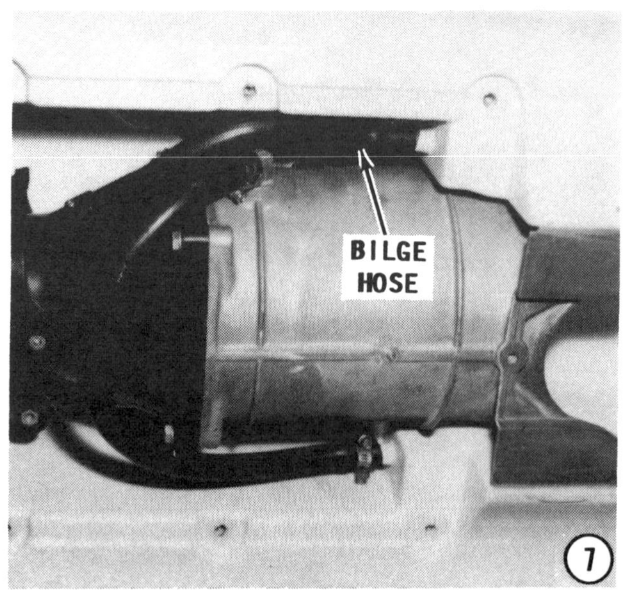

Tighten the hose clamp **SECURELY.** If the hose should blow off, the supply of cooling water to the engine would cease instantly and the engine would overheat.

6- Pull back the steering cable end and snap the cable end onto the steering ball joint on the steering arm of the nozzle.

7- Connect the bilge hose to the fitting on the hull. Secure the hose with a hose clamp.

8- Apply a coating of Loctite Stud N' Bearing Mount Compound Coat to the threads of the (four) pump attaching bolts.

9- Install the cover over the jet pump. Secure the cover with the attaching bolts and tighten them alternately and evenly to a torque value of 16 ft lb (22Nm) -- 550 Series; 87 in lb (9.8Nm) -- 650 Series; 69 in lb (7.8Nm) -- 750, 900 and 1100 Series.

10- Apply a coating of Loctite Stud N' Bearing Mount Compound to the threads of the three rock grate attaching bolts.

Install the rock grate and tighten the bolts to a torque value of 87 in lb (9.8Nm).

Rotate the craft upright. Install the battery and connect the cables to the proper terminals. Install or fill the fuel tank.

Fill the oil injection tank (on all Models other than the 550 Series). The oil injection system **MUST** be bled while operating the engine on a premix as directed in Section 6-9 of Chapter 6 -- **FUEL.**

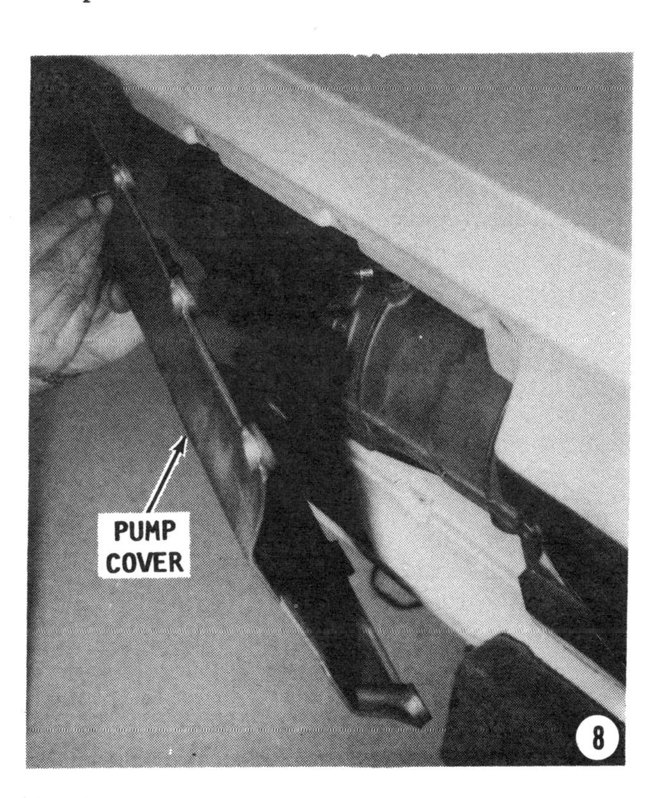

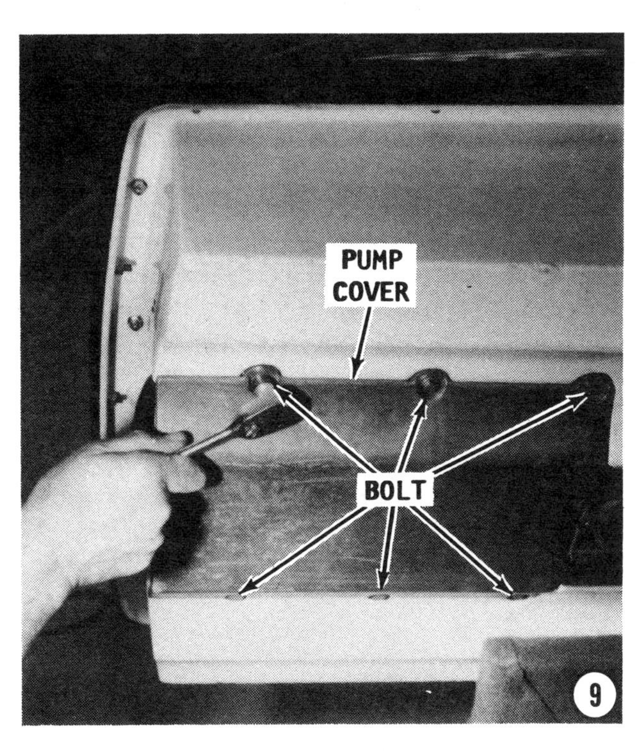

Check Completed Work

Move the craft to a body of water or connect a garden hose to the engine cooling water supply fitting or flush fitting on the cylinder head.

Start the engine and allow the rpm's to stabilize at idle speed **FOR JUST A FEW SECONDS,** and then turn the water on.

Adjust the water flow until a small trickle is discharged from the bypass outlet on the port side of the hull.

When the engine is to be shut down, turn the water off **FIRST** -- raise the aft portion of the hull -- **WHILE THE ENGINE IS OPERATING AT IDLE** -- "rev" the engine just a **COUPLE** times to clear water from the exhaust system -- and then shut it down.

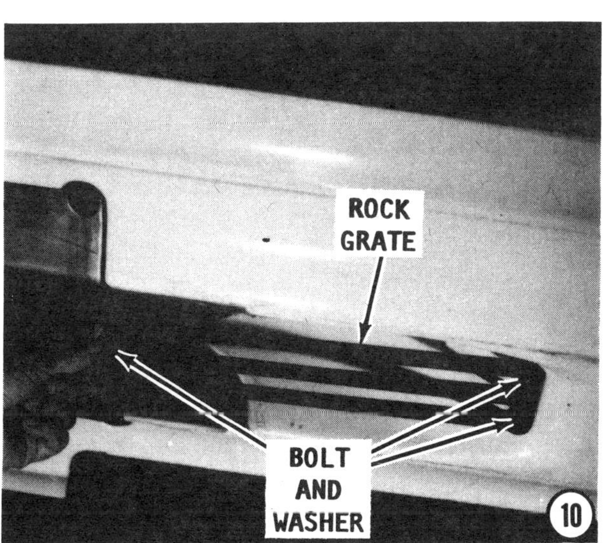

NEVER allow the engine to operate without cooling water for more than 15 seconds.

CAUTION

Water must circulate through the jet pump -- to and from the engine, anytime the engine is operating. Circulating water will prevent overheating -- which could cause damage to moving engine parts and possible engine seizure.

NEVER, AGAIN, NEVER operate the engine at high speed with a flush device attached. The engine, operating at high speed with such a device attached, would **RUNAWAY** from lack of a load on the impeller, causing extensive damage.

REVERSE SYSTEM

The reverse gate adjustment is covered in Chapter 11, Section 11-3.

10-6 BEARING HOUSING SERVICE

The driveshaft bearing housing serves as a support for the driveshaft and prevents water from entering the engine compartment from the pump intake.

Older type bearing housings consisted of two ball bearings separated by a collar with double oil seals on each side.

The forward oil seals both faced forward and the aft oil seals faced aft.

A very few 650 Series pumps had only a single bearing, with the same oil seal arrangement.

Newer type bearing housings have bushing type bearings with no seals and are maintenance free. The older type housings were equipped with lubrication fittings, which the manufacturer recommended lubricating every 25 hours of craft operation.

Only the older type serviceable housings are covered in this section.

If a newer type bearing housing needs replacement, follow the service procedures in the section to remove the housing, and then skip the disassembly and assembly procedures -- move directly to the installation procedures to install the housing.

Normally, the bearing housing never requires service, unless the oil seals have failed and allowed water to enter the bearing cavity.

SPECIAL WORDS ON SPECIAL TOOLS

The following list of special tools are required in order to perform some of the necessary work on the bearing housing. Some tasks become extremely difficult, if not impossible, without the special tools or some type of equivalent. Manufacturers' part numbers are given. In specific steps of the procedures -- whenever possible -- a special effort has been made to give some type of equivalent or homemade tool which may be substituted. However, the cost of a special tool may be offset by money saved if an expensive part is not damaged.

SPECIAL TOOLS FOR BEARING HOUSING SERVICE

550 Series
Mixed Flow Pump

Coupling Holder	P/N 57001-1230
Driveshaft Holder	P/N W56019-003
Adapter	P/N 57001-1231
Oil Seal and Bearing Remover	P/N 57001-1129

650 Series

Coupling Holder	P/N 57001-1230
Driveshaft Holder	P/N W56019-003
Adapter	P/N 57001-1231
Oil Seal and Bearing Remover	P/N 57001-1129

750 Series
900 Series
1100 Series

Coupling Holder	P/N 57001-1230
Driveshaft Holder	P/N 57001-1327
Adapter	P/N 57001-1231
Oil Seal and Bearing Remover	P/N 57001-1129

Bearing Housing Removal

Remove the battery, the water box muffler, and the fuel tank from the craft.

Remove the three bolts securing the coupling cover to the bearing housing.

Remove the engine, see Chapter 8.

Remove the jet pump, see Section 10-3.

All Series

Remove the four bolts securing the carrier bearing housing to the bulkhead. There may be one or more pieces of shim material behind the mounting holes (550 Series only). Save the pieces of shim material and make a note of their locations. Identifying the shim material, at this time, will **ENSURE** correct shaft alignment at the pump.

This bearing on all other Models covered in this manual has been redesigned, eliminating the need for shim material -- alignment purposes -- and a zirc fitting -- the fittings are sealed.

All Models
Except 550 Series

The driveshaft may be pulled forward out of the bearing housing.

550 Series

Cut away any silicone sealant from the edges of the housing flange and then pull the bearing housing, with the driveshaft attached, from the bulkhead.

Clean off all traces of silicone sealer from the housing flange and the hull.

Bearing Housing
and Driveshaft Disassembling
All Series

Refer to the accompanying illustrations and exploded diagram as an assist in disassembling and take note of the proper location of items in the assembly.

Bearing housing removed and ready for disassembling, if required.

Remove only the parts which will be replaced.

If the bearing inside the housing is satisfactory, do not remove it. If only the two sets of oil seals are to be replaced, the seals may be pulled from the housing and the bearing left in place.

Insert the pin in the adaptor into the hole in the driveshaft. Clamp the driveshaft holder tool around the adaptor and shaft. Insert the coupling holder into the metal coupler and unscrew the coupler from the threaded end of the driveshaft. Remove the washer.

SAFETY WORDS

The snap ring mentioned in the following step is under tremendous pressure when installed. Therefore, wear a pair of safety glasses and exercise extra **CARE** when removing the snap ring to prevent injury to self or others in the area.

Using a pair of internal snap ring pliers, remove the snap ring at the forward face of the housing. Make a setup in an arbor press to remove the seals and bearing from the housing. With the forward face down over a wooden support, press on the top seal to remove the two large forward seals, the bearing, and finally the two small aft oil seals.

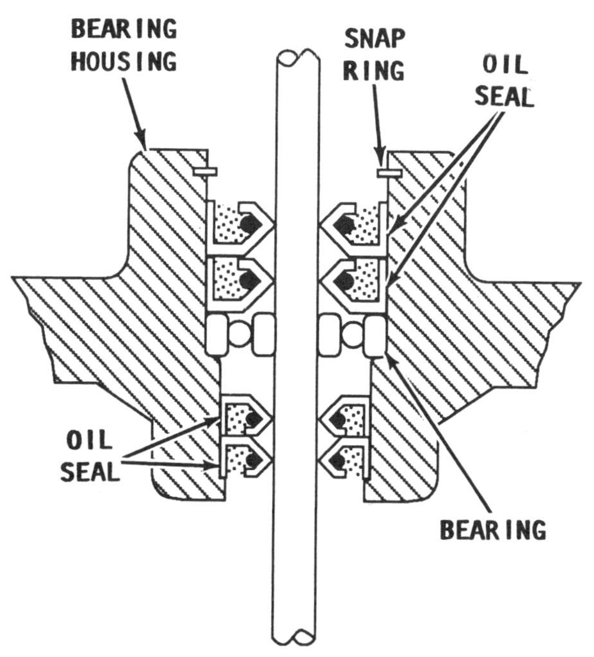

Cross-section line drawing of the bearing housing to identify the relationship of the various parts mentioned in the text.

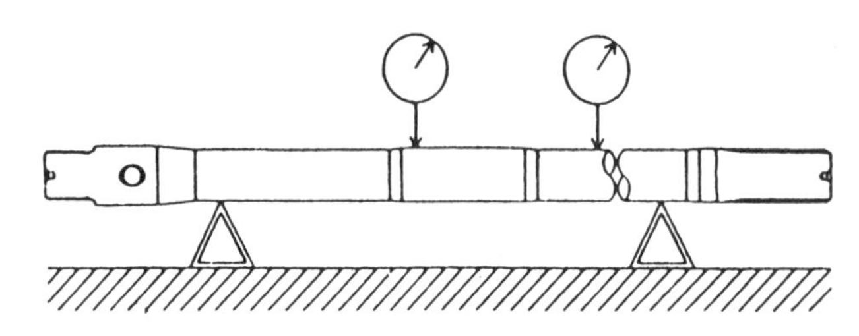

Typical of the process for testing drive shaft runout by placing the driveshaft on V blocks. Place a dial gauge against the shaft at each point. Turn the shaft slowly and note the difference between the highest and lowest reading, which is considered the runout.

CLEANING AND INSPECTING ALL SERIES

Place the driveshaft on a set of Vee blocks and install a dial indicator over the center of the shaft. Slowly rotate the shaft and observe the dial indicator. A shaft in satisfactory condition will have a runout less than 0.2mm. If the runout is more than 0.5mm, replace the shaft.

Next, position the dial indicator over the two bearing surfaces, and then the sleeve surface, if equipped. Measure the runout. A shaft in satisfactory condition will have runout less than 0.03mm. If the runout is more than 0.1mm, replace the driveshaft.

Rinse the bearings in a high flash point solvent and inspect them. Spin each bearing by hand -- never with compressed air -- and feel for rough spots. Inspect for water damage.

Bearing Housing and Driveshaft Assembling All Series

Pack the lips of all four **NEW** oil seals with Shell Alvania EP1 lubricant or equivalent multi-purpose water resistant lubricant.

Using a suitable mandrel, press the two small aft oil seals into the housing from the

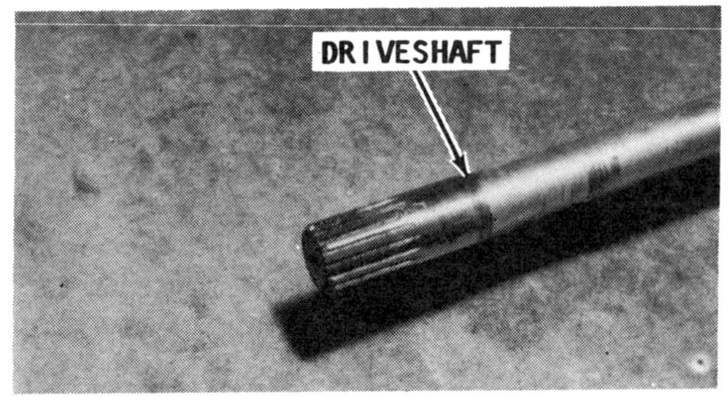

*Splined end of a Model **650** Series driveshaft. The splines index with the splines in the impeller shaft. If the splines become excessively worn or damaged, impeller rotation will be seriously restricted.*

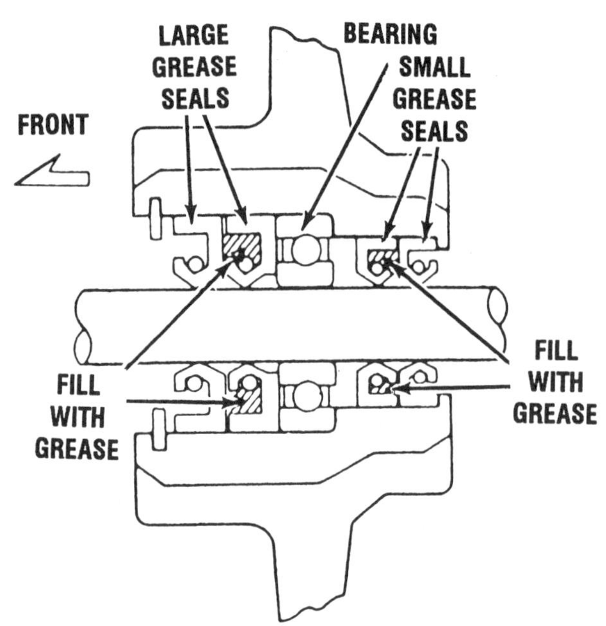

Line drawing of the bearing housing. The gaps between the sets of large and small seals should be filled with grease as shown.

forward side, with both seal lips facing **DOWN** -- aft -- after installation.

Press the bearing into the housing until it makes contact with a shoulder inside the housing.

Next, press in the two large oil seals, with the seal lips facing **UP** -- forward -- after installation. The seals are correctly installed when the groove for the snap ring is visible.

SAFETY WORDS

The snap ring mentioned in the following paragraph is under tremendous pressure while it is being installed. Therefore, wear a pair of safety glasses and exercise extra **CARE** when installing the snap ring to prevent injury to self or others in the area.

Install the snap ring.

Insert the pin in the adaptor into the hole in the driveshaft. Clamp the driveshaft holder tool around the adaptor and shaft. Place the washer over the threaded end of the driveshaft. Thread the coupler onto the driveshaft. Insert the coupling holder into the metal coupler and tighten the coupler to the threaded end of the driveshaft to a torque value of 29 ft lb (39 Nm).

Bearing Housing Installation All Series

Check to be sure the hull surface and the bearing housing surface are free of any old sealant.

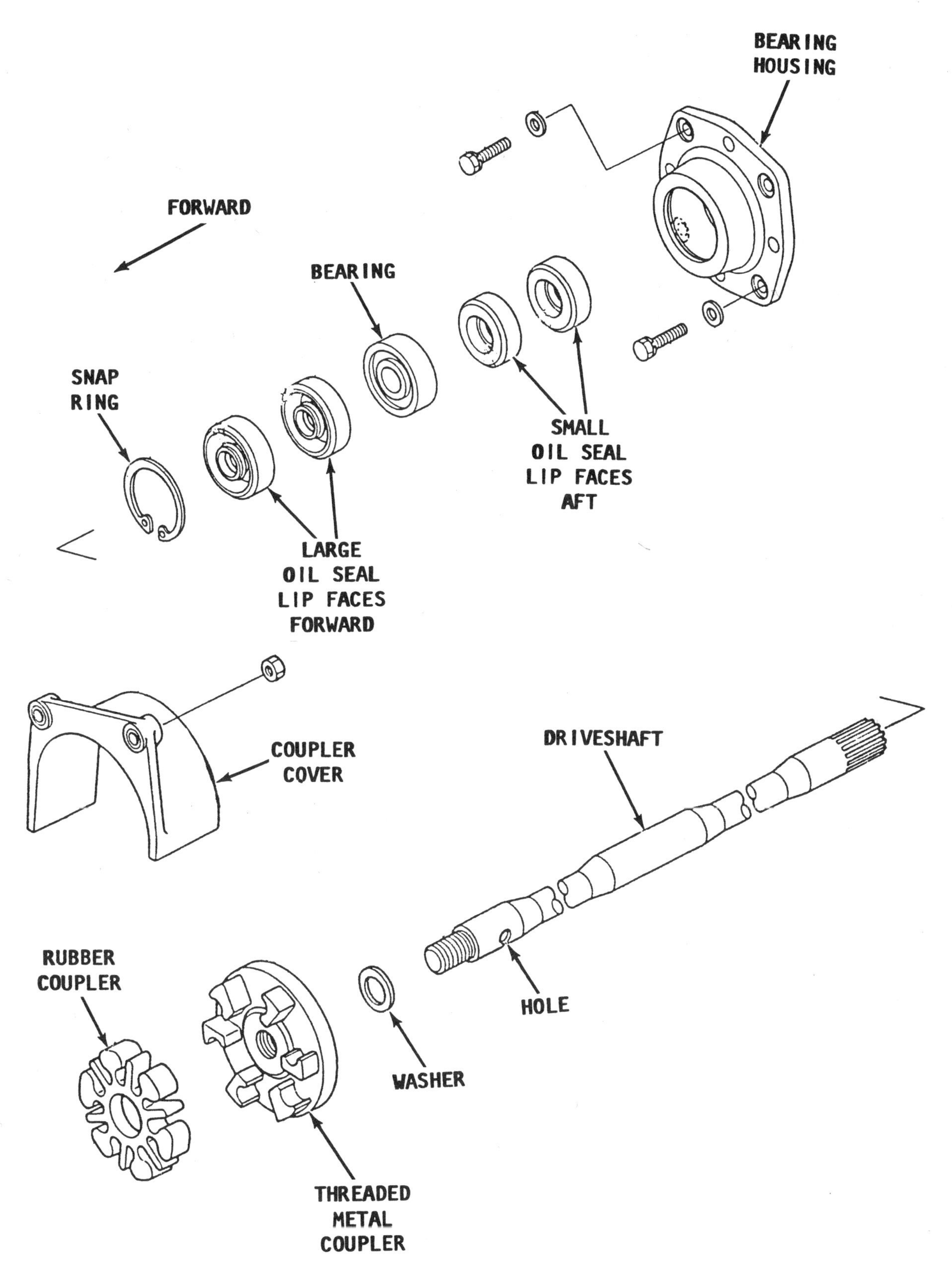

Exploded drawing of a typical bearing housing installed on all series watercraft covered in this manual. Major parts are identified.

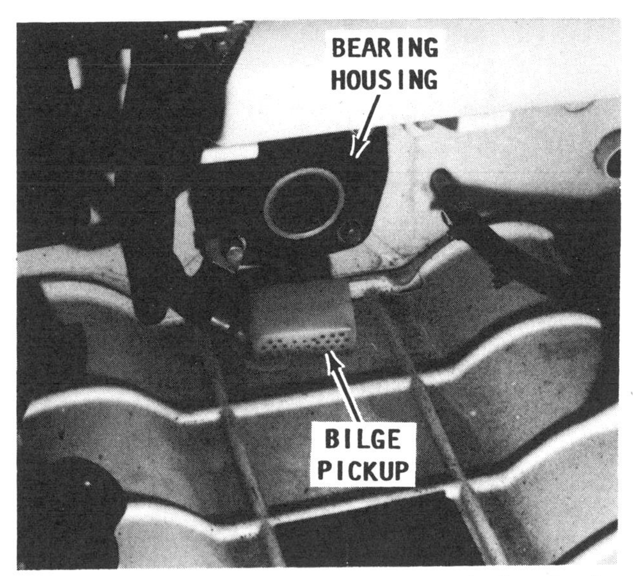

A new bearing housing in place ready for the jet pump and engine.

Position any shim material removed in disassembly in the exact original location. The purpose of the shim material is to **ENSURE** correct driveshaft alignment at the pump.

Next, measure the distance between the center of the driveshaft and the two side walls of the hull cavity. This distance must be equal on both sides. If the two measurements are not equal, shim material must be added or subtracted, in equal amounts, to the two left mounting corners/feet or the two right mounting corners/feet, so as not to disturb the vertical setting.

Once the shimming is completed, take **TIME** and **CARE** to identify the four new stacks of shim material and remove the bearing housing from the hull.

Once the jet pump is installed, the bilge and cooling hoses are attached to the proper fittings. The bilge hose is being attached in the above illustration.

Apply a bead of Loctite Superflex around the perimeter of the mounting flange, looping around the four bolt holes. Install the bearing housing with the correct stack of shim material behind each corner. Apply a coating of Loctite Lock N' Seal to the attaching bolt threads. Install the plain washers, lockwashers and bolts. Tighten the bolts to a torque value of 13.5 ft lb (19 Nm).

Insert the driveshaft into the bearing housing and index the threads into the impeller.

Position the rubber coupler inside the driveshaft metal coupler.

11
CONTROL ADJUSTMENTS

11-1 INTRODUCTION

Control adjustments on Kawasaki Personal Watercraft is limited to making adjustments to the steering cable, reverse cable, and trim system -- if so equipped. Normally, adjustments are not necessary for extended periods of time, unless the craft has experienced an accident or the control has been disconnected to service other components.

Detailed procedures to adjust the steering, reverse, and trim cables are presented in this short chapter. Adjustment of the throttle cable and the choke cable is thoroughly covered in Chapter 6, **FUEL**.

11-2 STEERING CABLE ADJUSTMENT ALL MODELS

If the steering nozzle is not positioned exactly in the center of the transom opening, with an equal clearance on both sides, the steering cable **MUST** be adjusted to prevent the craft from veering to port or starboard when the handle bar is positioned for "dead ahead".

Adjustment

1- Center the handle bar for a "dead ahead" course. Have an assistant prevent the handle bar from moving from this straight ahead position.

2- At the stern, measure the clearance, port and starboard, between the outside of the steering nozzle and the transom opening. Both measurements should be the same. If the measurements vary by more than 1/16", perform Steps 3 thru Step 7 to properly adjust the nozzle.

3- Raise the handle pole to the upright position. The steering cable is attached to a ball joint on the steering plate around the pole. Loosen the locknut on the steering cable.

4- Push the outer sleeve back from the ball joint and pry the cable end free.

5- With the handle bar still in the "dead ahead" position, use a tape measure to center the steering nozzle within the transom cavity.

6- Without disturbing the position of the handle bar or the position of the nozzle, rotate the cable end until the hole in the end aligns with the ball joint on the steering plate.

7- Push back the outer sleeve on the cable end, and then snap the cable end over the ball joint. Make another check of the steering nozzle position in relation to the handle still at "dead ahead". If the measurement, port and starboard has remained the same, tighten the locknut to hold the new adjustment.

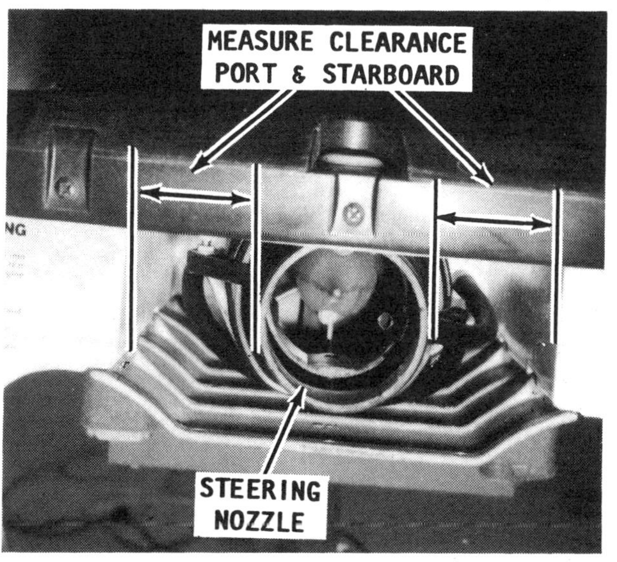

Close view of the steering nozzle to clearly indicate where the clearance measurements are to be taken when making a steering cable adjustment, on all models.

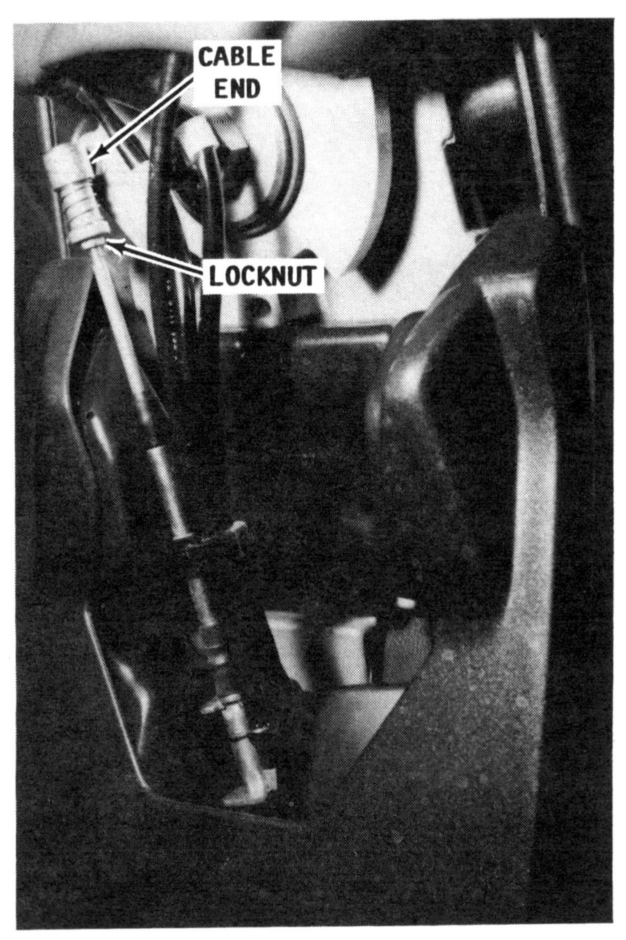

End fitting for the steering cable on a 550 and 650 Series Model.

8- A final check **MUST** be made with the craft in the water to ensure proper cable adjustment. Therefore, move the craft to a body of water or to a test tank.

9- Start the engine and allow it to warm to operating temperature.

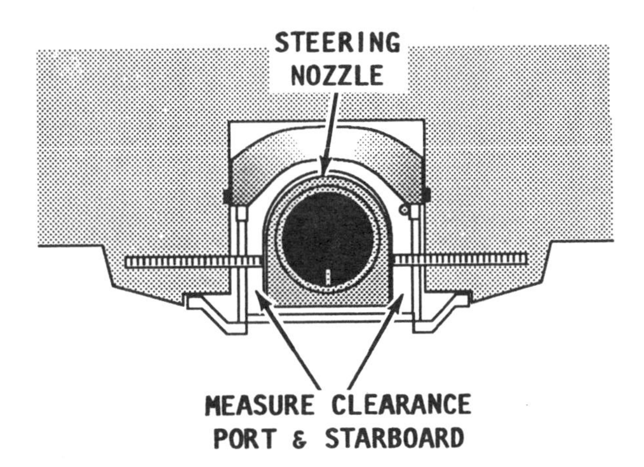

Line drawing to depict where the measurement between the nozzle and the edge of the transom opening is to be taken -- port and starboard -- when making a steering cable adjustment.

CAUTION

Water must circulate through the jet pump -- to and from the engine, anytime the engine is operating. Circulating water will prevent overheating -- which could cause damage to moving engine parts and possible engine seizure.

10- With the engine operating, hold the handlebar in the "dead ahead" position; advance the throttle and observe if the bow tends to veer to port or starboard. If the bow remains unmoved, the adjustment is accurate -- task completed.

11-3 REVERSE CABLE ADJUSTMENT

The position of the reverse bucket determines the direction of thrust from water expelled from the nozzle. If the bucket is not adjusted properly forward motion of the craft will be seriously hampered because some of the expelled water will be forced back under the craft -- as for going astern.

More importantly, improper adjustment of the reverse bucket can cause the bucket to be unintentionally lowered -- throwing the craft into reverse -- suddenly raising the bow up out of the water -- causing possible injury to operator and/or passengers.

A shifting cable is attached between the shift lever arm ball joint to the ball joint on the reverse bucket. To check adjustment of this cable, proceed as follows:

1- Move the shift lever to the full **FORWARD** position. The rubber damper on top of the reverse bucket should make contact with the pump housing.

2- Move the shift lever aft to the full **REVERSE** position. The two rubber dampers on the lower edge of the reverse bucket should make contact with the pump cover, as indicated in the accompanying illustration.

If the three dampers fail to make contact, as described, the shift cable **MUST** be adjusted at the reverse bucket pivot barrel.

3- Determine if the bucket is adjusted too high or too low.

POSITION WORDS

If the two lower dampers fail to make contact, but the upper damper makes contact, as it should -- the bucket is adjusted too high -- the cable must be shortened.

If the upper damper fails to make contact, but the lower dampers do, as they

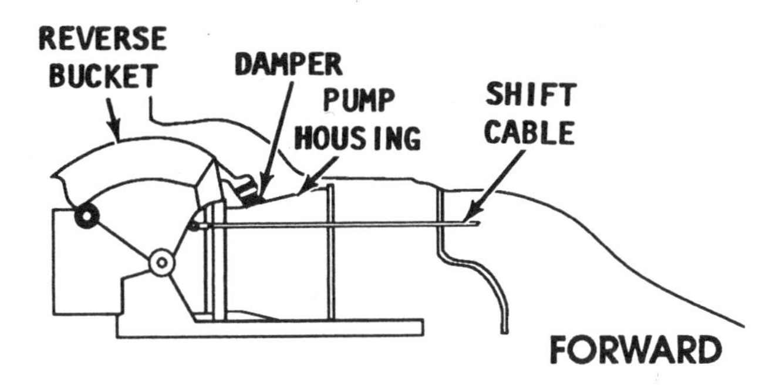

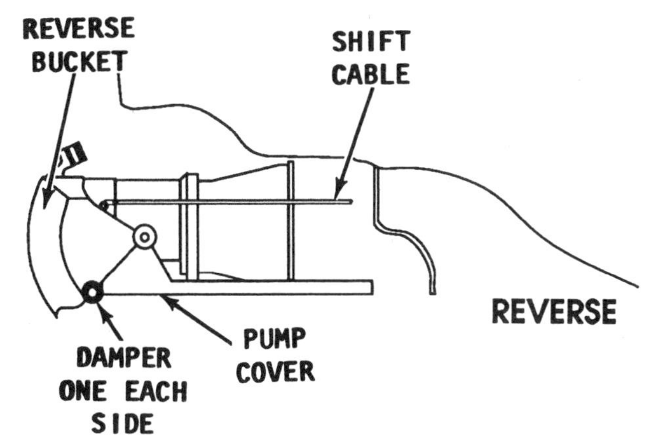

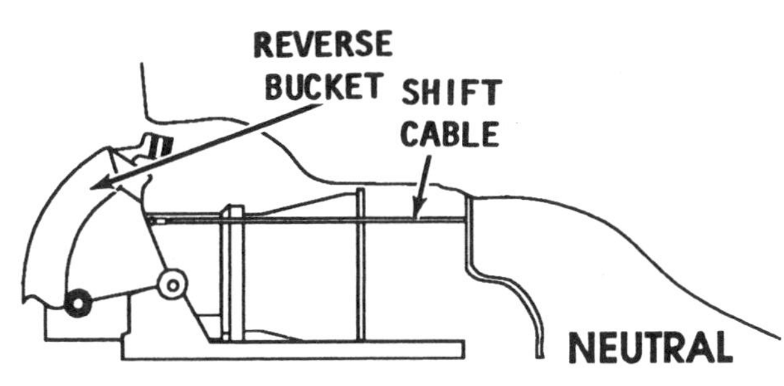

Line drawing to depict the reverse bucket in full ***FORWARD*** *(top),* ***REVERSE*** *(center), and in the full* ***NEUTRAL*** *position (bottom).*

should -- the bucket is adjusted too low -- the shift cable must be lengthened.

4- Pry the cable end free of the ball joint on the reverse bucket and loosen the locknut. Rotate the cable end **COUNTER-CLOCKWISE** to lengthen the cable or **CLOCKWISE** to shorten the cable, as necessary.

5- Push the outer sleeve down the cable and snap the cable end onto the ball joint.

6- Shift the reverse bucket again from full forward to full reverse and observe the dampers making contact -- on top for full forward -- on the bottom for full reverse. If the dampers make contact, as described, the adjustment is correct.

7- Tighten the locknut to hold the new adjustment.

SPECIAL WORDS

A final check **MUST** be made with the craft in the water to ensure proper adjustment has been attained.

8- Move the craft to a body of water or to a test tank. Start the engine and allow it to warm to operating temperature.

CAUTION

Water must circulate through the jet pump -- to and from the engine, anytime the engine is operating. Circulating water will prevent overheating -- which could cause damage to moving engine parts and possible engine seizure.

9- With the craft in the water; with the engine operating; and the shift lever in the **NEUTRAL** position; the craft **MUST** not move forward or sternward at any engine rpm.

An adjustment to the steering cable end at the reverse bucket may be made immediately while the craft is in the water.

EXCEPTION

If the three dampers fail to make contact, as described, but the craft does not move forward or sternward in the water with the engine operating, such misadjustment can be considered acceptable.

11-4 TRIM CABLE ADJUSTMENT

1- Turn the ignition switch **ON**.

2- Using the electric trim switch, center the trim meter needle at the level position.

3- Measure the distance from the pump

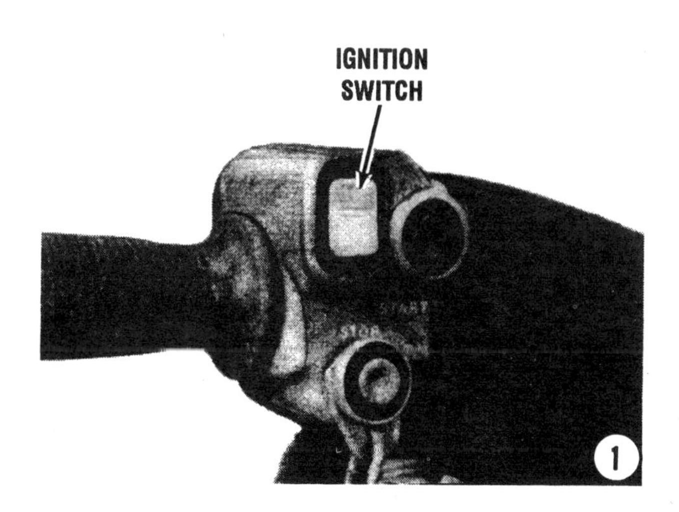

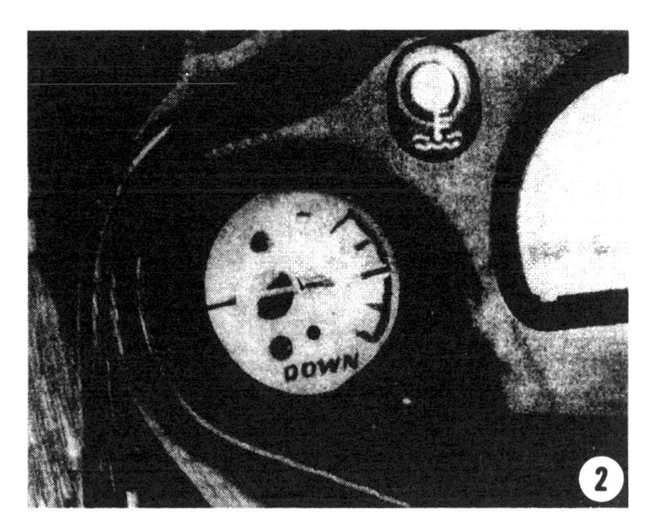

locations as shown in the illustration. Both distances should be equal. If they are not the same, adjustment is necessary, proceed with the next step.

4- Disconnect the trim cable ball joint by sliding the outer sleeve away from the post slightly, and lifting the cable from the ball. Rotate the ball joint on the cable to adjust the tilt angle. Connect the ball joint and check the cable adjustment again. Once adjustment is correct, tighten the trim cable locknut. Turn the ignition switch **OFF**.

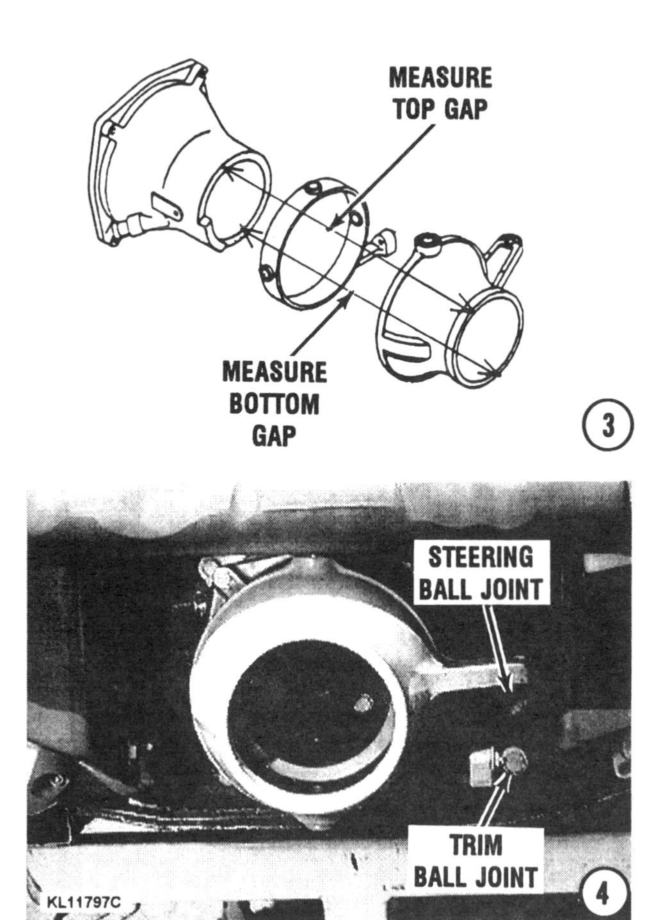

APPENDIX

METRIC CONVERSION CHART

LINEAR

inches	X 25.4	= millimetres (mm)
feet	X 0.3048	= metres (m)
yards	X 0.9144	= metres (m)
miles	X 1.6093	= kilometres (km)
inches	X 2.54	= centimetres (cm)

AREA

$inches^2$	X 645.16	= $millimetres^2$ (mm^2)
$inches^2$	X 6.452	= $centimetres^2$ (cm^2)
$feet^2$	X 0.0929	= $metres^2$ (m^2)
$yards^2$	X 0.8361	= $metres^2$ (m^2)
acres	X 0.4047	= hectares (10^4 m^2) (ha)
$miles^2$	X 2.590	= $kilometres^2$ (km^2)

VOLUME

$inches^3$	X 16387	= $millimetres^3$ (mm^3)
$inches^3$	X 16.387	= $centimetres^3$ (cm^3)
$inches^3$	X 0.01639	= litres (l)
quarts	X 0.94635	= litres (l)
gallons	X 3.7854	= litres (l)
$feet^3$	X 28.317	= litres (l)
$feet^3$	X 0.02832	= $metres^3$ (m^3)
fluid oz	X 29.60	= millilitres (ml)
$yards^3$	X 0.7646	= $metres^3$ (m^3)

MASS

ounces (av)	X 28.35	= grams (g)
pounds (av)	X 0.4536	= kilograms (kg)
tons (2000 lb)	X 907.18	= kilograms (kg)
tons (2000 lb)	X 0.90718	= metric tons (t)

FORCE

ounces - f (av)	X 0.278	= newtons (N)
pounds - f (av)	X 4.448	= newtons (N)
kilograms - f	X 9.807	= newtons (N)

ACCELERATION

$feet/sec^2$	X 0.3048	= $metres/sec^2$ (m/S^2)
$inches/sec^2$	X 0.0254	= $metres/sec^2$ (m/s^2)

ENERGY OR WORK (watt-second - joule - newton-metre)

foot-pounds	X 1.3558	= joules (j)
calories	X 4.187	= joules (j)
Btu	X 1055	= joules (j)
watt-hours	X 3500	= joules (j)
kilowatt - hrs	X 3.600	= megajoules (MJ)

FUEL ECONOMY AND FUEL CONSUMPTION

miles/gal	X 0.42514	= kilometres/litre (km/l)

Note:
235.2/(mi/gal) = litres/100km
235.2/(litres/100 km) = mi/gal

LIGHT

footcandles	X 10.76	= $lumens/metre^2$ (lm/m^2)

PRESSURE OR STRESS (newton/sq metre - pascal)

inches HG (60 F)	X 3.377	= kilopascals (kPa)
pounds/sq in	X 6.895	= kilopascals (kPa)
inches H_2O (60° F)	X 0.2488	= kilopascals (kPa)
bars	X 100	= kilopascals (kPa)
pounds/sq ft	X 47.88	= pascals (Pa)

POWER

horsepower	X 0.746	= kilowatts (kW)
ft-lbf/min	X 0.0226	= watts (W)

TORQUE

pound-inches	X 0.11299	= newton-metres (N·m)
pound-feet	X 1.3558	= newton-metres (N·m)

VELOCITY

miles/hour	X 1.6093	= kilometres/hour (km/h)
feet/sec	X 0.3048	= metres/sec (m/s)
kilometres/hr	X 0.27778	= metres/sec (m/s)
miles/hour	X 0.4470	= metres/sec (m/s)

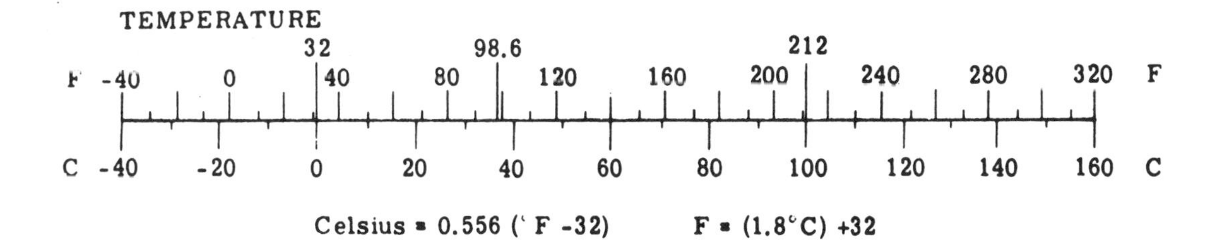

Celsius = 0.556 (°F -32)   F = (1.8°C) +32

RECOMMENDED TORQUE VALUES

SERIES:	JS750-A:SX	JS750-B:SXi	JT750-A:ST	JT750-B:STS	JT900-A:STX	JH900-A:ZXi	JH1100-A:ZXi	JT1100-A:STX
Flame Arr. Case Bolts	69 in lb (7.8 Nm)	69 in lb (7.8 Nm)	69 in lb (7.8 Nm)	69 in lb (7.8 Nm)	69 in lb (7.8 Nm)	69 in lb (7.8 Nm)	69 in lb (7.8 Nm)	69 in lb (7.8 Nm)
Intake Manifold Nuts	69 in lb (7.8 Nm)	87 in lb (9.8Nm)	87 in lb (9.8Nm)	87 in lb (98 Nm)	87 in lb (98 Nm)	87 in lb (98 Nm)	87 in lb (98 Nm)	87 in lb (98 Nm)
Exhaust Manifold Bolts/Nuts	14.5 ft lb (20 Nm)	13.5 ft lb (19 Nm)	13.5 ft lb (19 Nm)	13.5 ft lb (19 Nm)	14.5 ft lb (20 Nm)	14.5 ft lb (20 Nm)	14.5 ft lb (20 Nm)	14.5 ft lb (20 Nm)
Cylinder Head Nuts	22 ft lb (29 Nm)	22 ft lb (29 Nm)	22 ft lb (29 Nm)	22 ft lb (29 Nm)	22 ft lb (29 Nm)	22 ft lb (29 Nm)	22 ft lb (29 Nm)	22 ft lb (29 Nm)
Engine BED Mounting Bolts	27 ft lb (36 Nm)	27 ft lb (36 Nm)	27 ft lb (36 Nm)	27 ft lb (36 Nm)	27 ft lb (36 Nm)	27 ft lb (36 Nm)	27 ft lb (36 Nm)	27 ft lb (36 Nm)
Engine Mounting Bolts	27 ft lb (36 Nm)	33 ft lb (44 Nm)	33 ft lb (44 Nm)	33 ft lb (44 Nm)	33 ft lb (44 Nm)	33 ft lb (44 Nm)	33 ft lb (44 Nm)	33 ft lb (44 Nm)
Crankcase Large Bolts	22 ft lb (29 Nm)	22 ft lb (29 Nm)	22 ft lb (29 Nm)	22 ft lb (29 Nm)	22 ft lb (29 Nm)	22 ft lb (29 Nm)	22 ft lb (29 Nm)	22 ft lb (29 Nm)
Crankcase Small Bolts	69 in lb (7.8 Nm)	69 in lb (7.8 Nm)	69 in lb (7.8 Nm)	69 in lb (7.8 Nm)	69 in lb (7.8 Nm)	69 in lb (7.8 Nm)	78 in lb (8.8 Nm)	78 in lb (8.8 Nm)
Coupling (Engine Side)	72 ft lb (98 Nm)	72 ft lb (98 Nm)	72 ft lb (98 Nm)	72 ft lb (98 Nm)	98 ft lb (130 Nm)	72 ft lb (98 Nm)	94 ft lb (125 Nm)	94 ft lb (125 Nm)
Oil Pump Bolts	69 in lb (7.8 Nm)	69 in lb (7.8 Nm)	69 in lb (7.8 Nm)	69 in lb (7.8 Nm)	69 in lb (7.8 Nm)	69 in lb (7.8 Nm)	78 in lb (8.8 Nm)	78 in lb (8.8 Nm)
Flywheel Bolt	94 ft lb (125 Nm)	94 ft lb (125 Nm)	94 ft lb (125 Nm)	94 ft lb (125 Nm)	94 ft lb (125 Nm)	94 ft lb (125 Nm)	94 ft lb (125 Nm)	94 ft lb (125 Nm)
Flywheel Cover Bolts	69 in lb (7.8 Nm)	69 in lb (7.8 Nm)	69 in lb (7.8 Nm)	69 in lb (7.8 Nm)	69 in lb (7.8 Nm)	69 in lb (7.8 Nm)	78 in lb (8.8 Nm)	78 in lb (8.8 Nm)

RECOMMENDED TORQUE VALUES

DESCRIP:	JS550-C:SX	JF650-A:X2	JF650-B:TS	JL650-A:SC	JS650-B:SX	JH750-A:SS	JH750-B:Xi	JH750-C:ZXi	JH750-D:XiR
Flame Arr. Case Bolts	N/A	N/A	N/A	N/A	N/A	69 in lb (7.8 Nm)	69 in lb (7.8 Nm)	69 in lb (7.8 Nm)	69 in lb (7.8 Nm)
Intake Manifold Nuts	69 in lb (7.8 Nm)	69 in lb (7.8 Nm)	69 in lb (7.8 Nm)	69 in lb (7.8 Nm)	69 in lb (7.8 Nm)	69 in lb (7.8 Nm)	69 in lb (7.8 Nm)	87 in lb (9.8Nm)	69 in lb (7.8 Nm)
Exhaust Manifold Bolts/Nuts	52 in lb (6 Nm)	18 ft lb (25 Nm)	18 ft lb (25 Nm)	18 ft lb (25 Nm)	18 ft lb (25 Nm)	14.5 ft lb (20 Nm)	14.5 ft lb (20 Nm)	14.5 ft lb (20 Nm)	14.5 ft lb (20 Nm)
Cylinder Head Nuts	18 ft lb (25 Nm)	22 ft lb (29 Nm)	22 ft lb (29 Nm)	22 ft lb (29 Nm)	22 ft lb (29 Nm)	22 ft lb (29 Nm)	22 ft lb (29 Nm)	22 ft lb (29 Nm)	22 ft lb (29 Nm)
Engine BED Mounting Bolts	27 ft lb (37 Nm)	14.5 (20 Nm)	14.5 (20 Nm)	14.5 (20 Nm)	14.5 (20 Nm)	27 ft lb (36 Nm)	27 ft lb (36 Nm)	27 ft lb (36 Nm)	27 ft lb (36 Nm)
Engine Mounting Bolts	35 ft lb (48 Nm)	27 ft lb (36 Nm)	27 ft lb (36 Nm)	27 ft lb (36 Nm)	27 ft lb (36 Nm)	27 ft lb (36 Nm)	27 ft lb (36 Nm)	36 ft lb (49 Nm)	27 ft lb (36 Nm)
Crankcase Large Bolts	18 ft lb (25 Nm)	22 ft lb (29 Nm)	22 ft lb (29 Nm)	22 ft lb (29 Nm)	22 ft lb (29 Nm)	22 ft lb (29 Nm)	22 ft lb (29 Nm)	22 ft lb (29 Nm)	22 ft lb (29 Nm)
Crankcase Small Bolts	69 in lb (7.8 Nm)	69 in lb (7.8 Nm)	69 in lb (7.8 Nm)	69 in lb (7.8 Nm)	69 in lb (7.8 Nm)	69 in lb (7.8 Nm)	69 in lb (7.8 Nm)	69 in lb (7.8 Nm)	69 in lb (7.8 Nm)
Coupling (Engine Side)	40 ft lb (54 Nm)	72 ft lb (98 Nm)	72 ft lb (98 Nm)	72 ft lb (98 Nm)	72 ft lb (98 Nm)	72 ft lb (98 Nm)	72 ft lb (98 Nm)	72 ft lb (98 Nm)	72 ft lb (98 Nm)
Oil Pump Bolts	N/A	69 in lb (7.8 Nm)	69 in lb (7.8 Nm)	69 in lb (7.8 Nm)	69 in lb (7.8 Nm)	69 in lb (7.8 Nm)	69 in lb (7.8 Nm)	69 in lb (7.8 Nm)	69 in lb (7.8 Nm)
Flywheel Bolt	N/A	72 ft lb (98 Nm)	72 ft lb (98 Nm)	72 ft lb (98 Nm)	72 ft lb (98 Nm)	94 ft lb (125 Nm)	94 ft lb (125 Nm)	94 ft lb (125 Nm)	94 ft lb (125 Nm)
Flywheel Cover Bolts	12 ft lb (16 Nm)	69 in lb (7.8 Nm)	69 in lb (7.8 Nm)	69 in lb (7.8 Nm)	69 in lb (7.8 Nm)	69 in lb (7.8 Nm)	69 in lb (7.8 Nm)	69 in lb (7.8 Nm)	69 in lb (7.8 Nm)

ENGINE SPECIFICATIONS AND

SERIES	CYL.	DISPL. cc	BORE mm	STROKE mm	IDLE RPM IN WATER	IDLE RPM OUT OF WATER
JS550-C:SX 1992-1994	2	530	75	60	1500 ±100	1900 ± 100
JF650-A:X2 1992-1995	2	635	76	70	1250 ± 100	1800 ± 100
JF650-B:TS 1992-1996	2	635	76	70	1250 ± 100	1800 ± 100
JL650-A:SC 1992-1995	2	635	76	70	1250 ± 100	1800 ± 100
JS650-B:SX 1992-1993	2	635	76	70	1250 ± 100	1800 ± 100
JH750-A:SS 1992-1995	2	743	80	74	1250 ± 100	1700 ± 100
JH750-B:Xi 1992-1995	2	743	80	74	1250 ± 100	1700 ± 100
JH750-C:ZXi 1995 & on	2	743	80	74	1250 ± 100	1700 ± 100
JH750-D:XiR 1994	2	743	80	74	1250 ± 100	1700 ± 100
JS750-A:SX 1992-1995	2	743	80	74	1250 ± 100	1700 ± 100
JS750-B:SXi 1995-1996	2	743	80	74	1250 ± 100	1700 ± 100
JT750-A:ST 1994-1995	2	743	80	74	1250 ± 100	1700 ± 100
JT750-B:STS 1995 & on	2	743	80	74	1250 ± 100	1700 ± 100
JH900-A:ZXi 1995 & on	3	891	73	71	1250 ± 100	1700 ± 100
JT900-A:STX 1997	3	891	73	71	1250 ± 100	1800 ± 100
JH1100-A:ZXi 1996 & on	3	1071	80	71	1250 ± 100	1800 ± 100
JT1100-A:STX 1997	3	1071	80	71	1250 ± 100	1800 ± 100

TUNEUP ADJUSTMENTS

MFG. COMPRESSION RANGE PSI	SPARK PLUG NGK	TIMING DEGREES @ RPM BTDC	CARB. TYPE KIEHIN	LOW SPEED SCREW TURNS OUT	HIGH SPEED SCREW TURNS OUT
78 - 125	BR8ES	21° @ 6,000 rpm	CDK-38	1/1/16	1.0
125 - 192	BR7ES	17° @ 6,000 rpm	CDK-34	1.0	5/8
125 - 192	BR7ES	17° @ 6,000 rpm	CDK-34	1.0	5/8
125 - 192	BR7ES	17° @ 6,000 rpm	CDK-34	1.0	5/8
125 - 192	BR7ES	17° @ 6,000 rpm	CDK-34	1.0	5/8
129 - 199	BR8ES	16° @ 2,500 rpm	(2) CDK-40	1.0	1.0
129 - 199	BR8ES	N/A	(2) CDK-40	N/A	N/A
83 - 135	BR8ES	20.2° @ 4,000 rpm	(2) CDK-40	3/4 ± 1/4	1-1/4 ± 1/4
129 - 199	BR8ES	20° @ 4,000 rpm	(2) CDK-40	1 ± 1/4	3/4 ± 1/4
121 - 187	BR8ES	16° @ 2,500 rpm	(2) CDK-40	7/8	7/8
121 - 187	BR8ES	N/A	(2) CDK-40	N/A	N/A
129 - 199	BR8ES	16° @ 2,500 rpm	(2) CDK-40	1 ± 1/4	1 ± 1/4
129 - 199	BR8ES	16° @ 2,500 rpm	(2) CDK-40	1 ± 1/4	1 ± 1/4
83 - 135	BR9ES	25° @ 3,000 rpm	(3) CDK-38	7/8 ± 1/4	1.0 ± 1/4
83 - 135	BR9ES	25° @ 3,000 rpm	(3) CDK-38	1.0 ± 1/4	1.0 ± 1/4
95 - 151	BR9ES	27° @ 3,000 rpm	(3) CDK-38	1-1/8 ± 1/4	3/4 ± 1/4
95 - 151	BR9ES	27° @ 3,000 rpm	(3) CDK-38	1-1/8 ± 1/4	3/4 ± 1/4

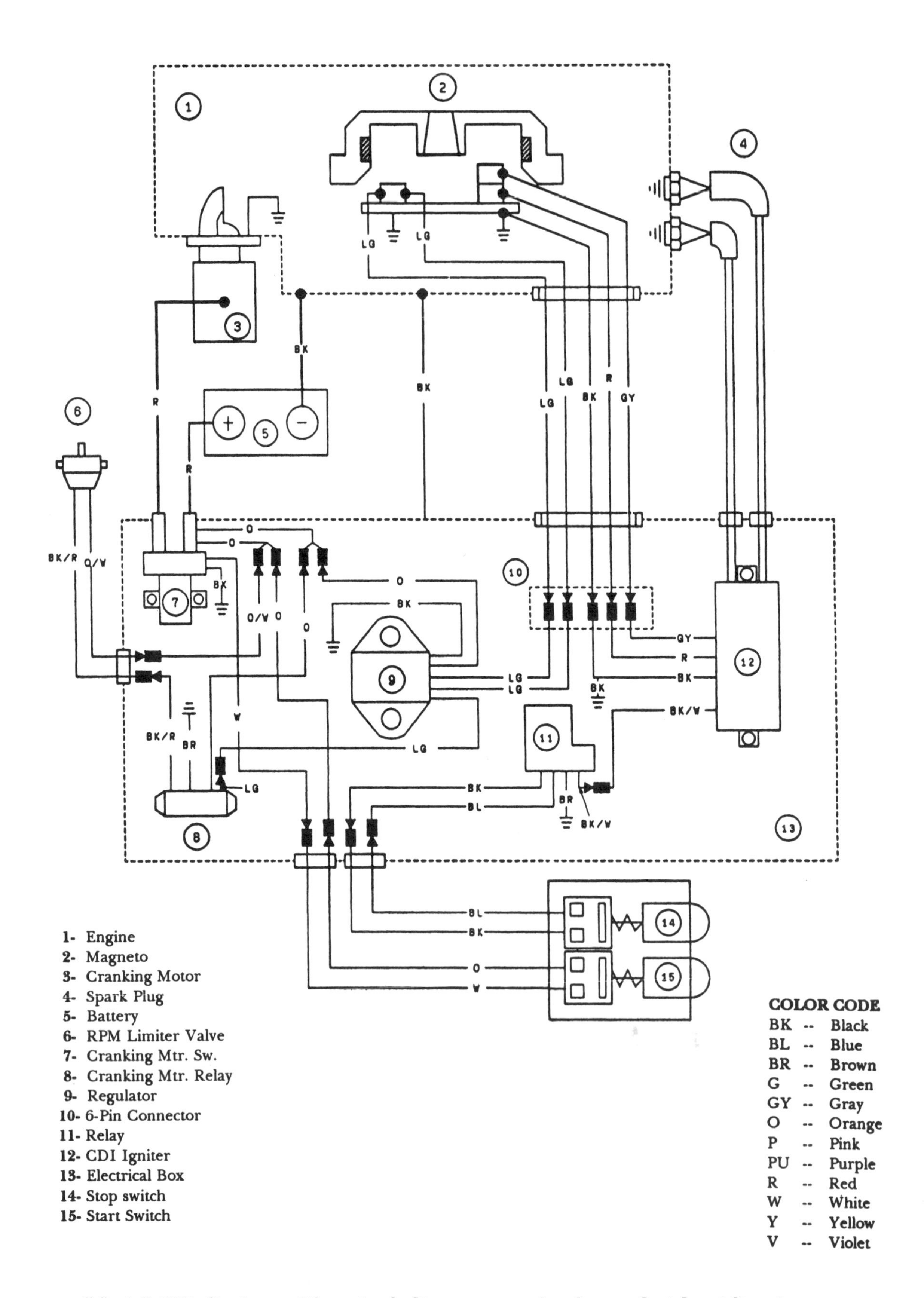

__Model 550 Series__ -- Electrical diagram and color code identification.

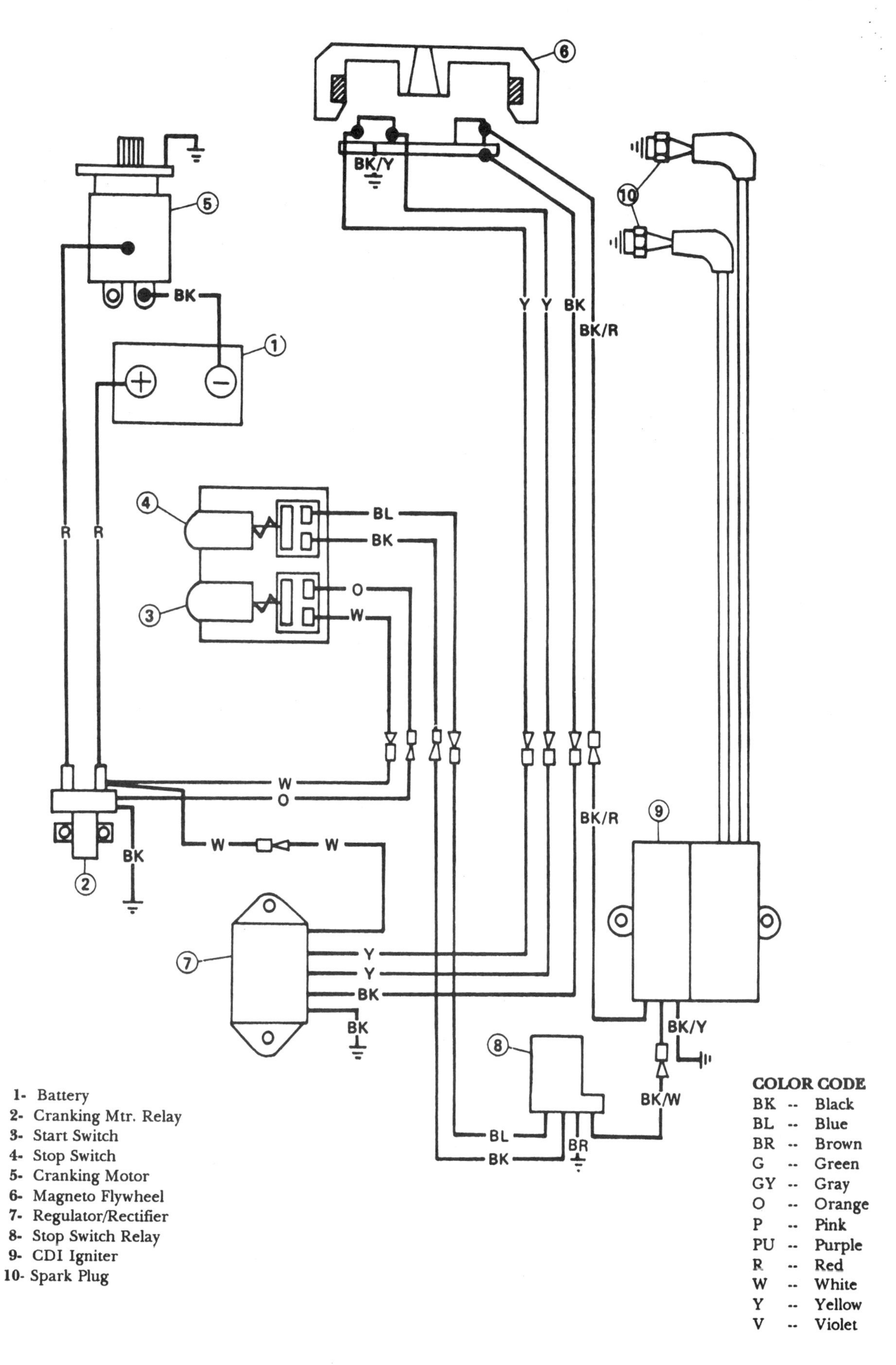

***Model 650 Series** -- Electrical diagram and color code identification.*

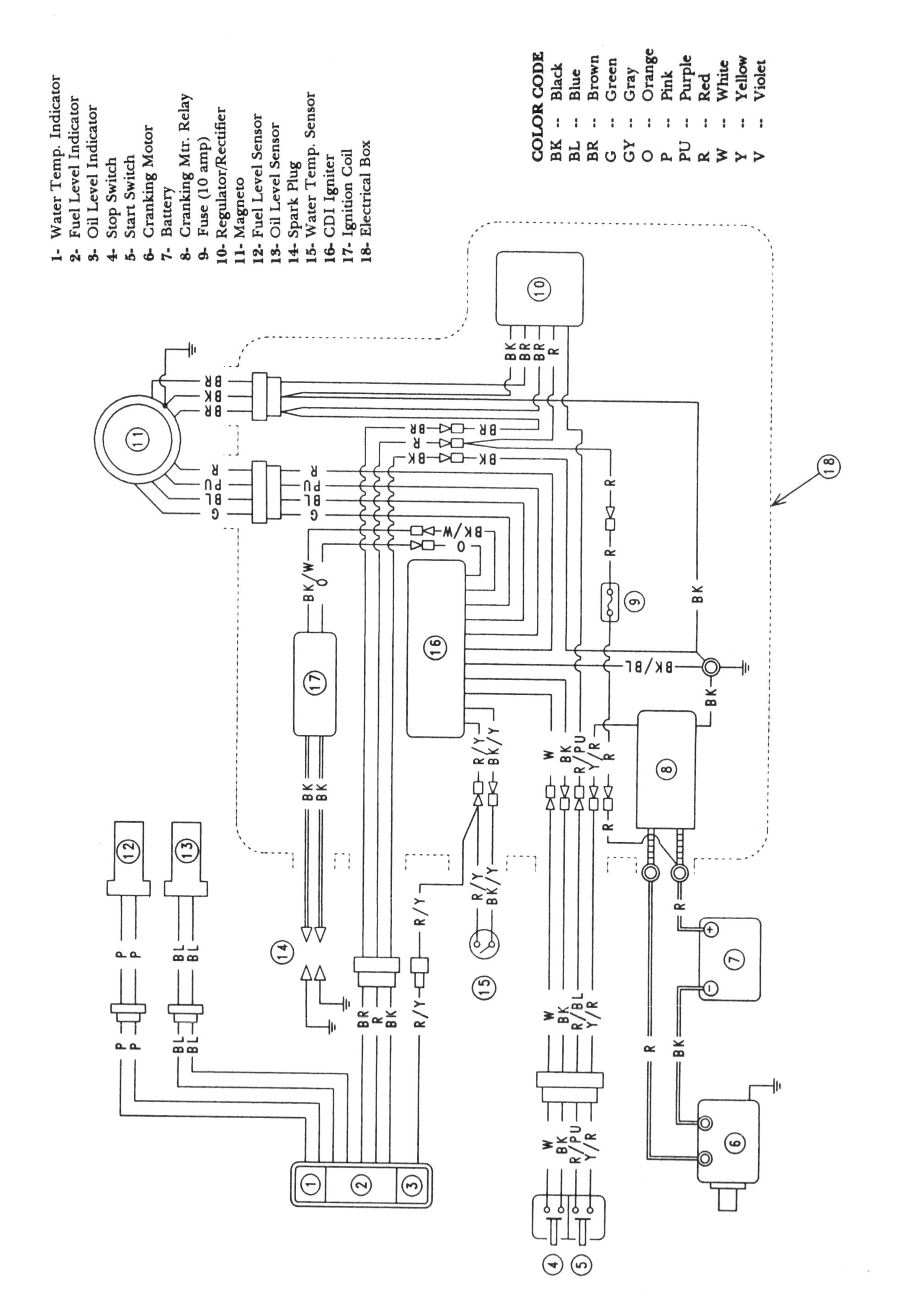

***Standard Model 750 Series** -- Electrical diagram and color code identification.*

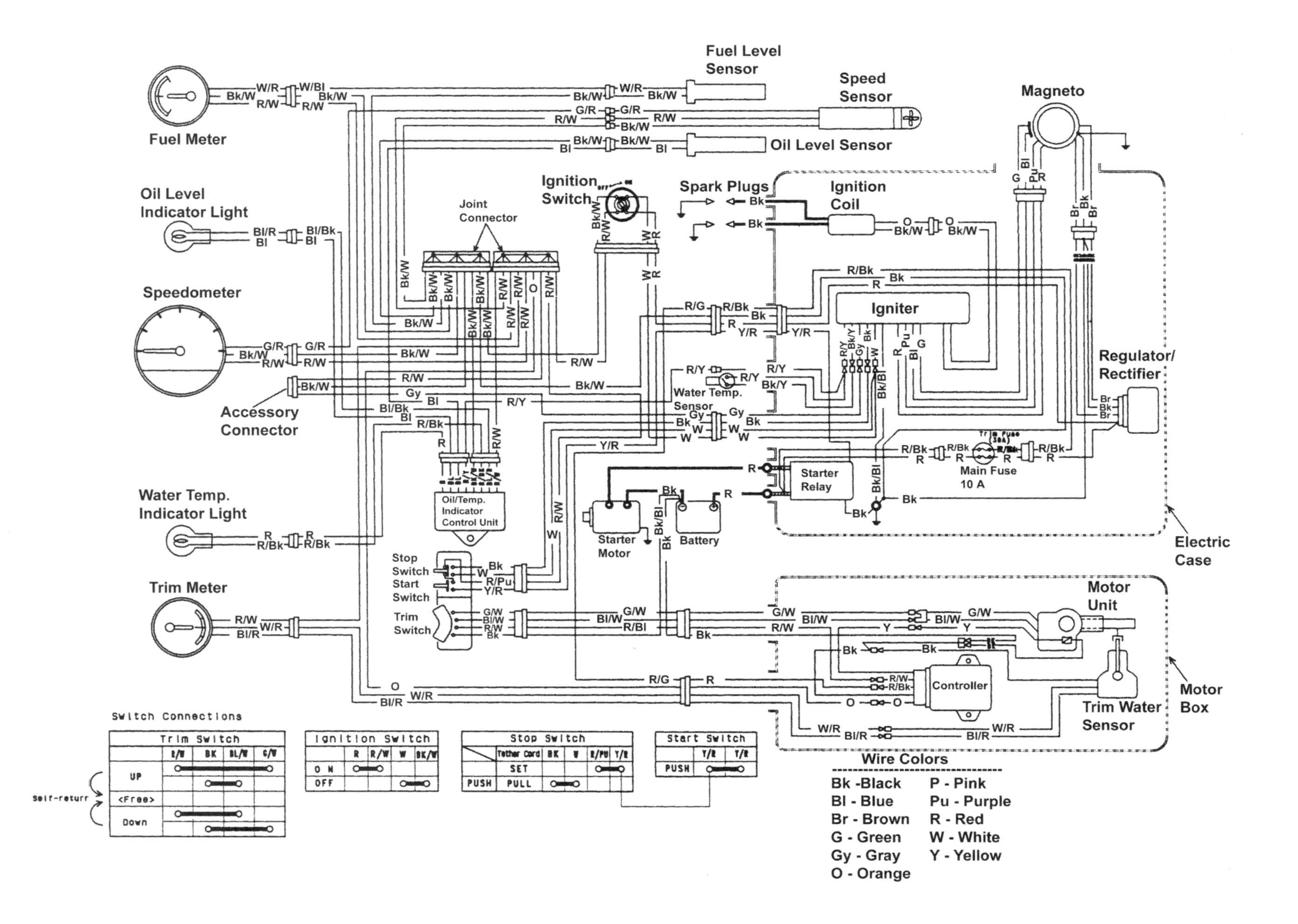

Model 750 High Performance Series – Electrical diagram and color code identification

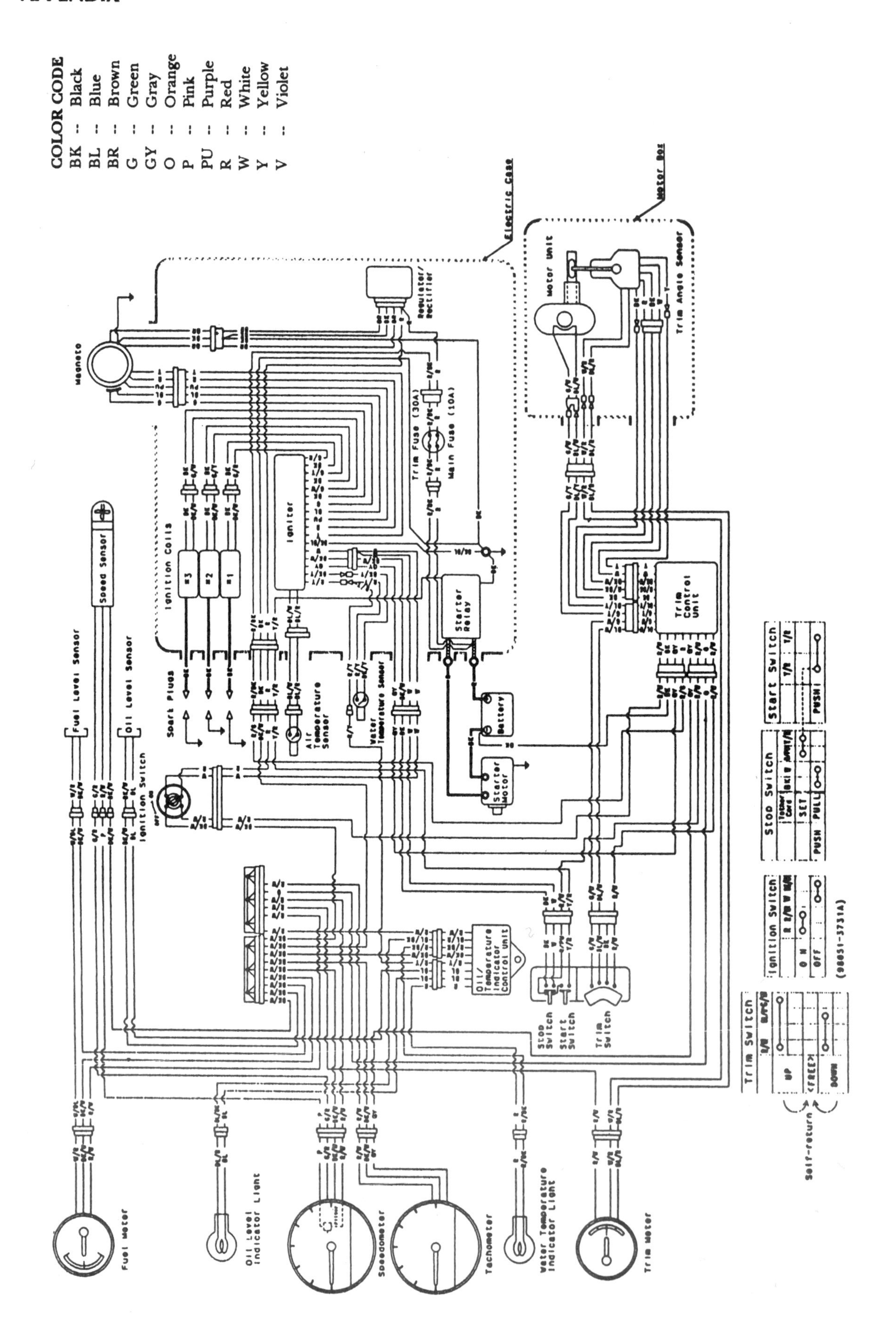

Model 900 High Performance Series -- Electrical diagram and color code identification.

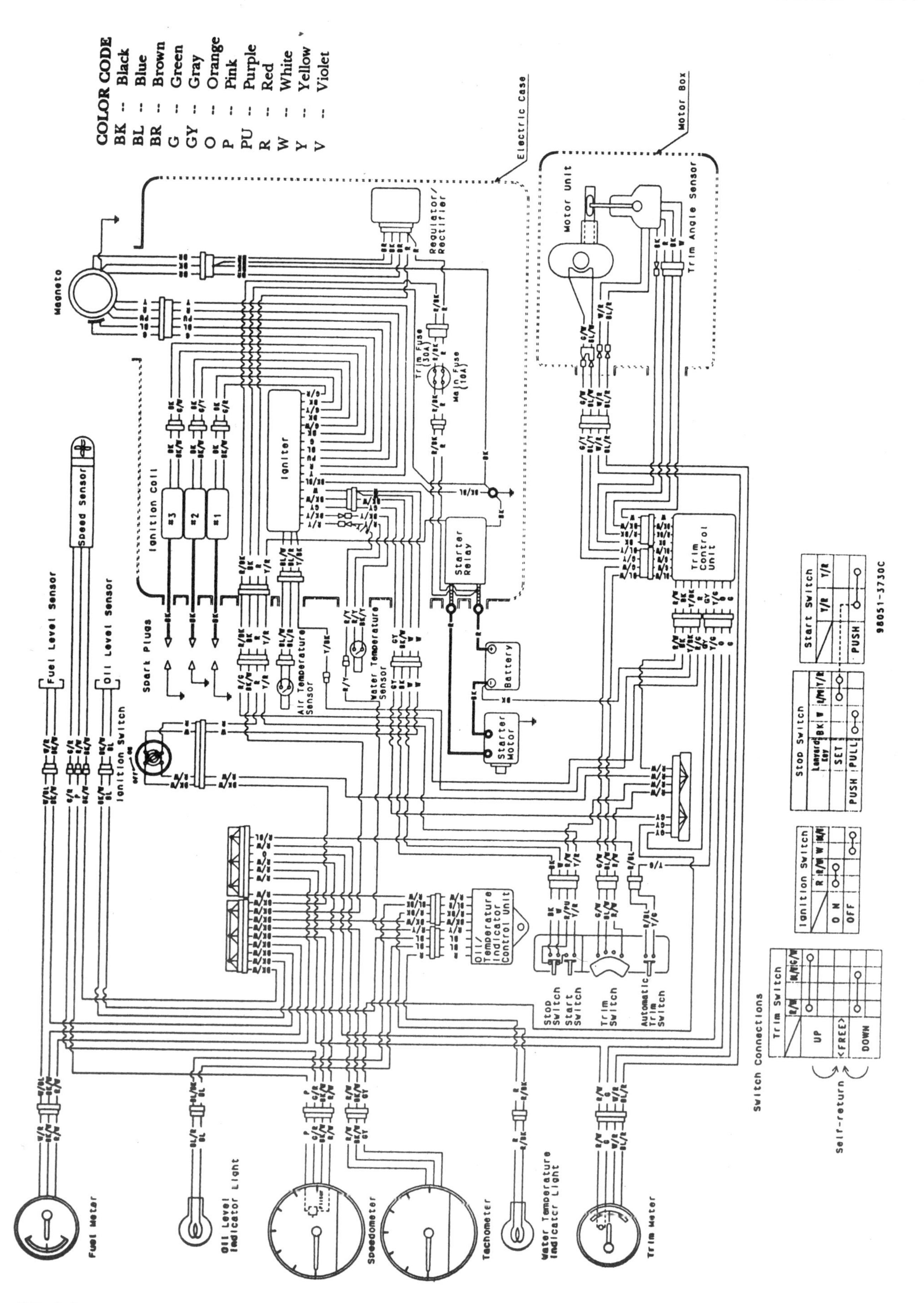

Model 1100 High Performance Series *-- Electrcial diagram and color code identification.*

NOTES & NUMBERS

NOTES & NUMBERS

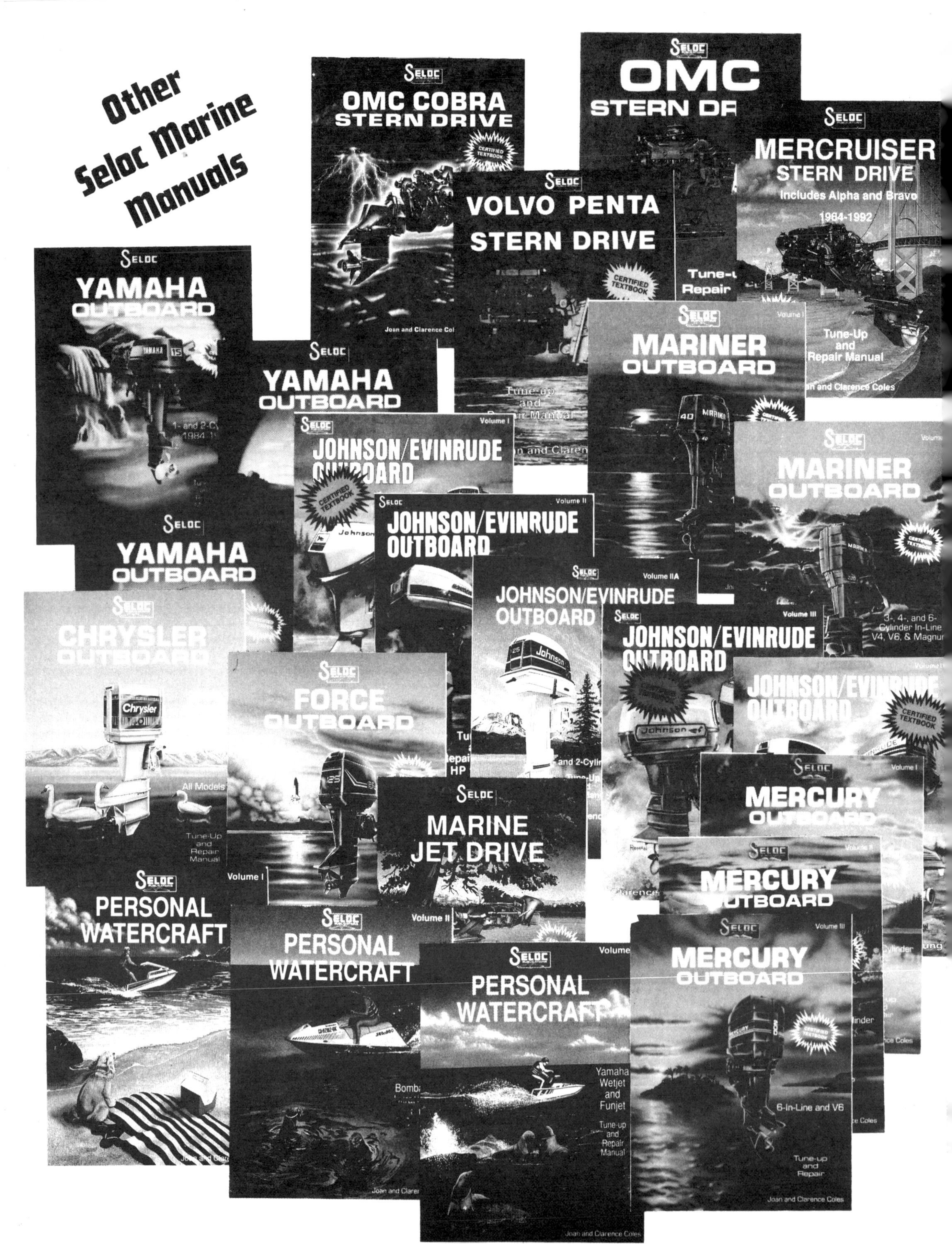

New titles are constantly being produced and the updating work on existing manuals never ceases.
All manuals contain complete detailed instructions, specifications, and wiring diagrams.